ELEMENTS OF MICROWAVE ENGINEERING

ELLIS HORWOOD SERIES IN ELECTRICAL AND ELECTRONIC ENGINEERING

Series Editors: PETER BRANDON, Emeritus Professor of Electrical and Electronic Engineering, University of Cambridge
P. R. ADBY, Department of Electronic and Electrical Engineering, University of London King's College

APPLIED CIRCUIT THEORY: Matrix and Computer Methods
P. R. ADBY, University of London King's College
BOND GRAPHS FOR MODELLING ENGINEERING SYSTEMS
ALAN BLUNDELL, Coventry (Lanchester) Polytechnic
NOISE IN ELECTRONIC DEVICES AND SYSTEMS
M. J. BUCKINGHAM, Royal Aircraft Establishment, Farnborough, Hampshire
DIGITAL AND MICROPROCESSOR ENGINEERING
S. J. CAHILL, Ulster Polytechnic
DIFFRACTION THEORY AND ANTENNAS
R. H. CLARKE and JOHN BROWN, Imperial College of Science and Technology, University of London
PARALLEL ARRAY PROCESSING
P. G. DUCKSBURY, Systems Designers Scientific, Gloucester
INTRODUCTION TO MICROCOMPUTER ENGINEERING
D. A. FRASER, *et al.*, Chelsea College, University of London
THEORY OF NON-LINEAR AND PARAMETRIC CIRCUITS
F. KOURIL and K. VRBA, University of Brno, Czechoslovakia
ELEMENTARY ELECTRIC POWER AND MACHINES
P. G. McLAREN, University Engineering Department, Cambridge
MICROWAVE PHASE MODULATORS
T. MORAWSKI and J. MODELSKI, Warsaw Technical University
CONTROL SYSTEMS WITH TIME DELAYS
A. W. OLBROT, University of Warsaw
POST-DIGITAL ELECTRONICS
F. R. PETTIT, Computing Teaching Centre, Oxford
OPTICAL AND OPTOELECTRONIC DEVICES AND SYSTEMS
DOUGLAS A. ROSS, University of Colorado at Denver
HANDBOOK OF RECTIFIER CIRCUITS
G. J. SCOLES, formerly English Electric Valve Company, Chelmsford, Essex
PRINCIPLES OF COMPUTER COMMUNICATION NETWORK DESIGN
J. SEIDLER, Institute of Fundamentals of Informatics, Polish Academy of Sciences

ELECTRONIC AND COMMUNICATION ENGINEERING

TELECOMMUNICATIONS TECHNOLOGY
R. L. BREWSTER, University of Aston in Birmingham
SATELLITE BROADCASTING SYSTEMS: Planning and Design
J. N. SLATER, Independent Broadcasting Authority, Winchester and L. A. TRINOGGA, Leeds Polytechnic
SIGNAL CODING AND PROCESSING: An Introduction Based on Video Systems
J. G. WADE, Plymouth Polytechnic

ELEMENTS OF MICROWAVE ENGINEERING

R. CHATTERJEE, Ph.D.
Department of Electrical Communication Engineering
Indian Institute of Science, Bangalore

ELLIS HORWOOD LIMITED
Publishers · Chichester

Halsted Press: a division of
JOHN WILEY & SONS
New York · Chichester · Brisbane · Toronto

First published in India by Afflilated East-West Press Private Limited
This Edition published in 1986 by
ELLIS HORWOOD LIMITED
Market Cross House, Cooper Street, Chichester, West Sussex, PO19 1EB, England

The publisher's colophon is reproduced from James Gillison's drawing of the ancient Market Cross, Chichester.

Distributors:
Australia and New Zealand:
Jacaranda-Wiley Ltd., Jacaranda Press,
JOHN WILEY & SONS INC.
GPO Box 859, Brisbane, Queensland 4001, Australia
Canada:
JOHN WILEY & SONS CANADA LIMITED
22 Worcester Road, Rexdale, Ontario, Canada
Europe and Africa:
JOHN WILEY & SONS LIMITED
Baffins Lane, Chichester, West Sussex, England
North and South America and the rest of the world (except in India):
Halsted Press: a division of
JOHN WILEY & SONS
605 Third Avenue, New York, NY 10158, USA

British Library Cataloguing in Publication Data
Chatterjee, Rajeswani
Elements of Microwave Engineering —
(Ellis Horwood series in electrical and electronic engineering)
1. Microwave devices
I. Title
621.381'3 TK7876

ISBN 0-7458-0060-2 (Ellis Horwood Limited — Library Edn.)
ISBN 0-7458-0069-6 (Ellis Horwood Limited — Student Edn.)
ISBN 0-470-20311-0 (Halsted Press)

Printed in Great Britain by Butler & Tanner, Frome, Somerset

To
(late) Dr. Jagadish Chandra Bose
the Indian scientist who did pioneering work
in microwave engineering in India in the nineteenth century

Contents

Preface

This text treats a subject that stems from Maxwell's mathematical formulation of the basic laws of electricity and magnetism, which led to the theoretical prediction of the reality of electromagnetic waves and which is of both theoretical elegance and growing practical utility. It is the outgrowth of a series of lectures on electromagnetic theory and applications and microwave engineering delivered over the last several years at the Indian Institute of Science, Bangalore. Most topics included here have also been successfully used in intensive courses specially organized for the professional engineers engaged in design, and research and development.

The volume is intended for the undergraduate in electronics, and electrical communication and electrical engineering as a comprehensive, up-to-date text, covering the fundamental principles of the various aspects of microwave engineering. It would be equally useful for the graduate in physics specializing in electronics.

Throughout, the theoretical principles and the analytical results have been presented in a form an average student can grasp and put into practical applications. The treatment is simple yet rigorous enough, and includes many worked examples illustrating the various analytical methods. No prior knowledge of mathematics, beyond that acquired by the engineering undergraduate or the graduate in physics, is required, and the chapters have been arranged on this basis. Owing to the limitations of space, the theoretical discussion and the more detailed parts of the applications have been curtailed to a certain extent in some places, but nothing of fundamental importance to a microwave engineer has been omitted. The reader wishing to supplement his knowledge of the subject is advised to consult the References/Suggested Reading provided.

The explosive growth of microwave engineering in the last few years makes it almost impossible to give a sufficiently thorough survey of all problems. An attempt has therefore been made to stress the basic scientific principles and to present in one volume several topics so that the reader can develop a good overall view of the subject. (A treatment of the advanced topics is given in the author's "Microwave Engineering: Special Topics" to be shortly published.) It is hoped that the book would benefit all those for whom it is intended.

I wish to thank Professor S. Dhawan, former Director, Professor S. Ramakrishna, former Chairman, Division of Electrical Sciences, and Professor S. Nagaraja, former Chairman, Department of Electrical Communication Engineering (all of the Indian Institute of Science) for their constant help and encouragement. I am indebted to the University Grants Commission for sponsoring a project for writing this text. I should also like to express my gratitude to Mr. N. Govindaraju and Mr. Varadaraja Iyengar for typing the manuscript, to Mr. R. Vijayendra for drawing the diagrams, and to Miss A. V. Ashajayanti, (late) Miss R. Sudha, and Dr. (Miss) Parveen Fatima Wahid for their assistance in preparing the manuscript.

October 1985
Bangalore

R. CHATTERJEE

1 Introduction

1.1 Microwave Frequency Range

The discovery of Maxwell that light, by its very nature, is electromagnetic, was the starting point for the evolution of the concept of an electromagnetic spectrum that extends from d-c to γ-rays. The term *microwave frequencies* is very commonly used for those wavelengths measured from 30 cm–0.3 mm which correspond to the frequency range 10^9–10^{12} Hz. Since a large number of electronic communication systems utilize the space propagation path, and since a certain bandwidth is required for each transmission, the frequency spectrum of interest to communication engineers has become an international resource. According to the International Radio Consultative Committee (CCIR), the frequency ranges are as designated in Table 1.1. For convenience, the frequencies are also often designated in terms of bands (see Table 1.2).

1.2 Historical Resumé of Early Work on Microwaves

Hertz (1893) conducted a series of experiments at $\lambda = 66$ cm, with a transmitter consisting of a parabolic mirror antenna which was fed by a dipole excited by spark discharges produced by an

Table 1.1 Designation of Frequency Ranges

Frequency (f)	*Wavelength (λ)*	*Designation*
<30 kHz	>10 km	VLF
30–300 kHz	10–1 km	LF
0.3–3 MHz	1–0.1 km	MF
3–30 MHz	100–10 m	HF
30–300 MHz	10–1 m	VHF
300–3000 MHz	1–0.1 m	UHF
3–30 GHz	100–10 mm	SHF
30–300 GHz	10–1 mm	EHF

induction coil, and with a receiver comprising a similar antenna with a dipole whose output was passed on to a spark-gap detector placed behind the mirror. These early experiments established beyond doubt *action at a distance* and proved that this action was communicated to a distance

Table 1.2 Bands for Frequency Ranges

Band designation	*Frequency range* (MHz)
P	225–390
L	390–1550
S	1550–5200
X	5200–10,900
K	10,900–36,000
Q	36,000–46,000
V	45,000–56,000

by wave motion. This justified Maxwell's theoretical prediction that the waves responsible for optical phenomena are electromagnetic. Hertz's work on reflection, diffraction, polarization, and measurement of wavelength by interference technique may be said to have led to the discovery of radio frequency optics, where the phenomenology of optics can be represented by microwaves. Righi (1897) performed many quasi-optical experiments at *X*- and *S*-bands and thus firmly laid the foundation of microwave optics. Lodge and Howard (1889) constructed a cylindrical lens of pitch. Lodge (1897, and 1898 and 1899) was also successful in establishing the mode property of propagation in a hollow tube and transmission of signals through space without wires. Bose (1895, 1897, and 1898a–1898c) conducted several microwave experiments at 5 mm with apparatus of his own design such as microwave spectrometers, diffraction gratings, polarimeters, spark generators, and coherer detectors. For a description of these experiments, see Ramsay (1958). Thus, it is evident that the pioneering work before 1900 by Hertz, Lodge, and Bose laid the foundations of modern microwave engineering.

1.3 Correspondence between Field and Circuit Concepts

Since the wavelengths at microwave frequencies are of the same order of magnitude as the dimensions of circuit devices, and the time of propagation of electrical effects from one part of the circuit to the other is comparable to the period of oscillating currents and charges, conventional circuit concepts of currents and voltages need to be replaced by field concepts. At microwave frequencies, the difficulty in applying circuit concepts is obvious when the potential difference between two points ordinarily means the line integral of the electric field strength, namely,

$$\int E \cdot ds,$$

taken at one instant of time, along some paths joining the two points. This concept is unique and useful only if the value of this line integral is independent of the path. But if the path length is not small compared to the wavelength, the line integral is not, in general, independent of the path, and hence the significance of the term *voltage* is lost. This suggests that, at microwave frequencies, we have to deal with electric and magnetic fields instead of with voltage and current.

Maxwell's field equations are generalizations of Faraday's laws of induction, Ampere's circuit law, and Gauss' law. These equations established that *magnetic flux source* does not exist. Hence, a close correspondence between circuit concepts and field concepts can be established: for example, the field equation $\nabla \times \boldsymbol{E} = -\partial \boldsymbol{B}/\partial t$ corresponds to the circuit equation $\Sigma v_n = -\partial \psi/\partial t$, ψ being the magnetic flux; the power flow given by the equation $\boldsymbol{P} = \boldsymbol{E} \times \boldsymbol{H}$ corresponds to the circuit concept of power, namely, $P = VI$.

Maxwell's field equations, their solutions, and their applications to several practical and useful problems form the subject matter of this text. The field and circuit concepts are used to study the characteristics of transmission lines, waveguides, and passive microwave components.

1.4 Some Useful Applications of Microwaves

Since the transit-time effects of electrons were the major limitations of the conventional high-frequency tubes, these conventional tubes could not succeed in the microwave region. These handicaps were overcome by introducing the concept of interaction of electron beams with electromagnetic fields, resulting in the development of magnetrons, klystrons, and travelling-wave tubes, which made the evolution of radar possible. These developments during World War II opened up new vistas for the extensive application of microwaves, not only to the technological fields such as defence, but to areas of civilian interest, e.g., microwave communication relay links, satellite communication, and domestic appliances, for instance, microwave ovens. Microwaves also find extensive application in pure scientific fields such as radio astronomy, spectroscopy, and materials research which led to the development of solid-state microwave generators, e.g., masers, coherent light generators such as lasers, and ferrite microwave devices.

The principles of microwave tubes and their modern solid-state counterparts are discussed in Chapter 10 in a language that can be understood not only by undergraduates in electronics and electrical communication engineering, but also by post-graduate students in physics.

Microwaves are currently used in India in the following areas: defence, post and telegraph, railways, civil aviation, space communication, police, and radio astronomy. It may therefore be stated that a comprehensive course in microwave engineering for the undergraduate and post-graduate level students should include the following topics:

(i) Maxwell's field equations and their solutions
(ii) Transmission lines and waveguides
(iii) Microwave networks
(iv) Microwave generators, including solid-state devices
(v) Microwave antennas
(vi) Microwave measurements
(vii) Other related topics.

All these topics are discussed in this text. The treatment of some of the topics may be found to be somewhat condensed due to limitations of space. However, we have endeavoured to emphasize the fundamental principles rather than details, with the conviction that, if the fundamental principles are properly grasped, the details can be learned by consulting the proper references.

REFERENCES

Bose, J. C., On a new electropolariscope, *The Electrician*, **36**, 291, 1895.

Bose, J. C., On the determination of the wavelength of electric radiation by a diffraction grating, *Proc. R. Soc.* (London), **60**, 167, 1897.

Bose, J. C., On the influence of the thickness of airspace on the total reflections of electric radiations, *The Electrician*, **42**, 154, 1898a.

Bose, J. C., On the rotation of plane of polarization of electric waves by a twisted structure, *Proc. R. Soc.* (London), **63**, 146, 1898b.

Bose, J. C., The refraction of electric waves, *Nature*, **57**, 353, 1898c.

Hertz, H., Electric Waves, Macmillan, London, 1893.

Lodge, O. J., The history of coherer principle, *The Electrician*, **40**, 87, 1897.

Lodge, O. J. and Howard, J. L., On the concentration of electric radiation by lenses, *Nature*, **40**, 94, 1889.

Lodge, O. J., Signalling through space without wires, *The Electrician*, **40**, 269, 305, 1898; **43**, 366, 402, 1899.

Ramsay, J. F., Microwave antenna and waveguide techniques before 1900, *Proc. I.R.E.*, **46**, 405, 1958.

Righi, A., L'Ottica delle Oscillazioni Electriche, N. Zanichelli, Bologna, 1897.

2 Mathematical Review

2.1 Vector Analysis

Vector is a name given to physical quantities such as force, velocity, and field intensity which can be defined uniquely only when their magnitudes and directions are specified. A vector is represented graphically by a directed segment $\overrightarrow{PQ}$ or $\boldsymbol{A}$ (see Fig. 2.1) whose length is proportional to

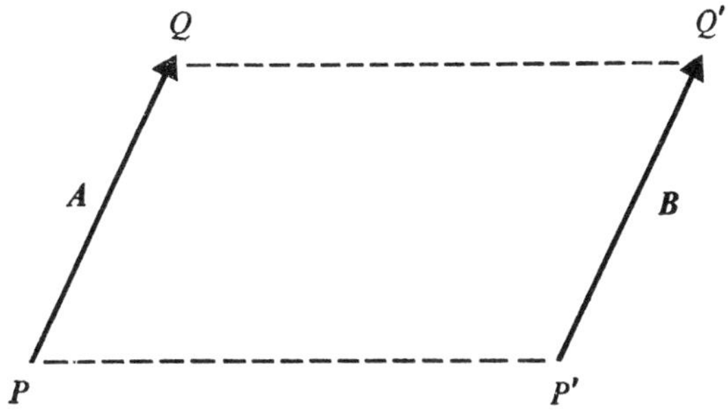

Fig. 2.1 Equal vectors.

the magnitude of the vector and whose direction is the same as that of $\overrightarrow{PQ}$. Two parallel vectors $\overrightarrow{PQ}\,(\boldsymbol{A})$ and $\overrightarrow{P'Q'}\,(\boldsymbol{B})$ are considered equal if they have the same magnitude and direction.

A *scalar* is a quantity which can be specified only by its magnitude and does not have a direction. Examples of scalar are temperature, speed, and mass.

The common laws of addition, subtraction, multiplication, and division which are applicable to scalars are not applicable to vectors. Throughout this text, therefore, we shall denote a vector by $\boldsymbol{F}$ and a scalar by F and the magnitude of $\boldsymbol{F}$ by $|F|$ or F.

Addition and Subtraction of Vectors

The sum of the vectors $\boldsymbol{A}$ and $\boldsymbol{B}$ is given by the diagonal of the parallelogram constructed with these vectors as the adjacent sides (see Fig. 2.2). Since $\overrightarrow{PQ} + \overrightarrow{QP} = 0$,

$$\overrightarrow{QP} = -\overrightarrow{PQ}.$$

Hence, $\boldsymbol{A} - \boldsymbol{B} = \overrightarrow{PQ} - \overrightarrow{PR} = \overrightarrow{PQ} + \overrightarrow{RP} = \overrightarrow{PQ} + \overrightarrow{PR'} = \overrightarrow{PS'} = \overrightarrow{RQ}$, as shown in Fig. 2.3.

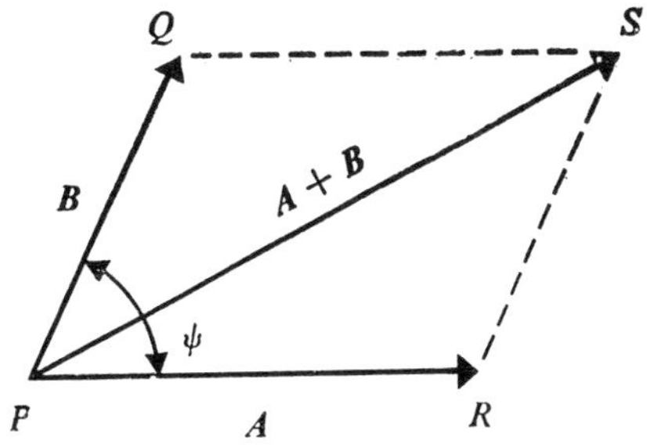

Fig. 2.2 Addition of two vectors.

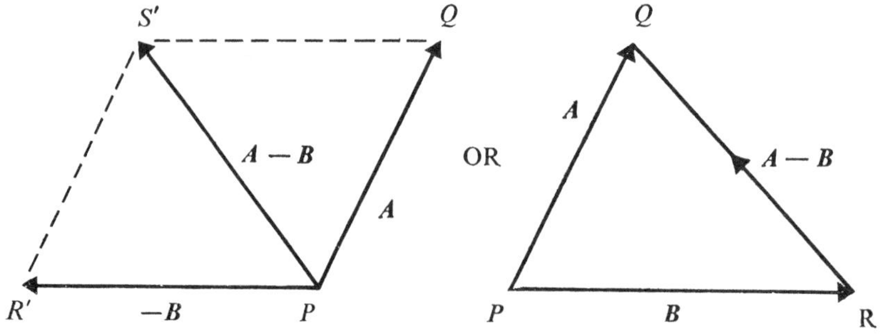

Fig. 2.3 Subtraction of two vectors.

Scalar and Vector Products

The *scalar product* of the vectors $\boldsymbol{A}$ and $\boldsymbol{B}$ is defined as the product of their magnitudes A and B and the cosine of the angle ψ between them. Thus

$$\boldsymbol{A} \cdot \boldsymbol{B} = AB \cos \psi. \tag{2.1}$$

If the scalar product of two vectors is zero, then they are perpendicular to each other.

Scalar multiplication obeys the commutative and distributive laws

$$\boldsymbol{A} \cdot \boldsymbol{B} = \boldsymbol{B} \cdot \boldsymbol{A}, \qquad (\boldsymbol{A} + \boldsymbol{B}) \cdot \boldsymbol{C} = \boldsymbol{A} \cdot \boldsymbol{C} + \boldsymbol{B} \cdot \boldsymbol{C}. \tag{2.2}$$

The component of a given vector $\boldsymbol{A}$ in a particular direction defined by a unit vector $\boldsymbol{u}$ in that direction is the scalar product of $\boldsymbol{A}$ and $\boldsymbol{u}$, that is, $\boldsymbol{A} \cdot \boldsymbol{u}$; or, in other words, the component is the projection of the vector $\boldsymbol{A}$ on $\boldsymbol{u}$. The cartesian components of a vector $\boldsymbol{A}$, drawn from a point $P(x_1, y_1, z_1)$ to a point $Q(x_2, y_2, z_2)$, along the positive directions of the x-, y-, and z-axis are given by $x_2 - x_1$, $y_2 - y_1$, and $z_2 - z_1$. If l is the length of the vector and α, β, γ are the angles the vector makes with the coordinate axes, then

$$PQ_x = A_x = x_2 - x_1 = l \cos \alpha, \tag{2.3}$$

$$PQ_y = A_y = y_2 - y_1 = l \cos \beta, \tag{2.4}$$

$$PQ_z = A_z = z_2 - z_1 = l \cos \gamma, \tag{2.5}$$

$$|A| = (A_x^2 + A_y^2 + A_z^2)^{1/2}. \tag{2.6}$$

The scalar product $\boldsymbol{A} \cdot \boldsymbol{B}$ of the two vectors $\boldsymbol{A}$ and $\boldsymbol{B}$ can be expressed as

$$\boldsymbol{A} \cdot \boldsymbol{B} = A_x B_x + A_y B_y + A_z B_z, \tag{2.7}$$

and the cosine of the angle ψ between the two vectors is given by

$$\cos \psi = \cos \alpha_A \cos \alpha_B + \cos \beta_A \cos \beta_B + \cos \gamma_A \cos \gamma_B, \tag{2.8}$$

where $(\alpha_A, \beta_A, \gamma_A)$ and $(\alpha_B, \beta_B, \gamma_B)$ are respectively the angles $\boldsymbol{A}$ and $\boldsymbol{B}$ make with the three coordinate axes x, y, and z.

The *vector product* $\boldsymbol{A} \times \boldsymbol{B}$ of the vectors $\boldsymbol{A}$ and $\boldsymbol{B}$ is defined as a vector perpendicular to both $\boldsymbol{A}$ and $\boldsymbol{B}$, pointing in the direction towards which a right-handed screw would advance if turned from $\boldsymbol{A}$ to $\boldsymbol{B}$ through the smaller angle (see Fig. 2.4). The magnitude of the vector product is the

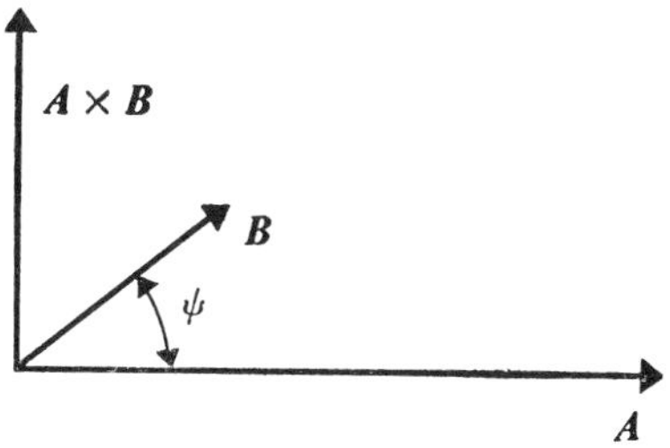

Fig. 2.4 Vector product.

product of the magnitudes of $\boldsymbol{A}$ and $\boldsymbol{B}$ and of the sine of the angle between them, that is, the area of the parallelogram constructed with $\boldsymbol{A}$ and $\boldsymbol{B}$ as the adjacent sides.

For vector products, we have

$$\boldsymbol{A}\times\boldsymbol{B} = -\boldsymbol{B}\times\boldsymbol{A}, \tag{2.9}$$

$$(\boldsymbol{A}+\boldsymbol{B})\times\boldsymbol{C} = \boldsymbol{A}\times\boldsymbol{C}+\boldsymbol{B}\times\boldsymbol{C}. \tag{2.10}$$

The components of the vector product $\boldsymbol{A}\times\boldsymbol{B}$ in the cartesian coordinates x, y, z are expressed as

$$(\boldsymbol{A}\times\boldsymbol{B})_x = A_yB_z - A_zB_y, \tag{2.11}$$

$$(\boldsymbol{A}\times\boldsymbol{B})_y = A_zB_x - A_xB_z, \tag{2.12}$$

$$(\boldsymbol{A}\times\boldsymbol{B})_z = A_xB_y - A_yB_x. \tag{2.13}$$

If $\boldsymbol{A}$ and $\boldsymbol{B}$ are expressed as

$$\boldsymbol{A} = \boldsymbol{u}_xA_x + \boldsymbol{u}_yA_y + \boldsymbol{u}_zA_z, \tag{2.14}$$

$$\boldsymbol{B} = \boldsymbol{u}_xB_x + \boldsymbol{u}_yB_y + \boldsymbol{u}_zB_z, \tag{2.15}$$

then

$$\boldsymbol{A}\times\boldsymbol{B} = \begin{vmatrix} \boldsymbol{u}_x & \boldsymbol{u}_y & \boldsymbol{u}_z \\ A_x & A_y & A_z \\ B_x & B_y & B_z \end{vmatrix}. \tag{2.16}$$

Functions of Position

A *function of position* or *point function* is a function $f(x, y, z)$ depending only on the position of points. The *loci* of equal values of a point function are called *level surfaces* or *contour surfaces*. Some level surfaces have special names, e.g., equipotential, isothermal, and isobar surfaces. Figure 2.5 illustrates how two-dimensional point functions may be represented graphically by drawing contour lines.

In Fig. 2.5a, the solid curves are the contour lines $x^2 + y^2 =$ constant r^2, and the dashed lines are the contour curves $y/x = \tan^{-1}\phi$ constant. In Fig. 2.5b, the solid curves are the contour lines $\log(\rho_1/\rho_2) = u$, where ρ_1 and ρ_2 are the distances from two fixed points A and B, and the dashed curves are the contour curves for $\theta =$ constant, θ being the angle made by BP with PA.

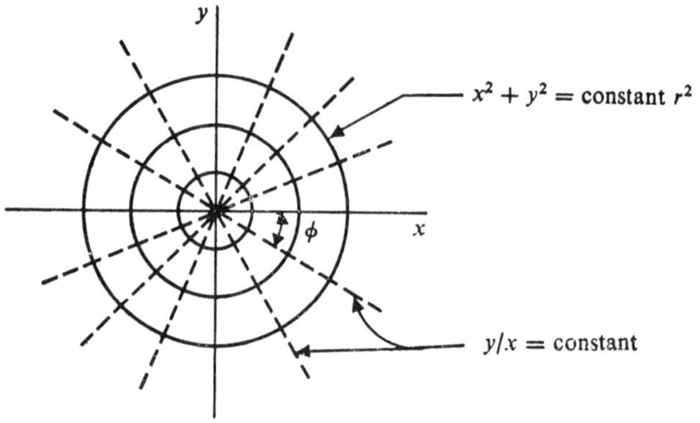

(a) For $y/x = \tan^{-1} \phi$ constant and $x^2 + y^2$ = constant

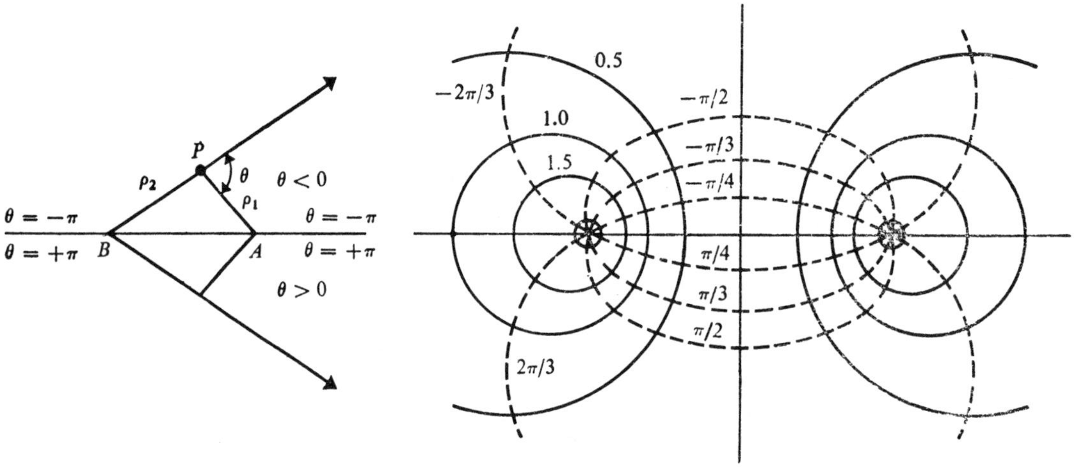

(b) For $\log(\rho_1/\rho_2)$ = constant

Fig. 2.5 Contour lines.

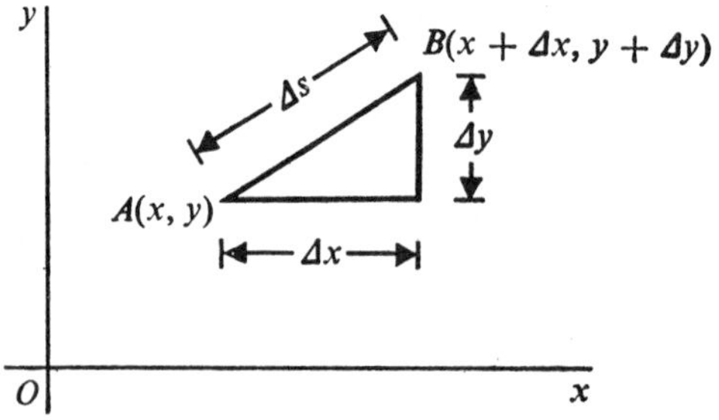

Fig. 2.6 Directional increments.

The rate of change of a point function depends not only on the position of a point but also on the particular direction of its travel. If ΔV is the change in the value of a point function $V(x, y)$ as we pass from a point $A(x, y)$ to a point $B(x + \Delta x, y + \Delta y)$ (see Fig. 2.6), and if Δs is the distance AB, then the ratio $\Delta V/(\Delta s)$ is the average rate of change of $V(x, y)$ in the direction AB. The limit of this ratio as B approaches A, while remaining on the same straight line, is the directional derivative of $V(x, y)$ in the direction AB. This derivative is denoted by $\partial V/\partial s$. If V is a function $V(x, y, z)$ of the cartesian coordinates x, y, z, then the partial derivatives $\partial V/\partial x$, $\partial V/\partial y$, $\partial V/\partial z$ are the directional derivatives along the three coordinate axes.

Gradient of Scalar Point Function $V(x, y, z)$

In three dimensions, the maximum rate of change of a scalar point function V at a point A is along the normal to the level surface through A, because V is constant along the level surface. But in two dimensions, V is a two-dimensional point function $V(x, y)$ (see Fig. 2.7). The gradient

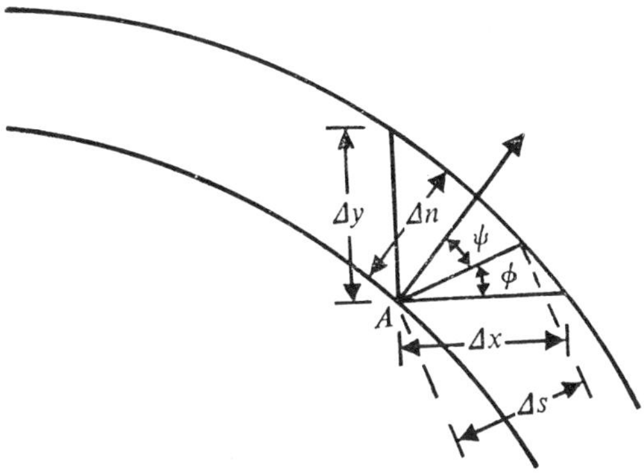

Fig. 2.7 Gradient.

of $V(x, y)$ at a point A is defined as a vector along the normal to the level surface through A, and is given by

$$\text{grad } V = \frac{\partial V}{\partial n}\boldsymbol{n}, \tag{2.17}$$

where $\boldsymbol{n}$ is the unit vector orthogonal to the level surface.

For an infinitesimal curvilinear triangle, we have

$$\Delta n = (\Delta s) \cos \psi, \tag{2.18}$$

where ψ is the angle between Δn and Δs, and therefore

$$\frac{\partial V}{\partial s} = \frac{\partial V}{\partial n} \cos \psi. \tag{2.19}$$

Hence, the directional derivative of a point function is the component of its gradient in that direction. In the cartesian coordinates, for two dimensions, grad V can be expressed as

$$\nabla V = \text{grad } V = \boldsymbol{u}_x \frac{\partial V}{\partial x} + \boldsymbol{u}_y \frac{\partial V}{\partial y}. \tag{2.20}$$

If V is a three-dimensional point function, then

$$\text{grad } V = \boldsymbol{u}_x \frac{\partial V}{\partial x} + \boldsymbol{u}_y \frac{\partial V}{\partial y} + \boldsymbol{u}_z \frac{\partial V}{\partial z}. \tag{2.21}$$

If α, β, γ are the angles made by the normal to the level surface through a point $A(x, y, z)$, then

$$\frac{\partial V}{\partial x} = \frac{\partial V}{\partial n} \cos \alpha, \qquad \frac{\partial V}{\partial y} = \frac{\partial V}{\partial n} \cos \beta, \qquad \frac{\partial V}{\partial z} = \frac{\partial V}{\partial n} \cos \gamma. \tag{2.22}$$

Hence

$$\frac{\partial V}{\partial n} = [(\frac{\partial V}{\partial x})^2 + (\frac{\partial V}{\partial y})^2 + (\frac{\partial V}{\partial z})^2]^{1/2}. \tag{2.23}$$

Also,

$$\frac{\partial V}{\partial n} = \frac{\partial V}{\partial x} \cos \alpha + \frac{\partial V}{\partial y} \cos \beta + \frac{\partial V}{\partial z} \cos \gamma. \tag{2.24}$$

Gradient of Complex Point Function

A complex point function is a function whose real and imaginary parts are point functions; for example,

$$V(x, y, z) = V_1(x, y, z) + jV_2(x, y, z). \tag{2.25}$$

There are four families of level surfaces for complex point functions: one family for the real part V_1, another family for the imaginary part V_2, yet a third family for the absolute value $(V_1^2 + V_2^2)^{1/2}$, and finally, a fourth family for the phase angle $\tan^{-1} (V_2/V_1)$.

The fourth family of surfaces, that is,

$$\tan^{-1} [\frac{V_2(x, y, z)}{V_1(x, y, z)}] = \psi = \text{constant}, \tag{2.26}$$

is called the equiphase surfaces. These surfaces are used in classifying the waves as plane, cylindrical, and spherical waves and so on.

The gradient of a complex function is defined as the complex vector whose components are the partial derivatives of the function.

Vector Point Function

A vector point function is a vector whose direction components are ordinary point functions.

Divergence of Vector Point Function

The flux of a vector point function $\boldsymbol{F}(x, y, z)$ through a surface S is defined as the surface integral

$$\Phi = \int_s F_n \, dS, \tag{2.27}$$

where F_n is the component of the vector $\boldsymbol{F}$ normal to the surface of integration. The *average divergence* of $\boldsymbol{F}$ is defined as the outward flux of $\boldsymbol{F}$ through a simply connected closed surface S divided by the volume V enclosed by S. The limit of the average divergence as S contracts to a point is the divergence of $\boldsymbol{F}$ at that point. Thus

$$\operatorname{div} \boldsymbol{F} = \lim_{V\to 0} \frac{\int_S F_n \, dS}{V} = \nabla \cdot \boldsymbol{F}. \tag{2.28}$$

Dividing the total volume V enclosed by the surface S into elementary cells (see Fig. 2.8), it can

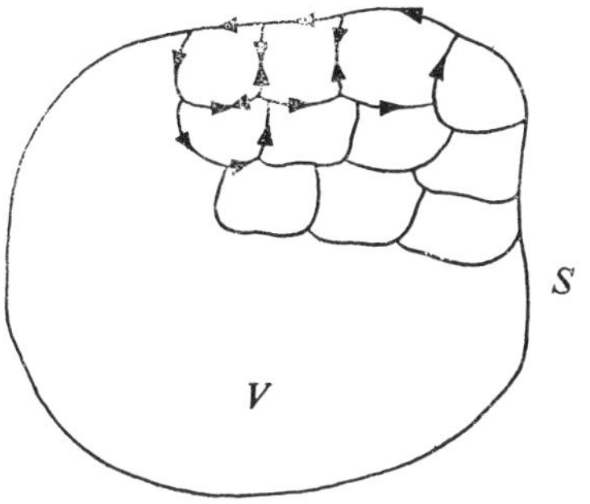

Fig. 2.8 Volume V enclosed by closed surface S.

be seen that the total flux of $\boldsymbol{F}$ across the surface is the sum of the fluxes through the boundaries of the elementary cells. However, the surface integrals over the common boundaries between two cells cancel each other. The flux through the boundary of a typical cell is div $\boldsymbol{F} \cdot dV$, where dV is the volume of the cell, and hence

$$\int_S F_n \, dS = \int_V \operatorname{div} \boldsymbol{F} \cdot dV \tag{2.29}$$

which is called *Gauss' theorem*.

The *surface divergence* of a two-dimensional vector point function $\boldsymbol{F}$ is defined as

$$\operatorname{div}' \boldsymbol{F} = \lim_{C\to 0} \frac{\oint_C F_n \, ds}{S} = \nabla' \cdot \boldsymbol{F}, \tag{2.30}$$

where C is the boundary of the elementary area S.

The *linear divergence* is merely the ordinary derivative.

The divergence of any vector point function is a scalar point function.

Line Integral, Circulation, and Curl

The *line integral* of a vector $\boldsymbol{F}$ along a path AB (see Fig. 2.9) is defined as the integral

$$\int_{(AB)} F_s \, ds,$$

where F_s is the tangential component of the vector and ds is an incremental length along the path. For example, if $\boldsymbol{F}$ is a force, then this integral represents the work done by $\boldsymbol{F}$ on a particle moving along the path AB. If the curve is closed, the line integral is called *circulation*. The circulation per unit area of an infinitely small loop is so oriented that it is maximum and is denoted by curl $\boldsymbol{F}$ or $\nabla \times \boldsymbol{F}$. It is a vector perpendicular to the plane of the loop. The relationship between the

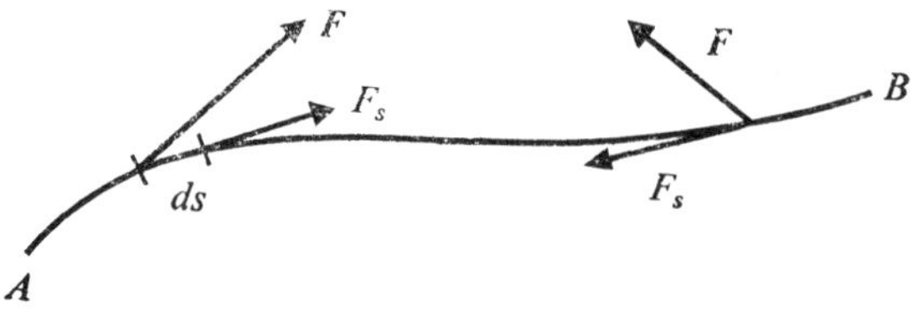

Fig. 2.9 Line integral.

positive direction of the circulation of $\boldsymbol{F}$ and that of curl $\boldsymbol{F}$ is illustrated in Fig. 2.10.

Let S be a surface bounded by a simple closed curve C, as shown in Fig. 2.11. If S is divided into a large number of small elementary areas, then the circulation of $\boldsymbol{F}$ along the boundary C of S is the sum of the circulations around the boundaries of the elements. However, the

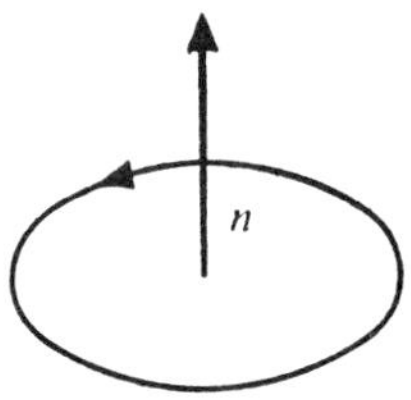

Fig. 2.10 Relationship between positive direction of circulation of $\boldsymbol{F}$ and that of curl $\boldsymbol{F}$.

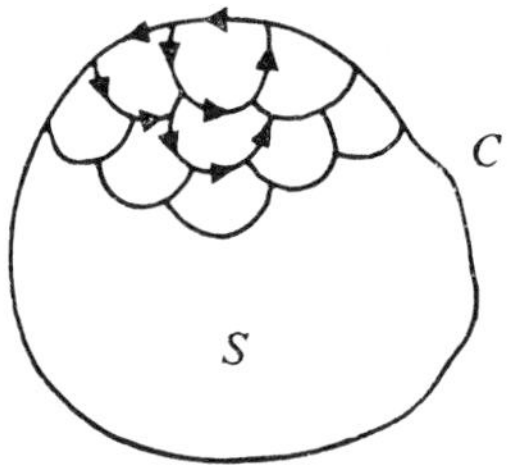

Fig. 2.11 Illustration of Stokes' theorem.

contributions due to the boundaries common to adjacent elements cancel out. Since the circulation around the boundary of each element is $\text{curl}_n\, \boldsymbol{F}\, dS$, we have

$$\oint_C F_s\, ds = \int_S \text{curl}_n\, \boldsymbol{F}\, dS \tag{2.31}$$

which is called *Stokes' theorem*.

Coordinate Systems

The most frequently used coordinate systems are the cartesian (or rectangular), cylindrical, and spherical systems. In these systems, a typical point P is denoted by (x, y, z), (ρ, ϕ, z), and (r, θ, ϕ), respectively (see Fig. 2.12). The relationships between the three coordinate systems are

$$\begin{aligned} &\rho = (x^2 + y^2)^{1/2}, \quad \phi = \tan^{-1}(y/x), \quad z = z, \\ &x = \rho\cos\phi, \quad y = \rho\sin\phi, \quad z = z, \end{aligned} \tag{2.32}$$

$$\begin{aligned} &r = (x^2 + y^2 + z^2)^{1/2}, \quad \theta = \tan^{-1}\left(\frac{x^2 + y^2}{z^2}\right)^{1/2}, \quad \phi = \tan^{-1}(y/x), \\ &x = r\sin\theta\cos\phi, \quad y = r\sin\theta\sin\phi, \quad z = r\cos\theta, \end{aligned} \tag{2.33}$$

$$\begin{aligned} &\rho = r\sin\theta, \quad \phi = \phi, \quad z = r\cos\theta, \\ &r = (\rho^2 + z^2)^{1/2}, \quad \phi = \phi, \quad \theta = \tan^{-1}(\rho/z). \end{aligned} \tag{2.34}$$

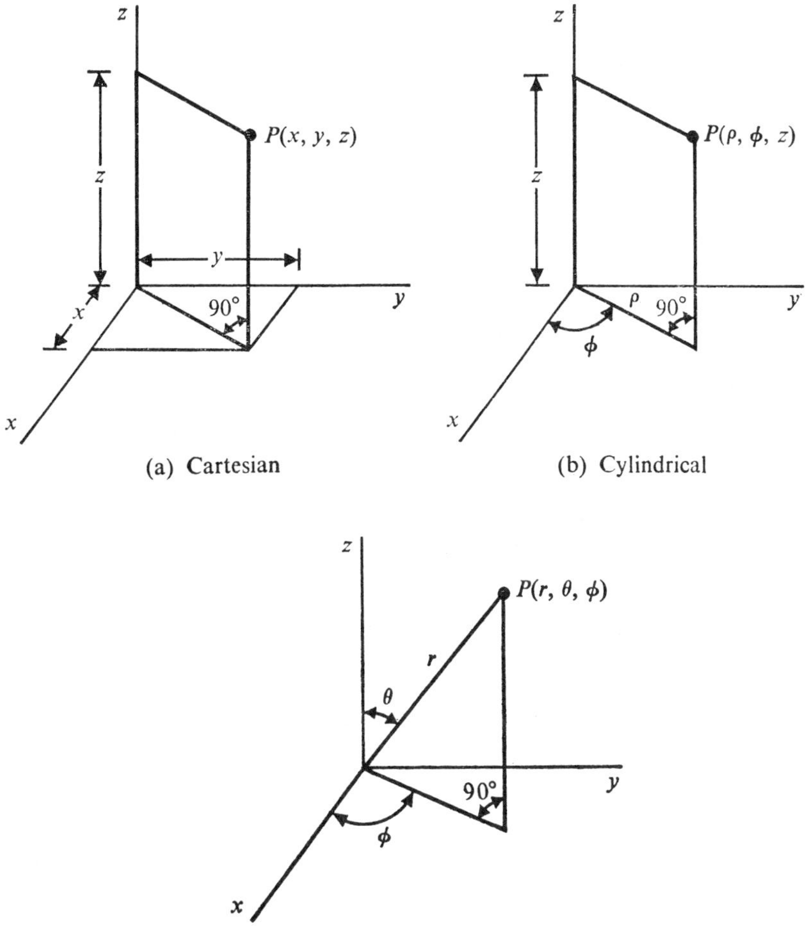

(a) Cartesian (b) Cylindrical

(c) Spherical

Fig. 2.12 Coordinate systems.

In a *general curvilinear system of coordinates*, a point $P(u, v, w)$ is specified as a point of intersection of the three coordinate surfaces

$$f_1(x, y, z) = u, \qquad f_2(x, y, z) = v, \qquad f_3(x, y, z) = w. \tag{2.35}$$

The lines of intersection of these coordinate surfaces are the coordinate lines. For example, the u-lines are the intersections of the v- and w-surface (see Fig. 2.13). If the coordinates are orthogonal, the differential distances along the coordinate lines are proportional to the differentials of the coordinates; and so

$$ds_u = e_1\, du, \qquad ds_v = e_2\, dv, \qquad ds_w = e_3\, dw, \tag{2.36}$$

$$ds = \text{a general element of length} = [(ds_u)^2 + (ds_v)^2 + (ds_w)^2]^{1/2} \tag{2.37}$$

$$= (e_1^2\, du^2 + e_2^2\, dv^2 + e_3^2\, dw^2)^{1/2}. \tag{2.38}$$

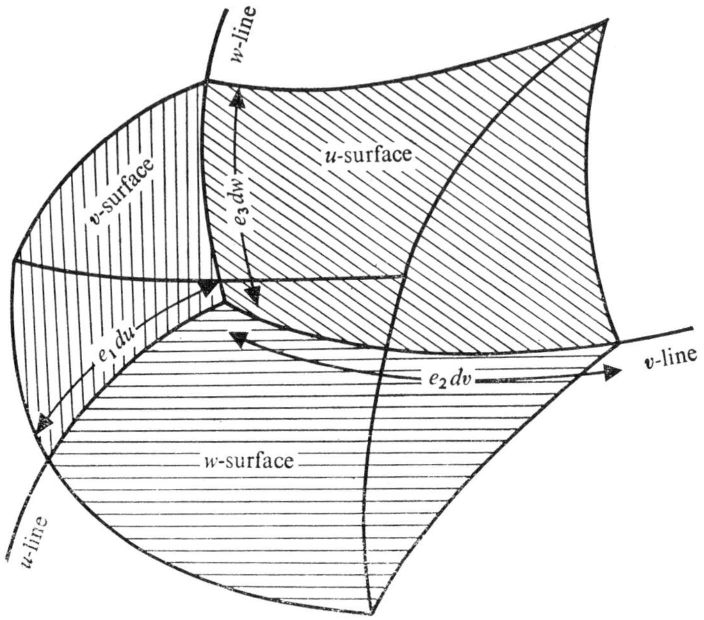

Fig. 2.13 Orthogonal curvilinear coordinates.

The elementary areas in the coordinate surfaces are

$$dS_u = ds_v\, ds_w, \qquad dS_v = ds_u\, ds_w, \qquad dS_w = ds_u\, ds_v, \tag{2.39}$$

$$dV = \text{elementary volume} = ds_u\, ds_v\, ds_w = e_1 e_2 e_3\, du\, dv\, dw. \tag{2.40}$$

The rectangular, cylindrical, and spherical coordinates are special cases of the orthogonal curvilinear coordinates (see Table 2.1).

Table 2.1 Rectangular, Cylindrical, and Spherical Coordinate Systems

Rectangular	*Cylindrical*	*Spherical*
$ds_x = dx$	$ds_\rho = d\rho$	$ds_r = dr$
$ds_y = dy$	$ds_\phi = \rho\, d\phi$	$ds_\theta = r\, d\theta$
$ds_z = dz$	$ds_z = dz$	$ds_\phi = r \sin\theta\, d\phi$
$ds^2 = dx^2 + dy^2 + dz^2$	$ds^2 = d\rho^2 + \rho^2\, d\phi^2 + dz^2$	$ds^2 = dr^2 + r^2\, d\theta^2 + r^2 \sin^2\theta\, d\phi^2$
$dV = dx\, dy\, dz$	$dV = \rho\, d\rho\, d\phi\, dz$	$dV = r^2\, dr\, d\theta \sin\theta\, d\phi$
$dS_x = dy\, dz$	$dS_\rho = \rho\, d\phi\, dz$	$dS_r = r^2 \sin\theta\, d\theta\, d\phi$
$dS_y = dz\, dx$	$dS_\phi = d\rho\, dz$	$dS_\theta = r \sin\theta\, dr\, d\phi$
$dS_z = dx\, dy$	$dS_z = \rho\, d\rho\, d\phi$	$dS_\phi = r\, dr\, d\theta$

Expressions for Gradient, Divergence, and Curl

By definition, the components of the gradient of a scalar point function V in the directions of the coordinate lines are the directional derivatives of V in those directions. Hence, in an orthogonal curvilinear system of coordinates, the components of grad V are

$$\frac{\partial V}{\partial s_u}, \qquad \frac{\partial V}{\partial s_v}, \qquad \frac{\partial V}{\partial s_w}.$$

Thus, in a general curvilinear coordinate system and in a rectangular, cylindrical, and spherical coordinate system, grad V is given by

$$\boldsymbol{u}\frac{1}{e_1}\frac{\partial V}{\partial u} + \boldsymbol{v}\frac{1}{e_2}\frac{\partial V}{\partial v} + \boldsymbol{w}\frac{1}{e_3}\frac{\partial V}{\partial w} = \text{grad } V, \tag{2.41}$$

$$\boldsymbol{u}_x\frac{\partial V}{\partial x} + \boldsymbol{u}_y\frac{\partial V}{\partial y} + \boldsymbol{u}_z\frac{\partial V}{\partial z} = \text{grad } V = \nabla V, \tag{2.42}$$

$$\boldsymbol{u}_\rho\frac{\partial V}{\partial \rho} + \boldsymbol{u}_\phi\frac{\partial V}{\rho\, \partial \phi} + \boldsymbol{u}_z\frac{\partial V}{\partial z} = \text{grad } V = \nabla V, \tag{2.43}$$

$$\boldsymbol{u}_r\frac{\partial V}{\partial r} + \boldsymbol{u}_\theta\frac{\partial V}{r\, \partial \theta} + \boldsymbol{u}_\phi\frac{1}{r \sin\theta}\frac{\partial V}{\partial \phi} = \text{grad } V = \nabla V. \tag{2.44}$$

To calculate the *divergence* of a vector point function $\boldsymbol{F}$ at a point P, we take an elementary cell about P, determine the flux through the surface of this cell, and divide the flux by its volume. The area of a u-surface about P is dS_u and the flux of $\boldsymbol{F}$ across this area is $F_u\, dS_u$. The rate of change of this flux in the u-direction is $\partial(F_u\, dS_u)/\partial u$. Hence, the residual flux of $\boldsymbol{F}$ over the two u-faces of the cell is

$$F_u\, dS_u + \frac{\partial}{\partial u}(F_u\, dS_u)\, du - F_u\, dS_u = \frac{\partial}{\partial u}(F_u\, dS_u)\, du.$$

Similarly, the residual flux of $\boldsymbol{F}$ over the two v-faces and the two w-faces can be calculated. Hence, the total outward flux of $\boldsymbol{F}$ from the cell is

$$\frac{\partial}{\partial u}(F_u\, dS_u)\, du + \frac{\partial}{\partial v}(F_v\, dS_v)\, dv + \frac{\partial}{\partial w}(F_w\, dS_w)\, dw. \tag{2.45}$$

Dividing this total outward flux by the volume $ds_u\, ds_v\, ds_w$ of the cell, we obtain, for div $\boldsymbol{F}$, the expression

$$\text{div } \boldsymbol{F} = \frac{\left[\frac{\partial}{\partial u}(F_u\, dS_u)\, du + \frac{\partial}{\partial v}(F_v\, dS_v)\, dv + \frac{\partial}{\partial w}(F_w\, dS_w)\, dw\right]}{ds_u\, ds_v\, ds_w} \tag{2.46}$$

$$= \frac{1}{e_1 e_2 e_3}\left[\frac{\partial}{\partial u}(e_2 e_3 F_u) + \frac{\partial}{\partial v}(e_1 e_3 F_v) + \frac{\partial}{\partial w}(e_1 e_2 F_w)\right]. \tag{2.47}$$

In the rectangular, cylindrical, and spherical coordinate systems, the expressions for div $\boldsymbol{F}$ are, respectively,

$$\text{div } \boldsymbol{F} = \frac{\partial F_x}{\partial x} + \frac{\partial F_y}{\partial y} + \frac{\partial F_z}{\partial z} = \nabla \cdot \boldsymbol{F}, \tag{2.48}$$

$$\text{div } \boldsymbol{F} = \frac{1}{\rho}\frac{\partial}{\partial \rho}(\rho F_\rho) + \frac{1}{\rho}(\frac{\partial F_\phi}{\partial \phi}) + \frac{\partial F_z}{\partial z} = \nabla \cdot \boldsymbol{F}, \tag{2.49}$$

$$\text{div } \boldsymbol{F} = \frac{1}{r^2 \sin \theta}[\sin \theta \frac{\partial}{\partial r}(r^2 F_r) + r\frac{\partial}{\partial \theta}(\sin \theta\ F_\theta) + r\frac{\partial F_\phi}{\partial \phi}] = \nabla \cdot \boldsymbol{F}. \tag{2.50}$$

To derive the expression for the curl of a vector $\boldsymbol{F}$ at a point P, its three components have to be calculated. By definition, the component $\text{curl}_u\, \boldsymbol{F}$ of curl $\boldsymbol{F}$ in the u-direction is the circulation of $\boldsymbol{F}$ per unit area in the u-surface passing through P (see Fig. 2.14). The circulation of F around the

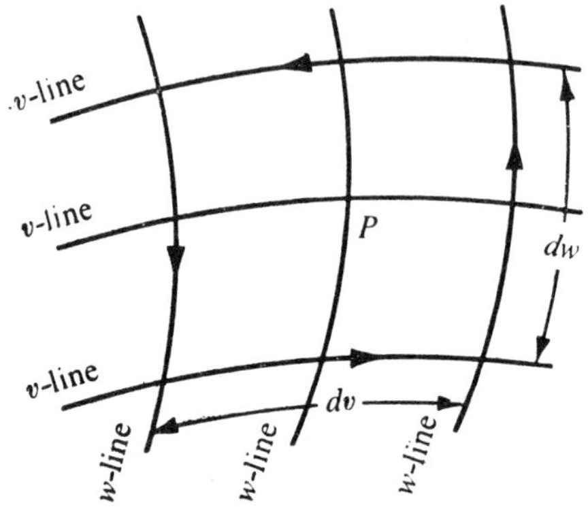

Fig. 2.14 Curl of vector.

boundary of an elementary area in the u-plane is

$$F_w\, ds_w + \frac{\partial}{\partial v}(F_w\, ds_w)\, dv - F_v\, ds_v - \frac{\partial}{\partial w}(F_v\, ds_v)\, dw - F_w\, ds_w + F_v\, ds_v$$

$$= \frac{\partial}{\partial v}(F_w\, ds_w)\, dv - \frac{\partial}{\partial w}(F_v\, ds_v)\, dw. \tag{2.51}$$

Hence

$$\text{curl}_u\, \boldsymbol{F} = \frac{\frac{\partial}{\partial v}(F_w\, ds_w)\, dv - \frac{\partial}{\partial w}(F_v\, ds_v)\, dw}{ds_v\, ds_w}$$

$$= \frac{\frac{\partial}{\partial v}(F_w e_3\, dw)\, dv - \frac{\partial}{\partial w}(F_v e_2\, dv)\, dw}{e_2 e_3\, dv\, dw} \tag{2.52}$$

$$= \frac{\frac{\partial}{\partial v}(e_3 F_w) - \frac{\partial}{\partial w}(e_2 F_v)}{e_2 e_3} = (\nabla \times \boldsymbol{F})_u. \tag{2.53}$$

Similarly,

$$\text{curl}_v\, \boldsymbol{F} = \frac{1}{e_3 e_1}[\frac{\partial}{\partial w}(e_1 F_u) - \frac{\partial}{\partial u}(e_3 F_w)] = (\nabla \times \boldsymbol{F})_v, \tag{2.54}$$

$$\text{curl}_w\, \boldsymbol{F} = \frac{1}{e_1 e_2}[\frac{\partial}{\partial u}(e_2 F_v) - \frac{\partial}{\partial v}(e_1 F_u)] = (\nabla \times \boldsymbol{F})_w. \tag{2.55}$$

In the three systems of coordinates, the expressions for the components of curl $\boldsymbol{F}$ are as follows. In the cartesian coordinates,

$$\text{curl}_x \boldsymbol{F} = \frac{\partial F_z}{\partial y} - \frac{\partial F_y}{\partial z} = (\nabla \times \boldsymbol{F})_x, \tag{2.56}$$

$$\text{curl}_y \boldsymbol{F} = \frac{\partial F_x}{\partial z} - \frac{\partial F_z}{\partial x} = (\nabla \times \boldsymbol{F})_y, \tag{2.57}$$

$$\text{curl}_z \boldsymbol{F} = \frac{\partial F_y}{\partial x} - \frac{\partial F_x}{\partial y} = (\nabla \times \boldsymbol{F})_z \tag{2.58}$$

or

$$\text{curl}\, \boldsymbol{F} = \begin{vmatrix} \boldsymbol{u}_x & \boldsymbol{u}_y & \boldsymbol{u}_z \\ \dfrac{\partial}{\partial x} & \dfrac{\partial}{\partial y} & \dfrac{\partial}{\partial z} \\ F_x & F_y & F_z \end{vmatrix} = \nabla \times \boldsymbol{F}. \tag{2.59}$$

In the cylindrical coordinates,

$$\text{curl}_\rho \boldsymbol{F} = \frac{1}{\rho}\frac{\partial F_z}{\partial \phi} - \frac{\partial F_\phi}{\partial z} = (\nabla \times \boldsymbol{F})_\rho, \tag{2.60}$$

$$\text{curl}_\phi \boldsymbol{F} = \frac{\partial F_\rho}{\partial z} - \frac{\partial F_z}{\partial \rho} = (\nabla \times \boldsymbol{F})_\phi, \tag{2.61}$$

$$\text{curl}_z \boldsymbol{F} = \frac{1}{\rho}[\frac{\partial}{\partial \rho}(\rho F_\phi) - \frac{\partial F_\rho}{\partial \phi}] = (\nabla \times \boldsymbol{F})_z. \tag{2.62}$$

In the spherical coordinates,

$$\text{curl}_r \boldsymbol{F} = \frac{1}{r \sin\theta}[\frac{\partial}{\partial \theta}(\sin\theta\, F_\phi) - \frac{\partial}{\partial \phi}(F_\theta)] = (\nabla \times \boldsymbol{F})_r, \tag{2.63}$$

$$\text{curl}_\theta \boldsymbol{F} = \frac{1}{r \sin\theta}[\frac{\partial}{\partial \phi}(F_r) - \sin\theta\, \frac{\partial}{\partial r}(rF_\phi)] = (\nabla \times \boldsymbol{F})_\theta, \tag{2.64}$$

$$\text{curl}_\phi \boldsymbol{F} = \frac{1}{r}[\frac{\partial}{\partial r}(rF_\theta) - \frac{\partial}{\partial \theta}(F_r)] = (\nabla \times \boldsymbol{F})_\phi. \tag{2.65}$$

Laplacian of Scalar Point Function

The Laplacian or second differential invariant of a scalar point function V is defined as the divergence of the gradient of the function or, symbolically,

$$\nabla^2 V = \Delta V = \text{div grad } V. \tag{2.66}$$

In the three coordinate systems,

$$\nabla^2 V = \frac{\partial^2 V}{\partial x^2} + \frac{\partial^2 V}{\partial y^2} + \frac{\partial^2 V}{\partial z^2}, \tag{2.67}$$

$$\nabla^2 V = \frac{1}{\rho}[\frac{\partial}{\partial \rho}(\rho \frac{\partial V}{\partial \rho}) + \frac{1}{\rho}\frac{\partial^2 V}{\partial \phi^2} + \rho\frac{\partial^2 V}{\partial z^2}], \tag{2.68}$$

$$\nabla^2 V = \frac{1}{r^2 \sin\theta}\left[\sin\theta \frac{\partial}{\partial r}\left(r^2 \frac{\partial V}{\partial r}\right) + \frac{\partial}{\partial \theta}\left(\sin\theta \frac{\partial V}{\partial \theta}\right) + \frac{1}{\sin\theta} \frac{\partial^2 V}{\partial \phi^2}\right]. \tag{2.69}$$

In the general orthogonal curvilinear coordinates,

$$\nabla^2 V = \frac{1}{e_1 e_2 e_3}\left[\frac{\partial}{\partial u}\left(\frac{e_2 e_3}{e_1} \frac{\partial V}{\partial u}\right) + \frac{\partial}{\partial v}\left(\frac{e_3 e_1}{e_2} \frac{\partial V}{\partial v}\right) + \frac{\partial}{\partial w}\left(\frac{e_1 e_2}{e_3} \frac{\partial V}{\partial w}\right)\right]. \tag{2.70}$$

Laplacian of Vector Point Function

The Laplacian of a vector point function $\boldsymbol{A}$ is defined as

$$\nabla^2 \boldsymbol{A} = \nabla(\nabla \cdot \boldsymbol{A}) - \nabla \times \nabla \times \boldsymbol{A} = \text{grad (div } \boldsymbol{A}) - \text{curl curl } \boldsymbol{A}. \tag{2.71}$$

In the rectangular coordinates,

$$\nabla^2 \boldsymbol{A} = \boldsymbol{u}_x \nabla^2 A_x + \boldsymbol{u}_y \nabla^2 A_y + \boldsymbol{u}_z \nabla^2 A_z. \tag{2.72}$$

In the cylindrical coordinates,

$$\begin{aligned} \nabla^2 \boldsymbol{A} = {} & \boldsymbol{u}_\rho\left[\left(\frac{\partial^2 A_\rho}{\partial \rho^2} + \frac{1}{\rho^2} \frac{\partial^2 A_\rho}{\partial \phi^2} + \frac{\partial^2 A_\rho}{\partial z^2} + \frac{1}{\rho} \frac{\partial A_\rho}{\partial \rho}\right) - \frac{2}{\rho^2} \frac{\partial A_\phi}{\partial \rho} - \frac{A_\rho}{\rho^2}\right] \\ & + \boldsymbol{u}_\phi\left[\left(\frac{\partial^2 A_\phi}{\partial \rho^2} + \frac{1}{\rho^2} \frac{\partial^2 A_\phi}{\partial \phi^2} + \frac{\partial^2 A_\phi}{\partial z^2} + \frac{1}{\rho} \frac{\partial A_\phi}{\partial \rho}\right) + \frac{2}{\rho^2} \frac{\partial A_\rho}{\partial \phi} - \frac{A_\phi}{\rho^2}\right] \\ & + \boldsymbol{u}_z\left(\frac{\partial^2 A_z}{\partial \rho^2} + \frac{1}{\rho^2} \frac{\partial^2 A_z}{\partial \phi^2} + \frac{\partial^2 A_z}{\partial z^2} + \frac{1}{\rho} \frac{\partial A_z}{\partial \rho}\right). \end{aligned} \tag{2.73}$$

In the spherical coordinates,

$$\begin{aligned} \nabla^2 \boldsymbol{A} = {} & \boldsymbol{u}_r\left[\left(\frac{\partial^2 A_r}{\partial r^2} + \frac{1}{r^2} \frac{\partial^2 A_r}{\partial \theta^2} + \frac{1}{r^2 \sin^2\theta} \frac{\partial^2 A_r}{\partial \phi^2} + \frac{2}{r} \frac{\partial A_r}{\partial r} + \frac{\cot\theta}{r^2} \frac{\partial A_r}{\partial \theta}\right)\right. \\ & \left. - \frac{2}{r^2 \sin\theta} \frac{\partial A_\phi}{\partial \phi} - \frac{2}{r^2} A_r - \frac{2}{r^2} \cot\theta \, A_\theta - \frac{2}{r^2} \frac{\partial A_\theta}{\partial \theta}\right] \\ & + \boldsymbol{u}_\theta\left[\left(\frac{\partial^2 A_\theta}{\partial r^2} + \frac{1}{r^2} \frac{\partial^2 A_\theta}{\partial \theta^2} + \frac{1}{r^2 \sin^2\theta} \frac{\partial^2 A_\theta}{\partial \phi^2} + \frac{2}{r} \frac{\partial A_\theta}{\partial r} + \frac{\cot\theta}{r^2} \frac{\partial A_\theta}{\partial \theta}\right)\right. \\ & \left. + \frac{2}{r^2} \frac{\partial A_r}{\partial \theta} - \frac{2}{r^2} \operatorname{cosec}\theta \cot\theta \frac{\partial A_\phi}{\partial \phi} - \frac{1}{r^2} \operatorname{cosec}^2\theta \, A_\theta\right] \\ & + \boldsymbol{u}_\phi\left[\left(\frac{\partial^2 A_\phi}{\partial r^2} + \frac{1}{r^2} \frac{\partial^2 A_\phi}{\partial \theta^2} + \frac{1}{r^2 \sin^2\theta} \frac{\partial^2 A_\phi}{\partial \phi^2} + \tfrac{1}{2} \frac{\partial A_\phi}{\partial r} + \frac{1}{r^2} \cot\theta \frac{\partial A_\phi}{\partial \theta}\right)\right. \\ & \left. + \frac{2}{r^2 \sin\theta} \frac{\partial A_r}{\partial \phi} + \frac{2}{r^2} \operatorname{cosec}\theta \cot\theta \frac{\partial A_\theta}{\partial \phi} - \frac{1}{r^2} \operatorname{cosec}^2\theta \, A_\phi\right]. \end{aligned} \tag{2.74}$$

Vector Identities

Some vector identities are as follows:

$$\nabla \times (\nabla V) = 0, \tag{2.75}$$

$$\nabla \times \nabla \times \boldsymbol{F} = \nabla(\nabla \cdot \boldsymbol{F}) - \nabla^2 \boldsymbol{F}, \tag{2.76}$$

$$\nabla \times (V\boldsymbol{F}) = V\,\nabla \times \boldsymbol{F} - \boldsymbol{F} \times (\nabla V), \tag{2.77}$$

$$\nabla \cdot (\nabla \times \boldsymbol{F}) = 0, \tag{2.78}$$

$$\nabla \cdot (V\boldsymbol{F}) = V(\nabla \cdot \boldsymbol{F}) + \boldsymbol{F} \cdot (\nabla V), \tag{2.79}$$

$$\nabla \cdot (\boldsymbol{F} \times \boldsymbol{G}) = \boldsymbol{G} \cdot \nabla \times \boldsymbol{F} - \boldsymbol{F} \cdot \nabla \times \boldsymbol{G}, \tag{2.80}$$

$$\int_V \boldsymbol{F} \cdot \nabla \times \boldsymbol{G}\, dV = \int_V \boldsymbol{G} \cdot \nabla \times \boldsymbol{F}\, dV + \oint_S \boldsymbol{n} \cdot (\boldsymbol{G} \times \boldsymbol{F})\, dS. \tag{2.81}$$

In Eq. (2.81), S is a closed surface enclosing the volume V.

Green's Theorem

From the vector identity

$$\nabla \cdot (V\boldsymbol{F}) = V(\nabla \cdot \boldsymbol{F}) + \boldsymbol{F} \cdot (\nabla V), \tag{2.82}$$

we have

$$\nabla \cdot (V\nabla g) = (\text{grad } V) \cdot (\text{grad } g) + V\nabla^2 g \tag{2.83}$$

by replacing $\boldsymbol{F}$ by the gradient of a scalar function g. Therefore, by the divergence theorem, we obtain

$$\int_V V\nabla^2 g + \int_V (\text{grad } V) \cdot (\text{grad } g)\, dV = \int_V V\nabla^2 g\, dV = \int_V V \cdot \nabla \cdot (\text{grad } g)\, dV \tag{2.84}$$

$$= \oint_S V \cdot \frac{\partial g}{\partial n}\, dS. \tag{2.85}$$

Interchanging V and g in Eq. (2.85), we obtain

$$\int_V g\nabla^2 V\, dV + \int_V (\text{grad } V) \cdot (\text{grad } g)\, dV = \oint_S g \cdot \frac{\partial V}{\partial n}\, dS. \tag{2.86}$$

Subtracting Eq. (2.84) from Eq. (2.86), we get

$$\int_V (V\nabla^2 g - g\nabla^2 V)\, dV = \oint_S \left(V\frac{\partial g}{\partial n} - g\frac{\partial V}{\partial n}\right) dS \tag{2.87}$$

which is called Green's formula or Green's theorem. The vector analogue of this theorem is

$$\int_V (\boldsymbol{F} \cdot \nabla \times \nabla \times \boldsymbol{G} - \boldsymbol{G} \cdot \nabla \times \nabla \times \boldsymbol{F})\, dV = \oint_S [\boldsymbol{G} \times (\nabla \times \boldsymbol{F}) - \boldsymbol{F} \times (\nabla \times \boldsymbol{G})] \cdot \boldsymbol{n}\, dS. \tag{2.88}$$

2.2 Dirac's Delta-Function

Dirac's delta-function is defined as an improper function existing in the neighbourhood of a point, say, r_0, and vanishing everywhere else on a one-dimensional space. Mathematically, $\delta(r - r_0)$ is defined such that

$$\int_\alpha^\beta \delta(r - r_0)\, dr = 1, \tag{2.89}$$

$$\delta(r - r_0) = 0 \quad \text{for } r \neq r_0, \tag{2.90}$$

where r_0 is in the interval (α, β) and $\delta(r - r_0)$ is the delta-function.

From definitions (2.89) and (2.90), it follows that, for a function $f(r)$ which is continuous at $r = r_0$,

$$\int_{\alpha}^{\beta} f(r)\delta(r - r_0) = f(r_0), \tag{2.91}$$

$$f(r)\delta(r - r_0) = f(r_0)\delta(r - r_0), \tag{2.92}$$

$$f(r)\delta'(r) = -f'(r)\delta(r), \tag{2.93}$$

$$f(r)\delta^n(r) = (-1)^n f^n(r)\delta(r), \tag{2.94}$$

$$\delta(r - r_0) = \frac{1}{2\pi}\int_{-\infty}^{\infty} \exp[-j\omega(r - r_0)]\, d\omega, \tag{2.95}$$

where $j = \sqrt{-1}$, and δ' and δ^n respectively denote the first and n-th derivative of δ.

The delta-function can be extended to more than one variable. For instance,

$$\delta(x - x_0)\delta(y - y_0) = 0 \quad \text{for } x \neq x_0,\ y \neq y_0, \tag{2.96}$$

$$\iint \delta(x - x_0)\delta(y - y_0)\, dx\, dy = 1, \tag{2.97}$$

$$\delta(x - x_0)\delta(y - y_0)\delta(z - z_0) = 0 \quad \text{for } x \neq x_0,\ y \neq y_0,\ z \neq z_0, \tag{2.98}$$

$$\iiint \delta(x - x_0)\delta(y - y_0)\delta(z - z_0)\, dx\, dy\, dz = 1. \tag{2.99}$$

The delta-function is used, for example, in applications of electromagnetic boundary conditions at a point source and a line source.

2.3 Matrices

Consider a lumped constant equivalent of a microwave network, as shown in Fig. 2.15. The

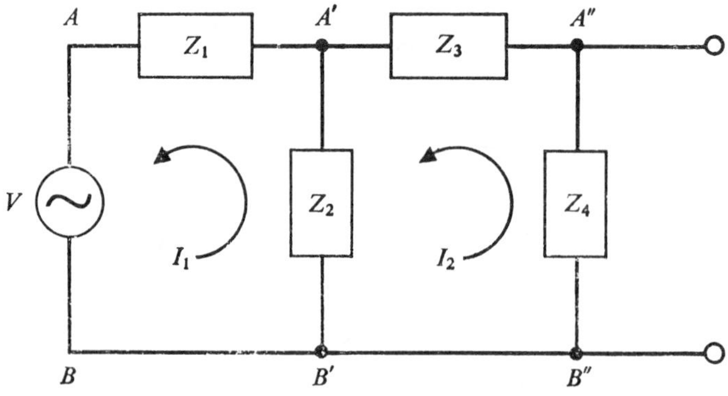

Fig. 2.15 Lumped constant equivalent of microwave network.

voltage equations for the two loops in the figure are

$$V = I_1(Z_1 + Z_2) - I_2Z_2 = Z_{11}I_1 + Z_{12}I_2, \tag{2.100}$$

$$0 = -I_1Z_2 + I_2(Z_2 + Z_3 + Z_4) = Z_{21}I_1 + Z_{22}I_2. \tag{2.101}$$

Equations (2.100) and (2.101) can be rewritten as

$$\begin{pmatrix} V \\ 0 \end{pmatrix} = \begin{pmatrix} Z_{11} & Z_{12} \\ Z_{21} & Z_{22} \end{pmatrix} \begin{pmatrix} I_1 \\ I_2 \end{pmatrix} \tag{2.102}$$

or

$$\hat{V} = \hat{Z}\hat{I}, \tag{2.103}$$

where

$$\hat{Z} = \begin{pmatrix} Z_{11} & Z_{12} \\ Z_{21} & Z_{22} \end{pmatrix}$$

is called the impedance matrix, and $\hat{V}$ and $\hat{I}$, referred to as the voltage and current matrices respectively, are given by

$$\hat{V} = \begin{pmatrix} V \\ 0 \end{pmatrix}, \qquad \hat{I} = \begin{pmatrix} I_1 \\ I_2 \end{pmatrix}. \tag{2.104}$$

These are one-column matrices and are also called vectors. Solving between Eqs. (2.100) and (2.101) for I_1 and I_2, we obtain

$$I_1 = \begin{vmatrix} V & Z_{12} \\ 0 & Z_{22} \end{vmatrix} \div \begin{vmatrix} Z_{11} & Z_{12} \\ Z_{21} & Z_{22} \end{vmatrix}, \tag{2.105}$$

$$I_2 = \begin{vmatrix} Z_{11} & V \\ Z_{21} & 0 \end{vmatrix} \div \begin{vmatrix} Z_{11} & Z_{12} \\ Z_{21} & Z_{22} \end{vmatrix}, \tag{2.106}$$

where

$$\begin{vmatrix} V & Z_{12} \\ 0 & Z_{22} \end{vmatrix}, \qquad \begin{vmatrix} Z_{11} & Z_{12} \\ Z_{21} & Z_{22} \end{vmatrix}$$

are determinants of the second order, and

$$|Z| = \begin{vmatrix} Z_{11} & Z_{12} \\ Z_{21} & Z_{22} \end{vmatrix} = Z_{11}Z_{22} - Z_{12}Z_{21} \ldots . \tag{2.107}$$

In general, for a network consisting of n loops,

$$\hat{V} = \begin{pmatrix} V_1 \\ V_2 \\ \vdots \\ V_n \end{pmatrix}, \qquad \hat{I} = \begin{pmatrix} I_1 \\ I_2 \\ \vdots \\ I_n \end{pmatrix}, \tag{2.108}$$

$$\hat{Z} = \begin{pmatrix} Z_{11} & Z_{12} & \ldots & Z_{1n} \\ Z_{21} & Z_{22} & \ldots & Z_{2n} \\ \vdots & \vdots & & \vdots \\ Z_{n1} & Z_{n2} & \ldots & Z_{nn} \end{pmatrix}. \tag{2.109}$$

Diagonal Matrix

A matrix $\hat{Z}$ is called diagonal if it is an $n \times n$ square matrix and all its elements except the diagonal elements are zero, that is,

$$\hat{Z} = \begin{pmatrix} Z_{11} & 0 & 0 & \dots & 0 \\ 0 & Z_{22} & 0 & \dots & 0 \\ \vdots & \vdots & \vdots & & \vdots \\ 0 & 0 & 0 & \dots & Z_{nn} \end{pmatrix}. \tag{2.110}$$

Such a diagonal matrix can represent a network consisting of several loops. In this network, there is no mutual impedance between any two loops. If

$$\hat{Z} = \begin{pmatrix} Z_0 & 0 & 0 & \dots & 0 \\ 0 & Z_0 & 0 & \dots & 0 \\ \vdots & \vdots & \vdots & & \vdots \\ 0 & 0 & 0 & \dots & Z_0 \end{pmatrix}, \tag{2.111}$$

then $\hat{Z}$ is called a *scalar matrix*, and it is a diagonal matrix whose elements are all equal. If

$$\boldsymbol{B} = \begin{pmatrix} B_{11} & B_{12} & \dots & B_{1n} \\ B_{21} & B_{22} & \dots & B_{2n} \\ \vdots & \vdots & & \vdots \\ B_{n1} & B_{n2} & \dots & B_{nn} \end{pmatrix}, \tag{2.112}$$

then

$$\hat{Z}\hat{B} = \begin{pmatrix} Z_0B_{11} & Z_0B_{12} & \dots & Z_0B_{1n} \\ Z_0B_{21} & Z_0B_{22} & \dots & Z_0B_{2n} \\ \vdots & \vdots & & \vdots \\ Z_0B_{n1} & Z_0B_{n2} & \dots & Z_0B_{nn} \end{pmatrix}. \tag{2.113}$$

This type of scalar multiplication of matrices results from a network that includes one or more transformers.

Identity Matrix

A scalar matrix whose diagonal elements are equal to unity is called an *identity matrix*, that is,

$$\hat{i} = \begin{pmatrix} 1 & 0 & 0 & \dots & 0 \\ 0 & 1 & 0 & \dots & 0 \\ \vdots & \vdots & \vdots & & \vdots \\ 0 & 0 & 0 & \dots & 1 \end{pmatrix}. \tag{2.114}$$

Also, $\hat{i}\hat{Z} = \hat{Z}$, that is, the product of any matrix with the identity matrix yields the matrix itself.

Multiplication of Two Matrices

The multiplication of two matrices is possible if and only if the number of columns in the first matrix is the same as the number of rows in the second matrix. For example, the product of the matrices $\hat{A}(m \times n)$ and $\hat{B}(n \times s)$ is

$$\hat{A}(m \times n) \times \hat{B}(n \times s) = \hat{C}(m \times s), \tag{2.115}$$

where

$$\hat{A} = \begin{pmatrix} a_{11} & a_{12} & \dots & a_{1n} \\ a_{21} & a_{22} & \dots & a_{2n} \\ \vdots & \vdots & & \vdots \\ a_{m1} & a_{m2} & \dots & a_{mn} \end{pmatrix}, \tag{2.116}$$

$$\hat{B} = \begin{pmatrix} b_{11} & b_{12} & \dots & b_{1s} \\ b_{21} & b_{22} & \dots & b_{2s} \\ \vdots & \vdots & & \vdots \\ b_{n1} & b_{n2} & \dots & b_{ns} \end{pmatrix}, \tag{2.117}$$

$$\hat{C} = \begin{pmatrix} c_{11} & c_{12} & \dots & c_{1s} \\ c_{21} & c_{22} & \dots & c_{2s} \\ \vdots & \vdots & & \vdots \\ c_{m1} & c_{m2} & \dots & c_{ms} \end{pmatrix}, \tag{2.118}$$

with

$$\begin{aligned} C_{11} &= a_{11}b_{11} + a_{12}b_{21} + a_{13}b_{31} + \dots + a_{1n}b_{n1} \\ C_{12} &= a_{11}b_{12} + a_{12}b_{22} + a_{13}b_{32} + \dots + a_{1n}b_{n2} \\ &\vdots \\ C_{ms} &= a_{m1}b_{1s} + a_{m2}b_{2s} + a_{m3}b_{3s} + \dots + a_{mn}b_{ns} \end{aligned} \tag{2.119}$$

or

$$c_{ij} = \sum_{k=1}^{n} a_{ik}b_{kj}. \tag{2.120}$$

Example The product

$$\begin{pmatrix} 2 & 3 & 4 \\ 1 & 2 & 3 \\ 4 & 5 & 6 \end{pmatrix} \times \begin{pmatrix} 1 & 2 \\ 3 & 4 \\ 5 & 6 \end{pmatrix} = \begin{pmatrix} 31 & 40 \\ 22 & 28 \\ 49 & 64 \end{pmatrix}. \tag{2.121}$$

Inverse Matrix

If $\hat{Z}$ is a square matrix and $\hat{i}$ the identity matrix, then the inverse matrix $\hat{Z}^{-1}$ is defined such that

$$\hat{Z} \times \hat{Z}^{-1} = \hat{i}. \tag{2.122}$$

It should be noted that $\hat{Z}^{-1}$ is also a square matrix.

Transpose of Matrix

If $\hat{Z}$ is a square matrix, then its transpose $\tilde{Z}$ is obtained by writing $\hat{Z}$ with the rows and columns interchanged. That is, if

$$\hat{Z} = \begin{pmatrix} Z_{11} & Z_{12} & \dots & Z_{1n} \\ Z_{21} & Z_{22} & \dots & Z_{2n} \\ \vdots & \vdots & & \vdots \\ Z_{n1} & Z_{n2} & \dots & Z_{nn} \end{pmatrix}, \tag{2.123}$$

then

$$\tilde{Z} = \begin{pmatrix} Z_{11} & Z_{21} & \dots & Z_{n1} \\ Z_{12} & Z_{22} & \dots & Z_{n2} \\ \vdots & \vdots & & \vdots \\ Z_{1n} & Z_{2n} & \dots & Z_{nn} \end{pmatrix}. \tag{2.124}$$

Adjoint of Matrix $\hat{Z}$

$\hat{Z}^{(a)}$ is called the adjoint matrix of $\hat{Z}$ if each element of $\hat{Z}^{(a)}$ is the cofactor of the corresponding element in $\tilde{Z}$. That is,

$$\hat{Z}^{(a)} = \begin{pmatrix} Z'_{11} & Z'_{21} & \dots & Z'_{n1} \\ Z'_{12} & Z'_{22} & \dots & Z'_{n2} \\ \vdots & \vdots & & \vdots \\ Z'_{1n} & Z'_{2n} & \dots & Z'_{nn} \end{pmatrix}, \tag{2.125}$$

where

$$Z'_{11} = \begin{pmatrix} Z_{22} & Z_{32} & \dots & Z_{n2} \\ Z_{23} & Z_{33} & \dots & Z_{n3} \\ \vdots & \vdots & & \vdots \\ Z_{2n} & Z_{3n} & \dots & Z_{nn} \end{pmatrix}. \dots \tag{2.126}$$

To get the inverse matrix $\hat{Z}^{-1}$ of $\hat{Z}$:

(i) Obtain $\tilde{Z}$.

(ii) Find the adjoint matrix $Z^{(a)}$ of $\tilde{Z}$.

(iii) Divide each element of $Z^{(a)}$ by the determinant of $\hat{Z}$, i.e., by $|Z|$.

Example Let

$$\hat{Z} = \begin{pmatrix} Z_{11} & Z_{12} & Z_{13} \\ Z_{21} & Z_{22} & Z_{23} \\ Z_{31} & Z_{32} & Z_{33} \end{pmatrix}. \tag{2.127}$$

Then

$$\tilde{Z} = \begin{pmatrix} Z_{11} & Z_{21} & Z_{31} \\ Z_{12} & Z_{22} & Z_{32} \\ Z_{13} & Z_{23} & Z_{33} \end{pmatrix}, \tag{2.128}$$

$$\hat{Z}^{-1} = \begin{pmatrix} \dfrac{Z_{22}Z_{33} - Z_{32}Z_{23}}{|Z|} & \dfrac{Z_{32}Z_{13} - Z_{12}Z_{33}}{|Z|} & \dfrac{Z_{12}Z_{23} - Z_{22}Z_{13}}{|Z|} \\ \dfrac{Z_{23}Z_{31} - Z_{21}Z_{33}}{|Z|} & \dfrac{Z_{11}Z_{33} - Z_{31}Z_{13}}{|Z|} & \dfrac{Z_{21}Z_{13} - Z_{11}Z_{23}}{|Z|} \\ \dfrac{Z_{21}Z_{32} - Z_{22}Z_{31}}{|Z|} & \dfrac{Z_{12}Z_{31} - Z_{11}Z_{32}}{|Z|} & \dfrac{Z_{11}Z_{22} - Z_{21}Z_{12}}{|Z|} \end{pmatrix}, \tag{2.129}$$

where

$$|Z| = \begin{vmatrix} Z_{11} & Z_{12} & Z_{13} \\ Z_{21} & Z_{22} & Z_{23} \\ Z_{31} & Z_{32} & Z_{33} \end{vmatrix}. \tag{2.130}$$

The use of matrices in the form of scattering matrices and impedance matrices is very useful in the analysis and synthesis of microwave circuits, especially in multiport waveguide junctions.

2.4 Green's Function

To obtain the field resulting from an electromagnetic source, usually the linear partial differential equation, called the wave equation, has to be solved. In many problems, the solution is found in the form of an infinite series such that each term of the series is, by itself, a solution of the differential equation. The coefficient of each such term is evaluated from the boundary and source conditions. If the series converges very slowly, then the computation of the solution becomes tedious. An alternative solution can be found in the form of an integral. Such a solution is called a closed-form solution. To obtain this solution, Green's function is used.

As an example, let us see how the current $I(t)$ can be determined in a circuit due to an electromotive force $V(t)$ applied across the terminals of the circuit. A unit impulse function is defined as in Fig. 2.16 such that the area of the step is one. Let $G(t, \hat{t})$ be the current at time t

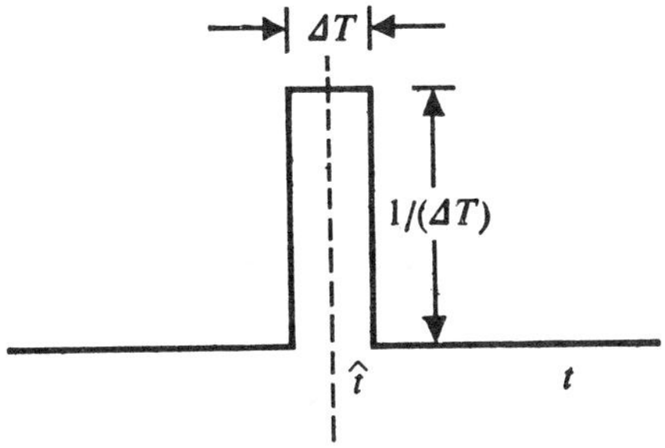

Fig. 2.16 Unit impulse function at $t = \hat{t}$.

due to a unit impulse of electromotive force at time $\hat{t}$. Then, $G(t, \hat{t})$ is called Green's function. Thus

$$I(t) = \int_{-\infty}^{t} V(\hat{t})G(t, \hat{t})\, d\hat{t}. \tag{2.131}$$

That is, $I(t)$ is the sum of all the contributions of current at time t due to the distributed source function V applied for all different times $\hat{t}$ up to t.

As another example, let us consider Green's function as a function of position instead of time. If $G(r, \hat{r})$ is the field at some point r, caused by a unit impulse source at the point $\hat{r}$, then the field $F(r)$ due to an arbitrary source distribution $\rho(\hat{r})$ over the entire range $\hat{r}$ is

$$F(r) = \int \rho(\hat{r})G(r, \hat{r})\, d\hat{r}, \tag{2.132}$$

the integral being taken over the entire range $\hat{r}$.

Properties of Green's Function

Green's function is a symmetric function of two sets of coordinates, namely, $(r, \hat{r})$ and $(t, \hat{t})$, and therefore it satisfies the reciprocity relation. Thus, if the response at r due to a unit source at $\hat{r}$ is $G(r, \hat{r})$, then the response at $\hat{r}$ due to a unit source at r is $G(\hat{r}, r)$, and

$$G(r, \hat{r}) = G(\hat{r}, r). \tag{2.133}$$

Making use of Green's function for the Helmholtz equation for a given geometry with homogeneous boundary conditions, it is possible to obtain a solution of the inhomogeneous equation

$$(\nabla^2 + \beta^2)\psi = \begin{cases} F & \text{at } P(x_0, y_0, z_0) \\ 0 & \text{elsewhere} \end{cases} \tag{2.134}$$

with a given homogeneous boundary condition such as $\psi = 0$, at any given surface S. Similarly, by employing Green's function, we may obtain a solution of the homogeneous Helmholtz equation

$$(\nabla^2 + \beta^2)\psi = 0 \tag{2.135}$$

with a given inhomogeneous boundary condition such as

$$\psi = \begin{cases} F & \text{on boundary} \\ 0 & \text{elsewhere} \end{cases} \tag{2.136}$$

Examples (i) In microwave engineering problems, an equation we frequently come across is

$$(\nabla^2 + \beta^2)\psi = -\rho(r), \tag{2.137}$$

where $\rho(r)$ is a distributed or discrete source. Here, the boundary condition is given by $\psi = 0$ on the closed bounded surface S. Let $G(r, \hat{r})$ be a Green's function that satisfies the inhomogeneous Helmholtz equation

$$\nabla^2 G(r, \hat{r}) + \beta^2 G(r, \hat{r}) = -\delta(r - \hat{r}) \tag{2.138}$$

for a unit point source at $\hat{r}$, where ∇^2 operates on r. Also, let G satisfy the boundary condition $G(r, \hat{r}) = 0$ when r lies on S. Assume that $\hat{r}$ is inside S. Then, multiplying Eq. (2.137) by $G(r, \hat{r})$ and Eq. (2.138) by $\psi(r)$ and interchanging the variables r and $\hat{r}$, we obtain

$$G(r, \hat{r})\nabla_0^2\psi(\hat{r}) - \psi(\hat{r})\nabla_0^2 G(r, \hat{r}) = -\rho(\hat{r})G(r, \hat{r}) + \psi(\hat{r})\delta(r - \hat{r}), \tag{2.139}$$

where ∇_0^2 operates on $\hat{r}$.

If V is the volume enclosed within a closed surface S, then integrating Eq. (2.139) over all the source coordinates (x, y, z) inside S, we obtain

$$\int_V [G(r, \hat{r})\nabla_0^2\psi(\hat{r}) - \psi(\hat{r})\nabla_0^2 G(r, \hat{r})]\, dV(\hat{r}) + \int_V \rho(\hat{r})G(r, \hat{r})\, dV(\hat{r})$$
$$= \begin{cases} \psi(r) & \text{if } r \text{ is inside } S \\ 0 & \text{if } r \text{ is outside } S \end{cases} \tag{2.140}$$

Using Green's theorem

$$\int_V (f\nabla^2 g - g\nabla^2 f)\, dV = \oint_S \left(f\frac{\partial g}{\partial n} - g\frac{\partial f}{\partial n}\right) dS, \tag{2.141}$$

we reduce Eq. (2.140) to

$$\oint_S \left[G(r, \hat{r})\frac{\partial\psi}{\partial n} - \psi(\hat{r})\frac{\partial}{\partial n}G(r, \hat{r})\right] dS(\hat{r}) + \int_V \rho(\hat{r})G(r, \hat{r})\, dV(\hat{r})$$
$$= \begin{cases} \psi(r) & \text{if } r \text{ is inside } S \\ 0 & \text{if } r \text{ is outside } S \end{cases} \tag{2.142}$$

For the inhomogeneous equation

$$(\nabla^2 + \beta^2)\psi = -\rho(r) \tag{2.143}$$

with $\rho \neq 0$, the homogeneous boundary condition is $\psi = 0$ on S, and for the inhomogeneous equation

$$\nabla^2 G(r, \hat{r}) + \beta^2 G(r, \hat{r}) = -\delta(r - \hat{r}), \tag{2.144}$$

the homogeneous boundary condition is $G(r, \hat{r}) = 0$ when r lies on S. Hence, in Eq. (2.142),

$$\oint_S \left[G(r, \hat{r})\frac{\partial\psi}{\partial n} - \psi(\hat{r})\frac{\partial}{\partial n}G(r, \hat{r})\right] dS(\hat{r}) = 0. \tag{2.145}$$

Therefore, from Eq. (2.142), we obtain

$$\int_V \rho(\hat{r})G(r, \hat{r})\, dV(\hat{r}) = \psi(r) \quad \text{if } r \text{ is inside } S. \tag{2.146}$$

Hence, the problem of finding a solution to the inhomogeneous equation (2.143) in $\psi(r)$ reduces to the problem of solving the similar but simpler inhomogeneous equation (2.146) in $G(r, \hat{r})$.

(ii) The same technique as used in Example (i) can be employed for solving a homogeneous

Helmholtz equation with inhomogeneous boundary conditions. Let ψ satisfy the equation

$$(\nabla^2 + \beta^2)\psi = 0 \tag{2.147}$$

such that $\psi = \psi_0(\hat{r}_S)$ on the surface S_0, and let G be a Green's function satisfying the equation

$$(\nabla^2 + \beta^2)G = -\delta(r - \hat{r}) \tag{2.148}$$

with $G = 0$ on S and S_0. Then, since $\rho = 0$, from Eq. (2.142), we have

$$\psi(r) = -\int_{S_0} \psi(\hat{r}_S)\frac{\partial}{\partial n}G(r, \hat{r}_S)\, dS_0\, (\hat{r}_S). \tag{2.149}$$

(iii) Green's function can also be used in solving time dependent inhomogeneous differential equations. Let ψ satisfy the scalar partial differential equation of the two variables r and t such that

$$\nabla^2\psi - \frac{1}{C^2}\frac{\partial^2\psi}{\partial t^2} = -q(r, t), \tag{2.150}$$

where both q and ψ are functions of r and t. If $G(r, t)$ satisfies the equation

$$\nabla^2 G - \frac{1}{C^2}\frac{\partial^2 G}{\partial t^2} = -\delta(r - \hat{r})\delta(t - \hat{t}), \tag{2.151}$$

then $G(r, t)$ is a Green's function that represents the response due to an impulse function at the instant $t = \hat{t}$, and is located at $r = \hat{r}$, and we may obtain a closed-form solution of $\psi(r, t)$ subject to appropriate boundary and initial conditions.

It is not always easy to determine the appropriate Green's function that satisfies all the necessary boundary conditions. However, for simple microwave problems, such as determining the electric and magnetic fields inside rectangular or cylindrical metal waveguides and cavities, a Green's function can be easily obtained.

(iv) To obtain Green's function for a rectangular metal waveguide having a cross-section of dimensions a and b, it is necessary to solve the equation

$$\nabla^2 G + \beta^2 G = -\delta(x - x')\delta(y - y')\delta(z - z'), \tag{2.152}$$

where G is a function of x, x', y, y', and z, z'. The boundary conditions are $G = 0$ at the guide walls, that is, at $x = 0$ and $x = a$, and at $y = 0$ and $y = b$. From these boundary conditions, it is possible to express intuitively a solution in the form

$$G = \sum_{m,n} \sin\left(\frac{m\pi x}{a}\right) \sin\left(\frac{m\pi x'}{a}\right) \sin\left(\frac{n\pi y}{b}\right) \sin\left(\frac{n\pi y'}{b}\right) \times F_{mn}(z, z'), \tag{2.153}$$

where $F_{mn}(z, z')$ is to be determined. Substituting Eq. (2.153) into the inhomogeneous wave equation (2.152), we get

$$\sum_{m,n} \sin\left(\frac{m\pi x}{a}\right) \sin\left(\frac{m\pi x'}{a}\right) \sin\left(\frac{n\pi y}{b}\right) \sin\left(\frac{n\pi y'}{b}\right)\left(\gamma_{mn}^2 F_{mn} + \frac{\partial^2}{\partial z'^2}F_{mn}\right)$$

$$= -\delta(x - x')\delta(y - y')\delta(z - z'), \tag{2.154}$$

where

$$\gamma_{mn}^2 = \beta^2 - (\frac{m\pi}{a})^2 - (\frac{n\pi}{b})^2. \tag{2.155}$$

Multiplying both sides of Eq. (2.154) by

$$\sin (\frac{m\pi x'}{a}) \sin (\frac{n\pi y'}{b})\, dx'\, dy'$$

and integrating between the limits $(a, 0)$ and $(b, 0)$, we obtain

$$(\frac{\partial^2}{\partial z'^2} + \gamma_{mn}^2)F_{mn} = -\frac{4}{ab}\delta(z - z'). \tag{2.156}$$

For a physically realizable field, F_{mn} should be bounded or

$$|F_{mn}| < \infty \quad \text{as } z \to \pm\infty.$$

Hence

$$F_{mn} = \frac{2}{\gamma_{mn}ab} \exp\,[-\gamma_{m}|(z - z')|]. \tag{2.157}$$

Therefore

$$G = \sum_{m,\,n} \sin (\frac{m\pi x}{a}) \sin (\frac{m\pi x'}{a}) \sin (\frac{n\pi y}{b}) \sin (\frac{n\pi y'}{b}) \frac{2}{\gamma_{mn}ab} \exp\,[-\gamma_m|(z - z')|]. \tag{2.158}$$

2.5 Differential Equations

Many physical problems, including microwave problems, can be represented by a second-order differential equation of the form

$$A(x, y)\frac{d^2y}{dx^2} + B(x, y)\frac{dy}{dx} + C(x, y)y + D(x, y) = 0. \tag{2.159}$$

For a linear differential equation, A, B, C, and D are functions of the independent variable x alone, and hence the general form of a linear second-order differential equation becomes

$$p(x)\frac{d^2y}{dx^2} + q(x)\frac{dy}{dx} + r(x)y + s(x) = 0. \tag{2.160}$$

If $s(x) = 0$, Eq. (2.160) becomes a homogeneous linear differential equation of the form

$$\frac{d^2y}{dx^2} + a(x)\frac{dy}{dx} + b(x)y = 0. \tag{2.161}$$

The solution of Eq. (2.161) in a closed form exists for only a few types of functions such as $a(x)$ and $b(x)$. For example, consider the equation

$$\frac{d^2y}{dx^2} - by = 0, \tag{2.162}$$

where b is a constant. If b is complex, the solution of Eq. (2.162) is of the form

$$y = P \exp(\sqrt{b}\,x) + Q \exp(-\sqrt{b}\,x). \tag{2.163}$$

If b is real and positive, the solution is of the form

$$y = P' \cosh(\sqrt{b}\,x) + Q' \sinh(\sqrt{b}\,x); \tag{2.164}$$

but if b is real and negative, then it is of the form

$$y = P_1 \cos(\sqrt{-b}\,x) + Q_1 \sin(\sqrt{-b}\,x), \tag{2.165}$$

where P, Q, P', Q', P_1, and Q_1 are all arbitrary constants of integration.

In general, a second-order differential equation has two independent solutions whose linear combination can represent the general solution.

If, in the differential equation (2.161), there is a singularity in any of the functions $a(x)$ and $b(x)$ at the point $x = x_0$, then at least one of the independent solutions is nonanalytic at the point, that is, the solution at the point $x = x_0$ cannot be expanded as a Taylor series.

If the solution of the homogeneous equation (2.161) is known, then the solution of the non-homogeneous equation (2.160) can be determined by a simple method. If the two independent solutions of Eq. (2.161) are y_1 and y_2, and if $s(x)/p(x) = -f(x)$, then it can be assumed that the solution of Eq. (2.160) is of the form

$$y = c(x)y_1 + d(x)y_2, \tag{2.166}$$

where $c(x)$ and $d(x)$ are two arbitrary functions of x and have to be determined. Thus

$$y' = \frac{dy}{dx} = cy_1' + c'y_1 + dy_2' + d'y_2. \tag{2.167}$$

Since $c(x)$ and $d(x)$ are arbitrary functions, we can impose some further restrictions on them. Let

$$c'y_1 + d'y_2 = 0. \tag{2.168}$$

Then

$$\frac{d^2y}{dx^2} = y'' = c'y_1' + cy_1'' + d'y_2' + dy_2'', \tag{2.169}$$

$$y'' + ay' + by = f(x) = c(y_1'' + ay_1' + by_1) + d(y_2'' + ay_2' + by_2) + c'y_1' + d'y_2'. \tag{2.170}$$

On the other hand, from Eq. (2.161), we get

$$y_1'' + ay_1' + by_1 = y_2'' + ay_2' + by_2 = 0.$$

Hence

$$c'y_1' + d'y_2' = f(x), \tag{2.171}$$

$$c'y_1 + d'y_2 = 0 \tag{2.172}$$

are two algebraic equations. Solving for c' and d' between them, we obtain

$$c' = \frac{-y_2}{y_1y_2' - y_2y_1'} f(x), \tag{2.173}$$

$$d' = \frac{y_1}{y_1 y_2' - y_2 y_1'} f(x). \tag{2.174}$$

The function

$$(y_1 y_2' - y_2 y_1') = (y_1 \frac{dy_2}{dx} - y_2 \frac{dy_1}{dx})$$

is called the Wronskian W of the solution. Hence, the solution of Eq. (2.160) is of the form

$$y = -y_1 \int \frac{f(x) y_2}{W} dx + y_2 \int \frac{f(x) y_1}{W} dx. \tag{2.175}$$

If $W = 0$, then solution (2.175) is meaningless. But if y_1 and y_2 are independent solutions, then $W \neq 0$.

Most microwave problems involve linear differential equations, but nonlinear equations do occur in problems such as nonuniform transmission lines.

2.6 Bessel Equation and its Solutions

The Bessel equation is a particular form of the homogeneous linear differential equation

$$\frac{d^2y}{dz^2} + a(z)\frac{dy}{dz} + b(z)y = 0. \tag{2.176}$$

It is of the form

$$\frac{d^2y}{dz^2} + \frac{1}{z}\frac{dy}{dz} + (1 - \frac{\nu^2}{z^2})y = 0 \tag{2.177}$$

whose solutions are represented by

$$y = A_1 J_\nu(z) + B_1 J_{-\nu}(z) \quad \text{for } \nu \neq \text{integer and any } z \tag{2.178}$$

or

$$y = A_2 J_\nu(z) + B_2 N_\nu(z) \quad \text{for any } \nu \text{ and any } z, \tag{2.179}$$

where A_1, A_2 and B_1, B_2 are arbitrary constants. $J_\nu(z)$ is called the Bessel function of the first kind of order ν; it can be represented by an infinite series

$$J_\nu(z) = \sum_{m=0}^{\infty} \frac{(-1)^m (z/2)^{\nu+2m}}{m!(\nu + m)!}, \tag{2.180}$$

where the generalized factorial is defined in terms of the gamma function as

$$p! = \Gamma(p + 1).$$

$J_{-\nu}(z)$ can be obtained by replacing ν by $-\nu$ in Eq. (2.180). $N_\nu(z)$ is called the Neumann function and is sometimes represented by the symbol $Y_\nu(z)$. It is also referred to as the Bessel function of the second kind and is related to $J_\nu(z)$ as

$$N_\nu(z) = \cot(\pi\nu) J_\nu(z) - \operatorname{cosec}(\nu\pi) J_{-\nu}(z), \tag{2.181}$$

where the right-hand side is to be interpreted as a limiting value when ν is an integer.

The Bessel equation (2.177) is satisfied also by the Hankel functions $H_\nu^{(1)}(z)$ and $H_\nu^{(2)}(z)$, and the solution may be written in the form

$$y = A_3 H_\nu^{(1)}(z) + B_3 H_\nu^{(2)}(z), \tag{2.182}$$

where A_3 and B_3 are arbitrary constants. $H_\nu^{(1)}(z)$ is called the Hankel function of the first kind and $H_\nu^{(2)}(z)$ the Hankel function of the second kind. They are related to $J_\nu(z)$ and $N_\nu(z)$ as

$$H_\nu^{(1)}(z) = J_\nu(z) + jN_\nu(z), \tag{2.183}$$

$$H_\nu^{(2)}(z) = J_\nu(z) - jN_\nu(z). \tag{2.184}$$

The functions $N_\nu(z)$, $H_\nu^{(1)}(z)$, and $H_\nu^{(2)}(z)$ have singularities at the origin $z = 0$, and hence their use is restricted to regions where $z = 0$ is excluded. On the other hand, $J_\nu(z)$ is finite at $z = 0$, and can be used in regions including $z = 0$. However, $J_\nu(z)$ has a branch point at $z = 0$. If ν is a positive integer n, then

$$J_{-n}(z) = (-1)^n J_n(z), \tag{2.185}$$

and hence only one solution, which is regular at $z = 0$, exists. As ν approaches the integer n,

$$N_n(z) = \lim N_\nu(z), \tag{2.186}$$

and this equation continues to be a solution of the Bessel equation (2.177).

A series expansion for $N_n(z)$ is

$$N_n(z) = -\frac{1}{\pi}\sum_{m=0}^{n-1}\frac{(n-m-1)!\,2^{n-2m}}{m!\,z^{n-2m}} + \frac{2}{\pi}(\ln z + C - \ln 2)J_n(z)$$
$$-\frac{1}{\pi}\sum_{m=1}^{\infty}\frac{(-1)^m z^{n+2m}}{2^{n+2m}m!\,(m+n)!}[\phi(m) + \phi(m+n)], \tag{2.187}$$

where the auxiliary function $\phi(m)$ is defined as

$$\phi(m) = 1 + \tfrac{1}{2} + \tfrac{1}{3} + \dots + \frac{1}{m}. \tag{2.188}$$

In the limit $\nu \to n$,

$$N_n(z) = \frac{1}{\pi}\left[\frac{\partial J_\nu}{\partial \nu} + (-1)^{n+1}\frac{\partial J_{-\nu}}{\partial \nu}\right]_{\nu=n}. \tag{2.189}$$

Using expression (2.189) and series (2.180), we obtain

$$N_0(z) = \frac{2}{\pi}(\ln z + C - \ln 2)J_0(z) + \frac{2}{\pi}\sum_{m=1}^{\infty}\frac{(-1)^m z^{2m}}{2^{2m}(m!)^2}\phi(m), \tag{2.190}$$

where C is Euler's constant $= 0.5772$.

In differentiating the power series, the logarithmic derivative of the generalized factorial function $\psi(z)$ is encountered. This function is

$$\psi(z) = \frac{d}{dz}\ln(z!) = \frac{1}{z!}\frac{d(z!)}{dz} = -C + \sum_{n=1}^{\infty}\left(\frac{1}{n} - \frac{1}{n+z}\right). \tag{2.191}$$

For integral and other values of z, we have (m being an integer)

$$\begin{aligned} \psi(0) &= -C, \\ \psi(m) &= -C + \phi(m), \\ \psi(-m) &= \infty, \\ \psi(-\tfrac{1}{2}) &= -C - 2 \ln 2 = -1.96351 \ldots, \\ \psi(-\tfrac{1}{2} \pm m) &= \psi(-\tfrac{1}{2}) + 2(1 + \tfrac{1}{3} + \tfrac{1}{5} + \ldots + \frac{1}{2m-1}). \end{aligned} \tag{2.192}$$

The following reduction formulae are useful:

$$\begin{aligned} \psi(z) &= \psi(z-1) + \frac{1}{z}, \\ \psi(-z) &= \psi(z-1) + \pi \cot(\pi z), \\ \psi(z - \tfrac{1}{2}) &= \psi(-z - \tfrac{1}{2}) + \pi \tan(\pi z). \end{aligned} \tag{2.193}$$

Examples (i) For certain types of propagation (TE modes) in a circular cylindrical metal waveguide, the magnetic field intensity H_z satisfies the partial differential equation

$$\nabla^2 H_z + \beta^2 H_z = 0. \tag{2.194}$$

Expanding $\nabla^2 H_z$ in the circular cylindrical coordinates, we get

$$\frac{1}{\rho}\left[\frac{\partial}{\partial\rho}\left(\rho\frac{\partial H_z}{\partial\rho}\right) + \frac{1}{\rho}\frac{\partial^2 H_z}{\partial\rho^2} + \rho\frac{\partial^2 H_z}{\partial z^2}\right] + \beta^2 H_z = 0. \tag{2.195}$$

Putting

$$H_z = R(\rho)\Phi(\phi)Z(z) \tag{2.196}$$

and using the method of separation of variables, we obtain three ordinary differential equations in R, Φ, and Z:

$$\rho^2\frac{d^2R}{d\rho^2} + \rho\frac{dR}{d\rho} + [(\gamma^2 + \beta^2)\rho^2 - n^2]R = 0, \tag{2.197}$$

$$\frac{d^2\Phi}{d\phi^2} + n^2\Phi = 0, \tag{2.198}$$

$$\frac{d^2Z}{dz^2} - \gamma^2 Z = 0, \tag{2.199}$$

where γ and n are separation constants. If we divide Eq. (2.197) by ρ^2 and substitute $s = (\gamma^2 + \beta^2)^{1/2}\rho$, we obtain the Bessel equation

$$\frac{d^2R}{ds^2} + \frac{1}{s}\frac{dR}{ds} + (1 - \frac{n^2}{s^2})R = 0 \tag{2.200}$$

which has the solution

$$R = A_n J_n(\sqrt{\gamma^2 + \beta^2}\,\rho) + B_n N_n(\sqrt{\gamma^2 + \beta^2}\,\rho). \tag{2.201}$$

Inside the waveguide, since the axis $\rho = 0$ is included, B_n is set equal to zero to avoid the singularity of $N_n(\sqrt{\gamma^2 + \beta^2}\rho)$ at $\rho = 0$.

(ii) Spherical Bessel functions. Consider a metal spherical cavity. The components of the electric or magnetic field intensity in such a cavity satisfy the same partial differential equation, namely,

$$\nabla^2\psi + \beta^2\psi = 0. \tag{2.202}$$

Expanding $\nabla^2\psi$ in the spherical polar coordinates, we obtain

$$\nabla^2\psi + \beta^2\psi = \frac{1}{r^2 \sin\theta}[\sin\theta \frac{\partial}{\partial r}(r^2\frac{\partial\psi}{\partial r}) + \frac{1}{\sin\theta}\frac{\partial^2\psi}{\partial\phi^2} + \frac{\partial}{\partial\theta}(\sin\theta \frac{\partial\psi}{\partial\theta})] + \beta^2\psi = 0. \tag{2.203}$$

Putting

$$\psi = R(r)\Theta(\theta)\Phi(\phi) \tag{2.204}$$

and using the method of separation of variables, we obtain the ordinary differential equation in R as

$$\frac{d^2(Rr)}{dr^2} + [\beta^2 - \frac{n(n+1)}{r^2}](Rr) = 0 \tag{2.205}$$

whose general solution is given by

$$Rr = (\frac{\pi r}{2})^{1/2}[\bar{A}_n J_{n+1/2}(\beta r) + \bar{B}_n N_{n+1/2}(\beta r)] \tag{2.206}$$

or

$$R = \bar{A}_n j_n(\beta r) + \bar{B}_n n_n(\beta r), \tag{2.207}$$

where

$$j_n(\beta r) = \sqrt{\frac{\pi}{2r}} J_{n+1/2}(\beta r), \tag{2.208}$$

$$n_n(\beta r) = \sqrt{\frac{\pi}{2r}} N_{n+1/2}(\beta r). \tag{2.209}$$

The functions $j_n(\beta r)$ and $n_n(\beta r)$ are called the spherical Bessel functions.

The spherical Hankel functions $h_n^{(1)}(\beta r)$ and $h_n^{(2)}(\beta r)$ are defined as

$$h_n^{(1)}(\beta r) = j_n(\beta r) + jn_n(\beta r), \tag{2.210}$$

$$h_n^{(2)}(\beta r) = j_n(\beta r) - jn_n(\beta r). \tag{2.211}$$

Asymptotic Values of Bessel and Hankel Functions

Though the numerical evaluation of the Bessel and Hankel functions is very tedious, the asymptotic values of these functions for very large or very small values of z compared with the order ν make the computation easy. For $z \gg \nu$ (usually $> 10\nu$),

$$J_\nu \approx \sqrt{\frac{2}{\pi z}} \cos(z - \frac{\nu\pi}{2} - \frac{\pi}{4}), \tag{2.212}$$

$$N_\nu \approx \sqrt{\frac{2}{\pi z}} \sin\left(z - \frac{\nu\pi}{2} - \frac{\pi}{4}\right), \tag{2.213}$$

$$H_\nu^{(1)} \approx \sqrt{\frac{2}{\pi z}}\, e^{j(z-\nu\pi/2-\pi/4)}, \tag{2.214}$$

$$H_\nu^{(2)} \approx \sqrt{\frac{2}{\pi z}}\, e^{-j(z-\nu\pi/2-\pi/4)}. \tag{2.215}$$

For z large (but not too large) compared to ν, though the asymptotic series are, in general, not convergent, the following asymptotic expressions are useful:

$$J_\nu(z) = \sqrt{\frac{2}{\pi z}} \cos\left(z - \frac{\nu\pi}{2} - \frac{\pi}{4}\right) \sum_{m=0}^{\infty} \frac{(-1)^m(\nu, 2m)}{(2z)^{2m}}$$

$$-\sin\left(z - \frac{\nu\pi}{2} - \frac{\pi}{4}\right) \sum_{m=0}^{\infty} \frac{(-1)^m(\nu, 2m+1)}{(2z)^{2m+1}}, \tag{2.216}$$

$$N_\nu(z) = \sqrt{\frac{2}{\pi z}} \left[\sin\left(z - \frac{\nu\pi}{2} - \frac{\pi}{4}\right) \sum_{m=0}^{\infty} \frac{(-1)^m(\nu, 2m)}{(2z)^{2m}}\right.$$

$$\left. + \cos\left(z - \frac{\nu\pi}{2} - \frac{\pi}{4}\right) \sum_{m=0}^{\infty} \frac{(-1)^m(\nu, 2m+1)}{(2z)^{2m+1}}\right], \tag{2.217}$$

$$H_\nu^{(1)}(z) = \sqrt{\frac{2}{\pi z}}\, e^{j(z-\nu\pi/2-\pi/4)} \sum_{m=0}^{\infty} \frac{(-1)^m(\nu, m)}{(2jz)^m}, \tag{2.218}$$

$$H_\nu^{(2)}(z) = \sqrt{\frac{2}{\pi z}}\, e^{-j(z-\nu\pi/2-\pi/4)} \sum_{m=0}^{\infty} \frac{(-1)^m(\nu, m)}{(2jz)^m}, \tag{2.219}$$

where (ν, k) is an even function of ν and is defined as

$$(\nu, k) = \frac{(4\nu^2 - 1)(4\nu^2 - 3^2) \ldots [4\nu^2 - (2k-1)^2]}{2^{2k} k!} \tag{2.220}$$

for any integer k. The asymptotic expressions for $J_{-\nu}$ and $N_{-\nu}$ can be obtained from Eqs. (2.212) and (2.213) by replacing ν by $-\nu$. Since (ν, k) is an even function of ν for any k,

$$(\nu, k) = (-\nu, k).$$

All these asymptotic expressions are valid for large values of $|z|$ even when z is a complex quantity $x + jy$, but the phase angle of z, i.e., $\tan^{-1}(y/x)$, should be $<\pi$. The error due to stopping at any term of the series for $J_\nu(z)$, $N_\nu(z)$, $H_n^{(1)}(z)$, and $H_n^{(2)}(z)$ is of the order of the magnitude of that term multiplied by $1/z$.

For $z \ll 1$, the asymptotic expressions are

$$J_\nu(z) \approx \frac{z^\nu}{2^\nu \Gamma(\nu + 1)}, \tag{2.221}$$

$$N_\nu(z) \approx \frac{-2^\nu \Gamma(\nu)}{\pi z^\nu}, \tag{2.222}$$

$$H_\nu^{(1)}(z) \approx J_\nu(z)\left[1 - \frac{j}{\pi\nu J_\nu^2(z)}\right], \tag{2.223}$$

$$H_\nu^{(2)}(z) \approx J_\nu(z)[1 + \frac{j}{\pi\nu J^2(z)}]. \tag{2.224}$$

In particular, for $z \ll 1$,

$$J_0(z) \approx 1, \qquad N_0(z) \approx \frac{2}{\pi} (\ln z + C - \ln 2), \tag{2.225}$$

where

$$C = \text{Euler's constant} = 0.5772. \tag{2.226}$$

For integral values of ν, i.e., for $\nu = n$,

$$N_n(z) = -\frac{1}{\pi} \sum_{m=0}^{n-1} \frac{(n-m-1)!\, 2^{n-2m}}{z^{n-2m} m!} + \frac{2}{\pi}(\ln z + C - \ln 2)J_n(z)$$

$$- \frac{1}{\pi} \sum_{m=0}^{\infty} \frac{(-1)^m z^{n+2m}}{2^{n+2m} m!\,(m+n)!}(Q_m + Q_{m+n}), \tag{2.227}$$

$$N_0(z) = \frac{2}{\pi}(\ln z + C - \ln 2)J_0(z) + \frac{2}{\pi} \sum_{m=1}^{\infty} \frac{(-1)^{m+1} z^{2m}}{2^{2m}(m!)^2} Q_m, \tag{2.228}$$

where

$$Q_m = 1 + \tfrac{1}{2} + \tfrac{1}{3} + \ldots + \frac{1}{m}. \tag{2.229}$$

Modified Bessel Functions

The modified Bessel functions are solutions of the differential equation

$$\frac{d^2y}{dz^2} + \frac{1}{z}\frac{dy}{dz} - (1 + \frac{\nu^2}{z^2})y = 0 \tag{2.230}$$

whose solution is of the form

$$y(z) = AI_\nu(z) + BK_\nu(z), \tag{2.231}$$

where A and B are arbitrary constants and $I_\nu(z)$ and $K_\nu(z)$ are respectively called the modified Bessel functions of the first and second kind. The function $I_\nu(z)$ is given by

$$I_\nu(z) = e^{-j\nu\pi/2} J_\nu(jz) \tag{2.232}$$

$$= \sum_{m=0}^{\infty} \frac{(z/2)^{\nu+2m}}{m!\,\Gamma(\nu + m + 1)}. \tag{2.233}$$

If ν is a positive integer n, then

$$I_{-n}(z) = I_n(z), \tag{2.234}$$

$$K_\nu(z) = \frac{\pi}{2}[I_{-\nu}(z) - I_\nu(z)] \operatorname{cosec}(\pi\nu) \quad \text{for all } \nu. \tag{2.235}$$

If ν is a complex quantity, whose absolute value of the phase angle is less than $3\pi/2$, then the asymptotic expansions of $K_\nu(z)$ and $I_\nu(z)$ for large values of z are

$$K_\nu(z) \approx \left(\frac{\pi}{2z}\right)^{1/2} e^{-z}\left[1 + \frac{4\nu^2 - 1}{1!\,8z} + \frac{(4\nu^2 - 1)(4\nu^2 - 3^2)}{2!\,(8z)^2} + \ldots\right], \tag{2.236}$$

$$I_\nu(z) \approx \frac{e^z}{\sqrt{2\pi z}} \sum_{m=0}^{\infty} \frac{(-1)^m(\nu, m)}{(2z)^m} + \frac{e^{-z+(\nu+1/2)\pi j}}{\sqrt{2\pi z}} \sum_{m=0}^{\infty} \frac{(\nu, m)}{(2z)^m}. \tag{2.237}$$

For small values of z, that is, for $z \ll 1$, the asymptotic expressions for $I_\nu(z)$ and $K_\nu(z)$ are

$$I_\nu(z) \approx \frac{z^\nu}{2^\nu \Gamma(\nu + 1)}, \tag{2.238}$$

$$K_\nu(z) \approx \frac{2^{\nu-1}\Gamma(\nu)}{z^\nu}. \tag{2.239}$$

In particular, for $z \ll 1$,

$$I_0(z) \approx 1, \tag{2.240}$$

$$K_0(z) = -(\ln z + C - \ln 2), \tag{2.241}$$

where $C = 0.5772 =$ Euler's constant.

Wronskians of Bessel Functions

The Wronskians of Bessel functions may be defined as

$$J_\nu(z)N_{\nu+1}(z) - J_{\nu+1}(z)N_\nu(z) = -\frac{2}{\pi z}, \tag{2.242}$$

$$J_\nu(z)N'_\nu(z) - N_\nu(z)J'_\nu(z) = \frac{2}{\pi z}, \tag{2.243}$$

$$J_\nu(z)N''_\nu(z) - N_\nu(z)J''_\nu(z) = \frac{-2}{\pi z^2}, \tag{2.244}$$

$$I_\nu(z)I'_{-\nu}(z) - I'_\nu(z)I_{-\nu}(z) = \frac{-2\sin(\nu\pi)}{\pi z}, \tag{2.245}$$

where the primes and double primes respectively denote the first and second derivatives with respect to the argument z.

Additional Formulae for Bessel Functions

Some of the important formulae for Bessel functions are

$$J_0\beta(\rho^2 + a^2 - 2\rho a \cos\phi)^{1/2} = \sum_{m=-\infty}^{\infty} J_m(\beta\rho)J_m(\beta a)e^{im\phi}, \tag{2.246}$$

$$J_\nu^2(z) = \sum_{m=0}^{\infty} \frac{(-1)^m(z/2)^{2\nu+2m}\Gamma(2\nu + 2m + 1)}{m!\,\Gamma(2\nu + m + 1)[\Gamma(\nu + m + 1)]^2}, \tag{2.247}$$

$$J_\nu(z)J_{-\nu}(z) = \sum_{m=0}^{\infty} \frac{(-1)^m(z/2)^{2m}2m!}{(m!)^2\Gamma(\nu + m + 1)\Gamma(-\nu + m + 1)}, \tag{2.248}$$

$$J_\mu(z)J_\nu(z) = \sum_{m=0}^{\infty} \frac{(-1)^m (z/2)^{\mu+\nu+2m}\Gamma(\nu+\mu+2m+1)}{m!\,\Gamma(\mu+m+1)\Gamma(\nu+m+1)\Gamma(\nu+\mu+m+1)}, \tag{2.249}$$

$$\int_0^z \frac{J_\mu(t)J_\nu(z-t)}{t(z-t)}\,dt = \left(\frac{1}{\mu}+\frac{1}{\nu}\right)J_{\nu+\mu}(z), \tag{2.250}$$

[where ν, μ are complex and Re $(\nu) > 0$, Re $(\mu) > 0$]

$$\int_0^z \sin(z-t)\frac{J_\mu(t)}{t}\,dt = \frac{2}{\mu}\sum_{m=0}^{\infty}(-1)^m J_{(\mu+2m+1)}(z), \tag{2.251}$$

$$\int_0^z \cos(z-t)J_0(t)\,dt = zJ_0(z), \tag{2.252}$$

$$\int_0^z \sin(z-t)J_1(t)\,dt = \sin z - zJ_0(z), \tag{2.253}$$

$$\int_0^\infty e^{-\alpha t}\,J_0(\beta t)\,dt = \frac{1}{\alpha^2+\beta^2}, \tag{2.254}$$

$$\int_0^z z'[J_n(\beta z')]^2\,dz' = \frac{z^2}{2}\left[\{J_n'(\beta z)\}^2 + \left(1-\frac{n^2}{\beta^2 z^2}\right)J_n^2(\beta z)\right] \tag{2.255}$$

$$= \frac{z^2}{2}[-J_{-n-1}(\beta z)J_{n+1}(\beta z) + J_n^2(\beta z)], \tag{2.256}$$

$$\int_0^z z[AJ_1(\alpha z)+BN_1(\alpha z)][AJ_1(\beta z)+BN_1(\beta z)]\,dz$$

$$= \frac{\beta z[AJ_1(\alpha z)+BN_1(\alpha z)][AJ_0(\beta z)+BN_0(\beta z)]}{\alpha^2-\beta^2} - \frac{\alpha z[AJ_0(\alpha z)+BN_0(\alpha z)][AJ_1(\beta z)+BN_1(\beta z)]}{\alpha^2-\beta^2}, \quad \alpha \neq \beta, \tag{2.257}$$

$$\int_0^1 zJ_n(\lambda_m z)J_n(\lambda_l z)\,dz = \begin{cases} 0 & \text{when } \lambda_m \neq \lambda_l \\ \tfrac{1}{2}J_{n+1}(\lambda_m) & \text{when } \lambda_m = \lambda_l \end{cases}$$

provided

$$J_n(\lambda_m) = J_n(\lambda_l) = 0. \tag{2.258}$$

Equation (2.258) describes the orthogonal property of the Bessel functions. A similar orthogonal property holds good for the spherical Bessel functions.

Some Important Recurrence Relations for Bessel Functions

All the four solutions of the Bessel equation, namely, $J_p(z)$, $N_p(z)$, $H_p^{(1)}(z)$, and $H_p^{(2)}(z)$ or, in general, $Z_p(z)$ satisfy the recurrence relations

$$Z_{p-1} + Z_{p+1} = \frac{2p}{z}Z_p, \tag{2.259}$$

$$\frac{dZ_p}{dz} = \tfrac{1}{2}Z_{p-1} - \tfrac{1}{2}Z_{p+1}, \tag{2.260}$$

$$\frac{d}{dz}[z^p Z_p(z)] = z^p Z_{p-1}, \tag{2.261}$$

$$\frac{d}{dz}[z^p Z_p(z)] = -z^{-p} Z_{p+1}. \tag{2.262}$$

2.7 Legendre Equation and its Solutions

The Legendre equation is another particular form of the general homogeneous linear differential equation

$$\frac{d^2y}{dz^2} + a(z)\frac{dy}{dz} + b(z)y = 0. \tag{2.263}$$

It is of the form

$$\frac{d^2u}{dz^2} - \frac{2z}{(1-z^2)}\frac{du}{dz} + \frac{\nu(\nu+1)}{(1-z^2)} = 0 \tag{2.264}$$

or

$$(1-z^2)\frac{d^2u}{dz^2} - 2z\frac{du}{dz} + \nu(\nu+1)u = 0. \tag{2.265}$$

The function $u = P_\nu(z)$, called the Legendre function of z of order ν, is a solution of the Legendre equation. When ν is an integer n, $P_n(z)$ becomes a polynomial of z, and is called the *Legendre polynomial* of degree n.

Using the binomial expansion of $(z^2 - 1)^n$, where n is a positive integer, we get

$$\frac{d^n}{dz^n}(z^2-1)^n = \frac{d^n}{dz^n}\Big[\sum_{m=0}^{n} (-1)^m \frac{n!}{m!\,(n-m)!} z^{2n-2m}\Big]. \tag{2.266}$$

The coefficients of negative power of z in expression (2.266) vanish. Substituting this expression in the Legendre equation (2.265), we can see that

$$P_n(z) = \frac{1}{2^n n!}\frac{d^n}{dz^n}(z^2-1)^n \tag{2.267}$$

is a solution.

For some particular values of n, say, 0, 1, 2, 3, 4, we have

$$\begin{aligned} &P_0(z) = 1, \qquad P_1(z) = z, \\ &P_2(z) = \tfrac{1}{2}(3z^2 - 1), \qquad P_3(z) = \tfrac{1}{2}(5z^2 - 3z), \\ &P_4(z) = \tfrac{1}{8}(35z^4 - 30z^2 + 3). \end{aligned} \tag{2.268}$$

Orthogonal Properties of Legendre Polynomials

The Legendre polynomials possess orthogonal properties just as trigonometric and Bessel functions possess these properties. For example,

$$\int_{-1}^{1} P_m(z)P_n(z)\,dz = \begin{cases} 0 & \text{for } m \neq n \\ \dfrac{2}{2n+1} & \text{for } m = n \end{cases} \tag{2.269}$$

Property (2.269) suggests the possibility of expressing an arbitrary function $f(z)$ by an infinite series involving the Legendre polynomials. Let

$$f(z) = \sum_{n=0}^{\infty} a_n P_n(z). \tag{2.270}$$

Then

$$a_n = \frac{2n+1}{2} \int_{-1}^{1} f(z) P_n(z)\, dz. \tag{2.271}$$

The series expression for $f(z)$ given by Eq. (2.270) is valid only when the integral in Eq. (2.271) exists within the interval $(-1, +1)$.

Recurrence Formulae for Legendre Polynomials

Some recurrence formulae are

$$(n+1)P_n(z) = P'_{n+1}(z) - zP'_n(z), \tag{2.272}$$

$$(2n+1)P_n(z) = P'_{n+1}(z) - P'_{n-1}(z), \tag{2.273}$$

$$(2n+1)zP_n(z) = (n+1)P_{n+1}(z) + nP_{n-1}(z), \tag{2.274}$$

$$nP_n(z) = zP'_n(z) - P'_{n-1}(z), \tag{2.275}$$

$$nzP_n(z) = (z^2 - 1)P'_n(z) + nP_{n-1}(z). \tag{2.276}$$

Legendre Functions of the Second Kind

Since the Legendre equation is of the second order, it should have two independent solutions. Hence, its general solution is of the form

$$u = AP_\nu(z) + BQ_\nu(z), \tag{2.277}$$

where A and B are arbitrary constants and $P_\nu(z)$ and $Q_\nu(z)$ are called the Legendre functions of the first kind and second kind, respectively. The function $Q_\nu(z)$ may be represented as a definite integral:

$$Q_\nu(z) = \frac{1}{2^{\nu+1}} \int_{-1}^{1} (1 - t^2)^\nu (z - t)^{-\nu-1}\, dt, \tag{2.278}$$

where ν may be a complex quantity with Re $(\nu + 1) \geqslant 0$. $Q_\nu(z)$ never becomes a polynomial of z even if ν is an integer n. If $P_n(z)$ is replaced by $Q_n(z)$, then the latter satisfies the same recurrence formulae as does the former, that is, it satisfies Eqs. (2.272)–(2.276).

If ν is not an integer, then the general solution of the Legendre equation (2.264) may also be written as

$$u(z) = A'P_\nu(z) + B'P_\nu(-z), \tag{2.279}$$

where A' and B' are arbitrary constants. If ν is an integer n, then $P_n(z)$ and $P_n(-z)$ are not independent and are related to each other as

$$P_n(-z) = (-1)^n P_n(z). \tag{2.280}$$

Associated Legendre Functions

The equations

$$P_\nu^m(z) = (1 - z^2)^{m/2}\frac{d^m P_\nu(z)}{dz^m}, \tag{2.281}$$

$$Q_\nu^m(z) = (1 - z^2)^{m/2}\frac{d^m Q_\nu(z)}{dz^m}, \tag{2.282}$$

where m is a positive integer and $-1 < z < 1$, are called respectively the associated Legendre functions of the first and second kind. They are two independent solutions of the equation

$$\sin\theta \frac{d}{d\theta}(\sin\theta \frac{d\Theta}{d\theta}) + [n(n+1)\sin^2\theta - m^2]\Theta = 0, \tag{2.283}$$

where $z = \cos\theta$, or of the equation

$$(1 - z^2)\frac{d^2u}{dz^2} - 2z\frac{du}{dz} + [n(n+1) - \frac{m^2}{1 - z^2}]u = 0. \tag{2.284}$$

Like $P_n(z)$ and $Q_n(z)$, $P_n^m(z)$ and $Q_n^m(z)$ too are orthogonal in the fundamental range $(-1, +1)$. Thus

$$\int_{-1}^{1} P_n^m(z)P_r^m(z)\, dz = \begin{cases} 0 & \text{for } r \neq n \\ \dfrac{2}{(2n+1)}\dfrac{(n+m)!}{n-m)!} & \text{for } r = n \end{cases} \tag{2.285}$$

The function

$$P_n^m(z) = 0 \quad \text{for } m > n, \tag{2.286}$$

becaues $P_n(z)$ is a polynomial of the n-th degree.

Singularities of Legendre Functions

The function $P_\nu(-z)$ has a singularity at $z = 1$; but $Q_n(z)$ has two singularities at $z = \pm 1$. These singular points of $Q_n(z)$ become evident from the expression for $Q_n(z)$ in terms of $P_n(z)$:

$$Q_n(\cos\theta) = P_n(\cos\theta)\ln\cot\frac{\theta}{2} - (P_{n-1}P_0 + \tfrac{1}{2}P_{n-2}P_1 + \tfrac{1}{3}P_{n-3}P_2 + \ldots + \frac{1}{n}P_0P_{n-1}), \tag{2.287}$$

where $\cos\theta = z$.

Recurrence Relations for Associated Legendre Functions

The recurrence relations for the associated Legendre functions are

$$(n - m + 1)P_{n+1}^m - (2n + 1)zP_n^m + (n + m)P_{n-1}^m = 0, \tag{2.288}$$

$$P_{n-1}^m = zP_n^m - (n - m + 1)\sqrt{1 - z^2}\, P_n^{m-1}, \tag{2.289}$$

$$P_{n+1}^m = zP_n^m + (n + m)\sqrt{1 - z^2}\, P_n^{m-1}, \tag{2.290}$$

$$\sqrt{1 - z^2}\, P_n^{m+1} = (n + m + 1)zP_n^m - (n - m + 1)P_{n+1}^m, \tag{2.291}$$

$$\sqrt{1-z^2}\,P_n^{m+1} = 2mzP_n^m - (n+m)(n-m+1)\sqrt{1-z^2}\,P_n^{m-1}, \tag{2.292}$$

$$\sqrt{1-z^2}\,P_n^m = \frac{1}{2n+1}(P_{n+1}^{m+1} - P_{n-1}^{m+1}), \tag{2.293}$$

$$\frac{m}{\sqrt{1-z^2}}P_n^m = \tfrac{1}{2}z[(n-m+1)(n+m)P_n^{m-1} + P_n^{m+1}] + m\sqrt{1-z^2}\,P_n^m, \tag{2.294}$$

$$(1-z^2)\frac{dP_n^m}{dz} = (n+1)zP_n^m - (n-m+1)P_{n+1}^m, \tag{2.295}$$

$$(1-z^2)\frac{dP_n^m}{dz} = (n+m)P_{n-1}^m - nzP_n^m, \tag{2.296}$$

$$\frac{dP_n^m}{d\theta} = -\sqrt{1-z^2}\frac{dP_n^m}{dz} = \tfrac{1}{2}[(n-m+1)(n+m)P_n^{m-1} - P_n^{m+1}]. \tag{2.297}$$

Relations (2.288)–(2.297) are also satisfied by $Q_n^m(z)$.

Orthogonal Properties of Associated Legendre Functions

The orthogonal properties of associated Legendre functions are

$$\int_{-1}^{1} P_n^m(z)P_l^m(z)\,dz = 0 \qquad \text{for } n \neq l,$$
$$\int_{-1}^{1} P_n^m(z)P_n^l(z)\frac{dz}{1-z^2} = 0 \quad \text{for } m \neq l, \tag{2.298}$$

$$\int_{-1}^{1} [P_n^m(z)]^2\,dz = \frac{2}{2n+1}\frac{(n+m)!}{(n-m)!}, \tag{2.299}$$

$$\int_{-1}^{1} [P_n^m(z)]^2\frac{dz}{1-z^2} = \frac{1}{m}\frac{(n+m)!}{(n-m)!}. \tag{2.300}$$

The Legendre functions and polynomials as well as the associated Legendre functions occur in solutions of electromagnetic field problems where the field has a spherical or partly spherical symmetry.

Example Consider the Laplace equation

$$\nabla^2\psi = 0. \tag{2.301}$$

In the spherical polar coordinates, its expansion is

$$\frac{1}{r^2\sin\theta}\left[\frac{\partial}{\partial r}\left(r^2\sin\theta\frac{\partial\psi}{\partial r}\right) + \frac{\partial}{\partial\theta}\left(\sin\theta\frac{\partial\psi}{\partial\theta}\right) + \frac{\partial}{\partial\phi}\left(\frac{1}{\sin\theta}\frac{\partial\psi}{\partial\phi}\right)\right] = 0. \tag{2.302}$$

Putting

$$\psi = R(r)\Theta(\theta)\Phi(\phi), \tag{2.303}$$

we obtain

$$\frac{d^2\Phi}{d\phi^2} + m^2\Phi = 0, \tag{2.304}$$

$$r^2\frac{d^2R}{dr^2} + 2r\frac{dR}{dr} - n(n+1)R = 0, \tag{2.305}$$

$$\sin\theta\frac{d}{d\theta}(\sin\theta\frac{d\Theta}{d\theta}) + [n(n+1)\sin^2\theta - m^2]\Theta = 0. \tag{2.306}$$

Putting $z = \cos\theta$, Eq. (2.306) becomes

$$(1-z^2)\frac{d}{dz}[(1-z^2)\frac{d\Theta}{dz}] + [(1-z^2)(n^2+n) - m^2]\Theta = 0. \tag{2.307}$$

Equation (2.307) is the associated Legendre equation whose two independent solutions are the associated Legendre functions $P_n^m(z)$ and $Q_n^m(z)$.

If ψ is independent of ϕ, then $m = 0$, and Eq. (2.307) becomes

$$(1-z^2)\frac{d^2\Theta}{dz^2} - 2z\frac{d\Theta}{dz} + n(n+1)\Theta = 0. \tag{2.308}$$

This is the Legendre equation which has two independent solutions, namely, $P_n(z)$ and $Q_n(z)$.

Incidentally, the solutions of Eqs. (2.304) and (2.305) in Φ and R, respectively, are of the form

$$\Phi = A\cos(m\phi) + B\sin(m\phi), \tag{2.309}$$

$$R = r^n. \tag{2.310}$$

Tesseral and Surface Harmonics

The functions $\cos(m\phi)P_n^m(\cos\theta)$ and $\sin(m\phi)P_n^m(\cos\theta)$ are called tesseral harmonics of the n-th degree and m-th order. As $P_n^m(\cos\theta) = 0$ for $m > n$, there are obviously $(2n+1)$ tesseral harmonics of the n-th degree. If the tesseral harmonics are multiplied by a set of arbitrary constants and summed, we obtain the spherical surface harmonics $Y_n(\theta, \phi)$ of degree n:

$$Y_n(\theta, \phi) = \sum_{m=0}^{n} [a_{nm}\cos(m\phi) + b_{nm}\sin(m\phi)]P_n^m(\cos\theta). \tag{2.311}$$

The tesseral harmonics form a complete system of orthogonal functions on the surface of a sphere (r = constant) because of the orthogonal properties of $\cos(m\phi)$, $\sin(m\phi)$, and $P_n^m(\cos\theta)$.

Let $g(\theta, \phi)$ be an arbitrary function on the surface of a sphere which, together with all its first and second derivatives, is continuous. Then, $g(\theta, \phi)$ can be represented by an absolutely convergent series of surface harmonics as

$$g(\theta, \phi) = \sum_{n=0}^{\infty} [a_{n0}P_n(\cos\theta) + \sum_{m=1}^{n} [a_{nm}\cos(m\phi) + b_{nm}\sin(m\phi)]P_n^m(\cos\theta)] \tag{2.312}$$

whose coefficients are determined by

$$a_{n0} = \frac{2n+1}{4\pi}\int_0^{2\pi}\int_0^{\pi} g(\theta, \phi)P_n(\cos\theta)\sin\theta\, d\theta\, d\phi, \tag{2.313}$$

$$a_{nm} = \frac{2n+1}{2\pi}\frac{(n-m)!}{(n+m)!}\int_0^{2\pi}\int_0^{\pi} g(\theta, \phi)P_n^m(\cos\theta)\cos(m\phi)\sin\theta\, d\theta\, d\phi, \tag{2.314}$$

$$b_{nm} = \frac{2n+1}{2\pi}\frac{(n-m)!}{(n+m)!}\int_0^{2\pi}\int_0^{\pi} g(\theta, \phi)P_n^m(\cos\theta)\sin(m\phi)\sin\theta\, d\theta\, d\phi. \tag{2.315}$$

Expression for $P_\nu^m(\cos\theta)$ for Nonintegral Values of ν

The expression for $P_\nu^m(\cos\theta)$ for nonintegral values of ν is

$$P_\nu^m(\cos\theta) = 2^m \cos\frac{(\nu+m)\pi}{2}\,\frac{\left(\frac{\nu+m-1}{2}\right)!}{\left(\frac{\nu-m}{2}\right)!\,(-\frac{1}{2})!}\sin^m\theta\, F\left(\frac{m+\nu+1}{2}, \frac{m-\nu}{2}; \tfrac{1}{2}; \cos^2\theta\right)$$

$$+ 2^{m+1}\sin\frac{(\nu+m)\pi}{2}\,\frac{\left(\frac{\nu+m}{2}\right)!}{\left(\frac{\nu-m-1}{2}\right)!\,(-\frac{1}{2})!}\sin^m\theta\cos\theta$$

$$\times F\left(\frac{m+\nu+2}{2}, \frac{m-\nu+1}{2}; \tfrac{3}{2}; \cos^2\theta\right), \tag{2.316}$$

where the hypergeometric function $F(\alpha, \beta; \gamma; x)$ is defined by the power series

$$F(\alpha, \beta; \gamma; x) = 1 + \frac{\alpha\cdot\beta}{1\cdot\gamma}x + \frac{\alpha(\alpha+1)\beta(\beta+1)}{1\cdot 2\gamma(\gamma+1)}x^2 + \dots. \tag{2.317}$$

This series is convergent for $|x| < 1$.

The function $P_\nu(\cos\theta)$ has a logarithmic singularity at $\theta = \pi$ and, in this neighbourhood, the series that is suitable for computing purposes is

$$P_\nu(\cos\theta) = \frac{\sin(\nu\pi)}{\pi}\sum_{s=0}^{\infty}\frac{(-1)^s(\nu+s)!}{(\nu-s)!}\left[2\ln\cos\frac{\theta}{2} + \psi(\nu+s) + \psi(\nu-s) - 2\psi(s)\right]$$

$$\times\frac{\cos^{2s}(\theta/2)}{(s!)^2} + \cos(\nu\pi)\sum_{s=0}^{\infty}\frac{(-1)^s(\nu+s)!}{(\nu-s)!\,(s!)^2}\cos^{2s}\left(\frac{\theta}{2}\right), \tag{2.318}$$

where

$$\psi(z) = \frac{d}{dz}\ln(z!) \tag{2.319}$$

$$= \frac{1}{z!}\frac{d(z!)}{dz} = -C + \sum_{n=1}^{\infty}\left(\frac{1}{n} - \frac{1}{n+z}\right) \tag{2.320}$$

and C = Euler's constant = 0.5772.

PROBLEMS

1 Evaluate

$$\int_{0,\,1}^{2,\,3}[(2xy-1)\,dx + (x^2+1)\,dy]$$

along the paths $y = x + 1$ and $y = x^2/2 + 1$.

2 Evaluate

$$\int_S (x+y+z)\,dS,$$

where S is the portion of the surface of the sphere $x^2 + y^2 + z^2 = a^2$ which lies in the first quadrant. Use the spherical polar coordinates.

3 If a particle is attracted towards the origin by a force proportional to the n-th power of the distance from the origin, show that the work done against this force in moving the particle from the point (x_0, y_0) to the point (x_1, y_1) is independent of the path, and find its amount.

4 Verify the divergence theorem for the function $(x^2\boldsymbol{u}_x + z\boldsymbol{u}_y + yz\boldsymbol{u}_z)$ over the cube whose vertices are (0, 0, 0), (1, 0, 0), (1, 1, 0), (0, 1, 0), (0, 0, 1), (1, 0, 1), (1, 1, 1), and (0, 1, 1).

5 Verify Stokes' theorem for the function $xy\boldsymbol{u}_x + yz\boldsymbol{u}_y + z^2\boldsymbol{u}_z$ for the cube of Problem 4.

6 If at each point of a closed surface S the vector $\boldsymbol{F}(x, y, z)$ is perpendicular to S, prove that

$$\int_V \nabla \times \boldsymbol{F}\, dV = 0.$$

7 Show that the number of terms in the expansion of a determinant of order n is $n!$.

8 If $|A| = |a_{ij}|$ is a determinant of order n with the property that $a_{ij} = -a_{ji}$ for all values of i and j, prove that $|A| = (-1)^n|A|$. What further conclusion can be drawn if n is odd?

9 Show that the area of the triangle whose vertices are the points (x_1, y_1), (x_2, y_2), (x_3, y_3) is given by

$$A = \pm\tfrac{1}{2}\begin{vmatrix} x_1 & y_1 & 1 \\ x_2 & y_2 & 1 \\ x_3 & y_3 & 1 \end{vmatrix}.$$

The plus or minus sign is chosen according to the consecutive numbering of the vertices of the triangle in the counter-clockwise or clockwise direction.

10 Show that

$$\begin{vmatrix} 1 & 1 & 1 & 1 \\ a_1 & a_2 & a_3 & a_4 \\ a_1^2 & a_2^2 & a_3^2 & a_4^2 \\ a_1^3 & a_2^3 & a_3^3 & a_4^3 \end{vmatrix} = (a_1 - a_2)(a_1 - a_3)(a_1 - a_4)(a_2 - a_3)(a_2 - a_4)(a_3 - a_4).$$

11 Multiply the matrices

$$\begin{pmatrix} 1 & 2 & -1 \\ 3 & 0 & 2 \end{pmatrix}$$

and

$$\begin{pmatrix} 3 & 1 \\ 1 & 3 \\ 2 & 0 \end{pmatrix}.$$

12 What is the adjoint of the matrix

$$\begin{pmatrix} -2 & 1 & 3 \\ 4 & 0 & -1 \\ 3 & 3 & 2 \end{pmatrix};$$

and what is its inverse?

13 If A and B are square matrices of the same order, show that $|AB| = |A|\,|B|$.

14 If A is a nonsingular matrix, show that $AB = 0$ implies $B = 0$.

15 Under what conditions, if any, does $AB = AC$ imply $B = C$?

16 If A and B are symmetric matrices of the same order, prove that the product AB is symmetric if and only if $AB = BA$.

17 Show that

(i) $$\frac{d[x^2 J_{\nu-1}(x) J_{\nu+1}(x)]}{dx} = 2x^2 J_\nu(x) \frac{dJ_\nu(x)}{dx},$$

(ii) $$4J_\nu''(x) = J_{\nu-2}(x) - 2J_\nu(x) + J_{\nu+2}(x),$$

(iii) $$J_\nu''(x) = [\frac{\nu(\nu+1)}{x^2} - 1]J_\nu(x) - \frac{J_{\nu-1}(x)}{x}.$$

18 Show that

$$J_0(x) = \frac{1}{\pi}\int_0^\pi \cos(x\cos\phi)\, d\phi.$$

19 By expanding the integrand into an infinite series and integrating term by term, show that

(i) $$\int_0^{\pi/2} J_0(x\cos\phi)\cos\phi\, d\phi = \frac{\sin x}{x},$$

(ii) $$\int_0^{\pi/2} J_1(x\cos\phi)\, d\phi = \frac{1-\cos x}{x}.$$

20 Show that

$$\int J_0(x)\,dx = 2[J_1(x) + J_3(x) + J_5(x) + \ldots]$$

$$= J_1(x) + \int \frac{J_1(x)}{x}\,dx$$

$$= J_1(x) + \frac{J_2(x)}{x} + 1\cdot 3\int \frac{J_2(x)}{x^2}\,dx$$

$$= J_1(x) + \frac{J_2(x)}{x} + \frac{1\cdot 3}{x^2}J_3(x) + 1\cdot 3\cdot 5\int \frac{J_3(x)}{x^3}\,dx$$

$$= J_1(x) + \frac{J_2(x)}{x} + \frac{1\cdot 3}{x^2}J_3(x) + \ldots + \frac{(2n-2)!\,J_n(x)}{2^{n-1}(n-1)!\,x^{n-1}} + \frac{(2n)!}{2^n_n!}\int \frac{J_n(x)}{x^n}\,dx.$$

21 Show that

(i) $$\int J_0(x)\cos x\,dx = xJ_0(x)\cos x + xJ_1(x)\sin x + c,$$

(ii) $$\int J_0(x)\sin x\,dx = xJ_0(x)\sin x - xJ_1(x)\cos x + c.$$

22 Express x^2 and x^3 as linear combinations of Legendre polynomials.

23 Show that both the Legendre polynomials with even subscripts and the Legendre polynomials with odd subscripts form orthogonal sets over the interval (0, 1).

3 Fundamental Electromagnetic Theory

We assume that the reader is familiar with the fundamental electromagnetic concepts. In our study, we shall use the metre-kilogram-second (MKS) system of units, proposed by Giorgi and popularly known as the MKS system. In this system, electromagnetic equations become simple and correspond closely to physical ideas or measurements, which are common in engineering laboratories.

3.1 Electric Charge and Coulomb's Law

The distinguishing characteristic of electricity is its corpuscular quality. The smallest particle of negative charge is the *electron* whose charge q_e is 1.602×10^{-19} coulomb. The *electric charge density* ρ is defined as the amount of electric charge contained within a unit volume. Though ρ can be represented both as a function of time and as a function of space coordinates, it need not be a continuous function.

The well-known law of forces, called *Coulomb's law* (see Fig. 3.1), refers to the force that exists between two electric charges q_1 and q_2. The force $\boldsymbol{F}$ between the two charges q_1 and q_2 separated by a distance r is given by

$$\boldsymbol{F} = \frac{q_1 q_2}{4\pi\epsilon r^2}\boldsymbol{u}_r, \tag{3.1}$$

where $\boldsymbol{F}$ is in newtons, r in metres, q_1 and q_2 are in coulombs, $\boldsymbol{u}_r$ is the unit vector along r, and ϵ is the permittivity of the medium in farad per metre. If q_1 and q_2 have the same sign, then the

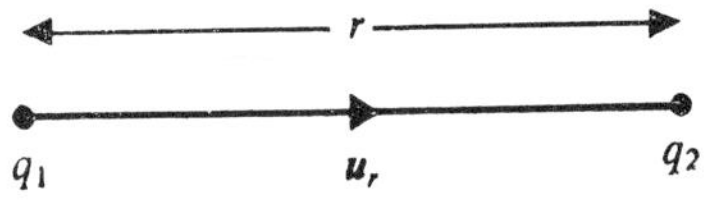

Fig. 3.1 Coulomb's law.

force is repelling; but if they have opposite signs, the force is attracting. $\epsilon = \epsilon_0\epsilon_r$, where ϵ_0 is the permittivity of free space and is given by

$$\epsilon_r = 8.854 \times 10^{-12} \approx \frac{1}{36\pi}10^{-9} \text{ farad/metre}$$

and ϵ_r is called the relative permittivity or dielectric constant of the medium.

3.2 Electric Intensity, Electric Displacement Density, and Electric Potential in Electrostatic Fields

An electrostatic field is produced by electric charges that do not change with time.

The *electric intensity* at a point is defined as the force on a unit positive charge placed at that point (assuming that this charge does not disturb the underlying field). Thus, the electric intensity $\boldsymbol{E}$ at a point distance r from a point charge q is obtained as

$$\boldsymbol{E} = \frac{q}{4\pi\epsilon r^2}\boldsymbol{u}_r \tag{3.2}$$

by putting $q_1 = q$ and $q_2 = 1$ in Coulomb's law. The MKS unit of electric intensity is the volt per metre. In general, the field $\boldsymbol{E}$ at any point in an electrostatic field may be due to a large number of charges distributed throughout space, and can be expressed as

$$\boldsymbol{E} = \sum_n \frac{q_n}{4\pi\epsilon r_n^2}\boldsymbol{u}_{r_n}, \tag{3.3}$$

where r_n is the distance of the point from the charge q_n and $\boldsymbol{u}_{r_n}$ is the unit vector in the direction $\boldsymbol{r}_n$.

The *electric displacement density* or *electric flux density* $\boldsymbol{D}$ is defined as

$$\boldsymbol{D} = \epsilon\boldsymbol{E}. \tag{3.4}$$

In an isotropic medium, ϵ is a scalar having only magnitude. However, in an anisotropic medium such as plasma, ϵ is a tensor of rank two with nine components. For example, in the rectangular coordinates,

$$\epsilon = \begin{pmatrix} \epsilon_{11} & \epsilon_{12} & \epsilon_{13} \\ \epsilon_{21} & \epsilon_{22} & \epsilon_{23} \\ \epsilon_{31} & \epsilon_{32} & \epsilon_{33} \end{pmatrix}, \tag{3.5}$$

and the relation $\boldsymbol{D} = \epsilon\boldsymbol{E}$ consists of three equations, namely,

$$\begin{aligned} D_x &= \epsilon_{11}E_x + \epsilon_{12}E_y + \epsilon_{13}E_z, \\ D_y &= \epsilon_{21}E_x + \epsilon_{22}E_y + \epsilon_{23}E_z, \\ D_z &= \epsilon_{31}E_x + \epsilon_{32}E_y + \epsilon_{33}E_z \end{aligned} \tag{3.6}$$

or, in matrix form,

$$\begin{pmatrix} D_x \\ D_y \\ D_z \end{pmatrix} = \begin{pmatrix} \epsilon_{11} & \epsilon_{12} & \epsilon_{13} \\ \epsilon_{21} & \epsilon_{22} & \epsilon_{23} \\ \epsilon_{31} & \epsilon_{32} & \epsilon_{33} \end{pmatrix}\begin{pmatrix} E_x \\ E_y \\ E_z \end{pmatrix}. \tag{3.7}$$

The MKS unit of electric displacement density is coulomb per square metre.

In an electrostatic field, the *electrostatic potential* V at a point is defined as the work done in moving a unit positive charge against the forces of the field from a point with zero potential (sometimes assumed to be at infinity) to the given point. Thus

$$V \text{ at point } P = \text{line integral} \int_{-\infty}^{P} \boldsymbol{E}\cdot d\boldsymbol{s}.$$

In practical problems, we are usually more concerned with potential differences. Thus, in Fig. 3.2, $V_B - V_A =$ the potential difference, or the electromotive force between the points A and

$B = -\int_A^B \boldsymbol{E}\cdot d\boldsymbol{s} =$ the work done in carrying a unit positive charge from A to B. An electrostatic field is a conservative field where no sources or sinks exist. Hence, the potential difference between

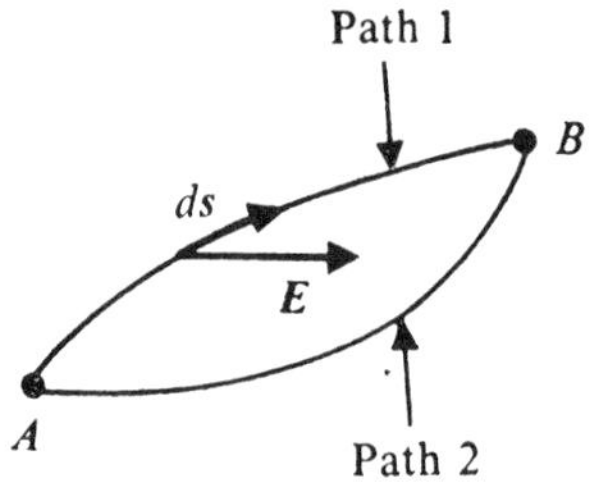

Fig. 3.2 Illustration of potential difference between points A and B.

the points A and B is independent of the path between A and B. Thus, in Fig. 3.2,

$$-\int_{A\ \text{path 1}}^{B} \boldsymbol{E}\cdot d\boldsymbol{s} = -\int_{A\ \text{path 2}}^{B} \boldsymbol{E}\cdot d\boldsymbol{s} \tag{3.8}$$

or

$$\oint_c \boldsymbol{E}\cdot d\boldsymbol{s} = 0,$$

where the integration is done over the closed path ABA along paths 1 and 2. This equation is true for any closed path. By Stokes' theorem,

$$\oint_c \boldsymbol{E}\cdot d\boldsymbol{s} = \int_s \text{curl } \boldsymbol{E}\, dS, \tag{3.9}$$

where the surface of integration is over a surface S enclosing the closed curve C on which the line integral is taken. Since this is true for any closed curve C enclosing a surface S, in an electrostatic field, curl $\boldsymbol{E} = 0$. Since the curl of the gradient of any scalar is always zero, we have

$$\boldsymbol{E} = -\text{grad } V = -\nabla V, \tag{3.10}$$

where V is a scalar. This V is the same as the electrostatic potential V, because the potential difference between two points, an infinitesimal distance ds apart, may be expressed as

$$dV = -\boldsymbol{E}\cdot d\boldsymbol{s} \tag{3.11a}$$

or

$$dV = -E_s\, ds, \tag{3.11b}$$

where E_s is the component of $\boldsymbol{E}$ in the direction $d\boldsymbol{s}$. Hence, $E_s = -dV/ds =$ the negative space derivative of V in the direction $d\boldsymbol{s}$. By the definition of the gradient of V,

$$\boldsymbol{E} = -\text{grad } V; \tag{3.12}$$

grad V is also known as the potential gradient. The unit of V is the volt.

3.3. Gauss' Law and Electric Flux

The *electric flux** ψ through a surface S is defined as $\int_S D_n \cdot dS$ and its unit is coulomb, where D_n is the component of $\boldsymbol{D}$ normal to the surface. Gauss' law states that the total displacement or electric flux through any closed surface surrounding the charges is equal to the amount of the charge enclosed.

Consider a point charge q located in a homogeneous isotropic medium whose permittivity is ϵ. The electric intensity at any point at a distance r from the charge q is

$$\boldsymbol{E} = \frac{q}{4\pi\epsilon r^2}\boldsymbol{u}_r, \tag{3.13}$$

and the displacement density $\boldsymbol{D}$ at the same point is

$$\boldsymbol{D} = \frac{q}{4\pi r^2}\boldsymbol{u}_r = \epsilon \boldsymbol{E}. \tag{3.14}$$

Let us consider the displacement or total electric flux through a closed surface enclosing the charge q (see Fig. 3.3). The displacement or electric flux through the element of the surface dS is

$$d\psi = D\, dS \cos\theta, \tag{3.15}$$

where θ is the angle between $\boldsymbol{D}$ and the normal to dS. If $d\Omega$ is the solid angle subtended by dS at q, then

$$dS \cos\theta = r^2\, d\Omega. \tag{3.16}$$

Hence

$$d\psi = Dr^2\, d\Omega. \tag{3.17}$$

The total displacement ψ through the closed surface S is therefore given by

$$\psi = \oint_S d\psi = \oint_S D_n r^2\, d\Omega = \oint_S \frac{q}{4\pi r^2} r^2\, d\Omega = \frac{q}{4\pi}\oint_S d\Omega = \frac{q}{4\pi}4\pi = q. \tag{3.18}$$

Thus, the total displacement through the closed surface S is equal to q, the charge enclosed. If

*The idea of electric displacement or electric flux can be understood by recalling Faraday's experiments with concentric spheres. In these experiments, a sphere with charge Q was placed within, but not touching a larger hollow sphere. The outer sphere was earthed momentarily, and then the inner sphere was removed. The charge remaining on the outer sphere was measured and was found to be equal and of opposite sign to the charge Q on the inner sphere, for all sizes of spheres and for all types of dielectrics between the spheres. Thus, it was concluded that there was an electric displacement from the charge on the inner sphere through the dielectric medium to the outer sphere, the amount of this displacement ψ being equal to Q (in coulombs). For an isolated charge q, the displacement per unit area or electric displacement density D at any point on a spherical surface of radius r centred at the isolated charge is $D = \psi/(4\pi r^2) = q/(4\pi r^2)$. The displacement per unit area depends on the direction of the area, and hence the displacement density $\boldsymbol{D}$ is a vector of magnitude $q/(4\pi r^2)$ and has the direction of the normal to the surface element which makes the displacement through the element of area a maximum. This direction is along the radial direction from the isolated charge q, and hence in the same direction as $\boldsymbol{E}$. Therefore, $\boldsymbol{D} = [q/(4\pi r^2)]\boldsymbol{u}_r = \epsilon \boldsymbol{E}$.

there are a number of charges within the volume enclosed, then

$$\oint_S D_n \, dS = \oint_S d\psi = \Sigma \, q. \tag{3.19}$$

Also, if the charge is continuously distributed throughout the volume with a charge density ρ

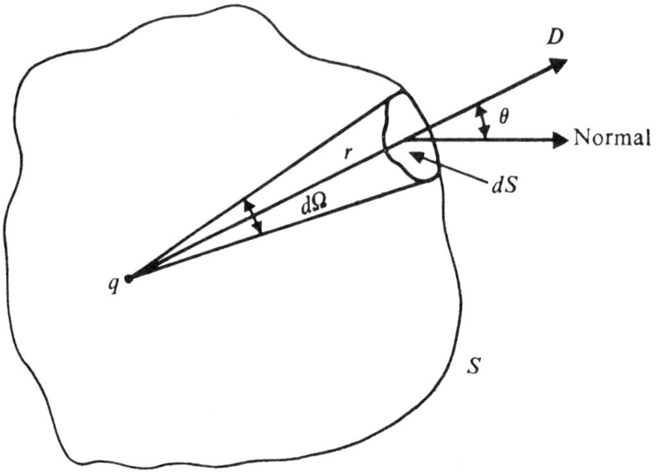

Fig. 3.3 Illustration of Gauss' law.

coulombs per cubic metre, then

$$\oint_S D_n \, dS = \int_V \rho \, dV, \tag{3.20}$$

where V is the volume enclosed by S.

3.4 Poisson's and Laplace's Equations

From Gauss' law,

$$\oint_S D_n \cdot dS = \int_V \rho \, dV, \tag{3.21}$$

where S is a closed surface enclosing a volume V. By Gauss' theorem,

$$\oint_S D_n \cdot dS = \int_V \operatorname{div} \boldsymbol{D} \, dV. \tag{3.22}$$

Therefore

$$\int_V \operatorname{div} \boldsymbol{D} \cdot dV = \int_V \rho \cdot dV. \tag{3.23}$$

Relation (3.23) is true for all closed surfaces S enclosing a volume V. Hence

$$\operatorname{div} \boldsymbol{D} = \rho \tag{3.24a}$$

or

$$\epsilon \operatorname{div} \boldsymbol{E} = \rho. \tag{3.24b}$$

However, $\boldsymbol{E} = -\operatorname{grad} V$. Thus

$$\epsilon \operatorname{div} (-\operatorname{grad} V) = \rho \tag{3.25}$$

or

$$-\nabla^2 V = \frac{\rho}{\epsilon}$$

or

$$\nabla^2 V = \frac{-\rho}{\epsilon}. \tag{3.26}$$

Relation (3.26) is known as Poisson's equation. In a region where $\rho = 0$,

$$\nabla^2 V = 0 \tag{3.27}$$

which is called Laplace's equation.

Many electrostatic problems can be solved by solving either Eq. (3.26) or Eq. (3.27) to satisfy the proper boundary conditions.

3.5 Conductors and Dielectrics, and Electric Current and Conductivity

Conductors are materials having many 'free' electrons, easily detachable from atoms. A *perfect dielectric* is a material which does not possess any detachable electron. Vacuum is the only physical example of a perfect dielectric.

In a conductor, the *electric current density* $\boldsymbol{J}$ is defined as the time rate of flow of electric charge per unit area normal to the lines of flow, and is proportional to the electric intensity $\boldsymbol{E}$. Thus

$$\boldsymbol{J} = \sigma \boldsymbol{E}, \tag{3.28}$$

where the coefficient of proportionality σ is called the *conductivity* of the material. Relation (3.28) is called the *microscopic Ohm's law*. By definition, the direction of the electric current coincides with that of the moving positive charge and is opposite to the direction of the moving negative charge. In practice, it is not possible to find a perfect dielectric or a perfect conductor. For a perfect conductor, σ should be equal to ∞. (The unit of current density $\boldsymbol{J}$ is ampere per square metre and the unit of conductivity is mho per metre. The reciprocal of conductivity is called resistivity and its unit is ohm-metre.) Examples of good conductors are silver ($\sigma = 6.2 \times 10^7$ mho/metre) and copper ($\sigma = 5.8 \times 10^7$ mho/metre).

A perfect dielectric material like vacuum has zero conductivity. But all other dielectric materials have finite conductivity. For example, mica, which is a good dielectric ($\epsilon_r = 5.6$–6.0), has a conductivity of 10^{-11}–10^{-15} mho/metre; distilled water has $\epsilon_r = 81.1$ and $\sigma = 2 \times 10^{-4}$ mho/metre. Table 3.1 gives the conductivity and permittivity of many commonly used materials.

Table 3.1 Conductivity of Materials

Material	*Conductivity* (σ) (mhos/metre)	*Type*
Aluminium (commercial)	3.54×10^7	Conductor
Carbon	3×10^4	Conductor
Constantin (60% Cu, 40% Ni)	0.20×10^7	Conductor
Copper (annealed)	5.8×10^7	Conductor
Copper (hard drawn)	5.65×10^7	Conductor
Duralumin	3×10^7	Conductor
Gold (pure)	4.10×10^7	Conductor
Graphite	10^5	Conductor
Cast iron	10^6	Conductor
Lead	0.48×10^7	Conductor
Magnesium	2.17×10^7	Conductor
Manganin (84% Cu, 12% Mn, 4% Ni)	0.23×10^7	Conductor
Mercury	0.1044×10^7	Conductor
Monel	0.24×10^7	Conductor
Nichrome	0.10×10^7	Conductor
Nickel	1.28×10^7	Conductor
Phosphor bronze	10^7	Conductor
Steel	$0.5\text{–}1.0 \times 10^7$	Conductor
Steel (manganese)	0.14×10^7	Conductor
Steel (silicon)	2×10^6	Conductor
Silver (99.98%, pure)	6.139×10^7	Conductor
Tellurium	$\approx 5 \times 10^2$	Conductor
Tin	9×10^6	Conductor
Tungsten	1.8×10^7	Conductor
Zinc	1.7×10^7	Conductor
Nb_3(Al-Ge) (superconductor)	$\approx\infty$	Conductor

Table 3.1 Conductivity of Materials (cont.)

Material	*Conductivity* (σ) (mhos/metre)	*Type*
Bakelite	10^{-8}–10^{-10}	Dielectric
Celluloid	10^{-8}	Dielectric
Ceresin wax	10^{-17}	Dielectric
Fibre (hard)	5×10^{-9}	Dielectric
Glass	10^{-12}	Dielectric
Glyptol	10^{-14}	Dielectric
Hard rubber	10^{-14}–10^{-16}	Dielectric
Marble	10^{-7}–10^{-9}	Dielectric
Mica	10^{-11}–10^{-15}	Dielectric
Paraffin	10^{-14}–10^{-15}	Dielectric
Porcelain	3×10^{-13}	Dielectric
Quartz (crystal)		
$\parallel$ to axis	10^{-12}	Dielectric
$\perp$ to axis	3×10^{-15}	Dielectric
Quartz (fused)	$\approx 10^{-17}$	Dielectric
Rosin	2×10^{-15}	Dielectric
Shellac	2×10^{-15}	Dielectric
Slate	10^{-6}	Dielectric
Sulphur	10^{-6}	Dielectric
Wood (paraffined)	10^{-15}	Dielectric
Alcohol		
ethyl	3.3×10^{-4}	Dielectric
methyl	7.1×10^{-4}	Dielectric
Petroleum	10^{-14}	Dielectric
Water (distilled)	2×10^{-4}	Dielectric

3.6 Capacitance and Stored Energy Density in Electrostatic Field

The capacitance (expressed in farad) between two conductors placed in a dielectric medium is defined as

$$C = \frac{Q}{V}, \tag{3.29}$$

where V is the voltage or potential difference between the conductors due to equal and opposite charges on them of magnitude Q. The capacitance of a single conductor is the capacitance between that conductor and a spherical shell of infinite radius.

When a condenser is charged so that there exists a voltage V between its plates, energy is stored and this energy can be converted into heat by discharging the condenser through a resistance. The amount of energy stored can be found by calculating the work done in charging the condenser. Since potential has been defined in terms of work done on unit positive charge, the work done in moving a small charge dQ against a potential difference V is $V\,dQ$. But the relation between the voltage V and capacitance C is

$$V = \frac{Q}{C}. \tag{3.30}$$

Therefore, the work done in increasing the charge on a condenser by an amount dQ is $(Q/C)\,dQ$. The total work W done in charging a condenser to Q coulombs is

$$W = \int_0^Q \frac{Q}{C}\,dQ = \tfrac{1}{2}\frac{Q^2}{C}. \tag{3.31}$$

Hence, the energy stored by a charged condenser is

$$\tfrac{1}{2}\frac{Q^2}{C} = \tfrac{1}{2}VQ = \tfrac{1}{2}V^2C. \tag{3.32}$$

This energy is said to be *associated with the electric charge on the conductors* or, alternatively, *associated with the electric field in the dielectric between the conductors.*

Using the concept that energy is associated with electric field and taking the example of a parallel-plate capacitor (see Fig. 3.4), whose plates of area A are separated by a distance d, we

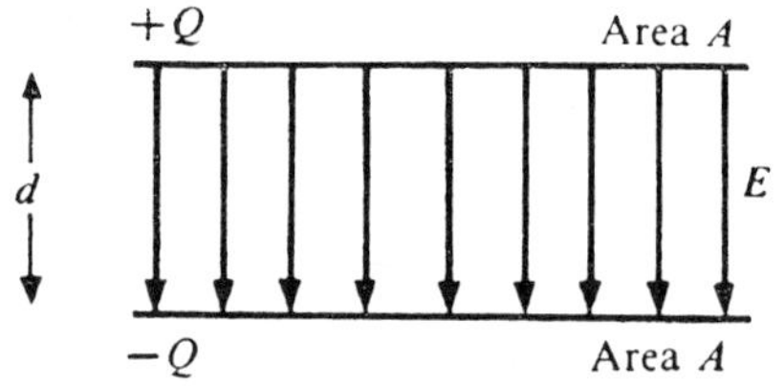

Fig. 3.4 Parallel-plate capacitor.

obtain the electric field E between the plates, which is uniform and has the value $E = V/d$, where V is the voltage or potential difference between the two plates. Then, the stored energy is

$$W_E = \frac{V^2C}{2} = \frac{E^2d^2\epsilon A}{2d} = \frac{\epsilon E^2}{2}(Ad) \tag{3.33}$$

because the capacitance

$$C = \frac{\epsilon A}{d}. \tag{3.34}$$

Since Ad is the volume between the two plates, the quantity $\epsilon E^2/2$ has the dimensions of energy per unit volume and is said to be the *stored energy density* of the electrostatic field.

3.7 Ohm's Law, Resistance, and Conductance

Consider a homogeneous conducting rod of length l and of cross-section S (see Fig. 3.5). The

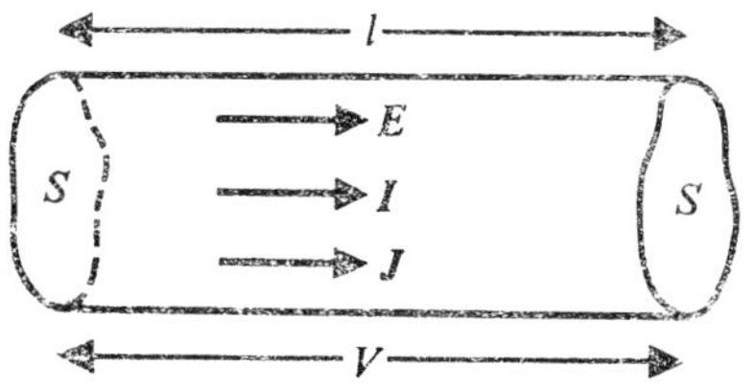

Fig. 3.5 Illustration of Ohm's law.

electric current I in the rod is SJ, where J is the electric current density. The electromotive force V or the electric potential difference over the length l is El. Therefore

$$I = SJ = S\sigma E = GV, \tag{3.35}$$

where

$$G = S\sigma/l, \tag{3.36}$$

$$V = RI \tag{3.37}$$

with

$$R = l/(\sigma S). \tag{3.38}$$

Equation (3.38) is called Ohm's law, R is the resistance of the rod, and G the conductance of the rod.

The current I can also be considered as the rate of change of charge with time or

$$I = Q/t. \tag{3.39}$$

Hence, the work done by V per second is

$$\frac{VQ}{t} = VI = GV^2 = \frac{V^2}{R} = \text{electric power}. \tag{3.40}$$

The work VI appears as heat. The electric power dissipated in heat per unit volume is, evidently, $JE = \sigma E^2$. The units of current and voltage are ampere and volt, and the unit of power or work done is watt.

3.8 Equation of Continuity, Conservation of Charge, and Relaxation Time

From the definition of current as a flow of charge across a surface, it follows that the surface integral of the normal component of J over a closed surface S must measure the loss of charge from the region V enclosed by S.

There is no experimental evidence to indicate that, under ordinary conditions, charge may be either created or destroyed in macroscopic amounts. Hence, we may write

$$\oint_S J_n \, dS = \oint_S \boldsymbol{J} \cdot \boldsymbol{n} \, dS = -\frac{d}{dt}\int_V \rho \, dV, \tag{3.41}$$

where $\boldsymbol{n}$ is the unit vector normal to the surface S. Relation (3.41) expresses the conservation of charge. The flow of charge across a surface can originate in two ways: (i) The surface S may be fixed in space, and the charge density ρ may be some function of time and of the coordinates, and (ii) the charge density may be invariable with time, and the surface moves in a specified manner. In the latter event, $\int_V \rho \, dV$ is a function of time. If, however, the surface is fixed and the integral convergent, we may replace d/dt by a partial derivative under the integral sign. Then

$$\oint_S \boldsymbol{J} \cdot \boldsymbol{n} \, dS = -\int_V \frac{\partial \rho}{\partial t} \, dV. \tag{3.42}$$

By Gauss' theorem,

$$\oint_S \boldsymbol{J} \cdot \boldsymbol{n} \, dS = \int_V \operatorname{div} \boldsymbol{J} \, dV. \tag{3.43}$$

Therefore

$$\int_V \operatorname{div} \boldsymbol{J} \, dV = -\int_V \frac{\partial \rho}{\partial t} \, dV \tag{3.44}$$

or

$$\operatorname{div} \boldsymbol{J} + \partial \rho / \partial t = 0 \tag{3.45}$$

which is called the equation of continuity. In a region of nonvanishing conductivity, $\boldsymbol{J} = \sigma \boldsymbol{E}$, and Eq. (3.45) becomes

$$\operatorname{div} (\sigma \boldsymbol{E}) + \partial \rho / \partial t = 0. \tag{3.46}$$

In any homogeneous medium,

$$\nabla \cdot \boldsymbol{E} = \rho / \epsilon. \tag{3.47}$$

Hence

$$\frac{\sigma \rho}{\epsilon} + \frac{\partial \rho}{\partial t} = 0 \tag{3.48}$$

or

$$\rho = \rho_0 e^{-(\sigma/\epsilon) t}, \tag{3.49}$$

where ρ_0 is the charge density at time $t = 0$. In a good conductor, σ is very large, and hence the initial charge distribution decays very fast exponentially with time at every point and in a manner wholly independent of the applied field. If the charge density is initially zero, it remains so subsequently. On the other hand, in a good dielectric, σ is very small, and the initial charge takes a long time to decay to a small value and, for all practical purposes, we may assume that the charge remains the same for a very long time. The time

$$\tau = \epsilon / \sigma \tag{3.50}$$

required for the charge at any point to decay to $1/e$ of its original value is called the relaxation time. In all good conductors, τ is exceedingly small. In sea water, τ is about 2×10^{-10} sec; even in such a poor conductor as distilled water, it is about 10^{-6} sec. In the best insulators, e.g., fused quartz, it may nevertheless have values exceeding 10^{6} sec.

3.9 Static Magnetic Field of Steady Electric Current

A static electric charge has an electric field, whereas an electric current possesses a magnetic field which can be detected with a magnetic compass. The direction of the magnetic field is the one indicated as "north" by the compass needle. A current carrying wire produces a magnetic field whose direction is related to that of the current by the right-hand rule. This magnetic field generates a force on another current carrying wire. (Wires carrying current in the same direction are attracted, whereas wires carrying current in opposite directions are repelled.)

3.10 Force on Current Element, Magnetic Flux Density

If $I\,dl$ is a current element placed in a magnetic field $\boldsymbol{B}$ (see Fig. 3.6), then the force on the current element is given by

$$\boldsymbol{F} = I\boldsymbol{B} \times d\boldsymbol{l}, \tag{3.51}$$

where $\boldsymbol{F}$ is the force, I the current, $\boldsymbol{B}$ the magnetic field, and $d\boldsymbol{l}$ is the length of the current ele-

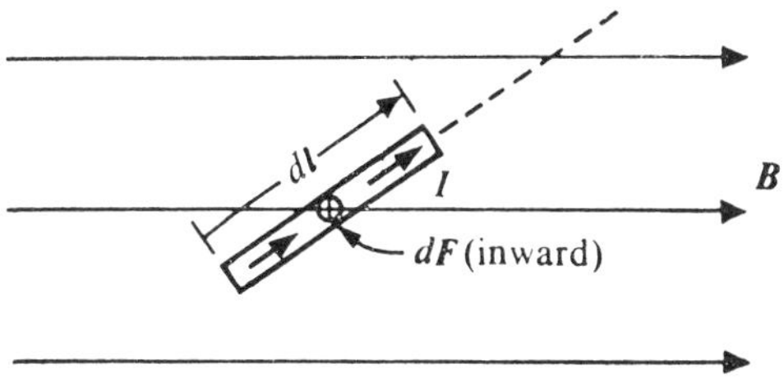

Fig. 3.6 Force exerted on current element by magnetic field.

ment (the direction of vector $d\boldsymbol{l}$ is the same as that of current flow). The force is maximum when $\boldsymbol{B}$ is normal to $d\boldsymbol{l}$. When $\boldsymbol{B}$ and $\boldsymbol{I}$ are at right angles,

$$F = BI\,dl \tag{3.52a}$$

or

$$B = F/(I\,dl). \tag{3.52b}$$

Hence, the *magnetic flux density* $\boldsymbol{B}$ may be defined as force per current moment, where $I\,dl$ is called the current element, and its direction is at right angles to the direction of $\boldsymbol{I}$. The unit of B is the newton per ampere metre or weber per square metre.

Electric charge, positive and negative, can be separated. The poles of a magnet, however, cannot be separated. Consequently, an isolated magnetic pole is not physically realizable, but its effect may be approximated by a pole of a very long magnetized needle. If Q_m is the strength of this pole, then the force on Q_m in a magnetic field $\boldsymbol{B}$ is given by

$$F = BQ_m \text{ newtons.} \tag{3.53}$$

The pole strength Q_m has the dimensions of current moment. Thus, $\boldsymbol{B}$ may also be defined as force per unit magnetic pole.

3.11 Force between Two Magnetic Poles or Two Current Elements, Coulomb's Law, and Permeability

The force between two magnetic poles Q_{m_1} and Q_{m_2} at a distance r from each other is

$$F = \frac{\mu Q_{m_1} Q_{m_2}}{r^2} \boldsymbol{u}_r, \tag{3.54}$$

where $\boldsymbol{u}_r$ is the unit vector in the direction of $\boldsymbol{r}$. The force between two current elements $I_1\, dl_1$ and $I_2\, dl_2$ is

$$F = \frac{\mu I_1 I_2}{4\pi r^2} dl_1 \times (dl_2 \times \boldsymbol{u}_r), \tag{3.55}$$

where μ is called the permeability of the medium and its unit is henry per metre. The permeability of free space is $\mu_0 = 4\pi \times 10^{-7}$ henry/metre. For any material, $\mu = \mu_0 \mu_r$, where μ_r is called the relative permeability. The relative permeability of diamagnetic materials (in which magnetization is opposed to the applied magnetic field), for example, bismuth, silver, lead, water, and copper, is slightly less than one. The relative permeability of paramagnetic materials (in which magnetization is in the same direction as the applied magnetic field) such as air, aluminium, and palladium is slightly more than one. For ferromagnetic materials (in which magnetization is in the same direction as the applied field), e.g., nickel, iron, cobalt, and their alloys, the relative permeability is much greater than one—for nickel 600; for mild steel 2000; and for superalloy 1,000,000. Table 3.2 gives the values of μ_r for a number of materials.

Table 3.2 Relative Permeability of Materials

Material	*Type*	*Relative permeability* (μ_r)
Supermalloy (5% Mo, 79% Ni)	Ferromagnetic	1×10^6
Purified iron (0.05% impurity)	Ferromagnetic	2×10^5
Mumetal (75% Ni, 5% Cu, 2% Cr)	Ferromagnetic	1×10^5
78 permalloy (78.5% Ni)	Ferromagnetic	1×10^5
Silicon iron (4% Si)	Ferromagnetic	7×10^3
Iron (0.2% impurity)	Ferromagnetic	5×10^3
Mild steel (0.2% C)	Ferromagnetic	2×10^3
Ferroxcube 3 (Mn-Zn-ferrite powder)	Ferromagnetic	1.5×10^3
Nickel	Ferromagnetic	600
Cobalt	Ferromagnetic	250
2–81 permalloy powder (2% Mo, 81% Ni)	Ferromagnetic	130

Table 3.2 Relative Permeability of Materials (cont.)

Material	*Type*	*Relative permeability* (μ_r)
Palladium	Paramagnetic	1.0008
Aluminium	Paramagnetic	1.00002
Air	Paramagnetic	1.0000004
Vacuum	Paramagnetic	1
Water	Diamagnetic	0.999991
Copper	Diamagnetic	0.999991
Lead	Diamagnetic	0.999983
Silver	Diamagnetic	0.99998
Bismuth	Diamagnetic	0.99983

3.12 Magnetic Flux over Closed Surface

The magnetic flux ϕ over a surface S is defined as

$$\phi = \int_S B_n \, dS = \int_S \boldsymbol{B} \cdot \boldsymbol{n} \, dS, \tag{3.56}$$

where B_n is the component of $\boldsymbol{B}$ normal to the surface. The magnetic flux over a closed surface is always zero, because the positive and negative magnetic poles cannot be separated. Thus

$$\oint_S B_n \cdot dS = 0. \tag{3.57}$$

By the divergence theorem,

$$\oint_S B_n \cdot dS = \int_V \text{div } \boldsymbol{B} \, dV, \tag{3.58}$$

where V is the volume enclosed by the closed surface S. Therefore

$$\text{div } \boldsymbol{B} = 0. \tag{3.59}$$

3.13 Magnetic Field Intensity, Ampere's Law, and Magnetomotive Force

The magnetic field intensity $\boldsymbol{H}$ at a point is defined as

$$\boldsymbol{H} = \frac{\boldsymbol{B}}{\mu}. \tag{3.60}$$

The unit of $\boldsymbol{H}$ is ampere per metre. Ampere's law states that the line integral of $\boldsymbol{H}$ around a closed path is equal to the electric current enclosed, or

$$\oint_c \boldsymbol{H} \cdot ds = I. \tag{3.61}$$

$\oint_c \boldsymbol{H} \cdot ds$ is called the *magnetomotive force* around the closed path. Or, in other words, Ampere's law may be stated as follows. The magnetomotive force around a closed path is equal to the electric current enclosed.

3.14 Magnetic Current, Magnetic Current Density, and Faraday's Law

Magnetic current is defined as the time rate of change of magnetic flux or, symbolically, magnetic current $= K = \partial\phi/\partial t$. Its unit is volt.

Magnetic current density $\boldsymbol{M}$ is defined as the time rate of change of the magnetic flux density $\boldsymbol{B}$ or, symbolically, $\boldsymbol{M} = \partial \boldsymbol{B}/\partial t$. Its unit is volt per square metre.

Faraday's law or the *first law of electromagnetic induction* may be stated as follows. *The electromotive force around a closed path is equal to the time rate of decrease of the total magnetic flux across a surface S enclosed by the closed path.* Or, symbolically,

$$\oint_c E_s \cdot ds = -\int_s M_n \cdot dS \tag{3.62a}$$

$$= -\int_s \frac{\partial B_n}{\partial t} \cdot dS. \tag{3.62b}$$

By Stokes' theorem,

$$\oint_c E_s \cdot ds = \int_s (\text{curl } \boldsymbol{E})_n \, dS. \tag{3.63}$$

Therefore

$$\int_s (\text{curl } \boldsymbol{E})_n \cdot dS = -\int_s \frac{\partial B_n}{\partial t} \cdot dS \tag{3.64}$$

or

$$\text{curl } \boldsymbol{E} = \frac{-\partial \boldsymbol{B}}{\partial t}. \tag{3.65}$$

3.15 Inductance and Flux Linkage

An inductor is a device for storing energy in a magnetic field, and may be regarded as the magnetic counterpart of a capacitor that stores electric energy. Examples of inductors are loops, coils, and solenoids.

The lines of a magnetic flux produced by a current in a solenoidal coil form closed loops as shown in Fig. 3.7. If all the magnetic flux lines link all the turns, then the total magnetic flux linkage Λ of the coil is equal to the total magnetic flux ϕ through the coil times the number of turns, or

$$\text{flux linkage} = \Lambda = N\phi \text{ weber-turns} \tag{3.66}$$

where N is the number of turns.

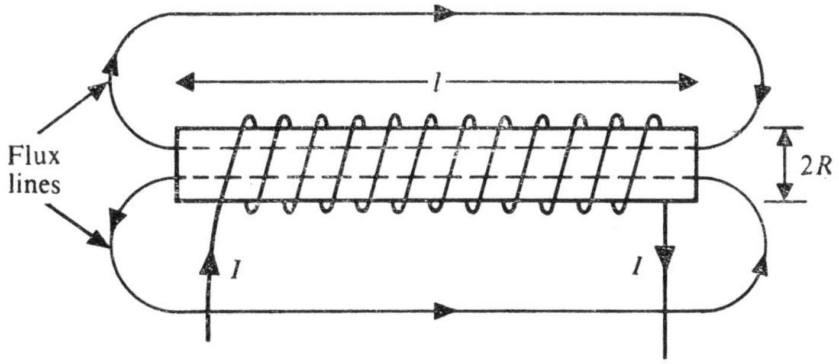

Fig. 3.7 Solenoidal coil.

By definition, the *inductance* L of an inductor is the ratio of its total magnetic flux linkage to the current I through the inductor, or

$$L = \frac{N\phi}{I} = \frac{\Lambda}{I}. \tag{3.67}$$

Definition (3.67) is satisfactory for a medium with a constant permeability such as air. However, for ferromagnetic materials, for which the permeability is not constant (since it varies with current), the inductance is defined as the ratio of the infinitesimal change in flux linkage to the infinitesimal change in current, or

$$L = d\Lambda/dI. \tag{3.68}$$

The unit of inductance is the henry.

3.16 Stored Energy Density in Magnetic Field

For a toroidal coil, as shown in Fig. 3.8, the magnetomotive force is

$$\mathcal{F} = nI, \tag{3.69}$$

where n is the number of turns. If D, the thickness of the core, is small as compared to R, the radius of the ring, then the radii of all circular paths C through the core are approximately equal to R_1 so that, at any point within the core,

$$H \approx \frac{\mathcal{F}}{2\pi R} = \frac{nI}{2\pi R} = \frac{nI}{l}, \tag{3.70}$$

where $l = 2\pi R$. The unit of H is ampere per metre.

The work done in establishing the current I against the electromotive force induced in the circuit by the increasing magnetic flux is given by Faraday's law. The energy thus transferred to the circuit is said to be stored in the magnetic field. The voltage V induced in the coil is given by

$$V = -n\frac{d\phi}{dt}. \tag{3.71}$$

In this equation, ϕ is the total magnetic flux through the coil and is given by $\phi = AB$, where A is

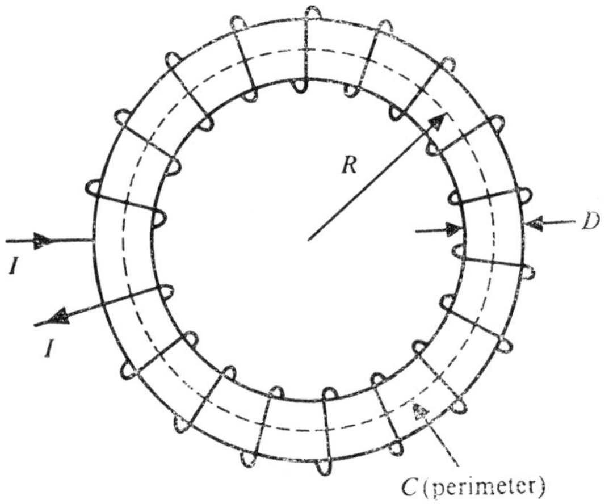

Fig. 3.8 Toroidal coil.

the cross-sectional area of the coil and B is the magnetic flux density defined as $B = \mu H$. Hence

$$V = -n\frac{d\phi}{dt} = -nA\frac{dB}{dt} \tag{3.72a}$$

$$= -n\mu A\frac{dH}{dt}. \tag{3.72b}$$

The work done in establishing the current I is

$$W = -\int_0^t VI\,dt \tag{3.73a}$$

$$= \int_0^t lAH\frac{dB}{dt}\,dt = \int_0^H \mu lAH\,dH$$

$$= lA\left(\frac{\mu H^2}{2}\right), \tag{3.73b}$$

where H is the magnetic field at time t. This W is the total energy stored in the field and, since lA is the volume of the region in which the magnetic field exists, we may conclude that $\mu H^2/2$ represents the stored energy density in the magnetic field.

3.17 Electric Displacement Current Density

From Gauss' law,

$$\text{div}\,\boldsymbol{D} = \rho, \tag{3.74}$$

and from the equation of continuity,

$$\text{div}\,\boldsymbol{J} = \frac{-\partial\rho}{\partial t}. \tag{3.75}$$

Hence

$$\frac{\partial}{\partial t}\operatorname{div}\boldsymbol{D} = \frac{\partial \rho}{\partial t} \tag{3.76}$$

or

$$\operatorname{div}\boldsymbol{J} = -\frac{\partial}{\partial t}\operatorname{div}\boldsymbol{D} = -\operatorname{div}\left(\frac{\partial \boldsymbol{D}}{\partial t}\right). \tag{3.77}$$

Therefore

$$\operatorname{div}\left(\boldsymbol{J} + \frac{\partial \boldsymbol{D}}{\partial t}\right) = 0. \tag{3.78}$$

Relation (3.78) holds good when there are time variations of currents and fields. The unit of $\boldsymbol{D}$ is coulomb per square metre and the unit of $\partial \boldsymbol{D}/\partial t$ is coulomb per square metre per second or ampere per square metre. Hence, $\partial \boldsymbol{D}/\partial t$ has the dimensions of electric current density and it exists in any region where the electric flux density $\boldsymbol{D}$ is changing with time. The idea of this displacement current density $\partial D/\partial t$ was first proposed by Maxwell.

3.18 Ampere's Law as Amended by Maxwell

Ampere's law as amended by Maxwell may be written as

$$\oint_c H_s \cdot ds = I = \int_s J_n\, dS = \int_s \boldsymbol{J} \cdot \boldsymbol{n}\, dS, \tag{3.79}$$

where $\boldsymbol{J} = \boldsymbol{J}_c + \boldsymbol{J}_d$, $\boldsymbol{J}_c$ being the conduction current density $\sigma \boldsymbol{E}$ and $\boldsymbol{J}_d$ the displacement current density $\partial \boldsymbol{D}/\partial t$. Thus

$$\oint_c H_s \cdot ds = \int_s \left(\sigma E_n + \frac{\partial \boldsymbol{D}_n}{\partial t}\right) dS. \tag{3.80}$$

By Stokes' theorem,

$$\oint_c H_s \cdot ds = \int_s (\operatorname{curl}\boldsymbol{H})_n \cdot dS. \tag{3.81}$$

Therefore

$$\int_s (\operatorname{curl}\boldsymbol{H})_n \cdot dS = \int_s \left(\sigma E_n + \frac{\partial \boldsymbol{D}_n}{\partial t}\right) \cdot dS \tag{3.82}$$

or

$$\operatorname{curl}\boldsymbol{H} = \sigma \boldsymbol{E} + \frac{\partial \boldsymbol{D}}{\partial t}. \tag{3.83}$$

Equation (3.83) is the *second law of electromagnetic induction.*

To obtain complete generality, a third term should be included in the current density in Eq. (3.83). This term is called the *convection electric current density*. If a cloud of electric charge is moving with a velocity $\boldsymbol{v}$, as in electron devices or batteries, and if the volume density of the electric charge is ρ, then the current density is called the convection electric current density. Or, symbolically,

$$J_{\text{convection}} = \rho \boldsymbol{v}. \tag{3.84}$$

Hence, a more general form of the second law of electromagnetic induction is

$$\text{curl } \boldsymbol{H} = \sigma \boldsymbol{E} + \frac{\partial \boldsymbol{D}}{\partial t} + \rho \boldsymbol{v}. \tag{3.85}$$

It should be noted that conduction current exists in good conductors, whereas displacement current exists in good dielectrics. In a material which is neither a good conductor nor a good dielectric, both conduction current and displacement current exist. Conduction current is proportional to the electric field intensity $\boldsymbol{E}$. Convection current, on the other hand, is not proportional to the electric field intensity $\boldsymbol{E}$. It usually exists in devices such as electron tubes, solid-state electron devices, and batteries; it can exist also in conductors, semiconductors, dielectrics, plasma, and ionosphere.

3.19 Force on Charged Particle in Electric Field, Electron Volt

The force $\boldsymbol{F}$ on a particle of charge q is given by

$$\boldsymbol{F} = q\boldsymbol{E}, \tag{3.86}$$

where $\boldsymbol{E}$ is the electric field intensity. If the mass of the particle is m, then

$$\boldsymbol{F} = m\boldsymbol{a}, \tag{3.87}$$

where $\boldsymbol{a}$, the acceleration of the particle, is equal to $d\boldsymbol{v}/dt$, $\boldsymbol{v}$ being the velocity. Therefore

$$\boldsymbol{v} = \boldsymbol{a}t = \frac{q}{m}\boldsymbol{E}t. \tag{3.88}$$

The electric field imparts energy to the charged particle. The energy acquired by the particle in moving from point 1 to point 2 is

$$W = m \int_1^2 \boldsymbol{a} \cdot d\boldsymbol{s} = -q \int_1^2 \boldsymbol{E} \cdot d\boldsymbol{s} = qV, \tag{3.89}$$

where V is the potential difference between points 1 and 2. The energy W is also given by

$$W = m \int_1^2 \boldsymbol{v} \cdot d\boldsymbol{v} = \tfrac{1}{2}m(v_2^2 - v_1^2), \tag{3.90}$$

where v_2 and v_1 are the velocities of the particle at points 1 and 2. Therefore

$$W = \tfrac{1}{2}m(v_2^2 - v_1^2) = qV. \tag{3.91}$$

If the particle starts from rest, $v_1 = 0$, and

$$W = qV = \tfrac{1}{2}mv^2, \tag{3.92}$$

where W is in joules, q in coulombs, V in volts, m in kilograms, and v is in metres per second. Equation (3.92) has dimensions of energy. From this equation,

$$v = (2qV/m)^{1/2}. \tag{3.93}$$

The energy acquired by an electron of charge 1.6×10^{-12} coulomb in 'falling' through a potential difference of 1 volt is 1.6×10^{-19} joule. This amount of energy is called one electron volt and is a convenient unit for designating the energies of particles. For an electron, $m = 0.91 \times 10^{-30}$ kg so that

$$v = 5.93 \times 10^5 \sqrt{V}. \tag{3.94}$$

3.20 Force on Charged Particle in Magnetic Field

The force $\boldsymbol{F}$ on a current element $I\ dl$ in a uniform magnetic field $\boldsymbol{B}$ is

$$\boldsymbol{F} = dl(\boldsymbol{I} \times \boldsymbol{B}). \tag{3.95}$$

The current $\boldsymbol{I}$ can be written as

$$\boldsymbol{I} = \boldsymbol{J}A = \rho \boldsymbol{v} A, \tag{3.96}$$

where $\boldsymbol{J}$ is the current density, $\boldsymbol{v}$ the velocity of the charged particles, ρ the charge density, and A is the cross-section. Hence

$$\boldsymbol{F} = \rho A\ dl\, \boldsymbol{v} \times \boldsymbol{B} = dq\, \boldsymbol{v} \times \boldsymbol{B}, \tag{3.97}$$

where dq is the charge in the volume $A\ dl$.

For a single particle of charge q moving with a velocity $\boldsymbol{v}$, the force is given by

$$\boldsymbol{F} = q\boldsymbol{v} \times \boldsymbol{B} = m\boldsymbol{a}, \tag{3.98}$$

where m is the mass and $\boldsymbol{a}$ the acceleration of the particle. Therefore

$$\boldsymbol{a} = \frac{q}{m}\boldsymbol{v} \times \boldsymbol{B}. \tag{3.99}$$

The acceleration $\boldsymbol{a}$ and the force $\boldsymbol{F}$ are both normal to $\boldsymbol{v}$ and to $\boldsymbol{B}$. If the particle is at rest, the force is zero; if it is moving in the same direction as that of $\boldsymbol{B}$, the force is again zero. If $\boldsymbol{v}$ is at right angles to $\boldsymbol{B}$, then the particle is acted upon by a force which is at right angles to $\boldsymbol{v}$ (see Fig. 3.9), and hence its path is a circle whose radius R is given by

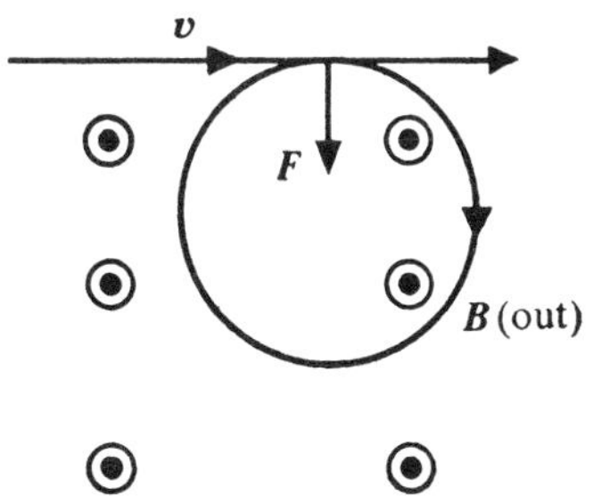

Fig. 3.9 Circular path of positively charged particle in magnetic field.

$$|\boldsymbol{F}| = |m\boldsymbol{a}| = qvB = \frac{mv^2}{R} \tag{3.100}$$

or

$$R = \frac{mv}{qB}.$$

3.21 Electromotive Force Induced by Moving Conductor in Magnetic Field

The force $\boldsymbol{F}$ on a particle of electric charge q moving with a velocity $\boldsymbol{v}$ in a magnetic field $\boldsymbol{B}$ is

$$\boldsymbol{F} = q\boldsymbol{v} \times \boldsymbol{B} \text{ newtons.} \tag{3.101}$$

If the charged particle is situated in a conducting wire moving with a velocity $\boldsymbol{v}$ in a magnetic field $\boldsymbol{B}$ (see Fig. 3.10), then

$$\boldsymbol{E} = \frac{\boldsymbol{F}}{q} = \boldsymbol{v} \times \boldsymbol{B} \tag{3.102}$$

is the force per unit positive charge or the electric field intensity induced in the conductor. The

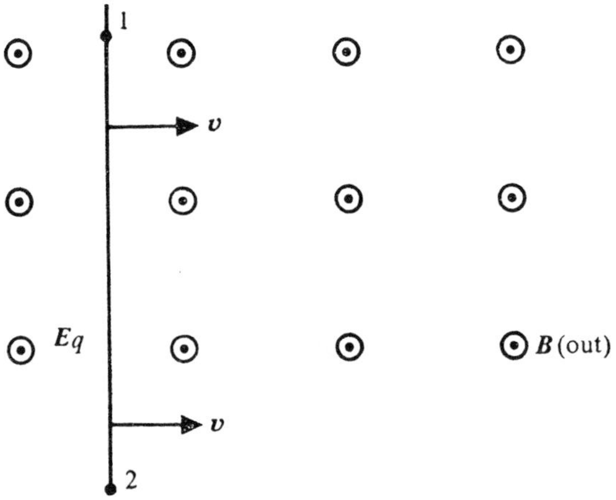

Fig. 3.10 emf induced in a conducting wire moving across a magnetic field.

electromotive force V induced between points 1 and 2 of the wire is

$$V = \int_1^2 \boldsymbol{E} \bullet d\boldsymbol{s} = \int_1^2 (\boldsymbol{v} \times \boldsymbol{B}) \bullet d\boldsymbol{s} \text{ volts.} \tag{3.103}$$

3.22 Generalized Form of the First Law of Electromagnetic Induction or Faraday's Law

If the magnetic flux density $\boldsymbol{B}$ is changing with time, and if a closed circuit C is in motion, then the total electromotive force induced in the closed circuit is given by

$$\oint_c \boldsymbol{E} \bullet d\boldsymbol{s} = V = \oint_c (\boldsymbol{v} \times \boldsymbol{B})\, d\boldsymbol{s} - \int_s \frac{\partial B_n}{\partial t}\, dS, \tag{3.104}$$

where S is a surface enclosed by C, or

$$\oint_c (\boldsymbol{E} - \boldsymbol{v} \times \boldsymbol{B}) \cdot ds = - \int_s \frac{\partial \boldsymbol{B}_n}{\partial t} \cdot dS. \tag{3.105}$$

By using Stokes' theorem, Eq. (3.105) becomes

$$\text{curl } \boldsymbol{E} - \text{curl } (\boldsymbol{v} \times \boldsymbol{B}) = - \frac{\partial \boldsymbol{B}}{\partial t}. \tag{3.106}$$

3.23 The Two Fundamental Electromagnetic Equations or Maxwell's Equations

The first and second laws of electromagnetic induction may be called the *two fundamental electromagnetic* (or *Maxwell's*) *equations.* These equations are

$$\text{curl } \boldsymbol{E} - \text{curl } (\boldsymbol{v} \times \boldsymbol{B}) = \frac{-\partial \boldsymbol{B}}{\partial t}, \tag{3.107}$$

$$\text{curl } \boldsymbol{H} = \sigma \boldsymbol{E} + \frac{\partial \boldsymbol{D}}{\partial t}. \tag{3.108}$$

In most problems of microwave engineering, it is not necessary to include the term curl $(\boldsymbol{v} \times \boldsymbol{B})$ which is due to motion. Hence, Eqs. (3.107) and (3.108) may be rewritten as

$$\text{curl } \boldsymbol{E} = - \frac{\partial \boldsymbol{B}}{\partial t}, \tag{3.109}$$

$$\text{curl } \boldsymbol{H} = \sigma \boldsymbol{E} + \frac{\partial \boldsymbol{D}}{\partial t}. \tag{3.110}$$

In Eqs. (3.109) and (3.110), the sources of the electromagnetic field, namely, the electric and magnetic currents, have been omitted. However, if the sources are included, the two fundamental electromagnetic equations may be written as

$$\text{curl } \boldsymbol{E} = \frac{-\partial \boldsymbol{B}}{\partial t} - \boldsymbol{M}^{\text{i}} \text{ volts/metre}^2, \tag{3.111}$$

$$\text{curl } \boldsymbol{H} = \sigma \boldsymbol{E} + \frac{\partial \boldsymbol{D}}{\partial t} + \boldsymbol{J}^{\text{i}} \text{ amperes/metre}^2, \tag{3.112}$$

where $\boldsymbol{M}^{\text{i}}$ and $\boldsymbol{J}^{\text{i}}$ may be called respectively the source or impressed magnetic and source or impressed electric current densities.

In integral form, the two equations may be written as

$$\oint_c \boldsymbol{E} \cdot ds = - \int_s \left(\frac{\partial \boldsymbol{B}_n}{\partial t} + \boldsymbol{M}_n^{\text{i}}\right) \cdot dS, \tag{3.113}$$

$$\oint_c \boldsymbol{H} \cdot ds = \int_s \left(\sigma E_n + \frac{\partial \boldsymbol{D}_n}{\partial t} + \boldsymbol{J}_n^{\text{i}}\right) \cdot dS. \tag{3.114}$$

In practical situations, we are mainly concerned with sources and fields varying harmonically with time. In such circumstances, Eqs. (3.113) and (3.114) become

$$\oint_c \boldsymbol{E} \cdot d\boldsymbol{s} = -\int_s (j\omega\mu \boldsymbol{H}_n + \boldsymbol{M}_n^i) \cdot d\boldsymbol{S} \text{ volts,} \tag{3.115}$$

$$\oint_c \boldsymbol{H} \cdot d\boldsymbol{s} = \int_s [(\sigma + j\omega\epsilon)\boldsymbol{E}_n + \boldsymbol{J}_n^i] \cdot d\boldsymbol{S} \text{ amperes,} \tag{3.116}$$

using the relations $\boldsymbol{B} = \mu\boldsymbol{H}$ and $\boldsymbol{D} = \epsilon\boldsymbol{E}$ and assuming that all quantities vary as exp $(j\omega t)$.

When Maxwell's equations are expressed in the form (3.115) and (3.116), they possess considerable symmetry: $\boldsymbol{E}$ and $\boldsymbol{H}$ correspond to each other, being expressed in volts per metre and amperes per metre, respectively; $\boldsymbol{D}$ and $\boldsymbol{B}$ correspond to each other, being expressed in coulombs or ampere-seconds per square metre and webers or volt-seconds per square metre, respectively; electric and magnetic currents correspond to each other, being measured in amperes and volts, respectively. Often, it is argued that $\boldsymbol{E}$ (force per unit positive charge) and $\boldsymbol{B}$ (force per unit electric current element) form one pair and $\boldsymbol{D}$ and $\boldsymbol{H}$ another. $\boldsymbol{D}$ and $\boldsymbol{H}$ are respectively determined from the charges and currents in the field, and represent lines of force resulting from the charges and currents. However, since electric and magnetic quantities are physically different, it is better to consider the similarity that arises from Maxwell's equations.

3.24 Currents across Closed Surface

We shall now show that the total electric and magnetic currents across a closed surface vanish. Applying the two fundamental electromagnetic equations

$$\oint_c E_s \cdot ds = -\int_s \left(\frac{\partial \boldsymbol{B}_n}{\partial t} + \boldsymbol{M}_n^i\right) \cdot d\boldsymbol{S}, \tag{3.117}$$

$$\oint_c H_s \cdot ds = \int_s \left(\sigma E_n + \frac{\partial \boldsymbol{D}_n}{\partial t} + \boldsymbol{J}_n^i\right) \cdot d\boldsymbol{S} \tag{3.118}$$

to the two parts S_1 and S_2 of a closed surface S, which has been divided into two parts by a

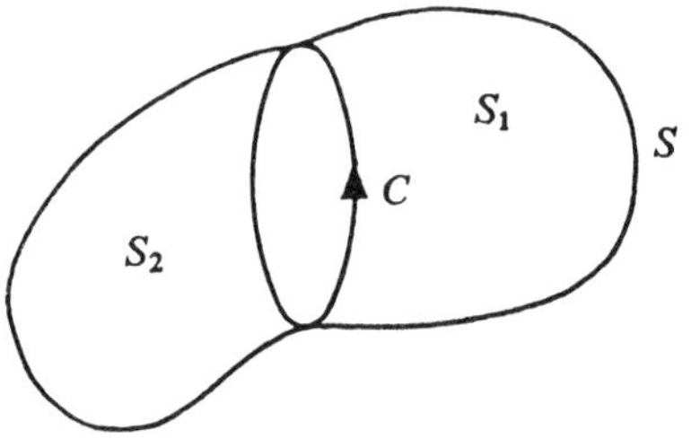

Fig. 3.11 Illustration of total currents across closed surface S.

closed curve C drawn on the surface (see Fig. 3.11), we obtain

$$-\int_{S_1} \left(\frac{\partial B_n}{\partial t} + M_n^i\right) dS - \int_{S_2} \left(\frac{\partial B_n}{\partial t} + M_n^i\right) dS = \oint_c E_s \cdot ds - \oint_c E_s \cdot ds = 0 \tag{3.119}$$

or

$$\oint_s \left(\frac{\partial B_n}{\partial t} + M_n^i\right) \cdot dS = 0. \tag{3.120}$$

Similarly,

$$\oint_s \left(\sigma E_n + \frac{\partial D_n}{\partial t} + J_n^i\right) \cdot dS = 0. \tag{3.121}$$

Equations (3.120) and (3.121) prove that the total electric and magnetic currents across a closed surface vanish. These equations can be rewritten as

$$\oint_s \frac{\partial B_n}{\partial t} \cdot dS = -\oint_s M_n^i \cdot dS = -K, \tag{3.122}$$

$$\oint_s \left(\sigma E_n + \frac{\partial D_n}{\partial t}\right) \cdot dS = -\oint_s J_n^i \cdot dS = -I, \tag{3.123}$$

where K and I are respectively the source magnetic and source electric currents flowing out of the closed surface S.

In perfect dielectrics, $\sigma = 0$. Substituting in Eq. (3.123) and integrating with respect to t, we obtain

$$\int_{-\infty}^{t} dt \oint_s \frac{\partial D_n}{\partial t} \cdot dS = -\int_{-\infty}^{t} I\, dt \tag{3.124}$$

or

$$\oint_s D_n \cdot dS = \oint_s \epsilon E_n \cdot dS = -\int_{-\infty}^{t} I\, dt = q, \tag{3.125}$$

where q is the total electric charge enclosed by the closed surface S. Equation (3.125) is called Gauss' theorem. Similarly, integrating Eq. (3.122) with respect to time, we get

$$\oint_s B_n \cdot dS = \oint_s \mu H_n \cdot dS = -\int_{-\infty}^{t} K\, dt = m, \tag{3.126}$$

where m is the total magnetic pole enclosed by S. But since positive and negative magnetic poles cannot be separated, m is always equal to zero. Hence

$$\oint_s B_n \cdot dS = 0. \tag{3.127}$$

It may be noted that the concept of the existence of $\boldsymbol{M}^i$ or source magnetic current is questionable. However, this is not true of source electric current $\boldsymbol{J}^i$ which can exist in the form of current in various types of generators of electric energy such as batteries, electric generators, and electron devices. Despite the uncertainty about the existence of source magnetic current, the concept of a fictitious source of magnetic current is very useful in practice. Besides, this concept makes the fundamental electromagnetic (or Maxwell's) equations mathematically symmetric and helps solve practical problems such as radiation from antennas.

3.25 Boundary Conditions in Electromagnetic Field

Applying the fundamental equations

$$\oint_c E_s \cdot ds = -\int_s \left(\mu \frac{\partial H_n}{\partial t} + M_n^i\right) \cdot dS, \tag{3.128}$$

$$\oint_c H_s \cdot ds = \int_s \left(\sigma E_n + \epsilon \frac{\partial E_n}{\partial t} + J_n^i\right) \cdot dS \tag{3.129}$$

to a rectangle $ABB'A'$ across a boundary S' between two media, as shown in Fig. 3.12, we get

$$E_t \cdot AB + E_n' \cdot BB' - E_t' \cdot A'B' - E_n \cdot AA' = \int_s \left(\frac{\mu \, \partial H_n}{\partial t} + M_n^i\right) \cdot dS, \tag{3.130}$$

where E_t and E_t' are respectively the tangential components of $\boldsymbol{E}$ along AB and $A'B'$ and E_n

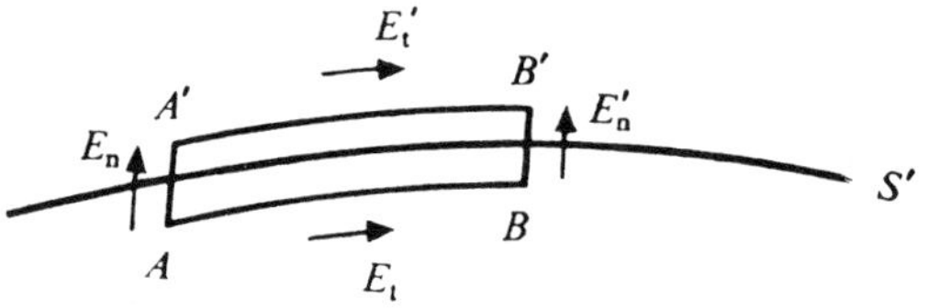

Fig. 3.12 Cross-section of boundary between two media and rectangle having two sides parallel to boundary and other two sides vanishingly small.

and E_n' are the normal components of E along AA' and BB', respectively. As $AA' \to 0$ and $BB' \to 0$,

$$E_t - E_t' = 0. \tag{3.131}$$

Similarly,

$$H_t - H_t' = 0. \tag{3.132}$$

Hence, the tangential components of $\boldsymbol{E}$ and $\boldsymbol{H}$ at the interface between the two media are continuous.

Since the circulation of the tangential component of $\boldsymbol{H}$ per unit area is the normal component of the electric current density $\boldsymbol{J}$, we can say that the normal component of $\boldsymbol{J}$ is also continuous across the boundary, or

$$J_n = J_n'. \tag{3.133}$$

Similarly,

$$M_n = M_n'. \tag{3.134}$$

For fields varying harmonically with time, that is, as $e^{j\omega t}$, the boundary conditions (3.133) and (3.134) become

$$(\sigma + j\omega\epsilon)E_n = (\sigma' + j\omega\epsilon')E_n', \tag{3.135}$$

$$\mu H_n = \mu' H_n'. \tag{3.136}$$

For static fields,

$$\sigma E_n = \sigma' E_n', \qquad \mu H_n = \mu' H_n'. \tag{3.137}$$

In perfect dielectrics ($\sigma = 0$), Eqs. (3.135) and (3.136) become

$$\epsilon E_n = \epsilon' E_n'$$

or

$$D_n = D_n' \tag{3.138}$$

and

$$\mu H_n = \mu' H_n'$$

or

$$B_n = B_n'. \tag{3.139}$$

This says that the normal components of $\boldsymbol{D}$ and $\boldsymbol{B}$ are continuous across the boundary.

In perfect conductors ($\sigma = \infty$), the electric intensity is zero for finite currents, and hence at the interface between a perfect conductor and a perfect dielectric,

$$E_t = 0$$

or

$$H_n = 0. \tag{3.140}$$

The concept of perfect conductors is valuable chiefly because it helps simplify mathematical calculations and provide approximations to solutions of problems involving good conductors.

3.26 Boundary Conditions in the Vicinity of Current Sheet

A current sheet is defined as an infinitely thin sheet carrying finite current (electric or magnetic) per unit length normal to the lines of flow. Figure 3.13 shows the cross-section of an electric

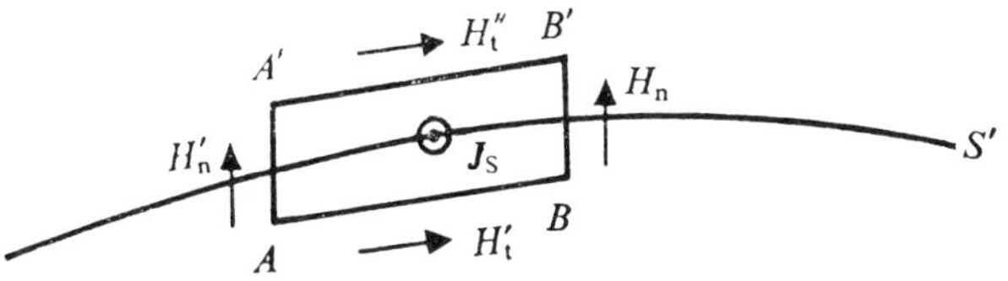

Fig. 3.13 Cross-section of boundary between two media.

current sheet whose linear current density J_S in amperes per metre is normal to the plane of the figure and is directed towards the reader.

Applying the equation

$$\oint_c H_s \, ds = \int_s \left(\sigma E_n + \epsilon \frac{\partial E_n}{\partial t}\right) dS + \int_s J_n^i \, dS \tag{3.141}$$

to the rectangle $ABB'A'$, we obtain

$$H'_t \cdot AB + H_n \cdot BB' - H''_t A'B' - H'_n \cdot AA' = \int_s (\sigma E_n + \epsilon \frac{\partial E_n}{\partial t}) \cdot dS + \int_s J^i_n \, dS. \quad (3.142)$$

As $AA' \to 0$ and $BB' \to 0$, relation (3.142) becomes

$$(H'_t - H''_t)AB = 0 + AB \cdot \lim_{AA' \to 0} AA' J^i_n. \quad (3.143)$$

If

$$\lim_{AA' \to 0} AA' \cdot J^i_n = J_S,$$

then

$$H'_t - H''_t = J_S. \quad (3.144)$$

In relation (3.144), the positive directions of the current density, the tangential component of ***H***, and the normal to the sheet form a right-handed triad. Since

$$\lim_{AA' \to 0} AA' M^i_n = 0,$$

applying the equation

$$\oint_c E_s \cdot ds = \int_s (\frac{\mu \partial H_n}{\partial t} + M^i_n) \, dS \quad (3.145)$$

to the same rectangle $ABB'A'$, we get

$$E'_t = E''_t. \quad (3.146)$$

Therefore, we can say that the tangential components of the electric field intensity in the vicinity of an electric current sheet of density J_S are continuous, whereas the tangential components of the magnetic field intensity are discontinuous by an amount equal to the linear current density J_S.

Similarly, for a magnetic current sheet density M_S volts per metre, we can show that

$$E'_t - E''_t = -M_S, \quad (3.147)$$

$$H'_t = H''_t. \quad (3.148)$$

In other words, there is a discontinuity in the tangential components of the electric field intensity

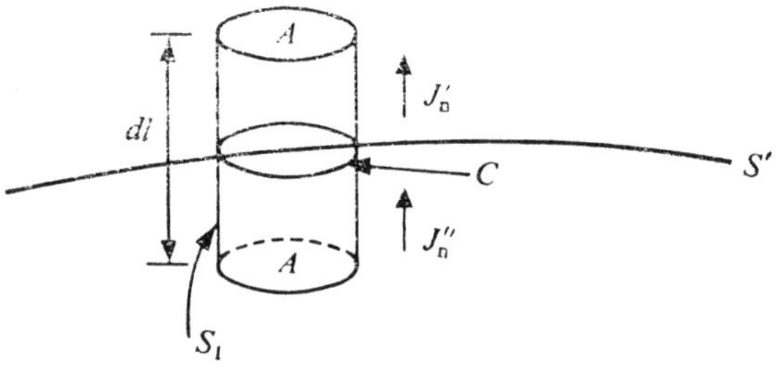

Fig. 3.14 Discontinuity in normal components of electric current densities in the vicinity of electric current sheet.

equal to the linear magnetic current density M_S, and the tangential components of the magnetic field intensity are continuous. The discontinuities in the tangential components of the field intensities imply discontinuities in the normal components of the current densities. Imagine a pill box with its broad faces infinitely close and parallel to the electric current sheet on its opposite sides, as illustrated in Fig. 3.14. Let A be the cross-section and dl the height of this pill box. If J_n' and J_n'' are the normal components of the electric current density J on either side of the electric current sheet, and if J_S is the surface current density on the electric current sheet, then using the theorem that the total electric current across the closed surface of the pill box is zero, we obtain

$$(J_n' - J_n'')A + \int_{S_1} J_n \, dS + \int_{A+A} J_n^i \, dS + \int_{S_1} J_n^i \cdot dS = 0, \tag{3.149}$$

where J^i is the source (or impressed) current inside the pill box. As $dl \to 0$,

$$\int_S J_n \cdot dS = 0, \qquad \int_{A+A} J_n^i \cdot dS = 0, \tag{3.150}$$

$$\int_{S_1} J_n^i \cdot dS = \lim_{dl \to 0} \oint_c (J_n^i \cdot dl) \cdot ds, \tag{3.151}$$

where c is the perimeter of the cross-section of the pill box. As $dl \to 0$ and $J_n^i \to \infty$, we have

$$\lim_{dl \to 0} J_n^i \cdot dl = J_S = \text{surface current density}, \tag{3.152}$$

and hence

$$(J_n' - J_n'')A = \oint_c \frac{\boldsymbol{J}_S \cdot \boldsymbol{ds}}{A}. \tag{3.153}$$

By definition, as $A \to 0$,

$$\frac{\oint_c \boldsymbol{J}_S \cdot \boldsymbol{ds}}{A} = \text{div}' \, J_S = \text{surface divergence of } J_S. \tag{3.154}$$

Therefore

$$J_n' - J_n'' = \text{div}' \, J_S = \text{surface divergence of } J_S. \tag{3.155}$$

Thus, the discontinuity in the normal component of the electric current density is equal to the surface divergence of the surface electric current density of the electric current sheet. Similarly, the discontinuity in the normal component of the magnetic current density is equal to the surface divergence of the magnetic surface current density of the magnetic current sheet.

3.27 Conditions in the Vicinity of Infinitely Thin Linear Current Filaments

For an infinitely thin electric current filament carrying current I amperes per metre of length in the z-direction, we have

$$H_\phi = \frac{I}{2\pi\rho}, \tag{3.156}$$

and for an infinitely thin magnetic current filament carrying current K amperes per metre of length in the z-direction,

$$E_\phi = \frac{K}{2\pi\rho}. \tag{3.157}$$

3.28 Energy Theorems, Poynting Vector

Starting with the fundamental equations of electromagnetic induction given by

$$\text{curl } \boldsymbol{E} = -\mu\frac{\partial \boldsymbol{H}}{\partial t} - \boldsymbol{M}^{\text{i}}, \tag{3.158}$$

$$\text{curl } \boldsymbol{H} = \sigma E + \epsilon\frac{\partial \boldsymbol{E}}{\partial t} + \boldsymbol{J}^{\text{i}} \tag{3.159}$$

and taking the scalar product of the first equation and $\boldsymbol{H}$ and subtracting from it the scalar product of the second equation and $\boldsymbol{E}$, we obtain

$$\boldsymbol{H}\cdot\text{curl } \boldsymbol{E} - \boldsymbol{E}\cdot\text{curl } \boldsymbol{H} = -\boldsymbol{M}^{\text{i}}\cdot\boldsymbol{H} - \boldsymbol{E}\cdot\boldsymbol{J}^{\text{i}} - \sigma E^2 - \mu\boldsymbol{H}\cdot\frac{\partial \boldsymbol{H}}{\partial t} - \epsilon\boldsymbol{E}\cdot\frac{\partial \boldsymbol{E}}{\partial t}. \tag{3.160}$$

Integrating over a volume V bounded by a closed surface S, and using the vector identity

$$\text{div } [\boldsymbol{F}\times\boldsymbol{G}] = \boldsymbol{G}\cdot\text{curl } \boldsymbol{F} - \boldsymbol{F}\cdot\text{curl } \boldsymbol{G} \tag{3.161}$$

and Gauss' theorem, we get

$$-\int_{\text{V}} \boldsymbol{E}\cdot\boldsymbol{J}^{\text{i}}\, dV - \int_{\text{V}} \boldsymbol{M}^{\text{i}}\cdot\boldsymbol{H}\cdot dV = \int_{\text{V}} \sigma E^2\, dV + \frac{\partial}{\partial t}\int_{\text{V}} \tfrac{1}{2}\epsilon E^2\, dV + \frac{\partial}{\partial t}\int_{\text{V}} \tfrac{1}{2}\mu H^2\, dV + \int_{\text{S}} (\boldsymbol{E}\times\boldsymbol{H})_{\text{n}}\, dS. \tag{3.162}$$

Integrating Eq. (3.162) with respect to t in the interval $(-\infty, t)$ and assuming that originally the space was field free, we obtain

$$-\int_{-\infty}^{t} dt \int_{\text{V}} (\boldsymbol{E}\cdot\boldsymbol{J}^{\text{i}} + \boldsymbol{M}^{\text{i}}\cdot\boldsymbol{H})\, dV = \int_{-\infty}^{t} dt \int_{\text{V}} \sigma E^2\, dV + \int_{\text{V}} (\tfrac{1}{2}\epsilon E^2 + \tfrac{1}{2}\mu H^2)\, dV + \int_{-\infty}^{t} dt \int_{\text{S}} (\boldsymbol{E}\times\boldsymbol{H})_{\text{n}}\, dS. \tag{3.163}$$

The left-hand side of Eq. (3.163) represents the total work performed by the impressed forces up to the instant t against the forces of the field in maintaining the impressed or source currents $\boldsymbol{J}^{\text{i}}$ and $\boldsymbol{M}^{\text{i}}$. In accordance with the principle of conservation of energy, we say that this work appears as electromagnetic energy, and we explain the various terms on the right-hand side of Eq. (3.163) as follows:

$$\int_{-\infty}^{t} dt \int_{\text{V}} \sigma E^2\, dV$$

is the total energy converted into heat,

$$\int_{\text{V}} \sigma E^2\, dV$$

being the rate at which the total energy is converted into heat; and

$$\int_V (\tfrac{1}{2}\epsilon E^2 + \tfrac{1}{2}\mu H^2)\, dV$$

is the total stored electric and magnetic energies within S,

$$\frac{\partial}{\partial t}\int_V (\tfrac{1}{2}\epsilon E^2 + \tfrac{1}{2}\mu H^2)\, dV$$

being the rate at which this total stored energy is changing with time. The last term in Eq. (3.163), namely,

$$\int_{-\infty}^{t} dt \int_S (\boldsymbol{E}\times\boldsymbol{H})_n\, dS$$

can be interpreted as the total flow of energy across S up to time t, and hence $\int_S (\boldsymbol{E}\times\boldsymbol{H})_n\, dS$ can be treated as the time rate of flow of energy or power across S. Hence, the vector $\boldsymbol{E}\times\boldsymbol{H} = \boldsymbol{P}$, called the *Poynting vector*, can be considered as a vector representing the time rate of energy flow or power per unit area. Its unit is watt per square metre.

The surface integral of the vector $\boldsymbol{P} = \boldsymbol{E}\times\boldsymbol{H}$ over a closed surface represents the difference between the energy contributed to the field inside S and the energy accounted for within S. On the other hand, it is also true that the value of this integral remains the same if the curl of an arbitrary vector $\boldsymbol{P}'$ is added to $\boldsymbol{P}$.

When the fields are varying harmonically with time, multiplying Eq. (3.158) by $\boldsymbol{H}^*$ and the conjugate of Eq. (3.159) by $\boldsymbol{E}$ and subtracting, we obtain

$$\begin{aligned}\boldsymbol{H}^*\cdot \text{curl}\, \boldsymbol{E} - \boldsymbol{E}\cdot\text{curl}\, \boldsymbol{H}^* &= -\boldsymbol{M}^i\cdot\boldsymbol{H}^* - \boldsymbol{E}\cdot\boldsymbol{J}^{i*} \\ &\quad -\sigma\boldsymbol{E}\cdot\boldsymbol{E}^* + j\omega\epsilon\boldsymbol{E}\cdot\boldsymbol{E}^* - j\omega\mu\boldsymbol{H}\cdot\boldsymbol{H}^*.\end{aligned} \tag{3.164}$$

Integrating Eq. (3.164) over a volume V bounded by a closed surface S, we get

$$\begin{aligned}-\tfrac{1}{2}\int_V (\boldsymbol{E}\cdot\boldsymbol{J}^{i*} + \boldsymbol{M}^i\cdot\boldsymbol{H}^*)\, dV &= \tfrac{1}{2}\int_V \sigma\boldsymbol{E}\cdot\boldsymbol{E}^*\, dV + \tfrac{1}{2}j\omega\int_V \mu\boldsymbol{H}\cdot\boldsymbol{H}^*\cdot dV \\ &\quad - \tfrac{1}{2}j\omega\int_V \epsilon\boldsymbol{E}\cdot\boldsymbol{E}^*\, dV + \tfrac{1}{2}\int_S (\boldsymbol{E}\times\boldsymbol{H}^*)_n\, dS.\end{aligned} \tag{3.165}$$

Let us take the real part of Eq. (3.165). The real part of the left-hand side of this equation is the average power spent by the impressed forces in sustaining the electromagnetic field. Some of this power is transformed into heat and this amount is given by

$$\tfrac{1}{2}\int_V \sigma\boldsymbol{E}\cdot\boldsymbol{E}^*\, dV.$$

The power that remains flows out of the volume across S, and the amount is represented by the real part of the term

$$\tfrac{1}{2}\int_S (\boldsymbol{E}\times\boldsymbol{H}^*)_n\, dS.$$

The term

$$\tfrac{1}{2}\int_V (\mu \boldsymbol{H}\cdot\boldsymbol{H}^* - \epsilon \boldsymbol{E}\cdot\boldsymbol{E}^*)\, dV$$

represents the difference between the average magnetic and electric power stored inside S, and it is equal to the imaginary part of the left-hand side of the equation. The term

$$\psi = \tfrac{1}{2}\int_S (\boldsymbol{E}\times\boldsymbol{H}^*)_n\, dS \tag{3.166}$$

is called the *complex power flow* across S. The vector $\boldsymbol{P} = \tfrac{1}{2}\boldsymbol{E}\times\boldsymbol{H}^*$ is the *complex Poynting vector*, and its real part is the average power flow per unit area.

If S is a perfect conductor, the tangential component of $\boldsymbol{E}$ on S vanishes, and hence there is no flow of energy across S and $\boldsymbol{P}$ is parallel to the surface. A perfectly conducting closed sheet separates space into two electromagnetically independent regions. A similar complete separation is provided by a closed surface on which the tangential component of $\boldsymbol{H}$ is zero. In the physical world, metals are good approximations to perfect conductors, but there are no good approximations to sheets on which H_t is zero except at zero (or nearly zero) frequency when substances with extremely high permeability have this property.

3.29 Normal and Surface Impedances of Surfaces

Only the tangential components of E and $\boldsymbol{H}$ contribute to the complex power flow ψ across a surface S. If u and v are orthogonal coordinates on S and if u, v, and n form a right-handed triplet of directions, then, from Eq. (3.166),

$$\psi = \tfrac{1}{2}\int_S (E_u H_v^* - E_v H_u^*)\, dS.$$

Putting

$$Z_{uv} = \frac{E_u}{H_v}, \qquad Z_{vu} = \frac{-E_v}{H_u}, \tag{3.167}$$

we obtain

$$\psi = \tfrac{1}{2}\int_S (Z_{vu} H_u H_u^* + Z_{uv} H_v H_v^*)\, dS. \tag{3.168}$$

If $Z_{uv} = Z_{vu} = Z_n$, then

$$\psi = \tfrac{1}{2}\int_S Z_n (H_u H_u^* + H_v H_v^*)\, dS. \tag{3.169}$$

Here, Z_n is called the *impedance normal to the surface S*.

Consider now a conducting surface of thickness t. The surface current density $\boldsymbol{J}_S$ in amperes per metre is equal to $\boldsymbol{J}t$, where $\boldsymbol{J}$ is the volume current density in amperes per square metre. If σ is the conductivity, then $\boldsymbol{J} = \sigma\boldsymbol{E}$, and hence $\boldsymbol{J}_S = \sigma t\boldsymbol{E}$. If t approaches zero and σ increases to a very large value so that the product $G = \sigma t$ remains constant, we have

$$\boldsymbol{J}_S = G\boldsymbol{E}, \qquad \boldsymbol{E} = R\boldsymbol{J}_S, \tag{3.170}$$

where G and R are called respectively the surface conductance (in mhos) and surface resistance (in ohms) of the sheet.

In any medium, if the displacement current density is considered in addition to the conduction current density, then for time-harmonic fields,

$$\boldsymbol{J} = j\omega\epsilon \boldsymbol{E} + \sigma \boldsymbol{E}, \tag{3.171}$$

and hence

$$J_S = (j\omega\epsilon + \sigma)tE. \tag{3.172}$$

As $t \to 0$ and σ and $\omega\epsilon$ become very large,

$$J_S = (G_S + jB_S)E = Y_S E, \tag{3.173}$$

$$\boldsymbol{E} = (R_S + jX_S)J_S = Z_S J_S, \tag{3.174}$$

where Y_S is called the surface admittance, G_S the surface conductance, B_S the surface susceptance, Z_S the surface impedance, R_S the surface resistance, and X_S the surface reactance.

A perfect conducting surface has zero normal impedance because the tangential component of $\boldsymbol{E}$ is zero, and hence it is called a *surface of zero impedance*. A *surface of infinite impedance* may be defined as one with zero tangential component of $\boldsymbol{H}$, and it can be approximated by a surface of high permeability only at zero frequency.

3.30 Primary and Secondary Electromagnetic Constants of a Medium

The conductivity σ, permittivity ϵ, and permeability μ are called the *primary electromagnetic constants of a medium*, in the sense that they appear directly in the formulation of the electromagnetic equations

$$\text{curl } \boldsymbol{E} = -\mu \frac{\partial \boldsymbol{H}}{\partial t} - \boldsymbol{M}^{\text{i}}, \tag{3.175}$$

$$\text{curl } \boldsymbol{H} = \sigma \boldsymbol{E} + \epsilon \frac{\partial \boldsymbol{E}}{\partial t} + \boldsymbol{J}^{\text{i}}. \tag{3.176}$$

For time-harmonic fields,

$$\text{curl } \boldsymbol{E} = -j\omega\mu \boldsymbol{H} - \boldsymbol{M}^{\text{i}}, \tag{3.177}$$

$$\text{curl } \boldsymbol{H} = \sigma \boldsymbol{E} + j\omega\epsilon \boldsymbol{E} + \boldsymbol{J}^{\text{i}}. \tag{3.178}$$

Compare Maxwell's equations (3.175) and (3.176) with the transmission line equations

$$\frac{dV}{dx} = -(R + j\omega L)I + E(x), \tag{3.179}$$

$$\frac{dI}{dx} = -(G + j\omega C)V, \tag{3.180}$$

where $E(x)$ is the voltage per unit length impressed along the line and in series with the line, V and I are the instantaneous transverse voltage across the line and the longitudinal electric current in the line, and R, L, G, and C are respectively the series resistance, series inductance, shunt conductance, and shunt capacitance per unit length. We then observe that the transmission line equations form a special one-dimensional case of Maxwell's equations, and the terminology of the transmission line may be extended to Maxwell's equations. Thus, we may call $j\omega\mu$ the distributed series impedance, σ the distributed shunt conductance, and $j\omega\epsilon$ the distributed shunt susceptance, and $(\sigma + j\omega\epsilon)$ the distributed shunt admittance per metre of the medium. The constants μ, σ, and ϵ may be called respectively the distributed series inductance, distributed shunt conductance, and distributed shunt capacitance per metre of the medium. In transmission line theory, two secondary constants are introduced, namely, the propagation constant

$$\gamma = [(R + j\omega L)(G + j\omega C)]^{1/2}$$

and the characteristic impedance

$$Z_0 = [(R + j\omega L)/(G + j\omega C)]^{1/2}.$$

Similarly, in the three-dimensional Maxwell's equations, we may introduce the two *secondary electromagnetic constants*, namely, the intrinsic propagation constant γ and the intrinsic impedance $Z_0 = \eta$, given by

$$\gamma = [j\omega\mu(\sigma + j\omega\epsilon)]^{1/2}, \tag{3.181}$$

$$\eta = Z_0 = \left(\frac{j\omega\mu}{\sigma + j\omega\epsilon}\right)^{1/2}. \tag{3.182}$$

The constants γ and η are independent of the geometry of the wave, and hence the adjective "intrinsic" or "characteristic" of the medium. The characteristic impedances of different types of waves contain η as a factor.

The primary constants are positive except when the frequencies are very high (at optical frequencies ϵ may be negative), and hence $\gamma = \alpha + j\beta$ and $\eta = \mathcal{R} + j\mathcal{X}$ are either in the first quadrant or in the third quadrant. The definitions of γ and η are made unambiguous if it is agreed that they always lie in the first quadrant or on its boundaries. For perfect conductors, $\sigma = \infty$, and hence both γ and η lie on the bisector of the first quadrant. For perfect dielectrics, $\sigma = 0$, and hence γ is on the positive imaginary axis and η on the positive real axis. In general, both γ and η are complex quantities:

$$\gamma = \alpha + j\beta, \tag{3.183}$$

$$\eta = \mathcal{R} + j\mathcal{X}, \tag{3.184}$$

where α, β, $\mathcal{R}$, $\mathcal{X}$ are called respectively the intrinsic attenuation constant, intrinsic phase constant, intrinsic resistance, and intrinsic reactance of the medium.

For perfect dielectrics,

$$\begin{aligned} &\gamma = j\beta, \qquad \beta = \omega\sqrt{\mu\epsilon} = \frac{\omega}{v} = \frac{2\pi}{\lambda}, \\ &\eta = \sqrt{\frac{\mu}{\epsilon}}, \qquad f\lambda = v, \\ &v = \frac{1}{\sqrt{\mu\epsilon}}, \qquad \lambda = \frac{2\pi}{\beta}, \qquad \mu = \frac{\eta}{v}, \qquad \epsilon = \frac{1}{\eta v}, \end{aligned} \tag{3.185}$$

where v and λ are called the intrinsic or characteristic velocity and wavelength in the medium. In general, the velocity and wavelength for different types of waves may have respectively the characteristic velocity and characteristic wavelength as factors.

For free space, the numerical values for various constants are

$$\eta_0 = \sqrt{\frac{\mu_0}{\epsilon_0}} = 376.7 \approx 377 \approx 120\pi \text{ ohms}, \tag{3.186}$$

$$v_0 = \frac{1}{\sqrt{\mu_0 \epsilon_0}} = 2.998 \times 10^8 \approx 3 \times 10^8 \text{ metres/second}, \tag{3.187}$$

$$\eta_0^{-1} = \sqrt{\frac{\epsilon_0}{\mu_0}} \approx \frac{1}{120\pi} \text{ mho}, \tag{3.188}$$

$$\mu_0 = 4\pi \times 10^{-7} = 1.257 \times 10^{-6} \text{ henry/metre}, \tag{3.189}$$

$$\epsilon_0 = 8.854 \times 10^{-12} \approx \frac{1}{36\pi} \times 10^{-9} \text{ farad/metre}. \tag{3.190}$$

For good conductors, $\sigma \gg \omega\epsilon$, and hence

$$\gamma = \sqrt{j\omega\mu\sigma}, \qquad \eta = \sqrt{\frac{j\omega\mu}{\sigma}}, \tag{3.191}$$

$$\alpha = \beta = \sqrt{\frac{\omega\mu\sigma}{2}}, \tag{3.192}$$

$$\mathcal{R} = \mathcal{X} = \sqrt{\frac{\omega\mu}{2\sigma}}. \tag{3.193}$$

The Q of a medium is defined as the ratio of the displacement current density to the conduction current density, and hence

$$Q = \frac{\omega\epsilon}{\sigma} = \frac{1}{D}, \tag{3.194}$$

where D is called the *dissipation factor*. For good dielectrics, $Q \gg 1$, and for good conductors, $Q \ll 1$. Table 3.3 gives the values of ϵ_r for different commonly used materials.

Table 3.3 Relative Permittivity of Materials

Material	*Relative permittivity* (ϵ_r)
Vacuum	1
Air (0°C)	
40 atmospheres	1.00059
80 atmospheres	1.0218
Carbon dioxide (0°C)	1.000985

Table 3.3 Relative Permittivity of Materials (cont.)

Material	*Relative permittivity* (ϵ_r)
Hydrogen (0°C)	1.000264
Water vapour (145°C)	1.00705
Styrofoam	1.03
Air (−191°C)	1.43
Paraffin	2.1
Plywood	2.1
Polystyrene	2.7
Paper	2.0–2.5
Rubber	2.3–4.0
Amber	2.7
Asphalt	2.68
Benzene	2.29
Petroleum	2.13
Amber	3.0
Linseed oil	3.35
Plexiglas	3.4
Dry sandy soil	3.4
Nylon (hard)	3.8
Wood	2.5–7.7
Sulphur (amorphous)	3.98
Sulphur (cast, fresh)	4.92
Guttapercha	3.3–4.9
Castor oil	4.67
Quartz	5.0
Bakelite	5.0
Mica	5.6–6.0
Lead glass	5.4–8.0
Formica	6.0
Flint glass	6.6–9.9

Table 3.3 Relative Permittivity of Materials (cont.)

Material	*Relative permittivity* (ϵ_r)
Alcohol	
amyl	16.0
ethyl	25.8
methyl	31.2
Ammonia	22.0
Glycerin (15°C)	56.2
Water (distilled)	81.1
Rutile (TiO_2)	89–173
Barium titanate ($BaTiO_3$)	1200
Barium strontium titanate ($2BaTiO_3 : 1SrTiO_3$)	10,000
Barium titanate zirconate ($4BaTiO_3 : 1BaZrO_3$)	13,000
Barium titanate stannate ($9BaTiO_3 : 1BaSnO_3$)	20,000

3.31 Electromagnetic Wave Equation in Dielectrics and Conductors

In Maxwell's equations

$$\text{curl } \boldsymbol{E} = -\mu \frac{\partial \boldsymbol{H}}{\partial t} - \boldsymbol{M}^{\text{i}}, \tag{3.195}$$

$$\text{curl } \boldsymbol{H} = \sigma \boldsymbol{E} + \epsilon \frac{\partial \boldsymbol{E}}{\partial t} + \boldsymbol{J}^{\text{i}} \tag{3.196}$$

if $\boldsymbol{M}^{\text{i}}$ and $\boldsymbol{J}^{\text{i}}$ are differentiable, then $\boldsymbol{E}$ and $\boldsymbol{H}$, which satisfy the nonhomogeneous wave equations, can be eliminated. In practice, however, $\boldsymbol{J}^{\text{i}}$ and $\boldsymbol{M}^{\text{i}}$ are usually discontinuous, and hence non-differentiable; in such cases, $\boldsymbol{E}$ and $\boldsymbol{H}$ cannot be eliminated without introducing auxiliary functions, also called potential functions.

In source-free regions, where $\boldsymbol{J}^{\text{i}} = \boldsymbol{M}^{\text{i}} = 0$, on the other hand, we can always eliminate either $\boldsymbol{E}$ or $\boldsymbol{H}$. In homogeneous source-free regions, we obtain the wave equations

$$\nabla^2 \boldsymbol{E} = \mu\epsilon \frac{\partial^2 \boldsymbol{E}}{\partial t^2} + \mu\sigma \frac{\partial \boldsymbol{E}}{\partial t}, \tag{3.197}$$

$$\nabla^2 \boldsymbol{H} = \mu\epsilon \frac{\partial^2 \boldsymbol{H}}{\partial t^2} + \mu\sigma \frac{\partial \boldsymbol{H}}{\partial t}. \tag{3.198}$$

In perfect dielectrics, $\sigma = 0$, and Eqs. (3.197) and (3.198) become

$$\begin{aligned} \nabla^2 \boldsymbol{E} &= \mu\epsilon \frac{\partial^2 \boldsymbol{E}}{\partial t^2}, \\ \nabla^2 \boldsymbol{H} &= \mu\epsilon \frac{\partial^2 \boldsymbol{H}}{\partial t^2}. \end{aligned} \tag{3.199}$$

For time-harmonic fields, the wave equations are

$$\nabla^2 \boldsymbol{E} = \gamma^2 \boldsymbol{E}, \tag{3.200}$$

$$\nabla^2 \boldsymbol{H} = \gamma^2 \boldsymbol{H}. \tag{3.201}$$

3.32 Solution of Wave Equation in Cartesian Coordinates in Homogeneous Dissipative and Nondissipative Regions: Uniform Plane Waves

In cartesian coordinates, the vector wave equation

$$\nabla^2 \boldsymbol{E} = \gamma^2 \boldsymbol{E} \tag{3.202}$$

reduces to three scalar wave equations, namely,

$$\begin{aligned} \nabla^2 E_x &= \gamma^2 E_x, \\ \nabla^2 E_y &= \gamma^2 E_y, \\ \nabla^2 E_z &= \gamma^2 E_z. \end{aligned} \tag{3.203}$$

A solution of any of the equations (3.203) is of the form

$$E_x = \exp\,(\pm \Gamma_x x \pm \Gamma_y y \pm \Gamma_z z), \tag{3.204}$$

where the propagation constants Γ_x, Γ_y, Γ_z in the x-, y-, z-coordinate axis respectively satisfy the condition

$$\Gamma_x^2 + \Gamma_y^2 + \Gamma_z^2 = j\omega\mu(\sigma + j\omega\epsilon). \tag{3.205}$$

In nondissipative media, we have $\sigma = 0$, and hence

$$\Gamma_x^2 + \Gamma_y^2 + \Gamma_z^2 = -\omega^2\mu\epsilon = -\beta^2 = -\frac{4\pi^2}{\lambda^2}. \tag{3.206}$$

Because of relation (3.206), only two of the propagation constants Γ_x, Γ_y, Γ_z are independent and depend on the distribution of sources producing the field. If the distribution of sources is uniform in planes parallel to the xy-plane, the field is uniform in these planes, and $\Gamma_x = \Gamma_y = 0$; then, the propagation constant Γ_z in the z-direction is equal to the intrinsic propagation constant γ. Such a wave is a uniform plane wave whose equiphase and equiamplitude planes are parallel to the xy-plane, and the expression for any of the components of the field is of the form

$$E_x = \exp\,(\pm \Gamma_z z) e^{j\omega t} = e^{\pm\gamma z} e^{j\omega t} = e^{\pm(\alpha + j\beta)z} e^{j\omega t}. \tag{3.207}$$

The positive signs in Eq. (3.207) give a wave with increasing amplitude as z increases. This is not possible in a physical situation unless there is a conversion of energy of some other form into electromagnetic energy (as seen in an electron device where the kinetic or potential energy of the charge carriers is converted into electromagnetic energy). Hence, it is customary to use the negative sign, and therefore

$$E_x = e^{-\alpha z} e^{j(\omega t - \beta z)} \tag{3.208}$$

which represents an attenuated uniform plane travelling in the positive z-direction with phase velocity v equal to ω/β. In a nondissipative or perfect dielectric medium, $\alpha = 0$, and hence there is no attenuation.

For a uniform plane wave, whose equiamplitude and equiphase planes are parallel to a plane

$$x \cos A + y \cos B + z \cos C = \text{constant},$$

where $\cos A$, $\cos B$, $\cos C$ are the direction cosines of a normal to the plane, the expression for any of the field components is of the form

$$E_x = e^{-\gamma(x \cos A + y \cos B + z \cos C)} = \exp(-\Gamma_x x - \Gamma_y y - \Gamma_z z) \tag{3.209}$$

so that the propagation constants along the three axes are

$$\Gamma_x = \gamma \cos A, \qquad \Gamma_y = \gamma \cos B, \qquad \Gamma_z = \gamma \cos C. \tag{3.210}$$

If the medium is nondissipative, $\sigma = 0$, and hence $\gamma = j\beta$. Therefore

$$\begin{aligned} &\Gamma_x = j\beta_x = j\beta \cos A \quad \text{or} \quad \beta_x = \beta \cos A, \\ &\beta_y = \beta \cos B, \qquad \beta_z = \beta \cos C. \end{aligned} \tag{3.211}$$

Hence

$$\begin{aligned} &\lambda_x = \frac{2\pi}{\beta_x} = \frac{2\pi}{\beta \cos A} = \frac{\lambda}{\cos A}, \\ &\lambda_y = \frac{\lambda}{\cos B}, \qquad \lambda_z = \frac{\lambda}{\cos C}, \end{aligned} \tag{3.212}$$

$$\begin{aligned} &v_x = \frac{\omega}{\beta_x} = \frac{\omega}{\beta \cos A} = \frac{v}{\cos A}, \\ &v_y = \frac{v}{\cos B}, \qquad v_z = \frac{v}{\cos C}, \end{aligned} \tag{3.213}$$

$$\beta_x^2 + \beta_y^2 + \beta_z^2 = \beta^2, \tag{3.214}$$

$$\frac{1}{\lambda_x^2} + \frac{1}{\lambda_y^2} + \frac{1}{\lambda_z^2} = \frac{1}{\lambda^2}, \tag{3.215}$$

$$\frac{1}{v_x^2} + \frac{1}{v_y^2} + \frac{1}{v_z^2} = \frac{1}{v^2}. \tag{3.216}$$

Thus, for uniform plane waves in a nondissipative medium, the phase velocities in various directions are always greater than the characteristic phase velocity, and the phase constants β_x, β_y, β_z are always less than the characteristic phase constant β of the medium. This property is true for a rectangular metal waveguide.

There are some types of physical situations, for example, for surface waves supported by surface-wave structures or slow-wave structures in nondissipative media, where the phase constant in some direction, say, the z-direction, is greater than β or, in other words, $\beta_z > \beta$. Then

$$\Gamma_x^2 + \Gamma_y^2 = -\beta^2 + \beta_z^2. \tag{3.217}$$

If $\Gamma_y = 0$ or there is no variation of the field in the y-direction, then

$$\Gamma_x^2 = -\beta^2 + \beta_z^2 > 0 \tag{3.218}$$

or Γ_x is a real quantity $\pm\alpha_x$.

Taking the negative value $-\alpha_x$ for a real physical situation, we find that there is attenuation in the x-direction. As $\beta_z > \beta$, $v_z < v$, the wave travels in the z-direction with a velocity less than the characteristic velocity of the medium, and is attenuated exponentially in the x-direction or, in other words, is evanescent in the x-direction. Such a wave is called a *surface* (or *slow*) *wave*. The attenuation is due to the boundary conditions, and not due to the loss in the medium. Any component of the field varies as $\exp(-\alpha_x x) \exp[j(\omega t - \beta_z z)]$ in a nondissipative medium. On the other hand, if $\beta_z < \beta$ and if $\Gamma_y = 0$, then

$$\Gamma_x^2 = \beta_z^2 - \beta^2 < 0 \tag{3.219}$$

or

$$\Gamma_x = j\beta_x, \tag{3.220}$$

and

$$v_z > v.$$

In this case, the wave travels in a direction making an angle $\theta = \tan^{-1}(\beta_z/\beta_x)$ with the z-axis and with a velocity v_z which is greater than the characteristic velocity v of the medium. Such a wave is called a *leaky* (or *fast*) *wave*, and any field component varies as $\exp[j(\omega t - \beta_x x - \beta_z z)]$.

In general, in a nondissipative medium, if the propagation constants Γ_x, Γ_y, Γ_z in the three directions are complex, then

$$\Gamma_x^2 + \Gamma_y^2 + \Gamma_z^2 = -\beta^2 \tag{3.221}$$

becomes

$$(\alpha_x + j\beta_x)^2 + (\alpha_y + j\beta_y)^2 + (\alpha_z + j\beta_z)^2 = -\beta^2 \tag{3.222}$$

or

$$\alpha_x^2 + \alpha_y^2 + \alpha_z^2 - \beta_x^2 - \beta_y^2 - \beta_z^2 = -\beta^2 \tag{3.223}$$

and

$$\alpha_x\beta_x + \alpha_y\beta_y + \alpha_z\beta_z = 0. \tag{3.224}$$

Equation (3.224) shows that the equiamplitude planes

$$\alpha_x x + \alpha_y y + \alpha_z z = \text{constant}$$

are perpendicular to the equiphase planes

$$\beta_x x + \beta_y y + \beta_z z = \text{constant}.$$

Thus, in nondissipative media, equiamplitude and equiphase planes either *coincide* with each other (as for uniform plane waves) or are *orthogonal* to each other. In the first instance, the waves are uniform on equiphase planes in the sense that E and H have constant values at all points of a given equiphase plane at a given instant; in the second case, the amplitude varies exponentially, the fastest variation being in the direction given by the direction components α_x, α_y, α_z. For example, for a surface (or slow) wave, any field component varies as

$$\exp(-\alpha_x x) \exp[j(\omega t - \beta_z z)],$$

and hence the equiphase planes are $z =$ constant and the equiamplitude planes are $x =$ constant, and these two families of planes are orthogonal to each other. For a leaky (or fast) wave, in general,

$$\Gamma_x = \alpha_x + j\beta_x, \qquad \Gamma_z = \alpha_z + j\beta_z, \qquad \Gamma_y = 0,$$

and hence the expression for any field component varies as

$$[\exp -(\alpha_x x - \alpha_z z) \exp \{j(\omega t - \beta_x x - \beta_z z)\}].$$

Therefore, the equiamplitude planes $\alpha_x x + \alpha_z z =$ constant are orthogonal to the equiphase planes $\beta_x x + \beta_z z =$ constant (see Fig. 3.15).

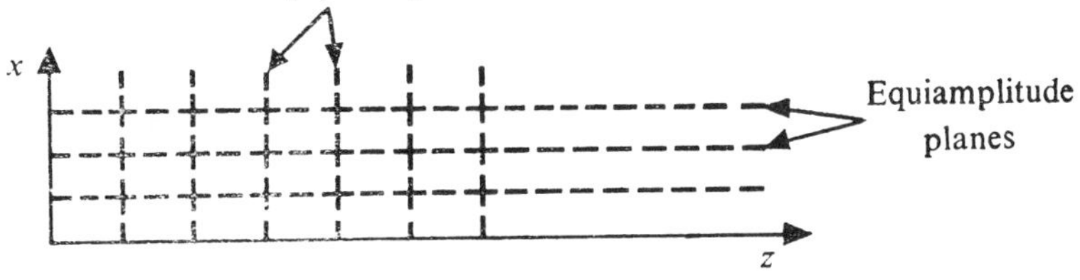

(a) Surface (or slow) waves ($\beta_z > \beta$, $v_z < v$)

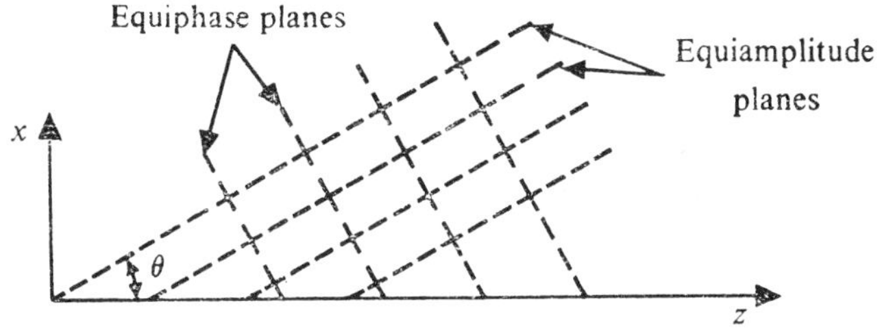

(b) Leaky (or fast) waves ($v_z > v$, $\beta_z < \beta$); [$\theta = \tan^{-1}(\beta_x/\beta_z)$]

Fig. 3.15 Illustration of surface (or slow) waves and leaky (or fast) waves.

In dissipative media, $\gamma^2 = (\alpha + j\beta)^2$, and hence

$$(\alpha_x + j\beta_x)^2 + (\alpha_y + j\beta_y)^2 + (\alpha_z + j\beta_z)^2 = (\alpha + j\beta)^2 \tag{3.225}$$

or

$$\alpha_x^2 + \alpha_y^2 + \alpha_z^2 - \beta_x^2 - \beta_y^2 - \beta_z^2 = \alpha^2 - \beta^2 \tag{3.226}$$

and

$$\alpha_x\beta_x + \alpha_y\beta_y + \alpha_z\beta_z = \alpha\beta = \tfrac{1}{2}\omega\mu\sigma. \tag{3.227}$$

Thus, equiamplitude planes are no longer perpendicular to equiphase planes.

The foregoing general conclusions on waves of exponential type of the form

$$E_x = \exp(-\Gamma_x x - \Gamma_y y - \Gamma_z z)e^{j\omega t} \tag{3.328}$$

have a broader significance than appears at first sight. The constant Γ_x represents the relative rate of change of E_x in the x-direction, and we have

$$\Gamma_x = -\frac{1}{E_x}\frac{\partial E_x}{\partial x}, \tag{3.229}$$

$$\Gamma_x^2 = \frac{1}{E_x}\frac{\partial^2 E_x}{\partial x^2}. \tag{3.230}$$

Equation (3.230) is also satisfied by $-\Gamma_x$. If the wave function is not exponential, we may still define Γ_x by Eq. (3.230) and, similarly, we may define Γ_y and Γ_z by

$$\Gamma_y^2 = \frac{1}{E_x}\frac{\partial^2 E_x}{\partial y^2}, \tag{3.231}$$

$$\Gamma_z^2 = \frac{1}{E_x}\frac{\partial^2 E_x}{\partial z^2}. \tag{3.232}$$

If the quantities Γ_x, Γ_y, and Γ_z vary slowly from one point to another, the solution of the wave equation will be approximately exponential, and the properties of exponential waves just discussed will be applicable in sufficiently small regions.

3.33 Waves at Interface between Conductors and Dielectrics

Consider a plane interface (the xy-plane) between air (substantially free space) above the plane and a conductor below the plane, as shown in Fig. 3.16. For an exponential wave of the form

$$F = \exp(-\Gamma_x x - \Gamma_y y - \Gamma_z z), \tag{3.233}$$

the propagation constants Γ_x and Γ_y in directions parallel to the boundary must be the same in both media so that the boundary conditions may be satisfied at all points of the air-conductor

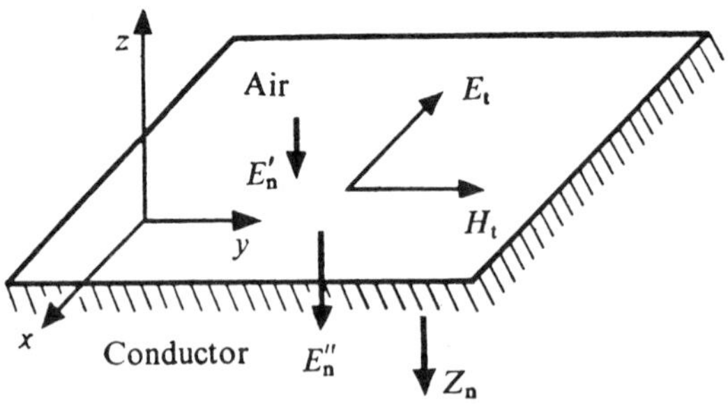

Fig. 3.16 Plane boundary between two semi-infinite media.

interface. This is obvious because F represents a component of either $\boldsymbol{E}$ or $\boldsymbol{H}$ parallel to the boundary; in this instance, F is continuous across the boundary and the continuity cannot be satisfied at all points unless Γ_x, Γ_y are the same on both sides of the boundary. The same condition applies also to the normal components of the current densities $(\sigma + j\omega\epsilon)E_z$ and $j\omega\mu H_z$. Thus, we have

$$\Gamma_x^2 + \Gamma_y^2 + \Gamma_{z,0}^2 = -\beta_0^2 \qquad \text{(in air)}, \tag{3.234}$$

$$\Gamma_x^2 + \Gamma_y^2 + \Gamma_z^2 = \gamma^2 \qquad \text{(in conductor)}, \tag{3.235}$$

where β_0 is the characteristic phase constant $\omega\sqrt{\mu_0\epsilon_0}$ of air and γ is the characteristic propagation constant of the conductor. Subtracting Eq. (3.234) from Eq. (3.235), we obtain

$$\Gamma_z^2 = \gamma^2 + \beta_0^2 + \Gamma_{z,0}^2. \tag{3.236}$$

The propagation constant γ in the conductor is very much larger than that in free space (or air). Hence, $\gamma^2 \gg \beta_0^2$. Also, $\Gamma_{z,0}$, which is the propagation constant in free space in the direction normal to the interface, is comparable to β_0 if the wave direction is nearly normal to the interface or it is much smaller than β_0 if the wave direction is nearly parallel to the interface. In other words, $\Gamma_{z,0} \approx \beta_0$ or $\Gamma_{z,0} \ll \beta_0$. Hence, $\Gamma_{z,0} \ll \gamma$. Therefore, from Eq. (3.236),

$$\Gamma_z \approx \gamma \tag{3.237}$$

or, in the conductor, the propagation constant normal to the interface is substantially equal to the intrinsic propagation constant of the conductor.

Since the current density normal to the interface is continuous, we have

$$\sigma E_n'' = j\omega\epsilon_0 E_n, \tag{3.238}$$

or

$$E_n'' = \frac{j\omega\epsilon_0}{\sigma} E_n', \tag{3.239}$$

where E_n' and E_n'' are the normal components of $\boldsymbol{E}$ in air and in the conductor, respectively. As σ for the conductor is very large, $E_n'' \ll E_n'$ or the normal component of $\boldsymbol{E}$ in the conductor is very much smaller than the normal component of $\boldsymbol{E}$ in air.

Even at moderately high frequencies, the attenuation constant in the conductor is large, and the field becomes quite small at rather short distances from the interface. Except at very low frequencies, the fields are confined largely to thin skins of conductors, and hence this effect is known as the *skin effect*.

The current density at the surface of the conductor is σE_t, where E_t is the tangential component of $\boldsymbol{E}$ at the interface, and elsewhere inside the conductor, it is $\sigma E_t e^{-\gamma z}$, where z is the normal distance from the surface. The total current per unit length normal to the lines of flow in a conductor is

$$J_S = \int_0^\infty \sigma E_t e^{-\gamma z}\, dz = \frac{\sigma}{\gamma} E_t = \frac{1}{\eta} E_t, \tag{3.240}$$

where $\eta = (j\omega\mu/\sigma)^{1/2}$ and $\gamma = \sqrt{j\omega\mu\sigma}$. On the other hand, at the surface of the conductor, the tangential component of $\boldsymbol{H}$, namely, H_t, is equal to J_S, and is given by

$$H_t = J_S = \frac{1}{\eta} E_t \tag{3.241}$$

or $E_t/H_t = Z_n$ = the normal impedance of the surface $= \eta$, which is the intrinsic impedance of the conductor.

By definition, E_t/J_S is the surface impedance Z_S of the interface. Therefore, Z_S of an interface between a good conductor and a good dielectric is equal to the normal impedance Z_n. The conductor may be replaced by a sheet, whose surface impedance is η adjacent to a sheet of infinite impedance, which would effectively exclude the space previously occupied by the conductor. Z_S and γ are given by

$$Z_S = \mathcal{R} + j\mathcal{X} = \eta = \left(\frac{j\omega\mu}{\sigma}\right)^{1/2} = \left(\frac{\omega\mu}{2\sigma}\right)^{1/2} + j\left(\frac{\omega\mu}{2\sigma}\right)^{1/2}, \tag{3.242}$$

$$\gamma = \alpha + j\beta = \sqrt{j\omega\mu\sigma} = \left(\frac{\omega\mu\sigma}{2}\right)^{1/2} + j\left(\frac{\omega\mu\sigma}{2}\right)^{1/2}. \tag{3.243}$$

Therefore

$$\mathcal{R} = \left(\frac{\omega\mu}{2\sigma}\right)^{1/2} = \frac{\alpha}{\sigma} = \frac{1}{\sigma t}, \tag{3.244}$$

where

$$t = \frac{1}{\alpha} = \left(\frac{2}{\omega\mu\sigma}\right)^{1/2}. \tag{3.245}$$

Hence, the surface resistance $\mathcal{R}$ is equal to the d-c resistance of a plate of thickness t, defined by the reciprocal of the attenuation constant. The thickness t is called the *skin depth* or *depth of penetration*, but the term should not be interpreted to mean that the rest of the conductor could be removed without changing its a-c resistance. The field reduces to $1/e$ of its value at the surface, at a distance equal to the skin depth, and the attenuation through the skin depth is only one neper.

3.34 Special Forms of Maxwell's Equations in Source-Free Regions in Different Systems of Coordinates

The fundamental electromagnetic (or Maxwell's) equations in source-free regions for time-harmonic fields are

$$\text{curl } \boldsymbol{E} = -j\omega\mu\boldsymbol{H}, \tag{3.246}$$

$$\text{curl } \boldsymbol{H} = j\omega\epsilon\boldsymbol{E} + \sigma\boldsymbol{E}. \tag{3.247}$$

In the cartesian coordinates, they become

$$\begin{aligned} \frac{\partial E_z}{\partial y} - \frac{\partial E_y}{\partial z} &= -j\omega\mu H_x, \\ \frac{\partial E_x}{\partial z} - \frac{\partial E_z}{\partial x} &= -j\omega\mu H_y, \\ \frac{\partial E_y}{\partial x} - \frac{\partial E_x}{\partial y} &= -j\omega\mu H_z, \end{aligned} \tag{3.248}$$

$$\frac{\partial H_z}{\partial y} - \frac{\partial H_y}{\partial z} = (\sigma + j\omega\epsilon)E_x,$$

$$\frac{\partial H_x}{\partial z} - \frac{\partial H_z}{\partial x} = (\sigma + j\omega\epsilon)E_y, \tag{3.249}$$

$$\frac{\partial H_y}{\partial x} - \frac{\partial H_x}{\partial y} = (\sigma + j\omega\epsilon)E_z.$$

In the cylindrical coordinates, they are of the form

$$\frac{\partial E_z}{\partial \phi} - \rho\frac{\partial E_\phi}{\partial z} = -j\omega\mu\rho H_\rho,$$

$$\frac{\partial E_\rho}{\partial z} - \frac{\partial E_z}{\partial \rho} = -j\omega\mu H_\phi, \tag{3.250}$$

$$\frac{\partial}{\partial \rho}(\rho E_\phi) - \frac{\partial E_\rho}{\partial \phi} = -j\omega\mu\rho H_z,$$

$$\frac{\partial H_z}{\partial \phi} - \rho\frac{\partial H_\phi}{\partial z} = (\sigma + j\omega\epsilon)\rho E_\rho,$$

$$\frac{\partial H_\rho}{\partial z} - \frac{\partial H_z}{\partial \rho} = (\sigma + j\omega\epsilon)E_\phi, \tag{3.251}$$

$$\frac{\partial}{\partial \rho}(\rho H_\phi) - \frac{\partial H_\rho}{\partial \phi} = (\sigma + j\omega\epsilon)\rho E_z.$$

In the spherical coordinates, they take the form

$$\frac{\partial}{\partial \theta}(\sin\theta\, E_\phi) - \frac{\partial E_\theta}{\partial \phi} = -j\omega\mu r \sin\theta\, H_r,$$

$$\frac{\partial}{\partial r}(rE_\theta) - \frac{\partial E_r}{\partial \theta} = -j\omega\mu r H_\phi, \tag{3.252}$$

$$\frac{\partial E_r}{\partial \phi} - \sin\theta\,\frac{\partial}{\partial r}(rE_\phi) = -j\omega\mu r \sin\theta\, H_\theta,$$

$$\frac{\partial}{\partial \theta}(\sin\theta\, H_\phi) - \frac{\partial H_\theta}{\partial \phi} = (\sigma + j\omega\epsilon)r \sin\theta\, E_r,$$

$$\frac{\partial}{\partial r}(rH_\theta) - \frac{\partial H_r}{\partial \theta} = (\sigma + j\omega\epsilon)rE_\phi, \tag{3.253}$$

$$\frac{\partial H_r}{\partial \phi} - \sin\theta\,\frac{\partial}{\partial r}(H_\phi r) = (\sigma + j\omega\epsilon)\sin\theta\, E_\theta.$$

3.35 Field Produced by Given Distribution of Currents in Infinite Homogeneous Medium: Vector and Scalar Wave Potentials

When the source (or impressed) currents are known throughout an infinite homogeneous medium, the field can be calculated fairly easily. If we obtain the field of a current element, then by the use of the principle of superposition, the field due to any distribution may be calculated.

The electromagnetic equations for harmonic fields are

$$\text{curl}\ \boldsymbol{E} = -j\omega\mu\boldsymbol{H} - \boldsymbol{M}^{\text{i}}, \tag{3.254}$$

$$\text{curl}\ \boldsymbol{H} = (\sigma + j\omega\epsilon)\boldsymbol{E} + \boldsymbol{J}^{\text{i}}. \tag{3.255}$$

If these equations are solved, the solution of the most general case can then be expressed in the form of a contour integral in the oscillation constant plane. If $\boldsymbol{M}^{\text{i}}$ is a continuous and differentiable function, then it is possible to eliminate $\boldsymbol{E}$ from Eq. (3.254) and obtain a second-order differential equation for $\boldsymbol{H}$. Similarly, if $\boldsymbol{J}^{\text{i}}$ is a continuous and differentiable function, we can eliminate $\boldsymbol{H}$ from Eq. (3.255) and derive a second-order differential equation for $\boldsymbol{E}$. However, in practical problems, $\boldsymbol{J}^{\text{i}}$ and $\boldsymbol{M}^{\text{i}}$ are localized and the regions occupied by them have sharp boundaries. When this happens, the equations can be solved by introducing a set of auxiliary functions, generally called potential functions. Let

$$\boldsymbol{E} = \boldsymbol{E}' + \boldsymbol{E}'', \qquad \boldsymbol{H} = \boldsymbol{H}' + \boldsymbol{H}'', \tag{3.256}$$

where $(\boldsymbol{E}', \boldsymbol{H}')$ and $(\boldsymbol{E}'', \boldsymbol{H}'')$ respectively are the solutions of

$$\text{curl}\ \boldsymbol{E}' = -j\omega\mu\boldsymbol{H}', \tag{3.257}$$

$$\text{curl}\ \boldsymbol{H}' = \boldsymbol{J}^{\text{i}} + (\sigma + j\omega\epsilon)\boldsymbol{E}' \tag{3.258}$$

and

$$\text{curl}\ \boldsymbol{E}'' = -\boldsymbol{M}^{\text{i}} - j\omega\mu\boldsymbol{H}'', \tag{3.259}$$

$$\text{curl}\ \boldsymbol{H}'' = (\sigma + j\omega\epsilon)\boldsymbol{E}''. \tag{3.260}$$

The field $(\boldsymbol{E}', \boldsymbol{H}')$ is produced by the electric current $\boldsymbol{J}^{\text{i}}$, and the field $(\boldsymbol{E}'', \boldsymbol{H}'')$ by the magnetic current $\boldsymbol{M}^{\text{i}}$. The sum of these fields satisfies Eqs. (3.254) and (3.255), and hence it is produced by $\boldsymbol{J}^{\text{i}}$ and $\boldsymbol{M}^{\text{i}}$.

Taking the divergence of each of the equations (3.257)–(3.260), we obtain

$$\text{div}\ \boldsymbol{H}' = 0, \tag{3.261}$$

$$\text{div}\ \boldsymbol{E}' = \frac{-\text{div}\ \boldsymbol{J}^{\text{i}}}{\sigma + j\omega\epsilon}, \tag{3.262}$$

$$\text{div}\ \boldsymbol{H}'' = \frac{-\text{div}\ \boldsymbol{M}^{\text{i}}}{j\omega\mu}, \tag{3.263}$$

$$\text{div}\ \boldsymbol{E}'' = 0. \tag{3.264}$$

Equations (3.262) and (3.263) require $\boldsymbol{J}^{\text{i}}$ and $\boldsymbol{M}^{\text{i}}$ to be continuous and differentiable. But one form of the solution to our problem is obtained without using these equations. In the other form of the solution which depends on them, to begin with we may assume $\boldsymbol{J}^{\text{i}}$ and $\boldsymbol{M}^{\text{i}}$ to be differentiable and then extend the results to include discontinuous distributions.

Equations (3.261) and (3.264) show that $\boldsymbol{H}'$ and $\boldsymbol{E}''$ can be represented as the curls of some vector point functions:

$$\boldsymbol{H}' = \text{curl}\ \boldsymbol{A}, \tag{3.265}$$

$$E'' = -\text{curl } \boldsymbol{F}. \tag{3.266}$$

Substituting Eqs. (3.265) and (3.266) in Eqs. (3.257) and (3.260), we obtain

$$E' = -j\omega\mu \boldsymbol{A} - \text{grad } V, \tag{3.267}$$

$$H'' = -(\sigma + j\omega\epsilon)\boldsymbol{F} - \text{grad } U, \tag{3.268}$$

where V and U are two new scalar point functions which are introduced because the equality of the curls of the two vectors does not imply that the vectors are identical. From Eqs. (3.267) and (3.268), and using Eqs. (3.258) and (3.259), we obtain

$$\text{curl curl } \boldsymbol{A} = \boldsymbol{J}^i - \gamma^2 \boldsymbol{A} - (\sigma + j\omega\epsilon) \text{ grad } V, \tag{3.269}$$

$$\text{curl curl } \boldsymbol{F} = \boldsymbol{M}^i - \gamma^2 \boldsymbol{F} - j\omega\mu \text{ grad } U. \tag{3.270}$$

Using the vector identity

$$\text{curl curl } \boldsymbol{A} = -\nabla^2 \boldsymbol{A} + \text{grad div } \boldsymbol{A}, \tag{3.271}$$

we obtain

$$\nabla^2 \boldsymbol{A} - \text{grad div } \boldsymbol{A} = -\boldsymbol{J}^i + \gamma^2 \boldsymbol{A} + (\sigma + j\omega\epsilon) \text{ grad } V, \tag{3.272}$$

$$\nabla^2 \boldsymbol{F} - \text{grad div } \boldsymbol{F} = -\boldsymbol{M}^i + \gamma^2 \boldsymbol{F} + j\omega\mu \text{ grad } U. \tag{3.273}$$

Thus, we have been able to express $\boldsymbol{E}$ and $\boldsymbol{H}$ in terms of the vectors $\boldsymbol{A}$ and $\boldsymbol{F}$ and the scalars V and U by means of Eqs. (3.272) and (3.273). The new functions, called auxiliary functions, are related to each other by Eqs. (3.265)–(3.268). If we add the gradients of arbitrary functions to $\boldsymbol{A}$ and $\boldsymbol{F}$, then Eqs. (3.265) and (3.266) remain unchanged. Hence, both $\boldsymbol{A}$ and $\boldsymbol{F}$ and V and U are arbitrary. Therefore, to suit our convenience, it is possible to impose further conditions on these functions. For instance, by putting

$$V = -\frac{\text{div } \boldsymbol{A}}{\sigma + j\omega\epsilon}, \tag{3.274}$$

$$U = -\frac{\text{div } \boldsymbol{F}}{j\omega\mu}, \tag{3.275}$$

Eqs. (3.272) and (3.273) become

$$\nabla^2 \boldsymbol{A} = \gamma^2 \boldsymbol{A} - \boldsymbol{J}^i, \tag{3.276}$$

$$\nabla^2 \boldsymbol{F} = \gamma^2 \boldsymbol{F} - \boldsymbol{M}^i. \tag{3.277}$$

When the auxiliary functions $\boldsymbol{A}$, $\boldsymbol{F}$, V, and U are specified as in Eqs. (3.274)–(3.277), they are called *wave potentials*. $\boldsymbol{A}$ is called the *magnetic vector potential*, $\boldsymbol{F}$ the *electric vector potential*, V the *electric scalar potential*, and U the *magnetic scalar potential*. H. A. Lorentz, while considering electromagnetic wave propagation in nondissipative media, introduced, for the first time, these wave potentials and called them *retarded potentials*. These wave potentials are not only *retarded* but also *attenuated*, and hence the term *wave potentials* is appropriate.

In terms of the wave potentials $\boldsymbol{A}$, $\boldsymbol{F}$, V, and U, the field vectors $\boldsymbol{E}$ and $\boldsymbol{H}$ can be expressed as

$$\boldsymbol{E} = \boldsymbol{E}' + \boldsymbol{E}'' = -j\omega\mu \boldsymbol{A} - \text{grad } V - \text{curl } \boldsymbol{F}, \tag{3.278}$$

$$\boldsymbol{H} = \boldsymbol{H}' + \boldsymbol{H}'' = \text{curl } \boldsymbol{A} - \text{grad } U - (\sigma + j\omega\epsilon)\boldsymbol{F}, \tag{3.279}$$

where V and U are defined by Eqs. (3.274) and (3.275) and $\boldsymbol{A}$ and $\boldsymbol{F}$ are the solutions of Eqs. (3.276) and (3.277).

If $\boldsymbol{J}^{\mathrm{i}}$ and $\boldsymbol{M}^{\mathrm{i}}$ are differentiable functions, V and U satisfy equations similar to those satisfied by $\boldsymbol{A}$ and $\boldsymbol{F}$. Taking the divergence of Eqs. (3.276) and (3.277) and substituting from Eqs. (3.274) and (3.275), we obtain

$$\nabla^2 V = \gamma^2 V + \frac{\text{div } \boldsymbol{J}^{\mathrm{i}}}{\sigma + j\omega\epsilon}, \tag{3.280}$$

$$\nabla^2 U = \gamma^2 U + \frac{\text{div } \boldsymbol{M}^{\mathrm{i}}}{j\omega\mu}. \tag{3.281}$$

In nondissipative media,

$$\text{div } \boldsymbol{J}^{\mathrm{i}} = -\frac{\partial \rho}{\partial t} = -j\omega\rho, \tag{3.282}$$

$$\text{div } \boldsymbol{M}^{\mathrm{i}} = -\frac{\partial m}{\partial t} = -j\omega m, \tag{3.283}$$

where ρ and m are the electric and magnetic volume charge densities. In fact, m is equal to zero. Consequently,

$$\nabla^2 V = -\beta^2 V - \frac{\rho}{\epsilon}, \tag{3.284}$$

$$\nabla^2 U = -\beta^2 U - \frac{m}{\mu} = -\beta^2 U. \tag{3.285}$$

For static fields, $\beta = \omega\sqrt{\mu\epsilon} = 0$. Hence

$$\nabla^2 V = -\frac{\rho}{\epsilon}, \tag{3.286}$$

$$\nabla^2 U = 0 \tag{3.287}$$

which are Poisson's and Laplace's equations, respectively.

3.36 Field of Electric Current Element

Let Il be an electric current element (see Fig. 3.17). Assume that the current I is uniform and

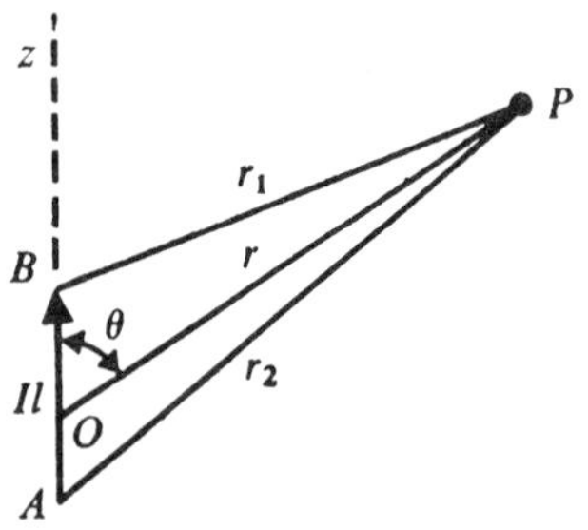

Fig. 3.17 Electric current element.

steady between the terminals A and B so that the entire current is forced to flow out of B into the external medium and then back into A. If the external medium is a perfect dielectric medium, then there is a concentration of positive electric charge at B and negative electric charge at A at the rate of I amperes or I coulombs per second. This results in an ever increasing electric field around the filament. The *moment* of the electric current element is defined as Il, the product of the current and length of the filament. If the current is along the z-axis and is centred at the origin of the coordinates, then the current density $\boldsymbol{J}$ at a point P at a radial distance r from the origin is given by

$$\boldsymbol{J} = \boldsymbol{u}_r J_r = \boldsymbol{u}_r \frac{I}{4\pi r^2} = \boldsymbol{u}_r \frac{\partial}{\partial r}\left(-\frac{I}{4\pi r}\right). \tag{3.288}$$

Assuming that the current element consists of two point sources, namely, $+I$ at B and $-I$ at A, the current density at a distant point P ($r \gg l$) is the gradient of the function

$$P = \frac{-I}{4\pi r_1} + \frac{I}{4\pi r_2} = \frac{+I}{4\pi}\left[\frac{-1}{r - (l/2)\cos\theta} + \frac{1}{r + (l/2)\cos\theta}\right] \approx \frac{-Il\cos\theta}{4\pi r^2}. \tag{3.289}$$

Hence

$$J_r = \frac{\partial P}{\partial r} = \frac{Il\cos\theta}{2\pi r^3}, \tag{3.290}$$

$$J_\theta = \frac{\partial P}{r\,\partial\theta} = \frac{Il\sin\theta}{4\pi r^3}. \tag{3.291}$$

In Fig. 3.18, the dashed lines represent the current flow lines and the solid lines show the magnetic field lines which are coaxial with the element. The magnetomotive force around the

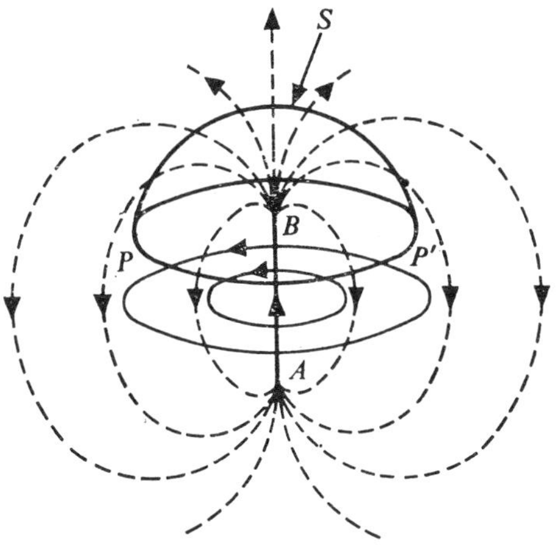

Fig. 3.18 Current flow lines, electric field lines, and magnetic field lines in the vicinity of current element.

circumference of a circle PP' coaxial with the element is equal to the total electric current $I(\theta)$

flowing through any surface S bounded by the circumference of this circle. If S is chosen as part of a sphere concentric with the origin, as shown in the figure, then we have

$$I(\theta) = \int_{\phi=0}^{2\pi} \int_{\theta=0}^{\theta} J_r r^2 \sin\theta \, d\theta \, d\phi = \int_{\phi=0}^{2\pi} \int_{\theta=0}^{\theta} \frac{Il \cos\theta}{2\pi r^3} r^2 \sin\theta \, d\theta \, d\phi = \frac{Il \sin^2\theta}{2r}. \tag{3.292}$$

Therefore

$$I(\theta) = \frac{Il \sin^2\theta}{2r} = H_\phi \cdot 2\pi r \sin\theta \tag{3.293}$$

or

$$H_\phi = \frac{Il \sin\theta}{4\pi r^2}. \tag{3.294}$$

But from Eq. (3.265),

$$\boldsymbol{H} = \text{curl } \boldsymbol{A},$$

and since

$$\nabla^2 \boldsymbol{A} = \gamma^2 \boldsymbol{A} - \boldsymbol{J}^i, \tag{3.295}$$

the component A_z of $\boldsymbol{A}$ depends only on the z-component of the source current $\boldsymbol{J}^i$. Since the current element is in the z-direction, $\boldsymbol{A}$ has only a z-component. Hence

$$H_\phi = \frac{-\partial A_z}{\partial \rho} = \frac{Il \sin\theta}{4\pi r^2}. \tag{3.296}$$

Since

$$r^2 = \rho^2 + z^2, \qquad \sin\theta = \rho/r = \frac{\rho}{\sqrt{\rho^2 + z^2}},$$

we have

$$\frac{\partial A_z}{\partial \rho} = \frac{-Il \sin\theta}{4\pi r^2} = \frac{-Il(\rho/r)}{\rho^2 + z^2} = \frac{-Il\rho}{(\rho^2 + z^2)^{3/2}}.$$

Therefore

$$A_z = \frac{Il}{4\pi(\rho^2 + z^2)^{1/2}} = \frac{Il}{4\pi r}. \tag{3.297}$$

If the current is a harmonic function of time, or $I = I_0 e^{j\omega t}$, then as the frequency ω approaches zero, the field of the alternating current element $I_0 l e^{j\omega t}$ must approach the values given by Eqs. (3.296) and (3.297) for a steady current element $I\,dl$. The magnetic vector potential $\boldsymbol{A}$ satisfies the equation

$$\nabla^2 \boldsymbol{A} = \gamma^2 \boldsymbol{A} \tag{3.298}$$

at all points external to the current element. And from Eq. (3.297), $\boldsymbol{A} = \boldsymbol{u}_z A_z$ must be independent of θ and ϕ, and hence

$$\frac{\partial}{\partial r}\left(r^2 \frac{\partial A_z}{\partial r}\right) = \gamma^2 r^2 A_z \tag{3.299}$$

whose solution is of the form

$$A_z = \frac{Pe^{-\gamma r}}{r} + \frac{Qe^{+\gamma r}}{r}. \tag{3.300}$$

In dissipative media ($\sigma \neq 0$), $[Q \exp(+\gamma r)]/r$ increases exponentially with the distance r from the element and it cannot therefore represent a physically realizable field. Hence, only the first term is retained, and

$$A_z = \frac{Pe^{-\gamma r}}{r} \tag{3.301}$$

approaches the value given by Eq. (3.297) for a steady current element if $P = Il/(4\pi)$. Hence

$$A_z = \frac{Ile^{-\gamma r}}{4\pi r}. \tag{3.302}$$

In a nondissipative medium, $\gamma = j\beta$, and hence

$$A_z = \frac{Ile^{-j\beta r}}{4\pi r}. \tag{3.303}$$

Therefore

$$\boldsymbol{A} = \boldsymbol{u}_z A_z = \boldsymbol{u}_z \frac{Ile^{-j\beta r}}{4\pi r}. \tag{3.304}$$

From Eq. (3.274),

$$V = \frac{-\operatorname{div} \boldsymbol{A}}{\sigma + j\omega\epsilon}, \tag{3.305}$$

$$V = -\frac{1}{\sigma + j\omega\epsilon}\frac{\partial A_z}{\partial z} = \frac{\eta Il}{4\pi r}\left(1 + \frac{1}{\gamma r}\right)e^{-\gamma r} \cos\theta, \tag{3.306}$$

where

$$\eta = \left(\frac{j\omega\mu}{\sigma + j\omega\epsilon}\right)^{1/2}.$$

$\boldsymbol{E}$ and $\boldsymbol{H}$ are now obtained from

$$\boldsymbol{E} = -j\omega\mu\boldsymbol{A} - \operatorname{grad} V, \tag{3.307}$$

$$\boldsymbol{H} = \operatorname{curl} \boldsymbol{A} \tag{3.308}$$

and have the components E_r, E_θ, H_ϕ given by

$$E_r = \frac{\eta Il}{2\pi r^2}\left(1 + \frac{1}{\gamma r}\right)e^{-\gamma r} \cos\theta, \tag{3.309}$$

$$E_\theta = \frac{j\omega\mu Il}{4\pi r}\left(1 + \frac{1}{\gamma r} + \frac{1}{\gamma^2 r^2}\right)e^{-\gamma r} \sin\theta, \tag{3.310}$$

$$H_\phi = \frac{\gamma Il}{4\pi r}\left(1 + \frac{1}{\gamma r}\right)e^{-\gamma r} \sin\theta. \tag{3.311}$$

In Eqs. (3.309)–(3.311), E_r has a $1/r^2$ term and a $1/r^3$ term; E_θ has a $1/r$ term, $1/r^2$ term, and $1/r^3$ term; and H_ϕ has a $1/r$ term and a $1/r^2$ term. The $1/r$ term is called the *radiation term*, the $1/r^2$ term the *induction term*, and the $1/r^3$ term the *electrostatic term*. At a large distance r, only the $1/r$ term predominates, and hence

$$H_\phi = \frac{jIl}{2\lambda r}e^{-j\beta r}\sin\theta, \qquad E_\theta = \eta H_\phi, \qquad E_r = 0.$$

Similarly, the field of a magnetic current element Kl can be obtained from the electric vector potential

$$\boldsymbol{F} = \boldsymbol{u}_z F_z = \boldsymbol{u}_z \frac{Kle^{-\gamma r}}{4\pi r}. \tag{3.312}$$

3.37 Radiation from Electric Current Element

The total power W radiated from an electric current element is calculated as the total flow of power across a large sphere concentric with the element:

$$W = \int_{\text{sphere}} P_r \cdot dS = \tfrac{1}{2}\int_{\phi=0}^{2\pi}\int_{\theta=0}^{\pi} E_\theta H_\phi^* r^2 \sin\theta\, d\theta\, d\phi, \tag{3.313}$$

where P_r is the radial component of the Poynting vector. W is given by

$$W = \frac{\eta I^2 l^2}{8\lambda^2}\int_0^{2\pi}\int_0^{\pi}\sin^3\theta\, d\theta\, d\phi = \tfrac{1}{3}\pi\eta\left(\frac{Il}{\lambda}\right)^2 \tag{3.314}$$

which, in free space, becomes equal to

$$40\pi^2\left(\frac{Il}{\lambda_0}\right)^2 = \frac{80\pi^2 l^2}{\lambda_0^2} I^2/2. \tag{3.315}$$

It should be noted here that, while integrating, only the $1/r$ terms in E_θ and H_ϕ contribute to this power. The other terms, namely, the $1/r^2$ and $1/r^3$ terms, contribute to the reactive power which is exchanged back and forth between the source and the field.

The radiated power can also be calculated from the work done by the electromotive force impressed on the current element. The electric intensity on the axis of the element in a nondissipative medium is given by

$$E_z = E_r = \frac{\eta Il}{2\pi}\left(\frac{1}{j\beta r^3} - \frac{j\beta}{2r} - \tfrac{1}{3}\beta^2 + \ldots\right). \tag{3.316}$$

The first two terms in Eq. (3.316) are in quadrature with I and on the average do not do any work; but the third term is 180° out of phase with I, and work is done against the field by the impressed electromotive force. Then, the in-phase component of this force is

$$\text{Re}\,(V^i) = -l\,\text{Re}\,(E_z) = \frac{\eta}{6\pi}\beta^2 l^2 I = \frac{2\pi\eta l^2 I}{3\lambda^2}. \tag{3.317}$$

The work done by this force per second is $2\pi\eta l^2 I^2/(2\times 3\lambda^2)$ which is equal to the total power radiated, W.

The ratio

$$R = \frac{\text{Re}(V^i)}{I} = \frac{2\pi\eta l^2}{3\lambda^2} = 80\pi^2\left(\frac{l}{\lambda_0}\right)^2 \text{ for free space} \tag{3.318}$$

is called the *radiation resistance* of the current element.

3.38 Electromagnetic Field Produced by Given Distribution of Applied Electric and Magnetic Currents

Any given distribution of applied electric or magnetic currents may be subdivided into elements, and the resulting field can be obtained by the superposition of the fields of individual elements. If an infinitesimal volume bounded by the lines of flow and two surfaces normal to them are considered, then the current I in this element is $\boldsymbol{J}^i\, dS$, where dS is the cross-section of the tube of current flow. If dl is the length of this tube, then the moment $I\,dl$ of the current element is $\boldsymbol{J}^i \cdot dS \cdot dl = \boldsymbol{J}^i \cdot dV$, where dV is the volume of the element (see Fig. 3.19).

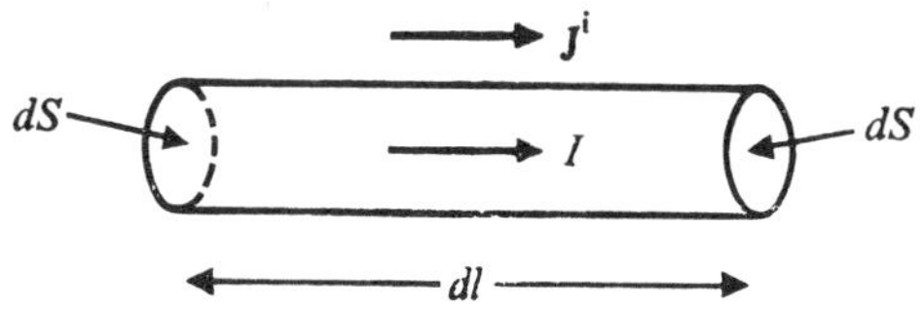

Fig. 3.19 Infinitesimal current element.

The magnetic vector potential due to the electric current distribution is

$$A = \int_V \frac{\boldsymbol{J}^i e^{-\gamma r}}{4\pi r}\, dV, \tag{3.319}$$

where r is the distance between a typical current element and a typical point in space, and is taken over the volume over which the current is distributed. Similarly, for a magnetic current distribution, the electric vector potential is

$$\boldsymbol{F} = \int_V \frac{\boldsymbol{M}^i e^{-\gamma r}}{4\pi r}\, dV. \tag{3.320}$$

The scalar potentials V and U may then be evaluated by using Eqs. (3.274) and (3.275), and hence $\boldsymbol{E}$ and $\boldsymbol{H}$ can be calculated from Eqs. (3.278) and (3.279). If, however, $\boldsymbol{J}^i$ and $\boldsymbol{M}^i$ are differentiable, then V and U can be computed from the equations

$$V = -\int_V \frac{\text{div}\, \boldsymbol{J}^i\, dV}{4\pi(\sigma + j\omega\epsilon)r}, \qquad U = \int_V \frac{\text{div}\, \boldsymbol{M}^i\, dV}{4\pi j\omega\mu r}. \tag{3.321}$$

In nondissipative media,

$$V = \int_V \frac{\rho e^{-j\beta r}}{4\pi\epsilon r}\, dV, \qquad U = \int_V \frac{m e^{-j\beta r}}{4\pi\epsilon r}\, dV. \tag{3.322}$$

3.39 Electromagnetic Field Produced by Impressed Currents Varying Arbitrarily with Time

In nondissipative media, the magnetic vector potential of a given electric current distribution varying harmonically with time (that is, varying as $e^{j\omega t}$) is

$$A = \int_V \frac{J^i e^{-j\beta r} e^{j\omega t}}{4\pi r}\, dV. \tag{3.323}$$

The phase of J^i is ωt, and the phase of the corresponding component of the vector potential is $\omega t - \beta r = \omega(t - r/v)$, where $v = \omega/\beta =$ the characteristic velocity of the medium. The time delay r/v is independent of the frequency; hence, all frequency components of a general function $J^i(x, y, z; t)$ are shifted equally on the time scale, and A will depend on $J^i(x, y, z; t - r/v)$. Thus we have

$$A(x, y, z; t) = \int_V \frac{J^i(x, y, z; t - r/v)}{4\pi r}\, dV. \tag{3.324}$$

In a dissipative medium, no simple formula such as the one given by Eq. (3.324) exists.

3.40 Field of Electric Current Element whose Current Varies Arbitrarily with Time

Consider an electric current element of length l along the z-axis at the origin, and suppose the current $I(t)$ is given by

$$\begin{aligned} I(t) &= 0 \quad \text{for } t \leqslant 0 \\ &= I(t) \quad \text{for } t > 0. \end{aligned} \tag{3.325}$$

Let dI/dt be finite. The charge $q(t)$ at the upper end is zero when $t \leqslant 0$, and is given by

$$q(t) = \int_0^t I(t)\, dt \quad \text{for } t > 0. \tag{3.326}$$

At the lower end the charge is $-q(t)$.

The field of the current element may now be evaluated. Both E and H consist of three parts: (E', H') depending only on the time derivative of the current, (E'', H'') depending on the current alone, and (E''', H''') depending on the charge. Thus

$$E = E' + E'' + E''', \tag{3.327}$$

$$H = H' + H'' + H''', \tag{3.328}$$

$$E'_\theta = \eta H'_\phi, \qquad H_\phi = \frac{lI'(t - r/v)}{4\pi v r} \sin\theta, \tag{3.329}$$

$$E'_r = 0, \qquad E''_r = 2E''_\theta \cot\theta, \tag{3.330}$$

$$E''_\theta = \eta H''_\phi, \qquad H''_\phi = \frac{lI(t - r/v)}{4\pi r^2} \sin\theta, \tag{3.331}$$

$$E'''_\theta = \frac{lq(t - r/v)}{4\pi\epsilon r^3} \sin\theta, \tag{3.332}$$

$$E_r''' = \frac{lq(t - r/v)}{2\pi\epsilon r^3} \cos\theta. \tag{3.333}$$

At sufficiently great distances, only the radiation field (E', H') is significant. The entire field is zero outside the spherical surface of radius vt with its centre at the element. This sphere is the wavefront of the wave emitted by the element and on it both (E'', H'') and (E''', H''') vanish. At the wavefront, $\boldsymbol{E}$ and $\boldsymbol{H}$ are perpendicular to the radius or are tangential to the wavefront.

3.41 Reflection of Electromagnetic Field at Boundary Surface

Let an infinite homogeneous medium be divided into two homogeneous parts by a surface S, so that one of these regions, say, region 2, is source-free (see Fig. 3.20). If the sources are distri-

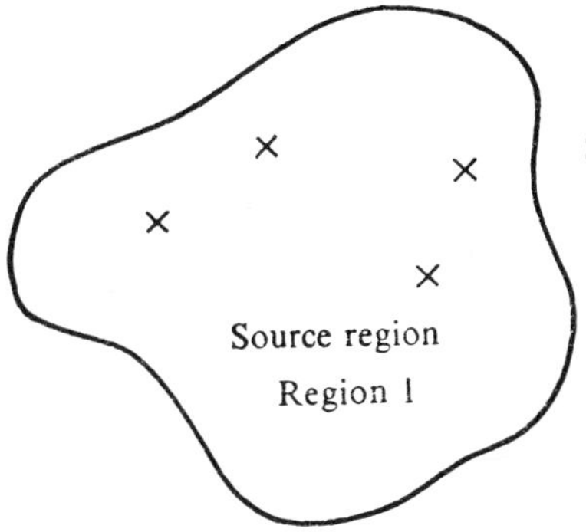

Fig. 3.20 Surface enclosing sources of field.

buted in both the regions, then the total field may be regarded as the superposition of the two fields, each produced by sources located in one region only. If the two regions have the same electromagnetic properties, the field of these sources can be found by using the results given in Section 3.35. But if the electromagnetic properties of the two regions are different, the field ($\boldsymbol{E}^{\mathrm{i}}$, $\boldsymbol{H}^{\mathrm{i}}$) thus obtained is not the actual field. In region 1 which has the sources, this field represents the primary field of the sources and is called the *impressed field*. To obtain the actual field in region 1, a field ($\boldsymbol{E}^{\mathrm{r}}$, $\boldsymbol{H}^{\mathrm{r}}$) has to be added to ($\boldsymbol{E}^{\mathrm{i}}$, $\boldsymbol{H}^{\mathrm{i}}$). The field ($\boldsymbol{E}^{\mathrm{r}}$, $\boldsymbol{H}^{\mathrm{r}}$) is then called the *reflected field* because it is reflected from the boundary surface S. This field has no sources in region 1, and hence it satisfies the homogeneous form of Maxwell's equations

$$\text{curl}\, \boldsymbol{E}^{\mathrm{r}} = -j\omega\mu_1 H^{\mathrm{r}}, \tag{3.334}$$

$$\text{curl}\, \boldsymbol{H}^{\mathrm{r}} = (\sigma_1 + j\omega\epsilon_1) E^{\mathrm{r}}, \tag{3.335}$$

where (μ_1, ϵ_1, σ_1) are the primary electromagnetic constants of region 1.

Let the actual field in region 2 be ($\boldsymbol{E}^{\mathrm{t}}$, $\boldsymbol{H}^{\mathrm{t}}$). This field is called the *transmitted* (or *refracted*) *field*; it too should satisfy the homogeneous Maxwell's equations

$$\text{curl}\, \boldsymbol{E}^{\mathrm{t}} = -j\omega\mu_2 \boldsymbol{H}^{\mathrm{t}}, \tag{3.336}$$

$$\text{curl}\, \boldsymbol{H}^{\mathrm{t}} = (\sigma_2 + j\omega\epsilon_2) \boldsymbol{E}^{\mathrm{t}}, \tag{3.337}$$

where (μ_2, ϵ_2, σ_2) are the primary electromagnetic constants of region 2.

At the boundary surface S of the two media, the tangential components of $\boldsymbol{E}$ and $\boldsymbol{H}$ are continuous. Thus

$$E_t^i + E_t^r = E_t^t, \tag{3.338}$$

$$H_t^i + H_t^r = H_t^t. \tag{3.339}$$

The set of equations (3.338) and (3.339) constitutes one formulation of the problem for determining the field of a given system of sources when the medium consists of two homogeneous regions. The method can be extended to any number of homogeneous regions.

If the boundary S is a perfectly conducting sheet, then the tangential component of $\boldsymbol{E}$ should vanish on S, or

$$E_t^i + E_t^r = 0$$

or

$$E_t^r = -E_t^i. \tag{3.340}$$

Such a sheet can support a finite electric current, and the tangential component of $\boldsymbol{H}$ is no longer continuous across S. In fact, the field in region 2 due to sources in region 1 is equal to zero. The component of $\boldsymbol{H}$ tangential to the surface S in region 1 represents the surface current density $\boldsymbol{J}_S$ on S. By the second law of electromagnetic induction, $\boldsymbol{J}_S$ is normal to the tangential component of $\boldsymbol{H}$. Therefore, if $\boldsymbol{n}$ is a unit normal to S, regarded as positive as it points into the source-free region 2, then

$$\boldsymbol{J}_S = (\boldsymbol{H}_t^i + \boldsymbol{H}_t^r) \times \boldsymbol{n}. \tag{3.341}$$

Since the vector product of $\boldsymbol{n}$ and the normal component of $\boldsymbol{H}$ is zero, the subscript "t" for tangential component may be dropped, and we may write

$$\boldsymbol{J}_S = (\boldsymbol{H}^i + \boldsymbol{H}^r) \times \boldsymbol{n}. \tag{3.342}$$

If the impedance Z_n normal to the boundary S is prescribed, then the relation between the tangential components in region 1 is

$$\boldsymbol{E}_t^i + \boldsymbol{E}_t^r = Z_n(\boldsymbol{H}_t^i + \boldsymbol{H}_t^r) \times \boldsymbol{n}. \tag{3.343}$$

The fields in waveguides and resonators are examples of reflection, and the total field is the sum of the impressed and reflected fields.

3.42 Induction Theorem

If we rewrite the set of Eqs. (3.338) and (3.339) as

$$E_t^t - E_t^r = E_t^i, \tag{3.344}$$

$$H_t^t - H_t^r = H_t^i, \tag{3.345}$$

we may define the *induced field* as the field that comprises the reflected field (E^r, H^r) in region 1 and the transmitted field (E^t, H^t) in region 2. The induced field satisfies the homogeneous Maxwell's equations everywhere except on the surface S. It can be shown that this field may be obtained from a distribution of sources on S as well as from the original sources. We may therefore write

$$M_S = (E_t^t - E_t^r) \times n = E_t^i \times n, \tag{3.346}$$

$$J_S = n \times (H_t^t - H_t^r) = n \times H_t^i. \tag{3.347}$$

Since the vector product of n and the normal component of the field is zero, we may write

$$M_S = E^i \times n, \tag{3.348}$$

$$J_S = n \times H^i. \tag{3.349}$$

These two equations show that the induced field, which is a combination of the reflected field (E^r, H^r) and the transmitted field (E^t, H^t), could be produced by the electric and magnetic current sheets J_S and M_S on the surface S given by Eqs. (3.346) and (3.347).

3.43 Equivalence Theorem

If both regions 1 and 2 separated by the surface S have the same electromagnetic constants, then the entire region is homogeneous. Hence, the reflected field is zero, and the transmitted field is the actual field in the homogeneous region due to the sources in region 1. But this transmitted field can also be calculated from a suitable distribution of electric and magnetic current sheets over the surface S. The required surface current densities of these sheets are

$$J_S = n \times H^t = n \times H^i, \tag{3.350}$$

$$M_S = E^t \times n = E^i \times n. \tag{3.351}$$

Thus, in a source-free region 1 bounded by a surfaces S, in order to compute the electromagnetic field, the source distribution outside S (in region 2) can be replaced by a distribution of surface electric and surface magnetic currents J_S and M_S over the surface S given by Eqs. (3.350) and (3.351). This is called the *equivalence theorem*, sometimes known as Love's equivalence principle or Schelkunoff's equivalence principle.

The equivalence theorem is a very powerful tool in solving many electromagnetic problems involving radiation from aperture-type antennas.

3.44 Hertz Potentials

Hertz[1] has shown that, under ordinary conditions, an electromagnetic field can be derived by any one of the two vector functions, π_h and π_e, which are called *Hertz vector potentials*.

In a homogeneous isotropic source-free region, $\nabla \cdot E = 0$. E may be derived from the curl of an auxiliary vector potential function which we shall call a magnetic type of Hertz vector potential. (The reason for using the adjective "magnetic" will be explained in Section 3.45.) Let

$$E = -j\omega\mu\nabla \times \pi_h, \tag{3.352}$$

where π_h is the magnetic Hertz vector potential. Substituting Eq. (3.352) in Maxwell's equations, we obtain

$$\nabla \times H = j\omega\epsilon E = +\omega^2\mu\epsilon\nabla \times \pi_h \tag{3.353}$$

[1] H. R. Hertz, *Ann. Physik*, **36**, 1, 1888.

so that

$$\boldsymbol{H} = k^2\boldsymbol{\pi}_h + \nabla\phi, \tag{3.354}$$

where $\omega^2\mu\epsilon = k^2$ and ϕ is an arbitrary scalar function. Substituting for $\boldsymbol{H}$ in the other Maxwell's equations, we have

$$\nabla\times\boldsymbol{E} = -j\omega\mu\nabla\times\nabla\times\boldsymbol{\pi}_h = -j\omega\mu\boldsymbol{H} = -j\omega\mu(k^2\boldsymbol{\pi}_h + \nabla\phi) \tag{3.355}$$

or

$$\nabla\times\nabla\times\boldsymbol{\pi}_h = \nabla\nabla\cdot\boldsymbol{\pi}_h - \nabla^2\boldsymbol{\pi}_h = k^2\boldsymbol{\pi}_h + \nabla\phi. \tag{3.356}$$

Since both ϕ and $\nabla\cdot\boldsymbol{\pi}_h$ are still arbitrary, we may make them satisfy a Lorentz type of condition, i.e., put $\nabla\cdot\boldsymbol{\pi}_h = \phi$, so that the equation for $\boldsymbol{\pi}_h$ becomes

$$\nabla^2\boldsymbol{\pi}_h + k^2\boldsymbol{\pi}_h = 0. \tag{3.357}$$

Since $\nabla\cdot\boldsymbol{H} = 0$, from Eq. (3.354),

$$\text{div}\,(\boldsymbol{H}) = k^2\,\text{div}\,\boldsymbol{\pi}_h + \nabla\cdot\nabla\phi = 0$$

or

$$\nabla^2\phi + k^2\phi = 0. \tag{3.358}$$

The fields $\boldsymbol{E}$ and $\boldsymbol{H}$ are now given by

$$\boldsymbol{E} = -j\omega\mu\nabla\times\boldsymbol{\pi}_h, \tag{3.359}$$

$$\boldsymbol{H} = k^2\boldsymbol{\pi}_h + \nabla\nabla\cdot\boldsymbol{\pi}_h = \nabla\times\nabla\times\boldsymbol{\pi}_h, \tag{3.360}$$

where $\boldsymbol{\pi}_h$ is a solution of the Helmholtz equation (3.357). In practice, we need not determine the scalar function ϕ explicitly since it is equal to $\nabla\cdot\boldsymbol{\pi}_h$ and has been eliminated from the equation determining the magnetic field $\boldsymbol{H}$.

We shall now introduce the electric Hertz vector potential $\boldsymbol{\pi}_e$ in the same way as we introduced the magnetic Hertz vector potential $\boldsymbol{\pi}_h$. Let

$$\boldsymbol{H} = j\omega\epsilon\nabla\times\boldsymbol{\pi}_e. \tag{3.361}$$

Thus, it can be shown that $\boldsymbol{\pi}_e$ is a solution of

$$\nabla^2\boldsymbol{\pi}_e + k^2\boldsymbol{\pi}_e = 0, \tag{3.362}$$

and that the electric field $\boldsymbol{E}$ is given by

$$\boldsymbol{E} = k^2\boldsymbol{\pi}_e + \nabla\nabla\cdot\boldsymbol{\pi}_e = \nabla\times\nabla\times\boldsymbol{\pi}_e. \tag{3.363}$$

[In a lossy dielectric medium, ϵ is replaced by $\epsilon - (j\sigma/\omega)$.]

3.45 Electric and Magnetic Polarization Sources for Hertz Vector Potentials

When an electric field is applied to a dielectric material, the electron orbits of the various atoms and molecules involved are perturbed, resulting in a dipole polarization $\boldsymbol{P}$ per unit volume. Some materials, called electrets, too have a permanent dipole polarization, but since these elementary dipoles are randomly oriented, there is no net resultant field as there is no applied external field. The application of an external field therefore tends to align the dipoles with the field, resulting in a decrease in the electric field intensity in the material. Thus, the relation between the electric displacement density $\boldsymbol{D}$, the electric field intensity $\boldsymbol{E}$, and the electric polarization vector $\boldsymbol{P}$, in the interior of a dielectric material, is

$$\boldsymbol{D} = \epsilon_0 \boldsymbol{E} + \boldsymbol{P}. \tag{3.364}$$

Here, $\boldsymbol{E}$ is the net field intensity in the dielectric, i.e., the vector sum of the applied field and the field arising from the dipole polarization. For many materials, $\boldsymbol{P}$ is collinear with and proportional to the applied field and hence proportional to $\boldsymbol{E}$. We may therefore write

$$\boldsymbol{P} = \chi_e \epsilon_0 \boldsymbol{E}, \tag{3.365}$$

where χ_e is called the *electric susceptibility* of the material, and is a dimensionless quantity. Consequently,

$$\boldsymbol{D} = \epsilon_0 \boldsymbol{E} + \boldsymbol{P} = \epsilon_0 (1 + \chi_e) \boldsymbol{E} = \epsilon \boldsymbol{E} \tag{3.366}$$

so that

$$\epsilon = (1 + \chi_e)\epsilon_0. \tag{3.367}$$

The relative permittivity or dielectric constant is

$$\epsilon_r = \frac{\epsilon}{\epsilon_0} = 1 + \chi_e. \tag{3.368}$$

Equation (3.367) is true only for materials having a high degree of symmetry in their crystal structure. Such materials are called *isotropic materials*.

In general, $\bar{\chi}_e$ is a dyadic or tensor quantity, and hence

$$\bar{\epsilon} = (\bar{I} + \bar{\chi}_e)\epsilon_0, \tag{3.369}$$

where $\bar{I}$ is the unit dyadic. This is somewhat analogous to the situation for materials having a permeability different from μ_0, $\boldsymbol{H}$ being defined by the relation

$$\boldsymbol{H} = \frac{\boldsymbol{B}}{\mu_0} - \boldsymbol{M}, \tag{3.370}$$

where $\boldsymbol{M}$ is the magnetic dipole polarization per unit volume. $\boldsymbol{M}$ is related to $\boldsymbol{H}$ as

$$\boldsymbol{M} = \chi_m \boldsymbol{H}, \tag{3.371}$$

where χ_m is the magnetic susceptibility of the medium. Equation (3.371) leads to

$$\mu = (1 + \chi_m)\mu_0. \tag{3.372}$$

As $\boldsymbol{P}$ is not always in the direction of $\boldsymbol{E}$ in the dielectric case, $\boldsymbol{M}$ is not always in the direction of $\boldsymbol{B}$ for materials having permeability different from μ_0, and hence χ_m and μ are, in general, dyadic quantities. Materials whose μ or ϵ are dyadics are called *anisotropic materials*.

If μ and ϵ are not functions of position, materials are said to be *homogeneous*; otherwise, they are called *nonhomogeneous*.

In materials with μ and ϵ as scalar functions of position, Maxwell's equations become

$$\nabla \times \boldsymbol{E} = -j\omega\mu_0 \boldsymbol{H} - j\omega\mu_0 \chi_m \boldsymbol{H}, \tag{3.373}$$

$$\nabla \times \boldsymbol{H} = j\omega\epsilon_0 \boldsymbol{E} + j\omega\epsilon_0 \chi_e \boldsymbol{E} + \boldsymbol{J}^i. \tag{3.374}$$

The term $j\omega\epsilon_0\chi_e\boldsymbol{E}$ may be regarded as an equivalent electric polarization current $\boldsymbol{J}_e$, and $j\omega\mu_0\chi_m\boldsymbol{H}$ may be considered as an equivalent magnetic polarization current $\boldsymbol{J}_m$.

Introducing the polarization vectors $\boldsymbol{P}$ and $\boldsymbol{M}$ in the divergence equations, we obtain

$$\nabla \cdot \boldsymbol{D} = \epsilon_0 \nabla \cdot \boldsymbol{E} + \nabla \cdot \boldsymbol{P} = \rho, \tag{3.375}$$

$$\nabla \cdot \boldsymbol{B} = \mu_0 \nabla \cdot \boldsymbol{H} + \mu_0 \nabla \cdot \boldsymbol{M} = 0. \tag{3.376}$$

Here, $\nabla \cdot \boldsymbol{P}$ and $\nabla \cdot \boldsymbol{M}$ may be interpreted as defining an electric polarization charge density $-\rho_e$ and a magnetic polarization charge density $-\rho_m$, respectively; or

$$-\nabla \cdot \boldsymbol{P} = \rho_e = -\epsilon_0 \nabla \cdot (\chi_e E) = -\epsilon_0 \chi_e \nabla \cdot \boldsymbol{E} - \epsilon_0 \boldsymbol{E} \cdot \nabla \chi_e, \tag{3.377}$$

$$-\nabla \cdot \boldsymbol{M} = \rho_m = -\nabla \cdot \chi_m \boldsymbol{H} = -\chi_m \nabla \cdot \boldsymbol{H} - \boldsymbol{H} \cdot \nabla \chi_m. \tag{3.378}$$

Now we shall show that the source functions for the Hertz vector potentials $\boldsymbol{\pi}_h$ and $\boldsymbol{\pi}_e$ are the magnetic and electric polarizations $\boldsymbol{M}$ and $\boldsymbol{P}$, respectively. If we put

$$\boldsymbol{E} = -j\omega\mu_0 \nabla \times \boldsymbol{\pi}_h, \tag{3.379}$$

then

$$\boldsymbol{H} = k_0^2 \boldsymbol{\pi}_h + \nabla\phi, \tag{3.380}$$

where $k_0^2 = \omega^2\mu_0\epsilon_0$. Hence, from

$$\nabla \times \boldsymbol{E} = -j\omega\mu_0 \boldsymbol{H} = -j\omega\mu_0 \nabla \times \nabla \times \boldsymbol{\pi}_h,$$

it follows that

$$\nabla \times \nabla \times \boldsymbol{\pi}_h = \frac{\mu_0 \boldsymbol{H}}{\mu_0} = \frac{\boldsymbol{B}}{\mu_0} = \boldsymbol{H} + \boldsymbol{M}$$

or

$$\nabla\nabla \cdot \boldsymbol{\pi}_h - \nabla^2 \boldsymbol{\pi}_h = k_0^2 \boldsymbol{\pi}_h + \nabla\phi + \boldsymbol{M} \tag{3.381}$$

since $\boldsymbol{B} = \boldsymbol{\mu}_0(\boldsymbol{H} + \boldsymbol{M})$. Putting

$$\phi = \nabla \cdot \boldsymbol{\pi}_h, \tag{3.382}$$

we obtain

$$\nabla^2 \boldsymbol{\pi}_h + k_0^2 \boldsymbol{\pi}_h = -\boldsymbol{M}. \tag{3.383}$$

Similarly, if we put

$$\boldsymbol{H} = j\omega\epsilon_0 \nabla \times \boldsymbol{\pi}_e, \tag{3.384}$$

then

$$\nabla^2 \boldsymbol{\pi}_e + k_0^2 \boldsymbol{\pi}_e = -\frac{\boldsymbol{P}}{\epsilon_0}, \tag{3.385}$$

$$E = k_0^2\pi_e + \nabla\nabla\bullet\pi_e = \nabla\times\nabla\times\pi_e - \frac{P}{\epsilon_0}. \tag{3.386}$$

Usually, it is more convenient to absorb M into the parameter μ and P into the parameter ϵ. Then

$$\nabla^2\pi_h + k^2\pi_h = 0, \qquad \nabla^2\pi_e + k^2\pi_e = 0,$$

where $k^2 = \omega^2\mu\epsilon$.

3.46 Babinet's Principle

Maxwell's equations in a source-free region are

$$\nabla\times E = -j\omega\mu H, \tag{3.387}$$

$$\nabla\times H = j\omega\epsilon E. \tag{3.388}$$

They exhibit a high degree of symmetry in the expressions for E and H. If a field (E_1, H_1) satisfies these equations, then a second field (E_2, H_2) obtained from it by the transformations

$$E_2 = \mp\left(\frac{\mu}{\epsilon}\right)^{1/2}H_1, \tag{3.389}$$

$$H_2 = \pm\left(\frac{\epsilon}{\mu}\right)^{1/2}E_1 \tag{3.390}$$

is also a solution of the source-free field equations.

Equations (3.387)–(3.390) are a statement of the duality property of the electromagnetic field and constitute the basis of *Babinet's principle*. We must, of course, consider also the effect of the transformations (3.389) and (3.390) on the boundary conditions to be satisfied by the new field. Clearly, since the roles of E and H are interchanged, we must interchange all electric walls $E_t = 0$ for magnetic walls $H_t = 0$ and magnetic walls for electric walls. At a boundary where the conditions on (E_1, H_1) specify that the tangential components be continuous across a surface with ϵ and μ equal to ϵ_1, μ_1 on one side and ϵ_2, μ_2 on the other side, the new boundary conditions on (E_2, H_2) are still the continuity of the tangential components, and this changes the relative amplitudes of the fields on the two sides of the discontinuity surface.

PROBLEMS

1 Using Gauss' law, find the total electric charge inside a cubical volume 2 m on a side situated in the positive octant with three edges coincident with the x-, y-, z-axis and one corner at the origin if the electric flux density D is given by

(i) $D = u_x(x + 4)$,

(ii) $D = u_y(y^2 + 2y + 3)$.

2 What electric charge would be required on the earth and the moon to balance their gravitational attraction if the charges were in the same ratio as their masses? (The masses of the earth and the moon are 6×10^{24} kg and 7×10^{22} kg, respectively, and their separation is 400 Mm. The gravitational constant is 6.7×10^{-11} Nm^2 kg^{-2}.)

3 Map the field lines and equipotential contours of a coaxial transmission line consisting of a circular inner conductor of diameter d, symmetrically located inside an outer conductor of square cross-section with an inner side dimension of $3d$.

4 A parallel-plate capacitor of area A is filled with a dielectric of permittivity

$$\epsilon = \epsilon_0\left(1 + \frac{\epsilon_r y}{d}\right),$$

where $y = 0$ at one plate and $y = d$ at the other plate. Find $\boldsymbol{E}$, $\boldsymbol{D}$, $\boldsymbol{P}$, and V as functions of distance y between the plates. Draw a graph of $\boldsymbol{E}$, $\boldsymbol{D}$, $\boldsymbol{P}$, V, and ϵ as functions of y. Find the capacitance.

5 Show that at a conductor-conductor boundary,

$$\frac{\sigma_1}{\sigma_2} = \frac{E_{n_2}}{E_{n_1}} = \frac{J_{t_1}}{J_{t_2}}.$$

6 A uniformly wound solenoid of 6000 turns is 2 m long by 15 mm in diameter. If the current $I = 3$ A, calculate the flux density $\boldsymbol{B}$ and the magnetic field intensity $\boldsymbol{H}$ at the centre of the solenoid.

7 A parallel-plate capacitor is filled with a dielectric of $\epsilon_r = 12$ and $\tan\delta = 0.003$. The plates are 250 mm square, and the distance between them is 5 mm. If 600 V rms at 2 MHz is applied to the capacitor, calculate the power dissipated as heat.

8 Show that the average Poynting vector of a circularly-polarized wave is twice that of a linearly-polarized wave if the maximum electric field $\boldsymbol{E}$ is the same for both waves.

9 A wave travelling in the z-direction is the resultant of two elliptically polarized waves, one with components

$$E_x = 4\cos\omega t, \qquad E_y = 3\sin\omega t$$

and the other with components

$$E_x = 3\cos\omega t, \qquad E_y = 5\sin\omega t.$$

For the resultant wave, calculate the axial ratio, and the angle between the major axis of the polarization ellipse and the positive axis. Does $\boldsymbol{E}$ rotate clockwise or counterclockwise?

10 Show that the instantaneous Poynting vector of a plane circularly-polarized travelling wave is a constant.

11 A half-space of air (medium 1) and a half-space of dielectric (medium 2) are separated by a 10 μm thick copper sheet which has constants

$$\sigma = 58 \text{ mmhos}, \qquad \mu_r = \epsilon_r = 1.$$

A plane linearly-polarized 3 GHz travelling wave in medium 1 (air) with electric field $E = 10$ Vm^{-1} rms is incident normally on the copper sheet. The constants for medium 2 are

$$\sigma = 0, \qquad \mu_r = 1, \qquad \epsilon_r = 5.$$

Determine the rms value of E

(i) inside the copper sheet adjacent to medium 1 (air),

(ii) inside the copper sheet adjacent to medium 2 (dielectric), and

(iii) at a distance of 2 m from the copper sheet in medium 2.

Also, determine H at these points.

12 Show that, if the boundary conditions are satisfied for the electric intensity and the field satisfies Maxwell's equations, the boundary conditions for the magnetic intensity are automatically satisfied, and vice versa.

13 A sheet of glass having a relative dielectric constant of 8 and negligible conductivity is coated with a silver plate. Show that at a frequency of 100 MHz, the surface impedance will be less for a 0.001 cm coating than for a 0.002 cm coating, and explain why this is so.

14 A particle with a negative charge of 10^{-17} coulombs and a mass of 10^{-26} kg is at rest in a field-free space. If a uniform electric field $E = 1$ kVm^{-1} is applied for 1 μs, find the velocity of the particle and the radius of curvature of the particle path if the particle enters a magnetic field $B = 2$ milliweber per square metre, with the velocity moving normal to $\boldsymbol{B}$.

15 Wherever the displacement current is negligible with respect to the conduction current ($\sigma/j\omega\epsilon \gg 1$), the field approximately satisfies the equations

$$\nabla \times \boldsymbol{E} + \mu\frac{\partial \boldsymbol{H}}{\partial t} = 0, \qquad \nabla \times \boldsymbol{H} = \boldsymbol{i}.$$

Show that to the same approximation,

$$\boldsymbol{E} = -\mu\frac{\partial \boldsymbol{A}}{\partial t}, \qquad \boldsymbol{H} = \nabla \times \boldsymbol{A},$$

$$\nabla^2 \boldsymbol{A} - \mu\sigma\frac{\partial \boldsymbol{A}}{\partial t} = 0, \qquad \nabla \cdot \boldsymbol{A} = 0,$$

and that the current density satisfies

$$\nabla^2 \boldsymbol{i} - \mu\sigma\frac{\partial \boldsymbol{i}}{\partial t} = 0, \qquad \nabla \cdot \boldsymbol{i} = 0.$$

The foregoing equations govern the distribution of current in metallic conductors at all frequencies in the radio spectrum and below.

16 A circular cylindrical wire of radius a, which is much greater than the skin depth δ, has a uniform electric field applied in the axial direction at its surface. Use the surface-impedance concept to find the total current on the wire. Show that the ratio of the a-c impedance of the wire to the d-c resistance is

$$\frac{Z_{\text{ac}}}{R_{\text{dc}}} = \frac{a\sigma}{2} Z_{\text{m}},$$

where Z_{m} is the surface impedance. Evaluate this ratio for copper at $f = 10^6$ cycles for $\sigma = 5.8 \times 10^7$ mhos/m, $a = 0.05$ cm, and $\mu = \mu_0$.

17 Show that, when the relaxation time for a material is small as compared with the period of the time-harmonic field, the displacement current may be neglected in comparison with the conduction current.

18 Show that, when a uniform plane wave is incident on an interface between two dielectric media, with its electric vector parallel to the plane of incidence (vertical or parallel polarization), the ratio of the reflected electric field E_r to the incident electric field E_i is given by

$$\frac{E_r}{E_i} = \frac{(\epsilon_2/\epsilon_1)\cos\theta_1 - [(\epsilon_2/\epsilon_1) - \sin^2\theta_1]^{1/2}}{(\epsilon_2/\epsilon_1)\cos\theta_1 + [(\epsilon_2/\epsilon_1) - \sin^2\theta_1]^{1/2}},$$

where ϵ_1 and ϵ_2 are the permittivities of the two media and θ_1 is the angle of incidence. Discuss the possibility of not obtaining any reflection. Show that, when Brewster's angle $\tan\theta_1 = \sqrt{\epsilon_2/\epsilon_1}$, there is no reflection. Find Brewster's angle when

$$\epsilon_1 = \epsilon_0, \qquad \epsilon_2 = 5\epsilon_0.$$

19 Obtain expressions for the standing-wave patterns of E and H when a uniform plane wave is incident normally at an interface between a perfect dielectric and a perfect conductor.

4 Solution of Scalar and Vector Wave Equations

4.1 Solution of Electromagnetic Field Problems in Microwave Engineering

Many of the problems which occur in microwave engineering, be they problems of microwave circuits, microwave propagation, or microwave devices including electron devices, entail the solution of the electromagnetic field. The general problem of the electromagnetic field may be defined as that of finding the solution of Maxwell's equations to obtain the field intensity as a function of space and time for a given physical system.

In this chapter, the general solutions of the scalar wave and vector wave equations are obtained in the three frequently used orthogonal coordinate systems, namely, the rectangular, cylindrical, and spherical systems. The general solution to the equations is a mathematical expression for the various types of waves which may exist in relation to a given coordinate system. This solution contains a number of arbitrary constants which are evaluated in such a manner as to make the field satisfy the boundary conditions for a given physical system.

Though a given field distribution may be expressed in any desired system of coordinates, the problem of satisfying the boundary conditions is greatly simplified by choosing a coordinate system that best suits the boundaries of the particular physical system. For example, the field inside a circular cylindrical metal waveguide is best expressed in the cylindrical coordinates, whereas the field inside a spherical metal resonator is aptly expressed in the spherical coordinates.

In Section 3.35, it was shown that the field produced by any given distribution of electric and magnetic currents in an infinite homogeneous medium can be expressed in terms of two vector and two scalar potentials, namely, the electric and magnetic vector potentials $\boldsymbol{F}$ and $\boldsymbol{A}$ and the electric and magnetic scalar potentials V and U. These vector and scalar potentials satisfy the inhomogeneous wave equations

$$\nabla^2 \boldsymbol{A} = \gamma^2 \boldsymbol{A} - \boldsymbol{J}^i, \qquad \nabla^2 \boldsymbol{F} = \gamma^2 \boldsymbol{F} - \boldsymbol{M}^i, \tag{4.1}$$

$$\nabla^2 V = \gamma^2 V + \frac{\operatorname{div} \boldsymbol{J}^i}{\sigma + j\omega\epsilon}, \qquad \nabla^2 U = \gamma^2 U + \frac{\operatorname{div} \boldsymbol{M}^i}{j\omega\mu}, \tag{4.2}$$

where $\boldsymbol{J}^i$ and $\boldsymbol{M}^i$ are respectively the source electric and source magnetic current densities.

In regions outside the current carrying regions, the homogeneous equations

$$\nabla^2 \boldsymbol{A} - \gamma^2 \boldsymbol{A} = 0, \tag{4.3}$$

$$\nabla^2 V - \gamma^2 V = 0 \tag{4.4}$$

and similar equations for $\boldsymbol{F}$ and U are satisfied. Equations (4.3) and (4.4) are called the homogeneous vector wave equation and homogeneous scalar wave equation, respectively.

The solutions of Eqs. (4.3) and (4.4) represent the fields which result from currents lying outside the given region. For example, to determine the electromagnetic field inside a metal waveguide, we may assume that the source currents are external to the region of interest and, for most practical purposes, it is enough to solve Eqs. (4.3) and (4.4).

4.2 Methods of Solving Homogeneous Wave Equations

The problem of solving the homogeneous scalar wave equation (4.4) is easier than solving the homogeneous vector wave equation (4.3). In the three systems of coordinates we are going to consider, namely, the rectangular, cylindrical, and spherical systems, it is possible to derive the solution of the vector wave equation in terms of the solution of the corresponding scalar wave equation.

Since the scalar wave equation is a second-order linear partial differential equation, it has two independent solutions. The usual method of solving this equation in the three systems of coordinates is known as the separation of variables. In this method, the equation is split into three linear ordinary second-order differential equations, corresponding to the three independent variables (or coordinates). Each one of these three second-order linear differential equations has two independent solutions. The general solution of the partial differential equation is the product of the solutions of the ordinary differential equations. For example, in the rectangular coordinates, let $X(x) = C_1\phi_1(x) + C_2\phi_2(x)$, $Y(y) = C_3\phi_3(y) + C_4\phi_4(y)$, $Z(z) = C_5\phi_5(z) + C_6\phi_6(z)$ be the solutions of the three ordinary differential equations. Then, the general solution of the partial differential equation is

$$f(x, y, z) = [C_1\phi_1(x) + C_2\phi_2(x)][C_3\phi_3(y) + C_4\phi_4(y)][C_5\phi_5(z) + C_6\phi_6(z)]. \tag{4.5}$$

This solution contains six arbitrary constants which must be determined so as to satisfy the boundary conditions. In most practical cases, the boundaries are relatively simple, and the constants can be easily determined.

For a given set of boundaries, the general solution of the wave equation shows that there is usually a very large, and often an infinite, number of possible field distributions which are called *modes*. The actual field distribution in a given physical system may consist of a single mode or two or more superposed modes or even an infinite number of modes. This depends not only on the boundary conditions, but also on the distribution of charges and currents at the exciting source. Although a number of modes may be required to describe the field, the field itself is single valued, i.e., there is one resultant electric field intensity and one resultant magnetic field intensity at a given point at a given instant of time. Though the solution of the wave equation provides all the modes which are physically possible within the given boundaries, it does not specify the existing modes.

To obtain the specific field set up by a given distribution of source currents, Eqs. (4.1) and (4.2) have to be solved so that the proper boundary conditions can be satisfied. This is more difficult than solving Eqs. (4.3) and (4.4). However, by knowing the method of excitation and the distribution of source currents, a good estimate can be made of the modes existing in the given region. Thus, in most practical situations, it is enough to solve the homogeneous equations (4.3) and (4.4).

4.3 Solution of Scalar Wave Equation in Rectangular Coordinates

Consider the scalar wave equation

$$\nabla^2 V = \gamma^2 V \tag{4.6}$$

applicable to a region not containing the source, and assume a time variation of $e^{j\omega t}$. The function γ may take imaginary or complex values, depending on the medium. When $\gamma = 0$, Eq. (4.6) reduces to Laplace's equation. Hence, the solution of the wave equation becomes the solution of Laplace's equation. In the rectangular coordinates, Eq. (4.6) becomes

$$\frac{\partial^2 V}{\partial x^2} + \frac{\partial^2 V}{\partial y^2} + \frac{\partial^2 V}{\partial z^2} = \gamma^2 V. \tag{4.7}$$

Using the method of separation of variables, let a solution of the form

$$V = X(x)Y(y)Z(z), \tag{4.8}$$

where $X(x)$ is a function of x, $Y(y)$ a function of y, and $Z(z)$ is a function of z, only. Now, substituting Eq. (4.8) in Eq. (4.7), we obtain

$$\frac{1}{X}\frac{d^2X}{dx^2} + \frac{1}{Y}\frac{d^2Y}{dy^2} + \frac{1}{Z}\frac{d^2Z}{dz^2} = \gamma^2. \tag{4.9}$$

Since each term in Eq. (4.9) is a function of only one of the variables x, y, z, and since their sum is a constant, the term must be equal to a constant. Therefore, we write

$$\frac{1}{X}\frac{d^2X}{dx^2} = a_x^2, \qquad \frac{1}{Y}\frac{d^2Y}{dy^2} = a_y^2, \qquad \frac{1}{Z}\frac{d^2Z}{dz^2} = a_z^2, \tag{4.10}$$

where

$$a_x^2 + a_y^2 + a_z^2 = \gamma^2. \tag{4.11}$$

Each equation in Eqs. (4.10) has a solution of the form

$$X = C_1 \exp(a_x x) + C_2 \exp(-a_x x). \tag{4.12}$$

Hence, the general solution of Eq. (4.7) is of the form

$$V(x, y, z) = [C_1 \exp(a_x x) + C_2 \exp(-a_x x)][C_3 \exp(a_y y) + C_4 \exp(-a_y y)] \times [C_5 \exp(a_z z) + C_6 \exp(-a_z z)], \tag{4.13}$$

where a_x, a_y, a_z (which may be real, imaginary, or complex) and $C_1, C_2, \ldots, C_6$ are determined from the boundary conditions. Some of the special cases have been discussed in Section 3.32.

4.4 Solution of Vector Wave Equation in Rectangular Coordinates

In the rectangular coordinates, the vector wave equation

$$\nabla^2 \boldsymbol{A} = \gamma^2 \boldsymbol{A} \tag{4.14}$$

becomes

$$\boldsymbol{u}_x \nabla^2 A_x + \boldsymbol{u}_y \nabla^2 A_y + \boldsymbol{u}_z \nabla^2 A_z = \gamma^2(\boldsymbol{u}_x A_x + \boldsymbol{u}_y A_y + \boldsymbol{u}_z A_z). \tag{4.15}$$

Equation (4.15) can be split into three scalar wave equations, namely,

$$\nabla^2 A_x = \gamma^2 A_x, \qquad \nabla^2 A_y = \gamma^2 A_y, \qquad \nabla^2 A_z = \gamma^2 A_z, \tag{4.16}$$

each of which has a solution of the form

$$A_x(x, y, z) = [C_1 \exp(a_x x) + C_2 \exp(-a_x x)][C_3 \exp(a_y y) + C_4 \exp(-a_y y)] \times [C_5 \exp(a_z z) + C_6 \exp(-a_z z)], \tag{4.17}$$

where $a_x^2 + a_y^2 + a_z^2 = \gamma^2$.

4.5 Solution of Scalar Wave Equation in Cylindrical Coordinates

In the cylindrical coordinates, the scalar wave equation (4.6) becomes

$$\frac{1}{\rho}\frac{\partial}{\partial \rho}\left(\rho\frac{\partial V}{\partial \rho}\right) + \frac{1}{\rho^2}\frac{\partial^2 V}{\partial \phi^2} + \frac{\partial^2 V}{\partial z^2} = \gamma^2 V. \tag{4.18}$$

To separate the variables here, let

$$V = R(\rho)\Phi(\phi)Z(z), \tag{4.19}$$

where $R(\rho)$, $\Phi(\phi)$, and $Z(z)$ are functions of ρ, ϕ, and z, only. Substituting Eq. (4.19) in Eq. (4.18), we obtain

$$\frac{1}{\rho R}\frac{d}{d\rho}\left(\rho\frac{dR}{d\rho}\right) + \frac{1}{\rho^2 \Phi}\frac{d^2\Phi}{d\phi^2} + \frac{1}{Z}\frac{d^2 Z}{dz^2} = \gamma^2. \tag{4.20}$$

The third term in Eq. (4.20) is a function of z only and may therefore be set equal to a constant a_z^2 so that

$$\frac{d^2 Z}{dz^2} = a_z^2 Z. \tag{4.21}$$

Equation (4.21) has a solution of the form

$$Z(z) = C_5 \exp(a_z z) + C_6 \exp(-a_z z). \tag{4.22}$$

Since the second term in Eq. (4.20) multiplied by ρ^2 is a function of ϕ only, it can be equated to a constant which may be represented by $-\nu^2$. Then, the equation for Φ becomes

$$\frac{d^2\Phi}{d\phi^2} = -\nu^2\Phi \tag{4.23}$$

whose solution is of the form

$$\Phi = C_3 \cos(\nu\phi) + C_4 \sin(\nu\phi). \tag{4.24}$$

Replacing the Φ-term in Eq. (4.20) by $-\nu^2$ and multiplying throughout by R, we obtain the equation for $R(\rho)$ as

$$\rho\frac{d}{d\rho}\left(\rho\frac{dR}{d\rho}\right) + [(a_z^2 - \gamma^2)\rho^2 - \nu^2]R = 0 \tag{4.25}$$

which is a form of Bessel's equation discussed in Section 2.6. It may be put in the standard form by letting $\lambda^2 = (a_z^2 - \gamma^2)$ and $x = \lambda\rho$, where x is a new variable (and not the x-coordinate).

Equation (4.25) then becomes

$$x^2\frac{d^2R}{dx^2} + x\frac{dR}{dx} + (x^2 - \nu^2)R = 0 \tag{4.26}$$

whose two independent solutions are $J_\nu(x)$ and $N_\nu(x)$, as discussed in Section 2.6. The Hankel functions $H_\nu^{(1)}(x)$ and $H_\nu^{(2)}(x)$ are also the solutions of Eq. (4.26).

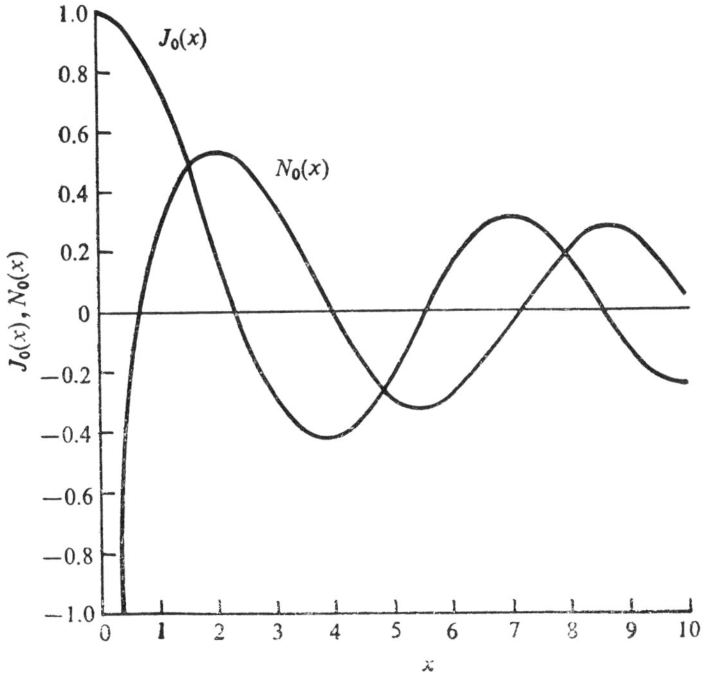

Fig. 4.1 Zero-order Bessel functions of the first and second kind.

The Bessel functions of the first and second kind are plotted for zero order in Fig. 4.1, and for orders 0–5 in Fig. 4.2.

$R(\rho)$ is related to $\Phi(\phi)$ since ν appears in both Eqs. (4.23) and (4.26). In many practical problems, for example, the cylindrical metal waveguide or the cylindrical dielectric waveguide, Φ should have a periodic variation with ϕ, having a period of 2π, because Φ must have the same value for $\phi = \delta$ and $\phi = \delta + 2\pi$. In such a case, ν must be an integer. But in some other problems, e.g., a metal sector of a cylinder, ν need not be an integer.

If ν is an integer n, then the general solution of the scalar wave equation (4.18) in the cylindrical coordinates is of the form

$$\begin{aligned} V(\rho, \phi, z) = {} & [C_1J_n(\lambda\rho) + C_2N_n(\lambda\rho)][C_3 \cos(n\phi) + C_4 \sin(n\phi)] \\ & \times [C_5 \exp(a_z z) + C_6 \exp(-a_z z)]. \end{aligned} \tag{4.27}$$

On the other hand, for nonintegral values of ν, we may use the solution

$$\begin{aligned} V(\rho, \phi, z) = {} & [C_1J_\nu(\lambda\rho) + C_2J_{-\nu}(\lambda\rho)][C_3 \cos(\nu\phi) + C_4 \sin(\nu\phi)] \\ & \times [C_5 \exp(a_z z) + C_6 \exp(-a_z z)]. \end{aligned} \tag{4.28}$$

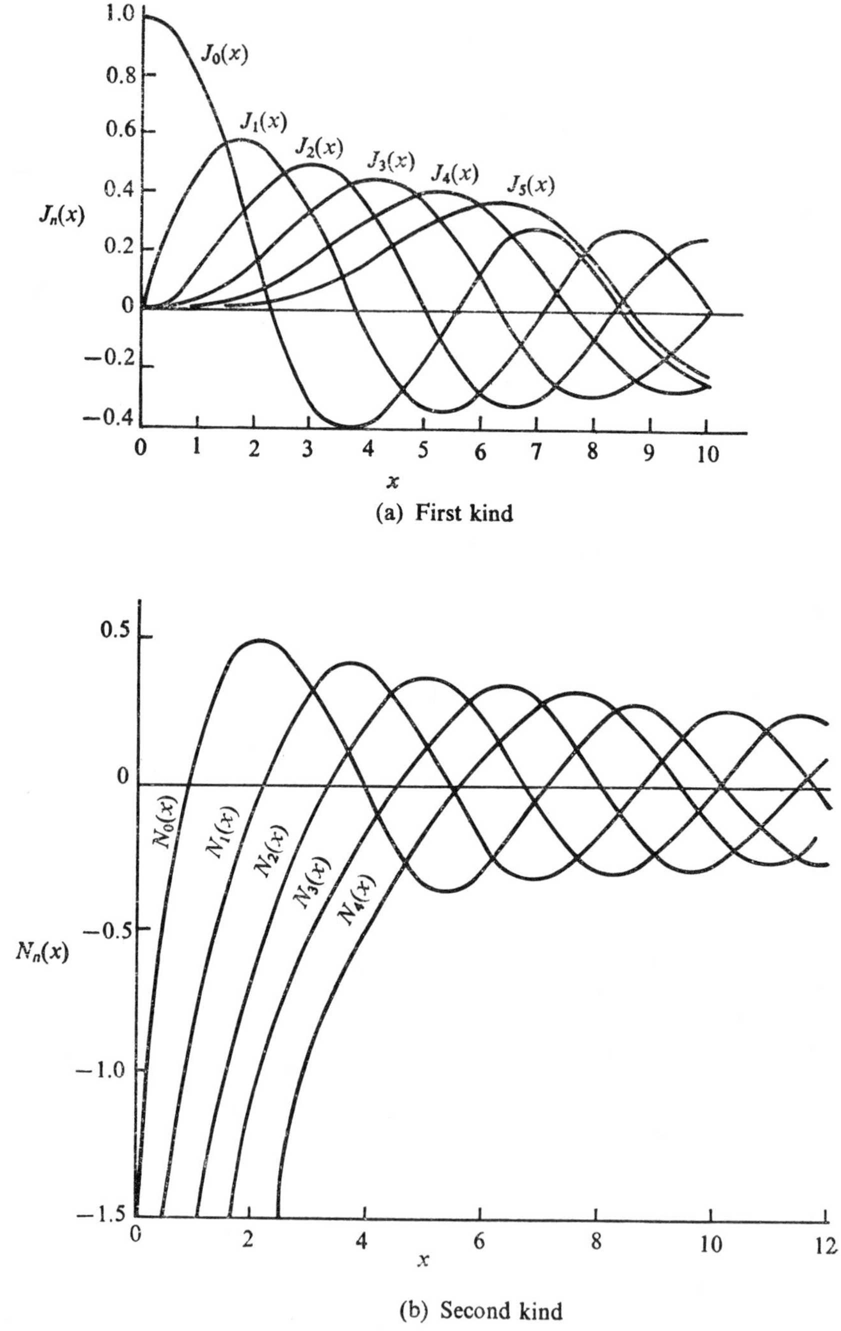

Fig. 4.2 Bessel functions for orders 0–5.

For some problems where the field has to decay with the radial distance ρ, for example, outside a dielectric cylinder, the Hankel functions $H_n^{(1)}(k\rho)$ and $H_n^{(2)}(k\rho)$ are used. For a problem where an azimuthal symmetry exists, $n = 0$ and $\partial/\partial\phi = 0$. If $\gamma = 0$, then the wave equation (4.20) becomes Laplace's equation. If a_z and γ are both zero, then Eq. (4.25) in R becomes

$$\rho \frac{d[\rho(\partial R/\partial \rho)]}{d\rho} - \nu^2 R = 0 \tag{4.29}$$

whose solution is of the form

$$R(\rho) = (C_1\rho^{\nu} + C_2\rho^{-\nu}). \tag{4.30}$$

Hence, the equation for $Z(z)$ becomes

$$\frac{d^2Z}{dz^2} = 0$$

whose solution is of the form

$$Z(z) = (C_5 z + C_6). \tag{4.31}$$

Then, V is of the form

$$V = (C_1\rho^{\nu} + C_2\rho^{-\nu})[C_3 \cos(\nu\phi) + C_4 \sin(\nu\phi)](C_5 z + C_6). \tag{4.32}$$

If a_z, γ, and ν are all zero, then V is of the form

$$V = (C_1 \ln \rho + C_2)(C_3\phi + C_4)(C_5 z + C_6). \tag{4.33}$$

The choice of the solution for V depends on the particular problem.

The properties of the Bessel, Neumann, and Hankel functions have been dealt with in detail in Section 2.6.

4.6 Solution of Vector Wave Equation in Cylindrical Coordinates

In the cylindrical coordinates, the vector wave equation

$$\nabla^2 \boldsymbol{A} = \gamma^2 \boldsymbol{A} \tag{4.34}$$

becomes

$$\boldsymbol{u}_\rho\left(\nabla^2 A_\rho - \frac{2}{\rho^2}\frac{\partial A_\phi}{\partial \rho} - \frac{A_\rho}{\rho^2}\right) + \boldsymbol{u}_\phi\left(\nabla^2 A_\phi + \frac{2}{\rho^2}\frac{\partial A_\rho}{\partial \phi} - \frac{A_\phi}{\rho^2}\right) + \boldsymbol{u}_z \nabla^2 A_z$$
$$= \gamma^2(\boldsymbol{u}_\rho A_\rho + \boldsymbol{u}_\phi A_\phi + \boldsymbol{u}_z A_z) \tag{4.35}$$

(see Eq. 2.73).

If the ρ-, ϕ-, and z-component in Eq. (4.35) are separated, we obtain

$$\nabla^2 A_\rho - \frac{2}{\rho^2}\frac{\partial A_\phi}{\partial \rho} - \frac{A_\rho}{\rho^2} = \gamma^2 A_\rho, \tag{4.36}$$

$$\nabla^2 A_\phi + \frac{2}{\rho^2}\frac{\partial A_\rho}{\partial \phi} - \frac{A_\phi}{\rho^2} = \gamma^2 A_\phi, \tag{4.37}$$

$$\nabla^2 A_z = \gamma^2 A_z. \tag{4.38}$$

Though it may be possible to eliminate A_ρ or A_ϕ from Eqs. (4.36) and (4.37) and obtain a differential equation for one of them, it is not so easy to solve the resulting differential equation for A_ρ or A_ϕ. However, Eq. (4.38) is the scalar wave equation for A_z and can be solved easily, as shown in Section 4.5. The most general form of the solution of A_z is

$$A_z = [C_1 J_\nu(\lambda\rho) + C_2 N_\nu(\lambda\rho)][C_3 \cos(\nu\phi) + C_4 \sin(\nu\phi)] \times [C_5 \exp(a_z z) + C_6 \exp(-a_z z)]. \tag{4.39}$$

From Eqs. (3.359), (3.360), (3.362), and (3.363), we have seen that in a homogeneous isotropic source-free region, it is possible to express the electric and magnetic field intensities $\boldsymbol{E}$ and $\boldsymbol{H}$ either in terms of the magnetic Hertz vector potential $\boldsymbol{\pi}_\mathrm{h}$ or in terms of the electric Hertz vector potential $\boldsymbol{\pi}_\mathrm{e}$, as follows:

$$\boldsymbol{E} = -j\omega\mu\nabla\times\boldsymbol{\pi}_\mathrm{h}, \tag{4.40}$$

$$\boldsymbol{H} = \nabla\times\nabla\times\boldsymbol{\pi}_\mathrm{h}, \tag{4.41}$$

$$\boldsymbol{H} = j\omega\epsilon\nabla\times\boldsymbol{\pi}_\mathrm{e}, \tag{4.42}$$

$$\boldsymbol{E} = \nabla\times\nabla\times\boldsymbol{\pi}_\mathrm{e}. \tag{4.43}$$

It may be noted that, in a lossy medium, ϵ is replaced by $[\epsilon - (j\sigma/\omega)]$.

Now let us calculate the cylindrical components of the electromagnetic field associated with the Hertz vector $\boldsymbol{\pi}_\mathrm{e}$ which has only a z-component $\pi_{\mathrm{e}z}$ so that $\pi_{\mathrm{e}\rho} = \pi_{\mathrm{e}\phi} = 0$. The cylindrical components of $\boldsymbol{E}$ and $\boldsymbol{H}$ of such a field, as derived from Eqs. (4.42) and (4.43), are

$$\begin{aligned} E_\rho &= \frac{\partial^2\pi_{\mathrm{e}z}}{\partial z\,\partial\rho}, \qquad E_\phi = \frac{1}{\rho}\frac{\partial^2\pi_{\mathrm{e}z}}{\partial z\,\partial\phi}, \\ E_z &= -\frac{1}{\rho}\Big[\frac{\partial}{\partial\rho}\Big(\rho\frac{\partial\pi_{\mathrm{e}z}}{\partial\rho}\Big) + \frac{\partial}{\partial\phi}\Big(\frac{1}{\rho}\frac{\partial\pi_{\mathrm{e}z}}{\partial\phi}\Big)\Big], \\ H_\rho &= j\omega\epsilon\frac{1}{\rho}\frac{\partial\pi_{\mathrm{e}z}}{\partial\phi}, \qquad H_\phi = -j\omega\epsilon\frac{\partial\pi_{\mathrm{e}z}}{\partial\rho}, \qquad H_z = 0. \end{aligned} \tag{4.44}$$

Equations (4.44) represent a partial electromagnetic field, which is characterized by the absence of the axial component H_z of the magnetic field. Since $\boldsymbol{\pi}_\mathrm{e}$ is the electric Hertz polarization potential, such a field is called a field of the *electric type* (E field) or a *transverse magnetic* (TM) field.

Similarly, by using Eqs. (4.40) and (4.41), a partial field can be derived from the magnetic Hertz polarization potential $\boldsymbol{\pi}_\mathrm{h}$ which has only a z-component $\pi_{\mathrm{h}z}$ so that $\pi_{\mathrm{h}\rho} = \pi_{\mathrm{h}\phi} = 0$. The cylindrical components of $\boldsymbol{E}$ and $\boldsymbol{H}$ of such a field are

$$\begin{aligned} E_\rho &= -\frac{j\omega\mu}{\rho}\frac{\partial\pi_{\mathrm{h}z}}{\partial\phi}, \qquad E_\phi = j\omega\mu\frac{\partial\pi_{\mathrm{h}z}}{\partial\rho}, \qquad E_z = 0, \\ H_\rho &= \frac{\partial^2\pi_{\mathrm{h}z}}{\partial z\,\partial\rho}, \qquad H_\phi = \frac{1}{\rho}\frac{\partial^2\pi_{\mathrm{h}z}}{\partial z\,\partial\phi}, \qquad H_z = -\frac{1}{\rho}\Big[\frac{\partial}{\partial\rho}\Big(\rho\frac{\partial\pi_{\mathrm{h}z}}{\partial\rho}\Big) + \frac{\partial}{\partial\phi}\Big(\frac{1}{\rho}\frac{\partial\pi_{\mathrm{h}z}}{\partial\phi}\Big)\Big]. \end{aligned} \tag{4.45}$$

A field represented by these equations is characterized by the absence of the axial component E_z of the electric field, and is derived from the magnetic Hertz vector potential $\boldsymbol{\pi}_\mathrm{h}$, and hence is called a field of the *magnetic type* (H field) or a *transverse electric* (TE) field.

π_{ez} and π_{hz} being the z-components of $\boldsymbol{\pi}_e$ and $\boldsymbol{\pi}_h$, they satisfy the scalar wave equation (4.38) in the cylindrical coordinates and have a solution of the form (4.39). Using relations (4.44) and (4.45), the components of the electric and magnetic fields $\boldsymbol{E}$ and $\boldsymbol{H}$ may be derived for transverse magnetic or transverse electric fields.

The electromagnetic field obtained by the superposition of the partial fields given by Eqs. (4.44) and (4.45) derived from π_{ez} and π_{hz} is of such generality that we can satisfy a prescribed set of boundary conditions on any of the coordinate surfaces, $z =$ constant, $\rho =$ constant, or $\phi =$ constant.

4.7 Solution of Scalar Wave Equation in Spherical Coordinates

In the spherical coordinates, the scalar wave equation becomes

$$\frac{1}{r^2}\frac{\partial[r^2(\partial V/\partial r)]}{\partial r} + \frac{1}{r^2 \sin\theta}\frac{\partial[\sin\theta\,(\partial V/\partial\theta)]}{\partial\theta} + \frac{1}{r^2\sin^2\theta}\frac{\partial^2 V}{\partial\phi^2} = \gamma^2 V. \tag{4.46}$$

Let us assume a solution of the form

$$V = R(r)P(\theta)\Phi(\phi). \tag{4.47}$$

Then, substituting Eq. (4.47) in Eq. (4.46), we obtain

$$\frac{1}{R}\frac{d[r^2(dR/dr)]}{dr} + \frac{1}{P\sin\theta}\frac{d[\sin\theta\,(dP/d\theta)]}{d\theta} + \frac{1}{\Phi\sin^2\theta}\frac{d^2\Phi}{d\phi^2} = \gamma^2 r^2. \tag{4.48}$$

The r-terms are contained in the first and last terms of Eq. (4.48), and hence we set them equal to a constant a_r^2. Thus

$$\frac{d[r^2(dR/dr)]}{dr} - (\gamma^2 r^2 + a_r^2)R = 0. \tag{4.49}$$

Using Eq. (4.49) in Eq. (4.48) and multiplying throughout by $\sin^2\theta$, we obtain

$$\frac{\sin\theta}{P}\frac{d[\sin\theta\,(dP/d\theta)]}{d\theta} + a_r^2\sin^2\theta + \frac{1}{\Phi}\frac{d^2\Phi}{d\phi^2} = 0. \tag{4.50}$$

The ϕ-term is equated to a constant $-m^2$, which yields

$$\frac{d^2\Phi}{d\phi^2} = -m^2\Phi. \tag{4.51}$$

Equation (4.51) has a solution of the form

$$\Phi = C_5\cos(m\phi) + C_6\sin(m\phi). \tag{4.52}$$

If the field is periodic in ϕ, then m is an integer n. Using Eq. (4.51) in Eq. (4.50) and multiplying throughout by $P/\sin^2\theta$, we obtain

$$\frac{d^2P}{d\theta^2} + \cot\theta\frac{dP}{d\theta} + (a_r^2 - \frac{m^2}{\sin^2\theta})P = 0 \tag{4.53}$$

which is the associated Legendre equation. Putting $a_r^2 = n(n+1)$, where n is usually an integer, and $x = \cos\theta$, Eq. (4.53) reduces to

$$(1 - x^2)\frac{d^2P}{dx^2} - 2x\frac{dP}{dx} + [n(n+1) - \frac{m^2}{1-x^2}]P = 0. \tag{4.54}$$

If the field has circular symmetry, then $m = 0$, and Eq. (4.54) reduces to

$$(1 - x^2)\frac{d^2P}{dx^2} - 2x\frac{dP}{dx} + n(n+1)P = 0 \tag{4.55}$$

which is called the Legendre equation. [The solutions of Eqs. (4.54) and (4.55) are given in Section 2.7.]

The two independent solutions of Eq. (4.54) are the associated Legendre functions $P_n^m(\cos\theta)$ and $Q_n^m(\cos\theta)$, and the solutions of Eq. (4.55) are the Legendre functions $P_n(\cos\theta)$ and $Q_n(\cos\theta)$. In the solutions of these equations by the series method, it is found that n must have integer values to avoid an infinite value of P. In most physical problems, the solutions of the wave equation which satisfy the proper boundary conditions can be obtained with integral values of n. However, in some physical problems, n may not be an integer; indeed, it may be a complex quantity.

The differential equation for R is Eq. (4.49) which can be reduced to the standard form of Bessel's equation by setting $k^2 = -\gamma^2$ and $x = kr$ and putting $a_r^2 = n(n+1)$. Then, Eq. (4.49) becomes

$$x^2\frac{d^2R}{dx^2} + 2x\frac{dR}{dx} + [x^2 - n(n+1)]R = 0. \tag{4.56}$$

Now, let $R = x^{-1/2}W$, where W is a new variable. Substituting for R in Eq. (4.56), we obtain

$$x^2\frac{d^2W}{dx^2} + x\frac{dW}{dx} + [x^2 - (n+\tfrac{1}{2})^2]W = 0. \tag{4.57}$$

Equation (4.57) is the Bessel equation of order $(n + 1/2)$, which has a general solution of the form

$$W = C_1 J_{n+1/2}(x) + C_2 J_{-(n+1/2)}(x). \tag{4.58}$$

Replacing x and W by the values given in Eq. (4.58) and choosing different constants, we may write the solution of Eq. (4.49) as

$$R = C_1\sqrt{\frac{\pi}{2kr}}J_{n+1/2}(kr) + C_2\sqrt{\frac{\pi}{2kr}}N_{n+1/2}(kr) = C_1 j_n(kr) + C_2 n_n(kr), \tag{4.59}$$

where $j_n(kr)$ and $n_n(kr)$ are the spherical Bessel functions defined in Section 2.6.

If $\gamma = 0$, the wave equation (4.46) becomes Laplace's equation, and the equation for R becomes

$$r^2\frac{d^2R}{dr^2} + 2r\frac{dR}{dr} - n(n+1)R = 0 \tag{4.60}$$

whose solution is of the form

$$R = C_1 r^n + C_2 r^{-(n+1)}. \tag{4.61}$$

The general solution of the wave equation (4.46) in the spherical coordinates may be written as

$$V = [C_1 j_n(kr) + C_2 n_n(kr)][C_3 P_n^m(\cos\theta) + C_4 Q_n^m(\cos\theta)][C_5 \cos(m\phi) + C_6 \sin(m\phi)]. \tag{4.62}$$

The following observations may be made on solution (4.62):

(i) In stationary or quasi-stationary fields, the spherical Bessel functions may be replaced by Eq. (4.61).

(ii) Integral values of n are required if infinite values of the Legendre functions are to be avoided. Also, m must be an integer if the field is periodic in ϕ with a period of 2π radians.

(iii) Since $Q_n^m(\cos\theta)$ has infinite value at $\theta = 0$ and $\theta = \pi$ radians, the solution is discarded if the field includes these regions.

(iv) For a circularly-symmetric field, $m = 0$, and the associated Legendre functions are replaced by the Legendre functions, and there is no variation with ϕ.

(v) In regions where the field has to decay with increasing r, for example, in the region external to a dielectric sphere, the spherical Hankel functions $h_n^{(1)}(kr)$ and $h_n^{(2)}(kr)$ are used instead of $j_n(kr)$ and $n_n(kr)$.

4.8 A Fundamental Set of Solutions of Vector Wave Equation: General Method of Solution

The vector wave equation is

$$\nabla^2 \boldsymbol{C} + k^2 \boldsymbol{C} = 0 \tag{4.63a}$$

or

$$\nabla\nabla\cdot\boldsymbol{C} - \nabla\times\nabla\times\boldsymbol{C} + k^2\boldsymbol{C} = 0. \tag{4.63b}$$

Here, $\boldsymbol{C}$ may represent any of the field vectors $\boldsymbol{E}$, $\boldsymbol{H}$, $\boldsymbol{D}$, $\boldsymbol{B}$, $\boldsymbol{A}$, $\boldsymbol{F}$, and $\boldsymbol{\pi}_e$ or $\boldsymbol{\pi}_h$, and $k^2 = -\gamma^2$ where γ is the propagation constant of the medium.

Let ψ be a solution of the scalar wave equation

$$\nabla^2\psi + k^2\psi = 0. \tag{4.64}$$

Then, it is easy to show that Eqs. (4.63) have three independent solutions $\boldsymbol{L}$, $\boldsymbol{M}$, and $\boldsymbol{N}$ given by

$$\boldsymbol{L} = \nabla\psi, \qquad \boldsymbol{M} = \nabla\times\boldsymbol{a}\psi, \qquad \boldsymbol{N} = \frac{1}{k}\nabla\times\boldsymbol{M}, \tag{4.65}$$

where $\boldsymbol{a}$ is a constant vector of unit length. If $\boldsymbol{C}$ is replaced by $\boldsymbol{L}$, $\boldsymbol{M}$, or $\boldsymbol{N}$, it is easy to show that Eq. (4.63b) is satisfied, provided Eq. (4.64) is also satisfied. Since $\boldsymbol{a}$ is a constant vector, it is also clear that

$$\boldsymbol{M} = \boldsymbol{L}\times\boldsymbol{a} = \frac{1}{k}\nabla\times\boldsymbol{N}. \tag{4.66}$$

Since $\boldsymbol{M} = \boldsymbol{L}\times\boldsymbol{a}$,

$$\boldsymbol{L}\cdot\boldsymbol{M} = 0. \tag{4.67}$$

Also,

$$\nabla\times\boldsymbol{L} = 0, \qquad \nabla\cdot\boldsymbol{L} = \nabla^2\psi = -k^2\psi, \tag{4.68}$$

$$\nabla \cdot \boldsymbol{M} = \nabla \cdot (\nabla \times \boldsymbol{a}\psi) = 0,$$
$$\nabla \cdot \boldsymbol{N} = \nabla \cdot (\frac{1}{k}\nabla \times \boldsymbol{M}) = 0. \tag{4.69}$$

The particular solutions of the scalar wave equation (4.64), which are finite, continuous, and single valued in a given domain, form a discrete set. Associated with each solution ψ_n of Eq. (4.64) are the three vector solutions $\boldsymbol{L}_n$, $\boldsymbol{M}_n$, and $\boldsymbol{N}_n$ of Eqs. (4.63), no two of which are collinear. Hence, any arbitrary wave function can be represented as a linear combination of the characteristic vector functions $\boldsymbol{L}_n$, $\boldsymbol{M}_n$, and $\boldsymbol{N}_n$. Further, it is possible to show that $\boldsymbol{L}_n$, $\boldsymbol{M}_n$, and $\boldsymbol{N}_n$ possess certain orthogonal properties, which will be demonstrated in due course, and the coefficients of the expansion can be determined. When the given function is solenoidal (its divergence is zero), it can be expressed only in terms of $\boldsymbol{M}_n$ and $\boldsymbol{N}_n$. If the function is not solenoidal, then $\boldsymbol{L}_n$ should also be included.

Since the electric field intensity $\boldsymbol{E}$ is proportional to the curl of the magnetic field intensity $\boldsymbol{H}$ and the magnetic field intensity $\boldsymbol{H}$ is proportional to the curl of the electric field intensity $\boldsymbol{E}$, these can be aptly represented by $\boldsymbol{M}$ and $\boldsymbol{N}$. If the time variation of the fields is taken as $e^{j\omega t}$, then

$$\nabla \times \boldsymbol{E} = -j\omega\mu\boldsymbol{H}, \qquad \nabla \times \boldsymbol{H} = (\sigma + j\omega\epsilon)\boldsymbol{E} = \frac{-k^2}{j\omega\mu}\boldsymbol{E}. \tag{4.70}$$

As $\nabla \cdot \boldsymbol{A} \neq 0$, the magnetic vector potential $\boldsymbol{A}$ can be represented as

$$\boldsymbol{A} = \frac{-j}{\omega}\sum_n (a_n\boldsymbol{M}_n + b_n\boldsymbol{N}_n + c_n\boldsymbol{L}_n), \tag{4.71}$$

where the coefficients a_n, b_n, and c_n are determined from the current distribution. From the relation $\boldsymbol{H} = \nabla \times \boldsymbol{A}$, we may write

$$\boldsymbol{E} = -\mu \sum_n (a_n\boldsymbol{M}_n + b_n\boldsymbol{N}_n), \tag{4.72}$$

$$\boldsymbol{H} = \frac{k}{j\omega}\sum_n (a_n\boldsymbol{N}_n + b_n\boldsymbol{M}_n). \tag{4.73}$$

The electric scalar potential V is given by [see Eq. (3.274)]

$$V = -\frac{\nabla \cdot \boldsymbol{A}}{\sigma + j\omega\epsilon} = \frac{j\omega\mu}{k^2}\nabla \cdot \boldsymbol{A}.$$

It can be written as

$$V = \sum_n c_n\psi_n. \tag{4.74}$$

The electric and magnetic Hertz potentials $\boldsymbol{\pi}_e$ and $\boldsymbol{\pi}_h$ are then given by

$$\boldsymbol{\pi}_e = -\frac{1}{k}\sum_n b_n\psi_n\boldsymbol{a}, \qquad \boldsymbol{\pi}_h = \frac{1}{j\omega}\sum_n a_n\psi_n\boldsymbol{a}. \tag{4.75}$$

4.9 Application of General Method of Solution of Vector Wave Equation to Cylindrical Coordinates

A general solution of the scalar wave equation

$$\nabla^2\psi + k^2\psi = 0 \tag{4.76}$$

in the cylindrical coordinates is given by

$$\psi_e = \cos(n\phi) Z_n(\lambda\rho) e^{jhz+j\omega t} \tag{4.77a}$$

or

$$\psi_o = \sin(n\phi) Z_n(\lambda\rho) e^{jhz+j\omega t}, \tag{4.77b}$$

where $Z_n(\lambda\rho)$ is a cylindrical Bessel function of any kind, namely, $J_n(\lambda\rho)$, $N_n(\lambda\rho)$, $H_n^{(1)}(\lambda\rho)$, and $H_n^{(2)}(\lambda\rho)$, and $\lambda^2 = h^2 - \gamma^2 = h^2 + k^2$, γ being the propagation constant in the medium. The subscripts e and o denote that the solution is an even or odd function of ϕ.

Setting

$$\boldsymbol{L}_{n\lambda} = \boldsymbol{l}_{n\lambda} e^{jhz+j\omega t}, \qquad \boldsymbol{M}_{n\lambda} = \boldsymbol{m}_{n\lambda} e^{jhz+j\omega t}, \qquad \boldsymbol{N}_{n\lambda} = \boldsymbol{n}_{n\lambda} e^{jhz+j\omega t} \tag{4.78}$$

and using the definitions of $\boldsymbol{L}$, $\boldsymbol{M}$, and $\boldsymbol{N}$, as given by Eqs. (4.65), and putting $\boldsymbol{a} = \boldsymbol{u}_z =$ a unit vector in the z-direction, we obtain

$$\boldsymbol{l}_{\substack{e\\o}n\lambda} = \frac{\partial}{\partial\rho}[Z_n(\lambda\rho)] \begin{matrix}\cos\\ \sin\end{matrix} (n\phi)\boldsymbol{u}_\rho \mp \frac{n}{\rho} Z_n(\lambda\rho) \begin{matrix}\sin\\ \cos\end{matrix} (n\phi)\boldsymbol{u}_\phi + jhZ_n(\lambda\rho) \begin{matrix}\cos\\ \sin\end{matrix} (n\phi)\boldsymbol{u}_z, \tag{4.79}$$

$$\boldsymbol{m}_{\substack{e\\o}n\lambda} = \mp \frac{n}{\rho} Z_n(\lambda\rho) \begin{matrix}\sin\\ \cos\end{matrix} (n\phi)\boldsymbol{u}_\rho - \frac{\partial}{\partial\rho}[Z_n(\lambda\rho)] \begin{matrix}\cos\\ \sin\end{matrix} (n\phi)\boldsymbol{u}_\phi, \tag{4.80}$$

$$\boldsymbol{n}_{\substack{e\\o}n\lambda} = \frac{jh}{k}\frac{\partial}{\partial\rho}[Z_n(\lambda\rho)] \begin{matrix}\cos\\ \sin\end{matrix} (n\phi)\boldsymbol{u}_\rho \mp \frac{jhn}{k\rho} Z_n(\lambda\rho) \begin{matrix}\sin\\ \cos\end{matrix} (n\phi)\boldsymbol{u}_\phi + \frac{\lambda^2}{k} Z_n(\lambda\rho)\boldsymbol{u}_z. \tag{4.81}$$

By using Eqs. (4.71), (4.72), and (4.73), the expressions for $\boldsymbol{A}$, $\boldsymbol{E}$, and $\boldsymbol{H}$ can be obtained. If $a_n = 0$, then

$$\boldsymbol{E} = -\mu \sum_n b_n \boldsymbol{N}_n, \qquad \boldsymbol{H} = \frac{k}{j\omega} \sum_n b_n \boldsymbol{M}_n, \tag{4.82}$$

and hence $\boldsymbol{H}$ has only the ρ- and ϕ-component and $H_z = 0$, whereas $\boldsymbol{E}$ has all the three components. This is true for a TM (or an E) field. If $b_n = 0$, then

$$\boldsymbol{E} = -\mu \sum_n a_n \boldsymbol{M}_n, \qquad \boldsymbol{H} = \frac{k}{j\omega} \sum_n a_n \boldsymbol{N}_n, \tag{4.83}$$

and hence $\boldsymbol{E}$ has only the ρ- and ϕ-component and $E_z = 0$, whereas $\boldsymbol{H}$ has all the three components. This is the case of a TE (or H). It can also be shown that

$$\boldsymbol{\pi}_e = -\frac{1}{k} \sum_n b_n \psi_n \boldsymbol{u}_z, \qquad \boldsymbol{\pi}_h = \frac{1}{i\omega} \sum_n a_n \psi_n \boldsymbol{u}_z. \tag{4.84}$$

It is evident from Eqs. (4.82)–(4.84) that a TM field is derived from $\boldsymbol{\pi}_e$, whereas a TE field is derived from $\boldsymbol{\pi}_h$ (see Section 4.6).

4.10 Solution of Vector Wave Equation in Spherical Coordinates

Let the solution of the scalar wave equation

$$\nabla^2\psi + k^2\psi = 0 \tag{4.85}$$

in the spherical coordinates be represented as

$$\psi_{\substack{e\\o}mn} = f_{\substack{e\\o}mn} e^{j\omega t},$$

where $f_{\substack{e\\o}mn}$ is given by

$$f_{\substack{e\\o}mn} = \begin{matrix}\cos\\ \sin\end{matrix}(m\phi) P_n^m(\cos\theta) z_n(kr). \tag{4.86}$$

Then, from Eqs. (4.65), if $\boldsymbol{L} = \boldsymbol{l}e^{j\omega t} = \nabla\psi$,

$$\boldsymbol{l}_{\substack{e\\o}mn} = \frac{\partial}{\partial r}[z_n(kr)] P_n^m(\cos\theta) \begin{matrix}\cos\\ \sin\end{matrix}(m\phi)\boldsymbol{u}_r + \frac{1}{r} z_n(kr) \frac{\partial}{\partial\theta} P_n^m(\cos\theta) \begin{matrix}\cos\\ \sin\end{matrix}(m\phi)\boldsymbol{u}_\theta$$

$$\mp \frac{m}{r\sin\theta} z_n(kr) P_n^m(\cos\theta) \begin{matrix}\sin\\ \cos\end{matrix}(m\phi)\boldsymbol{u}_\phi. \tag{4.87}$$

Then, $\boldsymbol{L}$ is a solution of the vector wave equation

$$\nabla^2\boldsymbol{C} + k^2\boldsymbol{C} = 0. \tag{4.88}$$

To obtain the other two independent solutions, namely, $\boldsymbol{M}$ and $\boldsymbol{N}$, we have to introduce a constant vector $\boldsymbol{a}$. Though such a procedure is possible in the spherical polar coordinates, it will not make $\boldsymbol{M}$ or $\boldsymbol{N}$ either normal or tangential over the entire surface of a sphere. On the other hand, if a radial vector $\boldsymbol{u}_r$ is used instead of a constant vector $\boldsymbol{a}$, it can be shown that a tangential solution $\boldsymbol{M}$ can be constructed from $\boldsymbol{u}_r$.

Let us try to find a solution of the vector wave equation (4.88) of the form

$$\boldsymbol{M} = \nabla \times \boldsymbol{u}_r v(r)\psi = \boldsymbol{L} \times \boldsymbol{u}_r v(r), \tag{4.89}$$

where $v(r)$ is an unknown scalar function of r. Then, the r-, θ-, and ϕ-component of M are

$$M_r = 0, \qquad M_\theta = \frac{1}{r\sin\theta}\frac{\partial}{\partial\phi}(v\psi), \qquad M_\phi = -\frac{1}{r}\frac{\partial}{\partial\phi}(v\psi). \tag{4.90}$$

Here, the divergence of $\boldsymbol{M}$ is zero, and hence

$$\nabla^2\boldsymbol{M} = \nabla\nabla\cdot\boldsymbol{M} - \nabla\times\nabla\times\boldsymbol{M} = -\nabla\times\nabla\times\boldsymbol{M}.$$

Therefore, Eq. (4.88) becomes

$$\nabla^2\boldsymbol{M} + k^2\boldsymbol{M} = -\nabla\times\nabla\times\boldsymbol{M} + k^2\boldsymbol{M} = 0. \tag{4.91}$$

If $(\nabla\times\nabla\times\boldsymbol{M} + k^2\boldsymbol{M})$ is expanded into the r-, θ- and ϕ-component, we find that the r-component is identically equal to zero, regardless of $v(r)$. The condition $\nabla\times\nabla\times\boldsymbol{M} - k^2\boldsymbol{M} = -\nabla^2\boldsymbol{M} + k^2\boldsymbol{M} = 0$ is satisfied on the θ- and ϕ-component if $v(r)$ satisfies the equation

$$\frac{\partial^2}{\partial r^2}(v\psi) + \frac{1}{r^2\sin\theta}\frac{\partial}{\partial\theta}\left[\sin\theta\frac{\partial}{\partial\theta}(v\psi)\right] + \frac{1}{r^2\sin^2\theta}\frac{\partial^2}{\partial\phi^2}(v\psi) + k^2 v\psi = 0. \tag{4.92}$$

If $v(r)$ is now chosen equal to r, Eq. (4.92) reduces to

$$\nabla^2\psi + k^2\psi = 0. \tag{4.93}$$

Hence, in the spherical coordinates, we can say that the vector wave equation (4.88) is satisfied by $\boldsymbol{L}$, $\boldsymbol{M}$, and $\boldsymbol{N}$ given by

$$\boldsymbol{L} = \nabla\psi, \qquad \boldsymbol{M} = \nabla\times\boldsymbol{r}\psi = \boldsymbol{L}\times\boldsymbol{r} = \frac{1}{k}\nabla\times\boldsymbol{N}. \tag{4.94}$$

The components of $\boldsymbol{M}$ are

$$M_r = 0, \qquad M_\theta = \frac{1}{\sin\theta}\frac{\partial\psi}{\partial\phi}, \qquad M_\phi = -\frac{\partial\psi}{\partial\theta}. \tag{4.95}$$

As $\boldsymbol{N} = \frac{1}{k}\nabla\times\boldsymbol{M}$, the components of $\boldsymbol{N}$ are

$$N_r = \frac{\partial^2(r\psi)}{k\partial r^2} + kr\psi, \qquad N_\theta = \frac{1}{kr}\frac{\partial^2(r\psi)}{\partial r\,\partial\theta}, \qquad N_\phi = \frac{1}{kr\sin\theta}\frac{\partial^2(r\psi)}{\partial r\,\partial\phi}. \tag{4.96}$$

Since $\psi = R(r)P(\theta)\Phi(\phi)$ from Eq. (4.47), and since $R(r)$ satisfies Eq. (4.56), by putting $x = kr$, Eq. (4.56) can be rewritten as

$$(k^2r^2)\frac{d^2R}{d(kr)^2} + 2kr\frac{dR}{d(kr)} + [k^2r^2 - n(n+1)]R = 0$$

or

$$r^2\frac{d^2R}{dr^2} + 2r\frac{dR}{dr} + [k^2r^2 - n(n+1)]R = 0$$

or

$$\frac{n(n+1)R}{kr} = \frac{r}{k}\frac{d^2R}{dr^2} + \frac{2}{k}\frac{dR}{dr} + krR = \frac{1}{k}\frac{d^2(rR)}{dr^2} + krR.$$

Hence, N_r reduces to the simpler form

$$N_r = \frac{n(n+1)\psi}{kr}. \tag{4.97}$$

Putting $\boldsymbol{M} = \boldsymbol{m}e^{j\omega t}$ and $\boldsymbol{N} = \boldsymbol{n}e^{j\omega t}$, we find that

$$\boldsymbol{m}_{\substack{e\\o}mn} = \mp\frac{m}{\sin\theta}z_n(kr)P_n^m(\cos\theta)\,{\sin \atop \cos}(m\phi)\boldsymbol{u}_\theta - z_n(kr)\frac{\partial}{\partial\theta}P_n^m(\cos\theta)\,{\cos \atop \sin}(m\phi)\boldsymbol{u}_\phi, \tag{4.98}$$

$$\boldsymbol{n}_{\substack{e\\o}mn} = \frac{n(n+1)}{kr}z_n(kr)P_n^m(\cos\theta)\,{\cos \atop \sin}(m\phi)\boldsymbol{u}_r + \frac{1}{kr}\frac{d}{dr}[rz_n(kr)]\frac{d}{d\theta}P_n^m(\cos\theta)\,{\cos \atop \sin}(m\phi)\boldsymbol{u}_\theta$$

$$\mp\frac{m}{kr\sin\theta}\frac{d}{dr}[rz_n(kr)]P_n^m(\cos\theta)\,{\sin \atop \cos}(m\phi)\boldsymbol{u}_\phi, \tag{4.99}$$

where $z_n(kr)$ is any one of the spherical Bessel functions $j_n(kr)$, $n_n(kr)$, $h_n^{(1)}(kr)$, and $h_n^{(2)}(kr)$.

From Eqs. (4.72) and (4.73),

$$\boldsymbol{E} = -\mu\sum_n(a_n\boldsymbol{M}_n + b_n\boldsymbol{N}_n), \tag{4.100}$$

$$H = \frac{k}{j\omega} \sum_n (a_n N_n + b_n M_n), \tag{4.101}$$

and E and H can be evaluated. The radial component M_r is always zero, and hence if all the coefficients a_n are zero, $H_r = 0$, which results in a TM field. If all the coefficients b_n are zero, then $E_r = 0$, resulting in a TE field.

By using Eqs. (4.100) and (4.101) and making all the coefficients a_n zero, the components of a TM field in the spherical polar coordinates may be written as

$$E_r = -\mu n(n+1) Y_{mn} \frac{z_n(kr)}{kr} e^{j\omega t}, \tag{4.102}$$

$$E_\theta = -\mu \frac{\partial Y_{mn}}{\partial \theta} \frac{1}{kr} [krz_n(kr)]' e^{j\omega t}, \tag{4.103}$$

$$E_\phi = \frac{-\mu}{\sin\theta} \frac{\partial Y_{mn}}{\partial \phi} \frac{1}{kr} [krz_n(kr)]' e^{j\omega t}, \tag{4.104}$$

$$H_r = 0, \tag{4.105}$$

$$H_\theta = \frac{k}{j\omega} \frac{1}{\sin\theta} \frac{\partial Y_{mn}}{\partial \phi} z_n(kr) e^{j\omega t}, \tag{4.106}$$

$$H_\phi = \frac{-k}{j\omega} \frac{\partial Y_{mn}}{\partial \theta} z_n(kr) e^{j\omega t}, \tag{4.107}$$

where

$$Y_{mn} = [A_{mn} \cos(m\phi) + B_{mn} \sin(m\phi)] P_n^m(\cos\theta). \tag{4.108}$$

Similarly, by making all the coefficients b_n zero, the components of a TE field in the spherical polar coordinates may be written as

$$E_r = 0, \tag{4.109}$$

$$E_\theta = \frac{-\mu}{\sin\theta} \frac{\partial Y_{mn}}{\partial \phi} z_n(kr) e^{j\omega t}, \tag{4.110}$$

$$E_\phi = \frac{\partial Y_{mn}}{\partial \theta} z_n(kr) e^{j\omega t}, \tag{4.111}$$

$$H_r = +\frac{n(n+1)}{j\omega} Y_{mn} \frac{1}{r} z_n(kr) e^{j\omega t}, \tag{4.112}$$

$$H_\theta = \frac{1}{j\omega} \frac{\partial Y_{mn}}{\partial \theta} \frac{1}{r} [krz_n(kr)]' e^{j\omega t}, \tag{4.113}$$

$$H_\phi = \frac{1}{j\omega} \frac{1}{\sin\theta} \frac{\partial Y_{mn}}{\partial \phi} \frac{1}{r} [krz_n(kr)]' e^{j\omega t}. \tag{4.114}$$

A field which is a superimposition of a TE field and a TM field can represent any general electromagnetic field in the spherical coordinates.

PROBLEMS

1 Consider a coaxial transmission line having an inner cylindrical conductor of radius a, an outer cylindrical conductor of radius b, and of length l. One end of the line is short circuited and a generator having a constant voltage V_1 is applied between the two conductors at the other end. Assuming V to be independent of ϕ, solve Laplace's equation, and obtain an expression for V at point (ρ, ϕ, z) as

$$V = V_1 \frac{z}{l} \ln \left(\frac{b/\rho}{b/a}\right).$$

2 If an uncharged conducting cylinder of radius R is placed in an originally uniform electric field E_0, obtain expressions for V in the medium around the cylinder. Draw the electric field lines and equipotentials.

3 If a dielectric sphere of radius b, having permittivity $\epsilon_2 = \epsilon_{r_2}\epsilon_0$, is placed in an originally uniform electric field E_0, show that the potential distribution outside the sphere is

$$V_e = E_0\left[\rho + \frac{b^3}{\rho^2}\left(\frac{\epsilon_{r_1} - \epsilon_{r_2}}{2\epsilon_{r_1} + \epsilon_{r_2}}\right)\right] \cos \theta$$

and the potential distribution inside the sphere is

$$V_i = \frac{3\epsilon_{r_1} E_0 \rho \cos \theta}{2\epsilon_{r_1} + \epsilon_{r_2}},$$

where ϵ_{r_1} = relative permittivity of medium outside the sphere.

4 If a dielectric sphere having permittivity ϵ_1 is placed in a dielectric medium of permittivity ϵ_0, write the expressions for the components of the electromagnetic field inside and outside the sphere. Applying the proper boundary conditions, obtain the characteristic equation.

5 In a region bounded by two perfectly conducting cones of infinite length of angles θ_1 and θ_2, write the expressions for the electromagnetic field. Apply the proper boundary conditions and obtain the characteristic equation.

5 Transmission Lines and Waveguides

5.1 Transmission of Electromagnetic Energy

In any electrical communication system, it is necessary to transfer electromagnetic energy from one place to another. The process of transferring such energy is usually called transmission which takes place through an unbounded or bounded medium. Examples of unbounded media are free space, ionosphere, troposphere, and under the ocean. Examples of bounded media are many, among which transmission lines or waveguides are the most important. These transmission lines or waveguides guide the electromagnetic waves from one point to another, without radiating energy. There are many types of transmission lines or waveguides, e.g., two-wire metal transmission lines, coaxial metal transmission lines, hollow metal waveguides, dielectric waveguides, surface-wave transmission lines, and strip transmission lines.

Most of the transmission lines and waveguides are cylindrical. This implies that the transmission of electromagnetic energy through such lines is along the direction of the axis of a cylinder. The cylinder may have any geometrical cross-section, e.g., circle, rectangle, ellipse, straight line, two or more circles, and concentric circles (see Fig. 5.1). The cross-sectional shape and electrical properties usually do not vary along the axis, which is generally chosen as the z-axis.

5.2 Classification of Transmission Lines and Waveguides Depending on Type of Wave Solution

While obtaining the solution for the electromagnetic field in and around a transmission line or waveguide, it is usual to omit the sources. The electric and magnetic field intensities $\boldsymbol{E}$ and $\boldsymbol{H}$ satisfy the homogeneous vector Helmholtz (or wave) equations

$$\nabla^2\boldsymbol{E} + k^2\boldsymbol{E} = 0, \qquad \nabla^2\boldsymbol{H} + k^2\boldsymbol{H} = 0, \tag{5.1}$$

where $k^2 = -\gamma^2$ with γ as the propagation constant of the medium. It is assumed that the electromagnetic wave is travelling along the axis of the cylindrical transmission line or waveguide, which is taken as the z-axis. Since the Helmholtz equations (5.1) are separable, it is possible to find solutions of the form $f(z)g(u, v)$, where $f(z)$ is a function of z only and $g(u, v)$ is a function of the transverse coordinates u and v. The coordinates u and v may be x and y, or ρ and ϕ, or other suitable transverse coordinates, depending on whether the cross-section is a rectangle, a straight line, a circle, or any other shape.

We have seen in Sections (4.3)–(4.6) that, in the rectangular or the cylindrical system of coordinates, the solution of the scalar wave equation as also the solution of the vector wave equation contains a factor of the form

$$C_5 \exp(a_z z) + C_6 \exp(-a_z z)$$

which gives the dependence on the z-coordinate. This is still true even if the cross-section of the

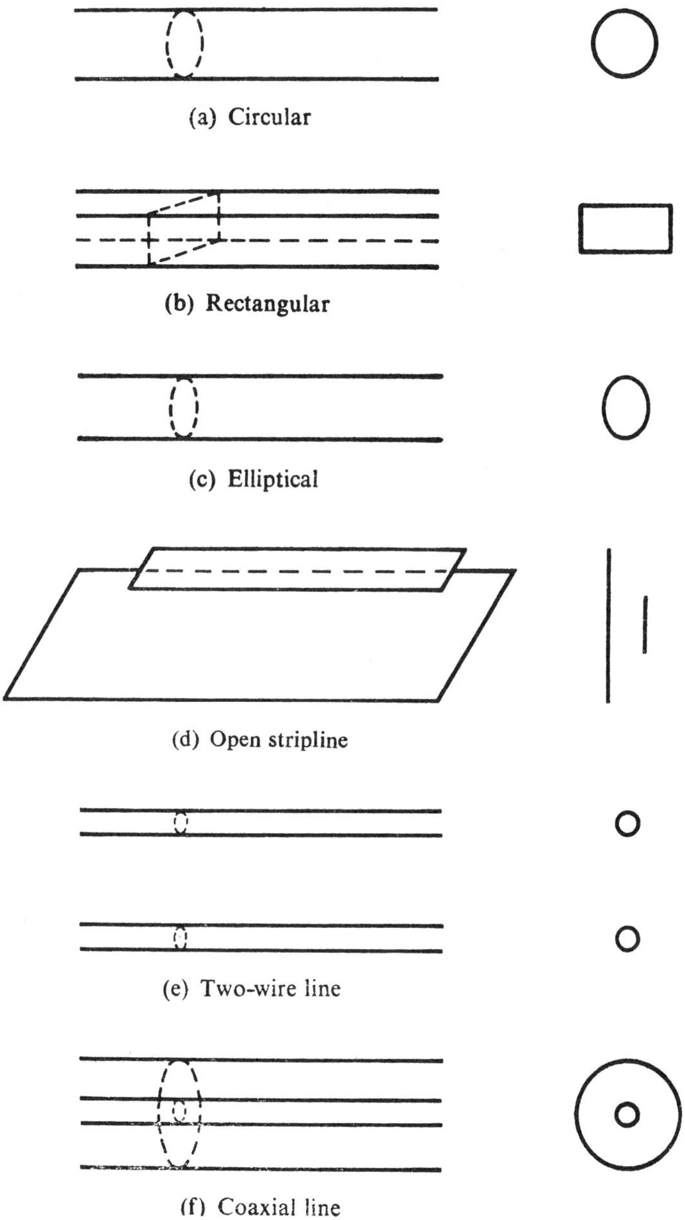

Fig. 5.1 Different types of transmission lines and waveguides.

cylinder is of a shape other than rectangle or circle. Hence, for a wave propagating along the z-axis, which is the axis of the cylinder, we may assume the z dependence to be exp $(\pm j\bar{\beta}z)$. Since the time dependence has been assumed as exp $(j\omega t)$, this assumption leads to wave solutions of the form exp $(j\omega t)$ or of the form cos $(\omega t \pm \bar{\beta}z)$ and sin $(\omega t \pm \bar{\beta}z)$. A wave propagating in the positive z-direction can be represented by exp $[j(\omega t - \bar{\beta}z)]$, and a wave propagating in the negative z-direction can be represented by exp $[+j(\omega t + \bar{\beta}z)]$. When there is attenuation, $j\bar{\beta}$ is replaced by

$\bar{\gamma} = \bar{\alpha} + j\bar{\beta}$. With an assumed z dependence exp $(-\bar{\beta}z)$, the del operator becomes

$$\nabla = \boldsymbol{u}_x\frac{\partial}{\partial x} + \boldsymbol{u}_y\frac{\partial}{\partial y} + \boldsymbol{u}_z\frac{\partial}{\partial z} = \nabla_t + \nabla_z = \nabla_t - j\bar{\beta}\boldsymbol{u}_z, \tag{5.2}$$

where

$$\nabla_t = \boldsymbol{u}_x\frac{\partial}{\partial x} + \boldsymbol{u}_y\frac{\partial}{\partial y} \quad \text{in rectangular coordinates}$$

$$= \boldsymbol{u}_\rho\frac{\partial}{\partial \rho} + \boldsymbol{u}_\phi\frac{1}{\rho}\frac{\partial}{\partial \phi} \quad \text{in cylindrical coordinates} \ldots .$$

Maxwell's equations can be considerably simplified by decomposing all fields into transverse and axial components and separating out the z dependence and by suppressing the time dependence $e^{j\omega t}$. Thus

$$\boldsymbol{E}(x, y, z) = \boldsymbol{E}_t(x, y, z) + \boldsymbol{E}_z(x, y, z) = [\boldsymbol{e}_t(x, y) + \boldsymbol{e}_z(x, y)] \exp(-j\bar{\beta}z), \tag{5.3}$$

$$\boldsymbol{H}(x, y, z) = \boldsymbol{H}_t(x, y, z) + \boldsymbol{H}_z(x, y, z) = [\boldsymbol{h}_t(x, y) + \boldsymbol{h}_z(x, y)] \exp(-j\bar{\beta}z), \tag{5.4}$$

where $\boldsymbol{E}_t$, $\boldsymbol{H}_t$ are the transverse components and $\boldsymbol{E}_z$, $\boldsymbol{H}_z$ are the axial components.

Maxwell's equations

$$\nabla \times \boldsymbol{E} = -j\omega\mu\boldsymbol{H}, \tag{5.5}$$

$$\nabla \times \boldsymbol{H} = j\omega\epsilon\boldsymbol{E} \tag{5.6}$$

become

$$(\nabla_t - j\bar{\beta}\boldsymbol{u}_z) \times (\boldsymbol{e}_t + \boldsymbol{e}_z) \exp(-j\bar{\beta}z) = -j\omega\mu(\boldsymbol{h}_t + \boldsymbol{h}_z) \exp(-j\bar{\beta}z), \tag{5.7}$$

$$(\nabla_t - j\bar{\beta}\boldsymbol{u}_z) \times (\boldsymbol{h}_t + \boldsymbol{h}_z) \exp(-j\bar{\beta}z) = j\omega\epsilon(\boldsymbol{e}_t + \boldsymbol{e}_z) \exp(-j\bar{\beta}z), \tag{5.8}$$

$$\boldsymbol{u}_z \times \boldsymbol{e}_z = 0, \qquad \nabla_t \times \boldsymbol{e}_z = -\boldsymbol{u}_z \times \nabla_t \boldsymbol{e}_z.$$

Also, $\nabla_t \times \boldsymbol{e}_t$ is directed along the z-axis only, and $\boldsymbol{u}_z \times \boldsymbol{e}_t$ and $\nabla_t \times \boldsymbol{e}_z$ have transverse components only. Hence, when the transverse and axial components of Eqs. (5.7) and (5.8) are equated, we obtain

$$\nabla_t \times \boldsymbol{e}_t = -j\omega\mu\boldsymbol{h}_z, \tag{5.9}$$

$$\nabla_t \times \boldsymbol{e}_z - j\bar{\beta}\boldsymbol{u}_z \times \boldsymbol{e}_t = -\boldsymbol{u}_z \times \nabla_t\boldsymbol{e}_z - j\bar{\beta}\boldsymbol{u}_z \times \boldsymbol{e}_t = -j\omega\mu\boldsymbol{h}_t, \tag{5.10}$$

$$\nabla_t \times \boldsymbol{h}_t = j\omega\epsilon\boldsymbol{e}_z, \tag{5.11}$$

$$\boldsymbol{u}_z \times \nabla_t\boldsymbol{h}_z + j\bar{\beta}\boldsymbol{u}_z \times \boldsymbol{h}_t = -j\omega\epsilon\boldsymbol{e}_t. \tag{5.12}$$

The divergence equation $\nabla \cdot \boldsymbol{B} = 0$ becomes

$$\nabla \cdot \boldsymbol{B} = \nabla \cdot \mu\boldsymbol{H} = (\nabla_t - j\bar{\beta}\boldsymbol{u}_z) \cdot (\boldsymbol{h}_t + \boldsymbol{h}_z) \exp(-j\bar{\beta}z)$$

$$= (\nabla_t \cdot \boldsymbol{h}_t - j\bar{\beta}\boldsymbol{u}_z \cdot \boldsymbol{h}_z)\mu \exp(-j\bar{\beta}z) = 0$$

or

$$\nabla_t \cdot \boldsymbol{h}_t = j\bar{\beta}\boldsymbol{h}_z. \tag{5.13}$$

Similarly, $\nabla \cdot \boldsymbol{D} = 0$ becomes

$$\nabla_t \cdot \boldsymbol{e}_t = j\beta e_z. \tag{5.14}$$

This reduced form of Maxwell's equations is very useful for the solution of many transmission line and waveguide problems.

For a large variety of practical transmission lines or waveguides, the boundary conditions can be satisfied by fields that do not have all the components of $\boldsymbol{E}$ and $\boldsymbol{H}$. In fact, the solution of interest for conducting transmission lines is a transverse electromagnetic (or TEM) wave with transverse components only, that is, with $E_z = H_z = 0$. For conducting waveguides, the solutions of interest are the transverse electric (or TE) waves with $E_z = 0$, but $H_z \neq 0$, and the transverse magnetic (or TM) waves with $H_z = 0$, but $E_z \neq 0$. The following classification is therefore of practical interest:

(i) Transverse electromagnetic (TEM) wave with $E_z = H_z = 0$.

(ii) Transverse electric (TE) wave with $E_z = 0$, but $H_z \neq 0$.

(iii) Transverse magnetic (TM) wave with $H_z = 0$, but $E_z \neq 0$.

There are, however, some practical waveguides, for example, dielectric waveguides, where a TE (or TM) wave does not by itself satisfy all the boundary conditions. In such cases, linear combinations of TE and TM waves, which may satisfy the boundary conditions, are used. Such waves are called *hybrid waves*. In some special cases, an infinite number of TE (or TM or hybrid) waves may have to be considered to satisfy the boundary conditions.

5.3 TEM Waves

For TEM waves, $e_z = h_z = 0$, and Eqs. (5.9)–(5.14) reduce to

$$\nabla_t \times \boldsymbol{e}_t = 0, \tag{5.15a}$$

$$\beta \boldsymbol{u}_z \times \boldsymbol{e}_t = \omega\mu \boldsymbol{h}_t, \tag{5.15b}$$

$$\nabla_t \times \boldsymbol{h}_t = 0, \tag{5.15c}$$

$$\beta \boldsymbol{u}_z \times \boldsymbol{h}_t = -\omega\epsilon \boldsymbol{e}_t, \tag{5.15d}$$

$$\nabla_t \cdot \boldsymbol{h}_t = 0, \tag{5.15e}$$

$$\nabla_t \cdot \boldsymbol{e}_t = 0. \tag{5.15f}$$

Equation (5.15a) means that the circulation or line integral of $\boldsymbol{e}_t$ around any closed path in the transverse plane is zero. This is so because there is no axial magnetic field H_z passing through such a contour. Hence, $\boldsymbol{e}_t$ can be expressed as the gradient of a scalar function, which may be called a potential function. Thus

$$\boldsymbol{e}_t = -\nabla_t \Phi. \tag{5.16}$$

Substituting Eq. (5.16) in Eq. (5.15f), we obtain

$$\nabla_t^2 \Phi = 0. \tag{5.17}$$

Equation (5.17) is the two-dimensional Laplace equation, where Φ is a function of the transverse coordinates u, v. Hence

$$\boldsymbol{E}_t = \boldsymbol{e}_t \exp(-j\beta z) = -\nabla_t \Phi \exp(-j\beta z). \tag{5.18}$$

But $\boldsymbol{E}_t$ also satisfies the three-dimensional wave equation

$$\nabla^2 \boldsymbol{E}_t + k^2 \boldsymbol{E}_t = 0. \tag{5.19}$$

Since $\nabla = \nabla_t - j\bar{\beta}\boldsymbol{u}_z$, $\nabla^2 = \nabla_t^2 - \bar{\beta}^2$, Eq. (5.19) reduces to

$$\nabla_t^2 \boldsymbol{E}_t + (k^2 - \bar{\beta}^2)\boldsymbol{E}_t = 0 \tag{5.20a}$$

or

$$\nabla_t[\nabla_t^2 \Phi + (k^2 - \bar{\beta}^2)\Phi] = 0. \tag{5.20b}$$

Comparing Eqs. (5.17) and (5.20b), it can be seen that $\bar{\beta} = \pm k$ for a TEM wave. This means that the axial propagation constant $\bar{\gamma} = j\bar{\beta}$ is the same as the intrinsic propagation constant $jk = \bar{\gamma}$ of the medium. For a medium with no losses, $\bar{\beta} = \beta = \omega\sqrt{\mu\epsilon}$ = intrinsic phase constant of the medium, and $\bar{v} = \omega/\bar{\beta} = 1/\sqrt{\mu\epsilon}$ = the intrinsic phase velocity of the medium. The transverse magnetic field $\boldsymbol{h}_t$ can be obtained from Eq. (5.15b) so that

$$\omega\mu \boldsymbol{h}_t = \bar{\beta}\boldsymbol{u}_z \times \boldsymbol{e}_t$$

or

$$\boldsymbol{u}_z \times \boldsymbol{e}_t = \frac{\omega\mu}{\bar{\beta}}\boldsymbol{h}_t = \frac{\omega\mu}{k}\boldsymbol{h}_t \tag{5.21}$$

and

$$k = \frac{\gamma}{j} = \frac{[j\omega\mu(\sigma + j\omega\epsilon)]^{1/2}}{j} = \frac{[\omega\mu(\sigma + j\omega\epsilon)]^{1/2}}{j}$$

so that

$$\frac{\omega\mu}{k} = \frac{\omega\mu\sqrt{j}}{[\omega\mu(\sigma + j\omega\epsilon)]^{1/2}} = \left(\frac{j\omega\mu}{\sigma + j\omega\epsilon}\right)^{1/2} = Z = \eta = \text{intrinsic impedance of the medium.}$$

Hence, for a TEM wave,

$$\frac{e_t}{h_t} = Z = \eta \tag{5.22}$$

which implies that the wave impedance looking in the z-direction is equal to the intrinsic impedance of the medium. This impedance is called the *characteristic impedance* of the transmission line supporting the TEM wave.

5.4 TE Waves

For TE waves, $E_z = 0$, but $H_z \neq 0$. The magnetic field $\boldsymbol{H}$ satisfies the wave equation

$$\nabla^2 \boldsymbol{H} + k^2 \boldsymbol{H} = 0. \tag{5.23}$$

By using Eqs. (5.2) and (5.4) in Eq. (5.23), the transverse and axial components can be separated. Thus

$$\nabla_t^2 h_z + k_c^2 h_z = 0, \tag{5.24}$$

$$\nabla_t^2 \boldsymbol{h} + k_c^2 \boldsymbol{h} = 0, \tag{5.25}$$

where

$$k_c^2 = k^2 - \beta^2. \tag{5.26}$$

k_c^2, which is called the eigenvalue, is determined from the appropriate boundary conditions, depending on the geometry of the waveguide.

As $\boldsymbol{e}_z = 0$, Maxwell's equations given by Eqs. (5.9)–(5.14) become

$$\nabla_t \times \boldsymbol{e}_t = -j\omega\mu \boldsymbol{h}_z, \tag{5.27a}$$

$$\beta \boldsymbol{u}_z \times \boldsymbol{e}_t = \omega\mu \boldsymbol{h}_t, \tag{5.27b}$$

$$\nabla_t \times \boldsymbol{h}_t = 0, \tag{5.27c}$$

$$\boldsymbol{u}_z \times \nabla_t h_z + j\beta u_z \times \boldsymbol{h}_t = -j\omega\epsilon \boldsymbol{e}_t, \tag{5.27d}$$

$$\nabla_t \bullet \boldsymbol{h}_t = j\beta \boldsymbol{h}_z, \tag{5.27e}$$

$$\nabla_t \bullet \boldsymbol{e}_t = 0. \tag{5.27f}$$

Taking the curl of Eq. (5.27c), we have

$$\nabla_t \times \nabla_t \times \boldsymbol{h}_t = \nabla_t \nabla_t \bullet \boldsymbol{h}_t - \nabla_t^2 \boldsymbol{h}_t = 0. \tag{5.28}$$

Using Eqs. (5.25) and (5.27e) in Eq. (5.28), we obtain

$$\boldsymbol{h}_t = \frac{-j\beta}{k_c^2} \nabla_t h_z. \tag{5.29}$$

From Eq. (5.27b),

$$\beta \boldsymbol{u}_z \times (\boldsymbol{u}_z \times \boldsymbol{e}_t) = \beta[(\boldsymbol{u}_z \bullet \boldsymbol{e}_t) \bullet \boldsymbol{u}_z - (\boldsymbol{u}_z \bullet \boldsymbol{u}_z) e_t] = -\beta \boldsymbol{e}_t = \omega\mu \boldsymbol{u}_z \times \boldsymbol{h}_t \tag{5.30}$$

or

$$\boldsymbol{e}_t = \frac{-\omega\mu}{\beta} \boldsymbol{u}_z \times \boldsymbol{h}_t = \frac{-k}{\beta} Z \boldsymbol{u}_z \times \boldsymbol{h}_t, \tag{5.31}$$

where

$$Z = \left(\frac{j\omega\mu}{\sigma + j\omega\epsilon}\right)^{1/2}.$$

The factor kZ/β has the dimension of impedance and is called wave impedance or characteristic impedance of TE (or H) modes, and can be designated Z_h so that

$$Z_h = \frac{kZ}{\beta}.$$

Thus, the solution for TE waves may be summarized as follows. Solve for h_z from Eq. (5.24) so that

$$\boldsymbol{h}_t = \frac{-j\beta}{k_c^2} \nabla_t h_z, \qquad \boldsymbol{e}_t = -Z_h \boldsymbol{u}_z \times \boldsymbol{h}_t,$$

where

$$\bar{\beta} = (k^2 - k_c^2)^{1/2}, \qquad Z_h = \frac{kZ}{\bar{\beta}},$$

k_c being determined from the boundary conditions.

5.5 TM Waves

For TM (or E) waves, $h_z = 0$, but $e_z \neq 0$. The derivations of the equations are similar to those of TE (or H) waves and may be summarized as follows. Obtain a solution for e_z which satisfies

$$\nabla_t^2 e_z + k_c^2 e_z = 0 \tag{5.32}$$

subject to the boundary conditions imposed. This determines the eigenvalue k_c^2. Then, the transverse fields are given by

$$\boldsymbol{E}_t = \boldsymbol{e}_t \exp(\mp j\bar{\beta}z) = \frac{-j\bar{\beta}}{k_c^2}\nabla_t e_z \exp(\mp j\bar{\beta}z), \tag{5.33}$$

$$\boldsymbol{H}_t = \pm \boldsymbol{h}_t \exp(\mp j\bar{\beta}z) = \pm Y_e \boldsymbol{u}_z \times \boldsymbol{e}_t \exp(\mp j\bar{\beta}z), \tag{5.34}$$

where $\bar{\beta} = (k^2 - k_c^2)^{1/2}$, and the wave admittance Y_e for TM waves is given by

$$Y_e = \frac{1}{Z_e} = \frac{k}{\bar{\beta}}Y = \frac{k}{\bar{\beta}Z}. \tag{5.35}$$

The dual nature of TE and TM waves is shown by the relation

$$Z_e Z_h = Z^2. \tag{5.36}$$

5.6 Transmission Lines: Field Analysis

A transmission line is usually defined as a guiding structure that supports the TEM wave, and a waveguide as a structure that does not support the TEM wave. Transmission lines which come under this definition include any cylindrical system with two or more conductors. Cross-sections of typical transmission lines are given in Fig. 5.2. Let us assume at first that the conductors are

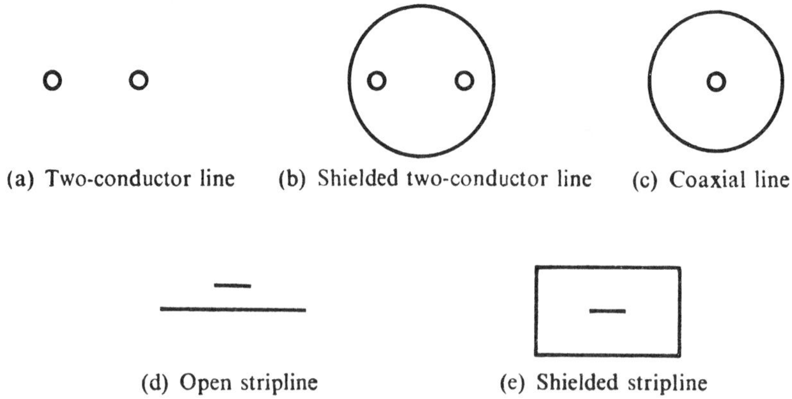

Fig. 5.2 Cross-sections of typical transmission lines.

perfectly conducting and that they are placed in air or free space so that $\epsilon = \epsilon_0$ and $\mu = \mu_0$. Later (in this chapter), we shall consider the effect of conductor losses.

Figure 5.3 shows the cross-section of a general two-conductor transmission line, together with the electric and magnetic field lines. Let $V_0/2$ and $-V_0/2$ be the potentials of the two conductors S_2 and S_1, respectively. For a TEM wave,

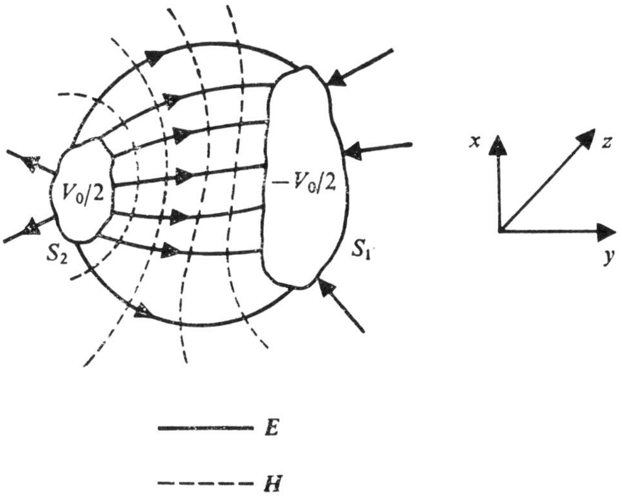

Fig. 5.3 Cross-section of general two-conductor transmission line showing electric and magnetic field lines.

$$\nabla_t^2 \Phi = 0, \tag{5.37}$$

where Φ is the scalar potential which is a function of the coordinates in the transverse plane. The solution of Eq. (5.37) should satisfy the boundary conditions

$$\Phi = \begin{cases} \dfrac{V_0}{2} & \text{on } S_2 \\ -\dfrac{V_0}{2} & \text{on } S_1 \end{cases} \tag{5.38}$$

If a solution for Φ is possible, a TEM mode field solution is also possible. Since Eq. (5.37) is Laplace's equation, the solution for Φ is an electrostatic problem, and the equation can be solved easily for simple geometries, as shown in Fig. 5.2.

If the propagation is in the z-direction, the fields are given by Eqs. (5.18) and (5.21) as

$$\boldsymbol{E} = \boldsymbol{E}_t = \boldsymbol{e}_t \exp(-jk_0z) = -\nabla_t \Phi \exp(-jk_0z), \tag{5.39}$$

$$\boldsymbol{H} = \boldsymbol{H}_t = Y_0 \boldsymbol{u}_z \times \boldsymbol{e}_t \exp(-jk_0z), \tag{5.40}$$

where

$$Y_0 = \frac{1}{Z_0} = \sqrt{\frac{\epsilon_0}{\mu_0}}.$$

The line integral of $\boldsymbol{e}_t$ between the two conductors is

$$\int_{S_1}^{S_2} \boldsymbol{e}_t \bullet d\boldsymbol{s} = \int_{S_1}^{S_2} -\nabla_t \Phi \bullet d\boldsymbol{s} = -\int_{S_1}^{S_2} \frac{d\Phi}{ds} ds = -[\Phi(S_2) - \Phi(S_1)] = -V_0. \tag{5.41}$$

Since the line integral of $\boldsymbol{e}_t$ between S_1 and S_2 is independent of the path chosen, because $\boldsymbol{e}_t$ is the gradient of a scalar potential, there is a unique voltage wave

$$V = V_0 e^{-j\beta z}. \tag{5.42}$$

If one conductor carries current I_0, then the other conductor carries current $-I_0$.

The line integral of $\boldsymbol{h}_t$ around one conductor, say, S_2, gives

$$\oint_{S_2} \boldsymbol{h}_t \bullet d\boldsymbol{s} = \oint_{S_2} J_S \, ds = I_0, \tag{5.43}$$

where $\boldsymbol{J}_S$ is the surface current density on the surface of the conductor. The boundary conditions on the surface of the conductors S_1 and S_2 are $\boldsymbol{n} \times \boldsymbol{e}_t = 0$ and $\boldsymbol{n} \times \boldsymbol{h}_t = \boldsymbol{J}_S$, where $\boldsymbol{n}$ is the outward unit normal vector. Since $\boldsymbol{n}$ and $\boldsymbol{h}_t$ are in the transverse plane, $\boldsymbol{J}_S$ is in the axial direction. Also, $\boldsymbol{n} \times \boldsymbol{e}_t = 0$ means that the electric field lines are normal to the surface of the conductors.

A unique current wave

$$I = I_0 e^{-j\beta z}$$

is associated with the magnetic field. Since Φ is a solution of Laplace's equation, it is independent of frequency, and hence the transverse fields $\boldsymbol{e}_t$ and $\boldsymbol{h}_t$ are also independent of frequency and are, in fact, the static fields that exist if the potential difference between the two conductors is V_0 and the currents in the two conductors are $+I_0$ and $-I_0$.

The electric and magnetic field lines are orthogonal to each other, and hence the magnetic field lines coincide with the equipotentials. As the electric and magnetic fields $\boldsymbol{e}_t$ and $\boldsymbol{h}_t$ are independent of frequency, transmission lines which can carry the TEM mode can be used at all frequencies—from very low frequencies to very high frequencies. However, such lines have other disadvantages such as radiation and attenuation losses at higher frequencies.

5.7 Microwave Transmission Lines

A conventional open two-wire conducting transmission line which is used at low frequencies is not very convenient at microwave frequencies for the reason that, as the frequency increases, there is more and more power loss by radiation. Figure 5.4 shows a parallel-wire transmission line in the $y'z'$-plane. Here, (r, θ, ϕ) are the spherical coordinates of the point of observation, P. The currents, as also the current densities in wires 1 and 2 at any position z, have equal magnitude but opposite signs. Let the current densities in wires 1 and 2 be $+\boldsymbol{J}$ and $-\boldsymbol{J}$, respectively. The magnetic potentials A_1 and A_2 at a point $P(r, \theta, \phi)$, due to these currents in wires 1 and 2, satisfy the inhomogeneous wave equations

$$\nabla^2 \boldsymbol{A}_1(r_1, t) - \mu\epsilon \frac{\partial^2 \boldsymbol{A}_1}{\partial t^2}(r_1, t) = -\boldsymbol{J}(z, t), \tag{5.44}$$

$$\nabla^2 \boldsymbol{A}_2(r_2, t) - \mu\epsilon \frac{\partial^2 \boldsymbol{A}_2}{\partial t^2}(r_2, t) = \boldsymbol{J}(z, t), \tag{5.45}$$

where $\boldsymbol{r}_1$ and $\boldsymbol{r}_2$ are the vectors joining the current elements $I\,dz'$ in wire 1 and $-I\,dz'$ in wire 2

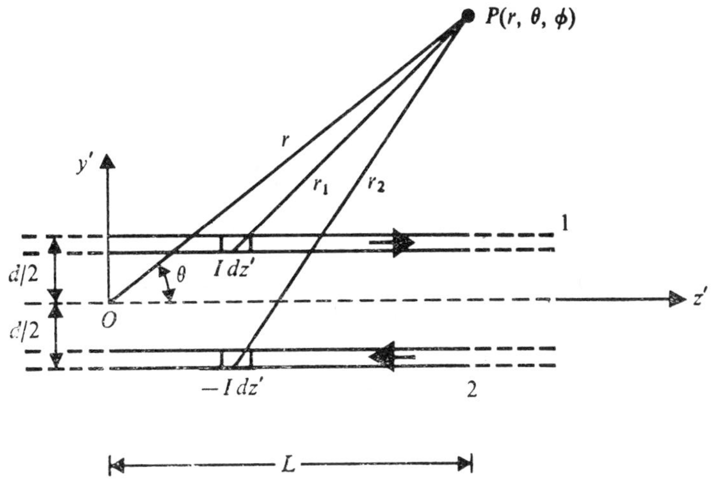

Fig. 5.4 Parallel-wire transmission line.

to the point P. Assuming the time variation as $e^{j\omega t}$, Eqs. (5.44) and (5.45) become

$$\nabla^2 \boldsymbol{A}_1 + \beta^2 \boldsymbol{A}_1 = -\boldsymbol{J}, \tag{5.46}$$

$$\nabla^2 \boldsymbol{A}_2 + \beta^2 \boldsymbol{A}_2 = \boldsymbol{J}, \tag{5.47}$$

where $\beta^2 = \omega^2 \mu\epsilon$.

The solutions of Eqs. (5.46) and (5.47) are of the form

$$\boldsymbol{A}_1 = \boldsymbol{u}_z \int_0^L \frac{I(z')\exp(-j\beta r_1)}{4\pi r_1}\,dz', \tag{5.48}$$

$$\boldsymbol{A}_2 = -\boldsymbol{u}_z \int_0^L \frac{I(z')\exp(-j\beta r_2)}{4\pi r_2}\,dz', \tag{5.49}$$

where $I(z')$ is the current per unit length of the wire and L is the length of the wire.

At low frequencies, the distance d of separation between the wires is very small compared to the wavelength, and $r_1 \approx r_2 \approx r$. When r is very much greater than a wavelength, $\boldsymbol{A}_1$ and $\boldsymbol{A}_2$ are very nearly equal in magnitude, but almost 180° out of phase. The resultant radiated field is then very nearly zero. At microwave frequencies, the distance d between the two devices becomes comparable with the wavelength. Therefore, for a large r, though $\boldsymbol{A}_1$ and $\boldsymbol{A}_2$ may be nearly equal in magnitude, their phase difference may differ very much from 180°. The total magnetic vector potential $\boldsymbol{A}$ at P is given by

$$\boldsymbol{A} = \boldsymbol{A}_1 + \boldsymbol{A}_2 = \boldsymbol{u}_z \frac{1}{4\pi} \int_0^L I(z')\left(\frac{\exp(-j\beta r_1)}{r_1} - \frac{\exp(-j\beta r_2)}{r_2}\right) dz'. \tag{5.50}$$

When r is very large compared with L and d,

$$r_1 \approx r - \frac{d}{2}\sin\theta\sin\phi - z'\cos\theta,$$

$$r_2 \approx r + \frac{d}{2} \sin \theta \sin \phi - z' \cos \theta,$$

where ϕ is defined by $y = r \sin \theta \sin \phi$. Hence

$$A = A_1(r) + A_2(r) \approx \boldsymbol{u}_z[\frac{j}{2\pi} \sin (\frac{\beta d}{2} \sin \theta \sin \phi) \frac{\exp(-j\beta r)}{r} \int_0^L I(z') \exp(j\beta z' \cos \theta)\, dz']. \tag{5.51}$$

If $\boldsymbol{E}$ and $\boldsymbol{H}$ are derived from $\boldsymbol{A}$, and if the Poynting vector is integrated over a large spherical surface completely enclosing the transmission line, it can be seen that the total radiated power flowing away from the line is nonzero. Hence, the efficiency of the transmission line is reduced, because the radiated power adds to the power loss in the line.

It can be seen from Eq. (5.51) that the radiated power can be reduced by making βd very small or, in other words, by making d/λ very small. But a very small d may also reduce the power handling capacity of the transmission line because of the possibility of arcing over the wires.

The foregoing analysis has shown that the open parallel-wire transmission line is not suitable at microwave frequencies.

By using a coaxial transmission line, the problem of radiation can be avoided. Air-filled coaxial lines are extensively used for transmission of microwave power. However, the electric and magnetic fields inside the coaxial line vary as $1/r$, where r is the radial distance, and hence the central conductor is located in a region of maximum electric and magnetic fields. This makes the ohmic loss at the central conductor very high. Therefore, when a very high efficiency of transmission is required, a coaxial line is not suitable.

Other transmission lines that carry the TEM mode and are used extensively include open striplines and shielded striplines.

Waveguides of various cross-sections bounded by highly conducting metal walls constitute a very large class of microwave transmission lines. Since these waveguides have a single conductor, they do not support the TEM mode, but support TE, TM, and hybrid modes. Similarly, other types of microwave transmission lines such as surface wavelines and dielectric waveguides also do not support the TEM mode.

5.8 Coaxial Transmission Line: TEM Mode

A coaxial transmission line, which is extensively used in microwave circuits, is formed by a pair of concentric circular cylinders, as shown in Fig. 5.5. Any point P in the line can be represented

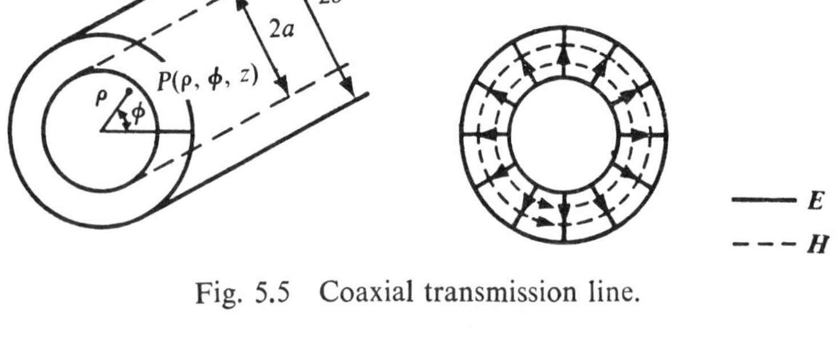

Fig. 5.5 Coaxial transmission line.

by the cylindrical coordinates (ρ, ϕ, z). Let a and b be the radii of the inner and outer cylinders, respectively, and the propagation of the wave be assumed to be in the z-direction.

The TEM wave is the dominant mode of propagation in the coaxial line. To derive the field components, the solution of the Laplace equation

$$\nabla_t^2 \Phi = 0 \tag{5.52}$$

in the transverse direction (ρ, ϕ) has to be obtained. In the cylindrical coordinates,

$$\nabla_t^2 \Phi = \frac{1}{\rho}\frac{\partial}{\partial \rho}\left(\rho \frac{\partial \Phi}{\partial \rho}\right) + \frac{1}{\rho^2}\frac{\partial^2 \Phi}{\partial \phi^2} = 0. \tag{5.53}$$

If the potential function Φ is independent of the coordinate ϕ, then

$$\frac{1}{\rho}\frac{\partial}{\partial \rho}\left(\rho \frac{\partial \Phi}{\partial \rho}\right) = 0 \tag{5.54}$$

whose solution is of the form

$$\Phi = C_1 \ln \rho + C_2. \tag{5.55}$$

If V_0 is the potential difference between the two concentric cylinders, then $\Phi = V_0$ at $r = a$, and $\Phi = 0$ at $r = b$. Therefore

$$V_0 = C_1 \ln a + C_2, \qquad 0 = C_1 \ln b + C_2.$$

Hence

$$C_2 = -C_1 \ln b, \qquad C_1 = \frac{V_0}{\ln (a/b)}. \tag{5.56}$$

Thus

$$\Phi = V_0 \frac{\ln (\rho/b)}{\ln (a/b)}. \tag{5.57}$$

The electric and magnetic fields of the TEM mode are given by Eqs. (5.39) and (5.40):

$$\boldsymbol{E} = -\boldsymbol{u}_\rho \frac{\partial \Phi}{\partial \rho} e^{-jkz} = \frac{-V_0}{\ln (a/b)} \frac{\boldsymbol{u}_\rho}{\rho} e^{-jkz} = \frac{V_0}{\ln (b/a)} \frac{\boldsymbol{u}_\rho}{\rho} e^{-jkz}, \tag{5.58}$$

$$\boldsymbol{H} = Y_0 \boldsymbol{u}_z \times \boldsymbol{e}_t e^{-jkz} = \frac{Y_0 V_0}{\ln (b/a)} \frac{\boldsymbol{u}_\phi}{\rho} e^{-jkz}, \tag{5.59}$$

where $Y_0 = \sqrt{\epsilon/\mu}$, $k = \omega\sqrt{\mu\epsilon}$. The voltage wave associated with the electric field is

$$V = V_0 e^{-jkz}. \tag{5.60}$$

The surface current density on the inner conductor is

$$J_S = \boldsymbol{n} \times \boldsymbol{H} = \boldsymbol{u}_\rho \times \boldsymbol{H} = \frac{Y_0 V_0}{\ln (b/a)} \frac{\boldsymbol{u}_z}{a} e^{-jkz}. \tag{5.61}$$

The total current I_0 is given by

$$I_0 = \frac{Y_0 V_0}{a \ln (b/a)} \int_0^{2\pi} a \, d\phi = \frac{Y_0 V_0 2\pi}{\ln (b/a)}. \tag{5.62}$$

Hence, Eq. (5.59) becomes

$$\boldsymbol{H} = \frac{\boldsymbol{u}_\phi}{2\pi\rho} I_0 e^{-jkz}, \tag{5.63}$$

where I_0 is the total current through the outer cylinder. The current wave associated with the magnetic field is

$$I = I_0 e^{-jkz}. \tag{5.64}$$

The power or rate of energy flow along the line is given by

$$P_{\mathrm{T}} = \tfrac{1}{2} \operatorname{Re} \int_{\rho=a}^{b} \int_{\phi=0}^{2\pi} \boldsymbol{E} \times \boldsymbol{H}^* \boldsymbol{u}_\rho \rho \, d\rho \, d\phi$$

$$= \tfrac{1}{2} \frac{Y_0 V_0^2}{[\ln (b/a)]^2} \int_{\rho=a}^{b} \int_{\phi=0}^{2\pi} \frac{d\phi \, d\rho}{\rho} = \frac{\pi Y_0 V_0^2}{\ln (b/a)} = \tfrac{1}{2} V_0 I_0 = \tfrac{1}{2} \operatorname{Re} (VI^*). \tag{5.65}$$

The ratio of V to I of an infinitely long line is defined as the characteristic impedance of the coaxial line given by

$$Z_c = \frac{1}{2\pi} \frac{1}{Y_0} \ln \left(\frac{b}{a}\right) = \frac{1}{2\pi} \sqrt{\frac{\mu}{\epsilon}} \ln \left(\frac{b}{a}\right) = \frac{1}{2\pi} Z_0 \ln \left(\frac{b}{a}\right), \tag{5.66}$$

where $Z_0 = \sqrt{\mu/\epsilon} = 1/Y_0 = \eta$.

The maximum electric field intensity in the TEM mode occurs at the surface of the inner conductor and is of the magnitude

$$|E_{\max}| = \frac{V_0}{a \ln (b/a)}. \tag{5.67}$$

For a constant outer radius b and voltage V_0, the value of $|E_{\max}|$ becomes minimum when $b/a = 2.72$. This value of b/a can be used to design a coaxial transmission line to transmit maximum amount of power without causing a dielectric breakdown which depends on the maximum electric field intensity within the two coaxial cylinders.

The total average power P_{T} flowing in the z-direction is given by Eq. (5.65). If the transmission line has losses, this expression has to be multiplied by $e^{-2\alpha z}$, where α is the attenuation constant, because both $\boldsymbol{E}$ and $\boldsymbol{H}$ are multiplied by $e^{-\alpha z}$. Then, P_{T} can be written as

$$P_{\mathrm{T}} = \frac{\pi V^2}{\eta \ln (b/a)} e^{-2\alpha z},$$

and hence

$$\frac{\partial P_{\mathrm{T}}}{\partial z} = -2\alpha \frac{\pi V^2}{\eta \ln (b/a)} e^{-2\alpha z} = -2\alpha P_{\mathrm{T}}. \tag{5.68}$$

The power loss $\partial P_{\mathrm{L}}/\partial z$ per unit length of the line, due to losses in the conducting walls, is given by

$$\frac{\partial P_{\mathrm{L}}}{\partial z} = \frac{\partial}{\partial z}\,\tfrac{1}{2}\int_{\text{walls}} \mathrm{Re}\,(Z_{\mathrm{m}})|H_{\mathrm{t}}|^2\,dS, \tag{5.69}$$

where Z_{m} is the surface impedance of the metal walls and H_{t} is the tangential magnetic field, and hence the surface current on the walls. For the coaxial line, this power loss per unit length should be calculated at the walls of both the inner and outer cylindrical conductors of radii a and b. Hence, the power loss is given by

$$\begin{aligned}\frac{\partial P_{\mathrm{L}a}}{\partial z} + \frac{\partial P_{\mathrm{L}b}}{\partial z} &= \tfrac{1}{2}\,\mathrm{Re}\,(Z_{\mathrm{m}a})\int_0^{2\pi} \frac{V_0^2}{[\ln(b/a)]^2}\,\frac{1}{\eta^2}\,\frac{d\phi}{a}\,e^{-2\alpha z} \\ &\quad + \tfrac{1}{2}\,\mathrm{Re}\,(Z_{\mathrm{m}b})\int_0^{2\pi} \frac{V_0^2}{[\ln(b/a)]^2}\,\frac{1}{\eta^2}\,\frac{d\phi}{b}\,e^{-2\alpha z} \\ &= \frac{\pi V_0^2}{[\ln(b/a)]^2}\,\frac{1}{\eta^2}\,e^{-2\alpha z}\left[\frac{\mathrm{Re}\,(Z_{\mathrm{m}a})}{a} + \frac{\mathrm{Re}\,(Z_{\mathrm{m}b})}{b}\right],\end{aligned} \tag{5.70}$$

where $Z_{\mathrm{m}a}$ and $Z_{\mathrm{m}b}$ are respectively the surface impedances of the inner and outer cylinders.

The power loss $[(\partial P_{\mathrm{L}a}/\partial z) + (\partial P_{\mathrm{L}b}/\partial z)]$ is equated to $\partial P_{\mathrm{T}}/\partial z$ to obtain the attenuation constant α of the line. The expression for α thus obtained is given by

$$\alpha = \left[\frac{\mathrm{Re}\,(Z_{\mathrm{m}a})}{2a} + \frac{\mathrm{Re}\,(Z_{\mathrm{m}b})}{2b}\right]\frac{1}{\eta\,\ln(b/a)}. \tag{5.71}$$

As usual, if the inner and outer conductors are made of copper and the dielectric is air, then

$$\alpha_{\mathrm{Cu}} = \frac{0.000006\sqrt{f}\,[1/(2a) + 1/(2b)]}{\ln(b/a)} = \frac{0.00036\sqrt{f}\,[1/(2a) + 1/(2b)]}{Z_{\mathrm{c}}}\ \text{db/ft}, \tag{5.72}$$

where f is the frequency in MHz, Z_{c} the characteristic impedance of the line in ohms, and a and b are measured in feet.

As already noted, a unique voltage V and a unique current I can be associated with any transmission line supporting the TEM mode. Also, the transverse fields of a TEM mode have the same transverse variation with the coordinates as that with static fields. For these reasons, the transmission line can be uniquely described as a distributed parameter electric network. Energy storage in the magnetic field is accounted for by the series inductance L per unit length, whereas energy storage in the electric field is due to the distributed shunt capacitance C per unit length. Power loss in the conductors is taken into account by a series resistance R per unit length. Finally, the power loss in the dielectric is accounted for by including a shunt conductance G per unit length.

Suitable definitions for the parameters L, C, R, and G based on the foregoing concepts are

$$L = \frac{\mu}{I_0 I_0^*}\int_{s\,\text{(cross-section)}} \boldsymbol{H}\cdot\boldsymbol{H}^*\cdot dS, \tag{5.73}$$

$$C = \frac{\epsilon}{V_0 V_0^*}\int_{s\,\text{(cross-section)}} \boldsymbol{E}\cdot\boldsymbol{E}^*\cdot dS, \tag{5.74}$$

$$R = \frac{R_{\mathrm{m}}}{I_0 I_0^*}\int_{s\,\text{(cross-section)}} \boldsymbol{H}\cdot\boldsymbol{H}^*\cdot dS, \tag{5.75}$$

$$G = \frac{\sigma}{V_0 V_0^*} \int_{s\,(\text{cross-section})} E \cdot E^* \cdot dS, \tag{5.76}$$

where $R_m = \text{Re}\,(Z_m)$ and σ is the conductivity of the dielectric medium.

For the coaxial line, the magnetic energy stored per unit length is $\mu \boldsymbol{H} \cdot \boldsymbol{H}^* = \mu(H^2/2)$, and is given by

$$W_H = \frac{\mu}{2} \int_{s\,(\text{cross-section})} |H_\phi|^2\, dS = \frac{\mu}{8\pi^2} I_0^2 \int_{\rho=a}^{b} \int_{\phi=0}^{2\pi} \frac{1}{\rho}\, d\rho\, d\phi = \frac{\mu}{4\pi} \ln\,(b/a) I_0^2. \tag{5.77}$$

This energy is equal to $\frac{1}{2} L I_0 I_0^*$ or $\frac{1}{2} L I_0^2$, and hence

$$L = \frac{\mu}{2\pi} \ln\,(b/a). \tag{5.78}$$

Similarly, the stored electric energy per unit length is given by

$$W_E = \frac{\epsilon}{2} \int_{s\,(\text{cross-section})} E \cdot E^*\, dS = \frac{\epsilon}{2} \int_{s\,(\text{cross-section})} E^2\, dS$$

$$= \frac{\epsilon}{2} \int_{s\,(\text{cross-section})} |E_\rho|^2\, dS = \frac{\epsilon V_0^2}{2[\ln\,(b/a)]^2} \int_{\rho=a}^{b} \int_{\phi=0}^{2\pi} \frac{d\rho}{\rho}\, d\phi = \frac{\epsilon V_0^2}{2 \ln\,(b/a)} 2\pi. \tag{5.79}$$

This energy is also equal to $\frac{1}{2} C V_0^2$, and hence the capacitance is

$$C = \frac{2\pi\epsilon}{\ln\,(b/a)}. \tag{5.80}$$

The series resistance R per unit length is given by

$$R = \frac{R_m}{I_0^2} \int_{\phi=0}^{2\pi} \int_{z=0}^{1} \frac{I_0^2}{(2\pi a)^2} a\, d\phi\, dz + \frac{R_m}{I_0^2} \int_{\phi=0}^{2\pi} \int_{z=0}^{1} \frac{I_0^2}{(2\pi b)^2} a\, d\phi\, dz = \frac{R_m}{2\pi} \left(\frac{1}{a} + \frac{1}{b}\right). \tag{5.81}$$

The shunt conductance G per unit length is given by

$$G = \frac{\sigma}{V_0^2} \int_{s\,(\text{cross-section})} E^2\, dS = \frac{2\pi\sigma}{\ln\,(b/a)}. \tag{5.82}$$

$R_m = \text{Re}\,(Z_m)$ is the surface resistance of the conductor and is equal to $[\omega\mu_c/(2\sigma_c)]^{1/2}$, where σ_c is the conductivity of the conducting walls and μ_c the permeability of the conducting walls. The permeability μ_c is usually equal to μ_0 for nonmagnetic conductors which are normally used for the coaxial lines.

In general, the characteristic impedance of the line is given by

$$Z_c = \left(\frac{R + j\omega L}{G + j\omega C}\right)^{1/2} \tag{5.83a}$$

and the propagation constant γ by

$$\gamma = [(R + j\omega L)(G + j\omega C)]^{1/2}. \tag{5.83b}$$

If

$$R = G = 0, \qquad Z_c = \sqrt{\frac{L}{C}} = \frac{1}{2\pi}\sqrt{\frac{\mu}{\epsilon}} \ln\left(\frac{b}{a}\right),$$

then

$$\gamma = j\omega\sqrt{LC} = j\omega\sqrt{\mu\epsilon} = j\beta = k.$$

Practical coaxial lines usually have a characteristic impedance of 50–52 ohms or 75 ohms. Sometimes, coaxial transmission lines have a flexible solid dielectric between the inner and outer conductors, and hence are flexible, and are used extensively in microwave circuits in short lengths. For long and high power coaxial lines, such as those used in television transmitting systems, rigid metal cylinders with air between the inner and outer cylinders are used to avoid additional dielectric loss and dielectric breakdown. A very long coaxial line can be obtained by joining several sections of identical coaxial lines. These joints cause mismatch and local standing waves, which can be reduced by proper design of the junctions. An example is given in Fig. 5.6 where, for an

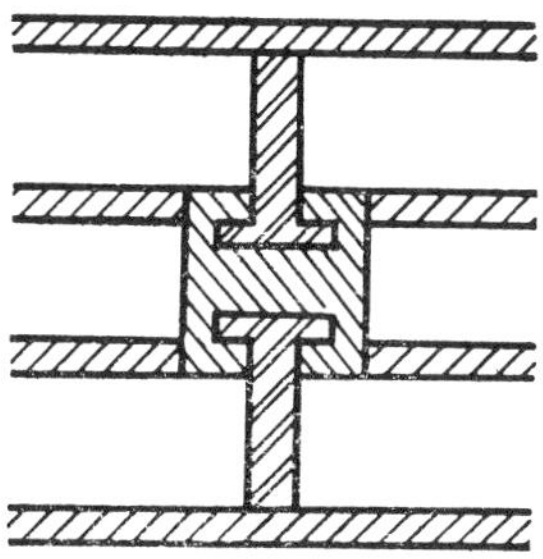

Fig. 5.6 Bead-supported coaxial-line junction.

air-filled rigid coaxial line, the capacitance due to the bead is compensated by the inductance resulting from the undercut.

5.9 TM and TE Modes in Coaxial Lines

Although the TEM mode is the dominant mode in a coaxial line, TE and TM modes also exist in it. The interest in studying the TE and TM modes lies in the design of a coaxial line that carries only the dominant TEM mode and does not carry the TE and TM modes. Also, such modes, especially the TM modes, may exist in the neighbourhood of a discontinuity.

For TM modes, e_z satisfies the scalar wave equation (see Section 4.5)

$$\nabla_t^2 e_z + k_c^2 e_z = 0, \tag{5.84}$$

where $E_z = e_z \exp(-j\bar{\beta}z)$ is the z-component of the electric field, and

$$\bar{\beta} = (k^2 - k_c^2)^{1/2}.$$

In the cylindrical coordinates, Eq. (5.84) becomes

$$\frac{1}{\rho}\frac{\partial}{\partial\rho}\left(\rho\frac{\partial e_z}{\partial\rho}\right) + \frac{1}{\rho^2}\frac{\partial^2 e_z}{\partial\phi^2} + k_c^2 e_z = 0. \tag{5.85}$$

From Section 4.5, it can be seen that a solution of Eq. (5.85) is of the form

$$e_z = [AJ_p(k_c\rho) + BN_p(k_c\rho)][Ce^{jp\phi} + De^{-jp\phi}], \tag{5.86}$$

where $p = 0, 1, 2, 3, \ldots$; both $J_p(k_c\rho)$ and $N_p(k_c\rho)$ are used because the axis $\rho = 0$ does not lie in the region $a < \rho < b$ where the field exists; and A, B, C, and D are arbitrary constants. Applying the boundary condition $E_z = 0$ at $\rho = a$ and $\rho = b$, we obtain

$$AJ_p(k_c a) + BN_p(k_c a) = CJ_p(k_c b) + DN_p(k_c b) = 0. \tag{5.87}$$

Therefore

$$-\frac{A}{B} = \frac{N_p(k_c a)}{J_p(k_c a)} = \frac{N_p(k_c b)}{J_p(k_c b)} \tag{5.88}$$

or

$$J_p(k_c a)N_p(k_c Ra) - N_p(k_c a)J_p(k_c Ra) = 0, \tag{5.89}$$

where $R = b/a$. Equation (5.89) is called the characteristic equation and its solutions give the value of k_c. It has an infinite number of solutions for k_c, which are designated k_{pq}, where q denotes the order of the root. The values of $k_{pq}a$ for different values of R are shown in Table 5.1.

Table 5.1 Values of $k_{pq}a$ for Different Values of p, q, R

R	$k_{01}a$	$k_{02}a$	$k_{11}a$	$k_{12}a$	$k_{21}a$	$k_{22}a$	$k_{03}a$	$k_{13}a$
1.2	15.70	31.41	15.73	31.43	15.81	31.47	47.12	47.13
1.5	6.27	12.56	6.32	12.59	6.47	12.66	18.85	18.86
2.0	3.12	6.27	3.20	6.31	3.41	6.43	9.42	9.44

Since

$$\bar{\beta} = (k^2 - k_c^2)^{1/2},$$

we can write

$$\bar{\beta}_{pq} = (k^2 - k_{pq}^2)^{1/2}, \qquad k^2 = \omega^2\mu\epsilon.$$

Therefore

$$\bar{\beta}_{pq} = (\omega^2\mu\epsilon - k_{pq}^2)^{1/2}. \tag{5.90}$$

For $\omega^2\mu\epsilon > k_{pq}^2$, $\bar{\beta}_{pq}$ is real, and hence

$$E_z = e_z \exp(-j\bar{\beta}z) = e_z \exp(-j\bar{\beta}_{pq}z)$$

represents a travelling wave in the z-direction, assuming a time variation $e^{j\omega t}$ which is suppressed in the expression for E_z. For $\omega^2\mu\epsilon < k_{pq}^2$, $\bar{\beta}_{pq}$ is imaginary and E_z represents an evanescent wave and thus there is no propagation.

$\omega_{c,pq} = k_{pq}/\sqrt{\mu\epsilon}$ is called the cut-off frequency of the TM_{pq} mode. This mode can be propagated by the coaxial line only if $\omega_c > k_{pq}/\sqrt{\mu\epsilon}$. The corresponding cut-off wavelength $\lambda_{c,pq}$ is

$$\lambda_{c,pq} = \frac{2\pi}{k_{pq}}. \tag{5.91}$$

The smallest value of k_{pq} for a given a occurs for $p = 0$, $q = 1$, as can be seen from Table 5.1.

The corresponding cut-off wavelength $\lambda_{c,pq}$ is equal to $2\pi/k_{01}$. For $R = 2.0$,

$$\lambda_{c,\,pq} = \frac{2\pi}{k_{01}} = \frac{6.28a}{3.12} \approx 2a \approx 2(b - a).$$

This relation holds good also for other values of R, as can be seen again from the table.

Since the propagation of the TM_{01} mode is only possible for $\lambda < \lambda_{c,\,01}$, the dimensions of the coaxial line to support the TM_{01} mode should be such that $2(b - a) > \lambda$, or $(b - a) > \lambda/2$. Hence, to prevent the TM_{01} mode, which is the lowest TM mode, from being propagated, the dimensions of the coaxial line should be selected so that $(b - a) < \lambda/2$. Most practical coaxial lines have $(b - a) \ll \lambda/2$. Once the lowest TM_{01} mode is prevented from propagating, the higher-order TM_{pq} modes too are certainly prevented from propagating, because the cut-off wavelengths for the higher-order modes are smaller than the cut-off wavelength for the TM_{01} mode. However, if there is a discontinuity in the line, then the TM waves can be excited and supported in a small neighbourhood of the discontinuity.

Similarly, an infinite set of TE modes too can exist in a coaxial line. For a TE mode, $E_z = 0$ and $H_z = h_z \exp(-j\beta z \neq 0)$, and h_z is a solution of the equation

$$\nabla_t^2 h_z + k_c'^2 h_z = 0. \tag{5.92}$$

A general solution of Eq. (5.92) is of the form

$$h_z = [A'J_p(k_c'\rho) + B'N_p(k_c'\rho)][C'e^{jp\phi} + D'e^{-jp\phi}]. \tag{5.93}$$

Hence, from Eqs. (5.29) and (5.31),

$$E_\phi = e_\phi \exp(-j\beta z) = \frac{-j\beta}{k_c'^2}\left(\frac{-k}{\beta}\right)\sqrt{\frac{\mu}{\epsilon}}\,\frac{\partial H_z}{\partial \rho}. \tag{5.94}$$

As E_ϕ is a tangential component of $\boldsymbol{E}$, $E_\phi = 0$ at $\rho = a$, $\rho = b$, and hence $\partial H_z/\partial\rho = 0$ or $\partial h_z/\partial\rho = 0$ at $\rho = a$, $\rho = b$. Therefore

$$A'\left[\frac{\partial}{\partial\rho}J_p(k_c'\rho)\right]_{\rho=a} + B'\left[\frac{\partial}{\partial\rho}N_p(k_c'\rho)\right]_{\rho=b} = 0. \tag{5.95}$$

Hence, the characteristic equation for TE modes is

$$\left[\frac{\partial}{\partial\rho}J_p(k_c'\rho)\right]_{\rho=a}\left[\frac{\partial}{\partial\rho}N_p(k_c'\rho)\right]_{\rho=b} - \left[\frac{\partial}{\partial\rho}N_p(k_c'\rho)\right]_{\rho=a}\left[\frac{\partial}{\partial\rho}J_p(k_c'\rho)\right]_{\rho=b} = 0. \tag{5.96}$$

The solution of Eq. (5.96) shows that, for each value of p, an infinite number of solutions k_{pq} exist. The lowest TE mode is the TE_{11} mode, and the cut-off wavelength of this mode is

$$\lambda_{c,\,11}' \approx \pi(a + b). \tag{5.97}$$

In order to prevent propagation of the TE_{11} mode in a coaxial line, the mean circumference of the coaxial cylinders must be smaller than the operating wavelength. This imposes a limitation on the spacing between the coaxial cylinders and their dimensions, which in turn limits the power handling capacity of the coaxial line.

5.10 Microstrip Transmission Lines

Microstrip transmission lines are extensively used in microwave circuits, because they can be fabricated easily by employing printed-circuit techniques. There are several types of microstrip-lines, some of which are given in Fig. 5.7. The shielded stripline shown in Fig. 5.7c is frequently

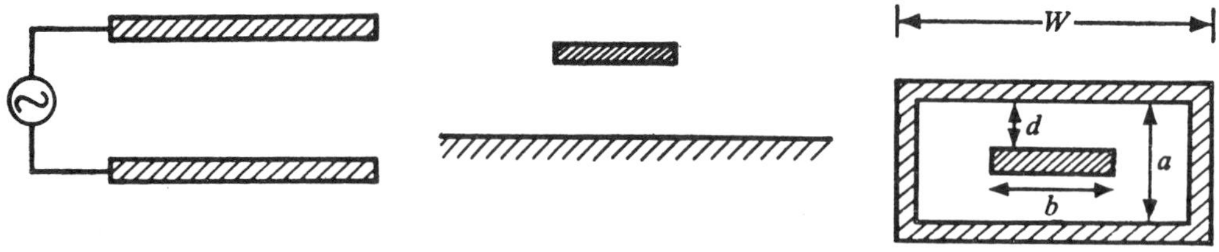

(a) Balanced line (b) Strip above ground plane (c) Shielded stripline

Fig. 5.7 Various types of strip transmission lines.

used in practice, because there is no loss by radiation, and hence the efficiency of transmission is higher in it than in the other two types shown in Figs. 5.7a and 5.7b.

Since the striplines have more than one conductor, they can support the TEM (or the dominant) mode such as the coaxial line. Hence, there is no cut-off frequency. They can also support the TE, TM, and hybrid modes. The statement that the TEM mode is supported by striplines is only an approximation. More rigorous field analysis by recent workers has shown that the modes are in fact hybrid. However, for a simple understanding of the behaviour of striplines, the TEM mode approximation is sufficient. This is assumed from the fact that two infinitely extending conducting parallel planes can support the TEM mode.

The shielded stripline illustrated in Fig. 5.7c may be considered as a deformed coaxial transmission line. If the width W of the outer conductor, which now becomes a shield around the inner conductor, is extended on both sides from a small value to infinity (see Fig. 5.8), the field configuration inside the line is only very slightly affected.

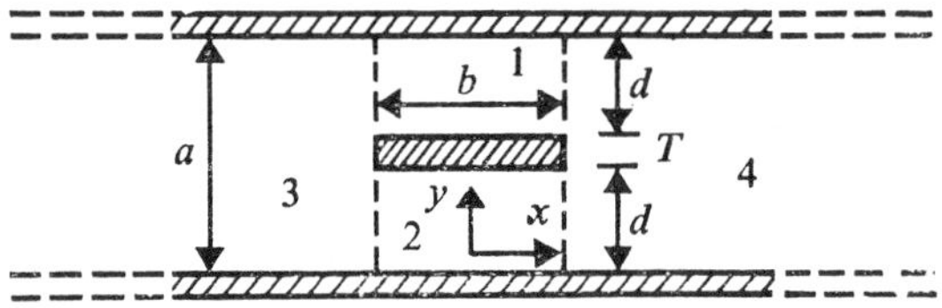

Fig. 5.8 Cross-section of shielded stripline between two infinite conducting plates.

The characteristic impedance and other characteristics of the shielded stripline can be determined from the field configuration inside the line. The field corresponding to the TEM mode can be derived from the scalar function Φ which satisfies the Laplace equation (see Section 5.3)

$$\nabla^2\Phi(x, y) = 0. \tag{5.98}$$

The solution of Eq. (5.98) for satisfying the proper boundary conditions can now be obtained. Taking $\Phi = 0$ at the outer conductor, the expressions for Φ which satisfy Laplace's equation in the different regions 1, 2, 3, and 4 can be written as

$$\Phi_1 = -\frac{V}{d}(y - a) + \sum_{m=1}^{\infty} A_m \cosh\left(\frac{m\pi x}{b}\right) \sin\left[\frac{m\pi}{d}(y - a + d)\right], \tag{5.99a}$$

$$\Phi_2 = \frac{Vy}{d} + \sum_{m=1}^{\infty} B_m \cosh\left(\frac{m\pi x}{b}\right) \sin\left(\frac{m\pi y}{d}\right), \tag{5.99b}$$

$$\Phi_3 = \sum_{m=1}^{\infty} C_m e^{m\pi x/b} \sin\left(\frac{m\pi y}{a}\right), \tag{5.99c}$$

$$\Phi_4 = \sum_{m=1}^{\infty} C_m e^{-m\pi x/b} \sin\left(\frac{m\pi y}{a}\right), \tag{5.99d}$$

where Φ_1, Φ_2, Φ_3, and Φ_4 are respectively the potential functions in regions 1, 2, 3, and 4. If W (see Fig. 5.7c) is not infinite, and the stripline is enclosed at $|x| = W/2$, the terms $e^{m\pi x/b}$ and $e^{-m\pi x/b}$ in Eqs. (5.99) have to be replaced by

$$\sinh\left[\frac{m\pi}{b}\left(\frac{W}{2} - |x|\right)\right].$$

The coefficients A_m, B_m, and C_m can be determined uniquely from the condition that the potential and its derivative $\partial\Phi/\partial x$ should be continuous across the common boundary $|x| = b/2$ for regions 1, 2, 3, and 4. The evaluation of these coefficients is straightforward though the expressions themselves are somewhat complicated. Even when we obtain expressions for the field components in this manner, the numerical evaluation of the circuit parameters such as inductance and capacitance per unit length and the characteristic impedance becomes rather tedious. We can, however, determine these quantities by considering the electrostatic field that might exist in the field configuration. In fact, the differential equation, which is Laplace's equation, and the boundary conditions to be satisfied by the potential functions for the electrostatic field, and the time-varying field in the TEM mode are identical.

The electrostatic field inside the shielded stripline can be obtained by using a conformal transformation such that the geometry of the line is simplified. If the field for this simple geometry is determined, the actual field in the stripline is obtained by inverse transformation. The transformation which can be conveniently used for this problem is of the Schwarz-Christoffel type. The characteristic impedance of a typical stripline with an inner conductor of finite thickness has been determined by several workers.

The derivation of an expression for the characteristic impedance of a stripline, in terms of the dimensions a, b, and T, is quite involved. If, however, the thickness of the inner conductor vanishes, an exact and simple expression for the characteristic impedance (as obtained by Oberhettinger and Magnus[1]) is given by

$$Z_0 = \frac{30\pi K(k)}{K(k')}, \tag{5.100}$$

where $K(k)$ is the complete elliptical integral of the first kind and

$$k = \operatorname{sech}\left(\frac{\pi b}{2a}\right), \qquad k' = \tanh\left(\frac{\pi b}{2a}\right).$$

For striplines with central conductors of nonzero thickness, the characteristic impedance varies considerably for different values of T/a. An approximate empirical formula valid for $0.05 < T/a$

[1]F. Oberhettinger and W. Magnus, Anwendung der elliptischen Funcktionen in Physik und Technik, Springer-Verlag, Berlin, 1949.

< 0.5 is given by

$$Z_c = \frac{94.15}{\sqrt{\epsilon_r}[\{b/(a-T)\} + 0.45 + 1.18(T/a)]} \text{ ohms,} \tag{5.101}$$

ϵ_r being the relative permittivity of the dielectric medium inside the stripline.

The stripline can also support TE and TM modes if the ground-plane spacing is large enough to support these modes. The ground-plane spacing must be greater than $\lambda/2$ to support these higher-order modes. Usually, the ground-plane spacing is kept $<\lambda/4$ so that the higher-order modes are avoided.

5.11 Rectangular Metal Waveguide

Figure 5.9 shows the section of a rectangular metal waveguide with conducting walls. Such a waveguide is used extensively in transmitting electromagnetic energy at microwave frequencies.

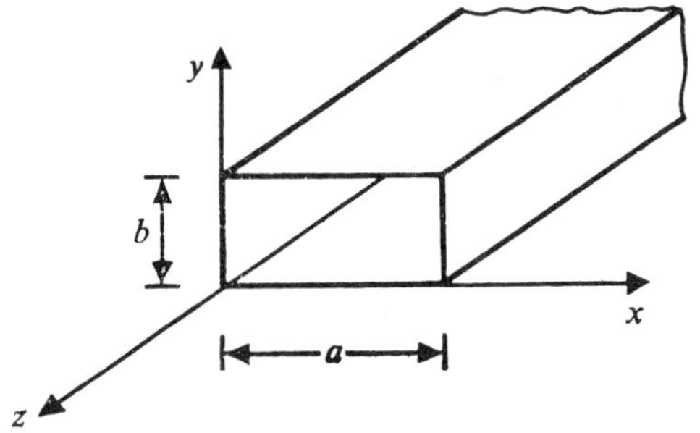

Fig. 5.9 Rectangular metal waveguide.

In any hollow metal waveguide with a closed cross-section, a transverse electric field can exist only if a time-varying axial magnetic field exists. Similarly, according to Maxwell's equations, a transverse magnetic field can exist only if either an axial displacement current or an axial conduction current is present. Since there is no central conductor in such a hollow metal waveguide, and since a TEM wave cannot have either an axial electric field or an axial magnetic field, the transverse electric and magnetic fields must be zero. Hence, a TEM wave cannot exist in a hollow metal waveguide of any closed cross-section.

Thus, the only types of waves that can be supported by a rectangular (or circular) metal hollow waveguide are the TE and TM waves, which we have discussed in Sections 5.4 and 5.5. The losses in the conducting walls and the dielectric inside the waveguide are initially neglected while studying the fields. The losses are taken into account when the attenuation in the waveguide is to be calculated.

The analysis of the loss-free waveguide shows that there is a double infinity of possible solutions for both TE and TM waves. These waves (or modes) are identified by integers m and n, e.g., TE_{mn} and TM_{mn}.

5.12 TE Waves or Modes in Rectangular Metal Waveguide

For TE (or H) modes, $E_z = e_z \exp(-j\beta z) = 0$, and all the other field components can be determined from $H_z = h_z \exp(-j\beta z)$ by using Eqs. (5.29) and (5.31) and assuming time dependence $e^{j\omega t}$.

h_z satisfies the two-dimensional wave equation

$$\nabla_t^2 h_z + k_c^2 h_z = 0. \tag{5.102}$$

In the rectangular coordinates, this equation becomes

$$\frac{\partial^2 h_z}{\partial x^2} + \frac{\partial^2 h_z}{\partial y^2} + k_c^2 h_z = 0 \tag{5.103}$$

whose solution is of the form (see Section 4.3)

$$h_z = f(x)g(y), \tag{5.104}$$

where

$$f(x) = A_1 \cos(k_x x) + A_2 \sin(k_x x), \tag{5.105}$$

$$g(y) = B_1 \cos(k_y y) + B_2 \sin(k_y y) \tag{5.106}$$

with

$$k_x^2 + k_y^2 = k_c^2, \tag{5.107}$$

$$k_c^2 = k^2 - \beta^2 \tag{5.108}$$

and A_1, A_2, B_1, B_2 as arbitrary constants. These constants, as well as the separation constants k_x, k_y, can be determined by considering the boundary conditions. The normal components of the magnetic field vanish at the perfectly conducting walls, and hence $\boldsymbol{n} \cdot \nabla_t h_z = 0$ at the walls, where $\boldsymbol{n}$ is a unit normal vector to the walls. Therefore

$$\frac{\partial h_z}{\partial x} = 0 \quad \text{at } x = 0, a, \tag{5.109}$$

$$\frac{\partial h_z}{\partial y} = 0 \quad \text{at } y = 0, b. \tag{5.110}$$

Substituting Eqs. (5.104)–(5.106) in Eqs. (5.109) and (5.110), we obtain

$$-k_x A_1 \sin(k_x x) + k_x A_2 \cos(k_x x) = 0 \quad \text{at } x = 0, a, \tag{5.111}$$

$$-k_y B_1 \sin(k_y y) + k_y B_2 \cos(k_y y) = 0 \quad \text{at } y = 0, b. \tag{5.112}$$

From Eqs. (5.111) and (5.112),

$$A_2 = B_2 = 0,$$

$$k_x = \frac{n\pi}{a} \qquad (n = 0, 1, 2, \ldots),$$

$$k_y = \frac{m\pi}{b} \qquad (m = 0, 1, 2, \ldots).$$

Hence

$$h_z = A_1 B_1 \cos(k_x x) \cos(k_y y) = A_{nm} \cos\left(\frac{n\pi}{a} x\right) \cos\left(\frac{m\pi}{b} y\right), \tag{5.113}$$

where $A_1B_1 = A_{nm}$. If $n = m = 0$, $h_z =$ constant, and all other components become zero. Hence, this gives a trivial solution which is of no practical interest. Equation (5.113) therefore gives a double infinity of solutions for h_z for $n = 0, 1, 2, \ldots, n = m \neq 0$. A particular set of values of n and m determines the nm-th TE (or TE_{nm} or H_{nm}) mode.

From Eqs. (5.107) and (5.108), we obtain

$$k^2 - \bar{\beta}_{nm}^2 = k_c^2 = k_x^2 + k_y^2 = \left(\frac{n\pi}{a}\right)^2 + \left(\frac{m\pi}{b}\right)^2 \tag{5.114}$$

or

$$\bar{\beta}_{nm}^2 = k^2 - \left(\frac{n\pi}{a}\right)^2 - \left(\frac{m\pi}{b}\right)^2. \tag{5.115}$$

If $\bar{\beta}_{nm}^2 > 0$ or < 0, then

$$k^2 = \omega^2\mu\epsilon > \text{ or } < \left(\frac{n\pi}{a}\right)^2 + \left(\frac{m\pi}{b}\right)^2 \tag{5.116}$$

or

$$\omega > \text{ or } < \left[\frac{(n\pi/a)^2 + (m\pi/b)^2}{\mu\epsilon}\right]^{1/2}$$

or

$$f > \text{ or } < \frac{c}{2\pi}\left[\left(\frac{n\pi}{a}\right)^2 + \left(\frac{m\pi}{b}\right)^2\right]^{1/2}. \tag{5.117}$$

The cut-off frequency is

$$f_{c,\,nm} = \frac{v}{2\pi}\left[\left(\frac{n\pi}{a}\right)^2 + \left(\frac{m\pi}{b}\right)^2\right]^{1/2}, \tag{5.118}$$

where $v = 1/\sqrt{\mu\epsilon}$ is the intrinsic phase velocity in the dielectric medium. For $f > f_c$, $\bar{\beta}$ is real, and hence there is propagation of the wave in the z-direction, with a velocity

$$v_{p,\,nm} = \frac{\omega}{\bar{\beta}} = \frac{\omega}{[\omega^2\mu\epsilon - (n\pi/a)^2 - (m\pi/b)^2]^{1/2}}.$$

For $f < f_{c,\,nm}$, $\bar{\beta}_{nm}$ is imaginary, and being evanescent the wave does not propagate in the z-direction. The phase velocity $v_{p,\,nm}$ is

$$v_{p,\,nm} = \frac{\omega}{[\omega^2\mu\epsilon - (n\pi/a)^2 - (m\pi/b)^2]^{1/2}} = \frac{v}{\left[1 - \dfrac{\{(n\pi/a)^2 + (m\pi/b)^2\}}{\omega^2\mu\epsilon}\right]^{1/2}} \tag{5.119}$$

and the cut-off wavelength $\lambda_{c,\,nm}$ is given by

$$\lambda_{c,\,nm} = \frac{2ab}{(n^2b^2 + m^2a^2)}. \tag{5.120}$$

Equation (5.119) shows that the phase velocity of the wave propagating in the z-direction is always greater than v, the intrinsic phase velocity of the dielectric medium.

The foregoing analysis has shown that the rectangular waveguide carrying the TE_{nm} mode behaves like a high-pass filter with a cut-off frequency $f_{c,\,nm}$.

A typical practical rectangular metal waveguide has dimensions a and b such that $a = 2b$. Therefore

$$\lambda_{c,nm} = \frac{2a}{(n^2 + 4m^2)^{1/2}}. \tag{5.121}$$

Hence, $\lambda_{c,10} = 2a$, $\lambda_{c,01} = a$, $\lambda_{c,11} = 2a/\sqrt{5}, \ldots$. It can be seen that the TE_{10} (or H_{10}) mode has the largest cut-off wavelength or the lowest cut-off frequency. The TE_{10} mode is called the *dominant mode* and is the one most commonly used in practice. If $a < \lambda < 2a$ or $c/(2a) < f < c/a$, then it can be seen that only the TE_{10} mode can be propagated. Therefore, one-octave band from $c/(2a)$ to c/a is the useful frequency band that can be propagated in this mode for a rectangular waveguide for which $a = 2b$. The remaining field components can be obtained by using Eq. (5.113) in Eqs. (5.29) and (5.31).

The field components of the TE_{nm} mode are

$$H_z = A_{nm} \cos\left(\frac{n\pi x}{a}\right) \cos\left(\frac{m\pi y}{b}\right) \exp(\mp j\bar{\beta}_{nm} z), \tag{5.122a}$$

$$H_x = \pm j\frac{\bar{\beta}_{nm}}{k^2_{c,nm}} A_{nm}\frac{n\pi}{a} \sin\left(\frac{n\pi x}{a}\right) \cos\left(\frac{m\pi y}{b}\right) \exp(\mp j\bar{\beta}_{nm} z), \tag{5.122b}$$

$$H_y = \pm j\frac{\bar{\beta}_{nm}}{k^2_{c,nm}} A_{nm}\frac{m\pi}{b} \cos\left(\frac{n\pi x}{a}\right) \sin\left(\frac{m\pi y}{b}\right) \exp(\mp j\bar{\beta}_{nm} z), \tag{5.122c}$$

$$E_x = Z_{h,nm} A_{nm} j\frac{\bar{\beta}_{nm}}{k^2_{c,nm}} \frac{m\pi}{b} \cos\left(\frac{n\pi x}{a}\right) \sin\left(\frac{m\pi y}{b}\right) \exp(\mp j\bar{\beta}_{nm} z), \tag{5.122d}$$

$$E_y = -Z_{h,nm} A_{nm}\frac{\bar{\beta}_{nm}}{k^2_{c,nm}} \frac{n\pi}{a} \sin\left(\frac{n\pi x}{a}\right) \cos\left(\frac{m\pi y}{b}\right) \exp(\mp j\bar{\beta}_{nm} z), \tag{5.122e}$$

where $Z_{h,nm}$ is the wave impedance for the TE_{nm} (or H_{nm}) mode and is equal to $(k/\bar{\beta}_{nm})Z_0$. Hence

$$Z_{h,nm} = \frac{\omega\sqrt{\mu\epsilon}}{\bar{\beta}_{nm}} \sqrt{\frac{\mu}{\epsilon}} = \frac{\omega\mu}{\bar{\beta}_{nm}}. \tag{5.123}$$

For $f < f_{c,nm}$, $\bar{\beta}_{nm}$ is imaginary, and hence $Z_{h,nm}$ is also imaginary, and so there is no propagation of the TE_{nm} mode. This indicates that there is no net energy flow with the evanescent mode. A general TE field with $E_z = 0$ can be described by a linear superposition of all the TE_{nm} modes.

5.13 Power Flow in Rectangular Metal Waveguide for TE Modes

For a propagating TE_{nm} mode, i.e., for $f > f_{c,nm}$, the power or rate of energy flow in the positive z-direction is given by

$$\begin{aligned} P_{nm} &= \tfrac{1}{2} \operatorname{Re} \int_{x=0}^{a} \int_{y=0}^{b} \boldsymbol{E} \times \boldsymbol{H}^* \cdot \boldsymbol{u}_z \, dx\, dy \\ &= \tfrac{1}{2} \operatorname{Re} \int_{x=0}^{a} \int_{y=0}^{b} (E_x H_y^* - E_y H_x^*)\, dx\, dy \\ &= \tfrac{1}{2} \operatorname{Re} (Z_{h,nm}) \int_{x=0}^{a} \int_{y=0}^{b} (H_y H_y^* + H_x H_x^*)\, dx\, dy. \end{aligned} \tag{5.124}$$

Substituting Eqs. (5.122) and (5.123) in Eq. (5.124) and noting that

$$\int_0^a \int_0^b \sin^2 (\frac{n\pi x}{a}) \cos^2 (\frac{m\pi y}{b})\, dx\, dy = \int_0^a \int_0^b \cos^2 (\frac{n\pi x}{a}) \sin^2 (\frac{m\pi y}{b})\, dx\, dy$$
$$= \frac{ab}{4} \quad \text{for } n \neq 0,\ m \neq 0$$
$$= \frac{ab}{2} \quad \text{for } n = 0,\ m = 0,$$

we obtain

$$P_{nm} = |A_{nm}|^2 \frac{ab}{\epsilon_{0n}\epsilon_{0m}} \frac{\bar{\beta}^2_{nm}}{k^4_{c,\,nm}} Z_{\text{h},\,nm}[(\frac{m\pi}{b})^2 + (\frac{n\pi}{a})^2] = \frac{|A_{nm}|^2 ab}{\epsilon_{0n}\epsilon_{0m}} (\frac{\bar{\beta}_{nm}}{k_{c,\,nm}})^2 Z_{\text{h},\,nm}, \tag{5.125}$$

where ϵ_{0m} is the Neumann factor which is equal to 1 for $m = 0$ and equal to 2 for $m > 0$.

If two TE modes, say, the TE_{nm} and TE_{rs} modes, are present simultaneously, then it is found that the power is the sum of the powers in the two modes, i.e., $P = P_{nm} + P_{rs}$. This is a general property of the loss-free waveguides, and results from the orthogonality of the eigenfunctions that describe the transverse variation of the fields when integrated over the guide cross-section. In the rectangular metal waveguide, the orthogonality property is given by

$$\int_0^a \begin{matrix}\sin\\ \cos\end{matrix} (\frac{n\pi x}{a}) \begin{matrix}\sin\\ \cos\end{matrix} (\frac{r\pi x}{a})\, dx = 0 \quad \text{for } n \neq r,$$
$$\int_0^a \begin{matrix}\sin\\ \cos\end{matrix} (\frac{m\pi y}{b}) \begin{matrix}\sin\\ \cos\end{matrix} (\frac{s\pi y}{b})\, dy = 0 \quad \text{for } m \neq s. \tag{5.126}$$

Even when small losses are present, the energy flow may be taken as that contributed by each individual mode, with negligible error in all cases except when two or more degenerate modes are present. Degenerate modes are modes for which the propagation constant $\bar{\gamma} = \bar{\alpha} + j\bar{\beta}$ is the same, and for these the presence of even small losses may result in strong coupling between the modes.

5.14 Attenuation in Rectangular Waveguides for TE Modes

The attenuation of electromagnetic waves in metal waveguides may result from one or more of the following causes:

(i) An impressed frequency less than the cut-off frequency
(ii) Losses in the dielectric
(iii) Losses in the metal guide walls.

If the frequency is less than the cut-off frequency, the field is evanescent, and the attenuation constant is given by

$$\bar{\alpha} = [(\frac{n\pi}{a})^2 + (\frac{m\pi}{b})^2 - \omega^2\mu\epsilon]^{1/2} = \omega(\mu\epsilon)^{1/2}\left[\frac{(\frac{n\pi}{a})^2 + (\frac{m\pi}{b})^2}{\omega^2\mu\epsilon} - 1\right]^{1/2}. \tag{5.127}$$

For frequencies very much less than the cut-off frequency, $\bar{\alpha}$ is nearly equal to

$$[(\frac{n\pi}{a})^2 + (\frac{m\pi}{b})^2]^{1/2}.$$

If the dielectric is lossy, then k is replaced by $j\gamma$, where γ is the complex propagation constant given by

$$\gamma = \sqrt{j\omega\mu(\sigma + j\omega\epsilon)} = \sqrt{-\omega^2\mu\epsilon[1 - j\sigma/(\omega\epsilon)]} = j\omega\sqrt{\mu\epsilon}[1 - j\sigma/(\omega\epsilon)]^{1/2}.$$

Then, the propagation constant γ_z in the z-direction in the waveguide is given by

$$\gamma_z^2 = \gamma^2 + (\frac{n\pi}{a})^2 + (\frac{m\pi}{b})^2 = j\omega\mu(\sigma + j\omega\epsilon) + (\frac{n\pi}{a})^2 + (\frac{m\pi}{b})^2$$

so that

$$\gamma_z = [(\frac{n\pi}{a})^2 + (\frac{m\pi}{b})^2 - \omega^2\mu\epsilon + j\omega\mu\sigma]^{1/2} = \bar{\alpha}_d + j\bar{\beta}$$

or, for a low-loss dielectric,

$$\bar{\beta} - j\bar{\alpha}_d = [\omega^2\mu\epsilon - (\frac{n\pi}{a})^2 - (\frac{m\pi}{b})^2 - j\omega\mu\sigma]^{1/2} = \omega\sqrt{\mu\epsilon}[1 - \frac{(n\pi/a)^2 + (m\pi/b)^2}{\omega^2\mu\epsilon} - \frac{j\sigma}{\omega\epsilon}]^{1/2}$$

$$= \omega\sqrt{\mu\epsilon}[1 - \frac{(n\pi/a)^2 + (m\pi/b)^2}{\omega^2\mu\epsilon}]^{1/2}\left[1 - \frac{j(\sigma/\omega\epsilon)}{1 - \frac{(n\pi/a)^2 + (m\pi/b)^2}{\omega^2\mu\epsilon}}\right]^{1/2}$$

$$\approx \omega\sqrt{\mu\epsilon}[1 - \frac{(n\pi/a)^2 + (m\pi/b)^2}{\omega^2\mu\epsilon}]^{1/2} - \frac{[j\sigma/(2\epsilon\omega)]\omega\sqrt{\mu\epsilon}}{[1 - \frac{(n\pi/a)^2 + (m\pi/b)^2}{\omega^2\mu\epsilon}]^{1/2}},$$

where $\bar{\alpha}_d$, which is the attenuation constant due to the losses in the low-loss dielectric medium, is given by

$$\bar{\alpha}_d = \frac{\sigma\sqrt{\mu/\epsilon}}{2[1 - \frac{(n\pi/a)^2 + (m\pi/b)^2}{\omega^2\mu\epsilon}]^{1/2}} = \frac{\sigma\sqrt{\mu/\epsilon}}{2(1 - \lambda^2/\lambda_c^2)}. \tag{5.128}$$

To find the attenuation constant due to the finite conductivity of the metal guide walls, we have to consider the loss of power to the walls as the modes propagate along the guide. Due to the loss in the walls, the propagation constant $j\bar{\beta}$ in the z-direction becomes $\bar{\alpha}_c + j\bar{\beta}$, where $\bar{\alpha}_c$ is the attenuation constant due to the losses in the walls. This $\bar{\alpha}_c$ is calculated in a way similar to that used for coaxial lines, and is given by

$$-\frac{dP}{dz} = 2\bar{\alpha}_c P = P_L, \tag{5.129}$$

where P is the power carried by the wave and P_L is the power loss of the waveguide in the walls per unit length.

We shall now calculate the attenuation constant $\bar{\alpha}_c$ for the dominant TE_{10} (or H_{10}) mode. From Eqs. (5.122), the field components for this mode are

$$H_z = h_z \exp(-j\bar{\beta}_{10}z) = A_{10}\cos(\frac{\pi x}{a})\exp(-j\bar{\beta}_{10}z), \tag{5.130a}$$

$$H_x = \frac{j\bar{\beta}_{10}}{k_{c,10}^2}A_{10}\frac{\pi}{a}\sin(\frac{\pi x}{a}), \tag{5.130b}$$

$$E_y = -Z_{h,\,10}A_{10}\frac{j\bar{\beta}_{10}}{k^2_{c,\,10}}\frac{\pi}{a}\sin\left(\frac{\pi x}{a}\right). \tag{5.130c}$$

The power flow along the guide is given by Eq. (5.125) as

$$P_{10} = |A_{10}|^2\frac{ab}{2}\left(\frac{\bar{\beta}_{10}}{k_{c,\,10}}\right)^2 Z_{h,\,10}. \tag{5.131}$$

The currents on the lossy walls are assumed to be the same as the loss-free currents, and are given by $\boldsymbol{J}_S = \boldsymbol{n}\times\boldsymbol{H}$, where $\boldsymbol{n}$ is an outward normal to the guide wall. Therefore, on the walls at $x = 0$, $x = a$ (see Fig. 5.9), the surface currents are

$$\begin{aligned} \boldsymbol{J}_S &= \boldsymbol{u}_x\times\boldsymbol{H} = -\boldsymbol{u}_y A_{10} \qquad \text{at } x = 0 \\ &= -\boldsymbol{u}_x\times\boldsymbol{H} = -\boldsymbol{u}_y A_{10} \quad \text{at } x = a. \end{aligned} \tag{5.132}$$

The currents on the upper and lower walls at $y = 0$, $y = b$ are

$$\begin{aligned} \boldsymbol{J}_S &= \boldsymbol{u}_y\times\boldsymbol{H} = -\boldsymbol{u}_z\frac{j\bar{\beta}_{10}}{k^2_{c,\,10}}A_{10}\frac{\pi}{a}\sin\left(\frac{\pi x}{a}\right) + \boldsymbol{u}_x A_{10}\cos\left(\frac{\pi x}{a}\right) \quad \text{at } y = 0 \\ &= -\boldsymbol{u}_y\times\boldsymbol{H} = \boldsymbol{u}_z\frac{j\bar{\beta}_{10}}{k^2_{c,\,10}}A_{10}\frac{\pi}{a}\sin\left(\frac{\pi x}{a}\right) - \boldsymbol{u}_x A_{10}\cos\left(\frac{\pi x}{a}\right) \quad \text{at } y = b. \end{aligned} \tag{5.133}$$

With a finite conductivity σ of the waveguide walls, the surface impedance of the walls is

$$Z_m = \frac{1+j}{\sigma\delta_s} = (1+j)R_m, \tag{5.134}$$

where δ_s is the skin depth. The power loss in the resistive part R_m of Z_m per unit length of guide is

$$\begin{aligned} P_L &= \frac{R_m}{2}\int_{\text{width of guide wall}} \boldsymbol{J}_S\cdot\boldsymbol{J}_S^*\cdot ds \\ &= \frac{R_m|A_{10}|^2}{2}\left[2\int_{y=0}^{b} dy + 2\int_{x=0}^{a}\frac{\bar{\beta}^2_{10}\pi^2}{k^4_{c,\,10}a^2}\sin^2\left(\frac{\pi x}{a}\right)dx + 2\int_{x=0}^{a}\cos^2\left(\frac{\pi x}{a}\right)dx\right]. \end{aligned} \tag{5.135}$$

Since $k_{c,\,10} = \pi/a$,

$$P_L = R_m|A_{10}|^2\left[b + \frac{a}{2}\left(\frac{\bar{\beta}_{10}}{k_{c,\,10}}\right)^2 + \frac{a}{2}\right]. \tag{5.136}$$

If P_0 is the power at $z = 0$, then $P_{10} = P_0 e^{-2\alpha z}$ is the power in the guide at any z. The rate of decrease of power propagated is

$$\frac{-dP_{10}}{dz} = 2\alpha P_{10} = P_L \tag{5.137}$$

and equals the power loss as given by Eq. (5.137). Hence, the attenuation constant $\bar{\alpha}_c$ for the TE_{10} (or H_{10}) mode is given by

$$\begin{aligned} \bar{\alpha}_c &= \frac{P_L}{2P_{10}} = \frac{R_m[(b + a/2)(\bar{\beta}_{10}/k_{c.\,10})^2 + a/2]}{(ab/2)(\bar{\beta}_{10}/k_{c,\,10})^2 Z_{h,\,10}} \\ &= \frac{R_m}{ab\bar{\beta}_{01}k_0 Z_0}(2bk^2_{c,\,10} + ak_0^2) \text{ nepers/metre.} \end{aligned} \tag{5.138}$$

The attenuation for other TE_{nm} modes is given in Table 5.2 which summarizes the solutions for the TE_{nm} and TM_{nm} modes. Figure 5.10 gives the attenuation for the TE_{10} mode in a rectangular copper waveguide as a function of frequency. To convert attenuation given in nepers to decibels, multiply it by 8.7.

Table 5.2 Properties of Modes in Rectangular Waveguides

Property	*TE modes*	*TM modes*
H_z	$\cos(\frac{n\pi x}{a})\cos(\frac{m\pi y}{b})\exp(-j\bar{\beta}_{nm}z)$	0
E_z	0	$\sin(\frac{n\pi x}{a})\sin(\frac{m\pi y}{b})\exp(-j\bar{\beta}_{nm}z)$
E_x	$Z_{h,nm}H_y$	$-\frac{j\bar{\beta}_{nm}n\pi}{ak_{c,nm}^2}\cos(\frac{n\pi x}{a})\sin(\frac{m\pi y}{b})\exp(-j\bar{\beta}_{nm}z)$
E_y	$-Z_{h,nm}H_x$	$-\frac{j\bar{\beta}_{nm}m\pi}{bk_{c,nm}^2}\sin(\frac{n\pi x}{a})\cos(\frac{m\pi y}{b})\exp(-j\bar{\beta}_{nm}z)$
H_x	$\frac{j\bar{\beta}_{nm}n\pi}{ak_{c,nm}^2}\sin(\frac{n\pi x}{a})\cos(\frac{m\pi y}{b})\exp(-j\bar{\beta}_{nm}z)$	$\frac{-E_y}{Z_{c,nm}}$
H_y	$\frac{j\bar{\beta}_{nm}m\pi}{bk_{c,nm}^2}\cos(\frac{n\pi x}{a})\sin(\frac{m\pi y}{b})\exp(-j\bar{\beta}_{nm}z)$	$\frac{E_x}{Z_{c,nm}}$
$Z_{h,nm}$	$(\frac{k_0}{\bar{\beta}_{nm}})Z_0$	
$Z_{e,nm}$		$(\frac{\bar{\beta}_{nm}}{k_0})Z_0$
$k_{c,nm}$	$[(\frac{n\pi}{a})^2+(\frac{m\pi}{b})^2]^{1/2}$	$[(\frac{n\pi}{a})^2+(\frac{m\pi}{b})^2]^{1/2}$
$\bar{\beta}_{nm}$	$(k^2-k_{c,nm}^2)^{1/2}$	$(k^2-k_{c,nm}^2)^{1/2}$
$\lambda_{c,nm}$	$\frac{2ab}{(n^2b^2+m^2a^2)^{1/2}}$	$\frac{2ab}{(n^2b^2+m^2a^2)^{1/2}}$
$\bar{\alpha}_c$	$\frac{2R_m}{bZ_0(1-k_{c,nm}^2/k_0^2)^{1/2}}(1+b/a)\frac{k_{c,nm}^2}{k_0^2} + \frac{b}{a}(\frac{\epsilon_{0m}}{2}-\frac{k_{c,nm}^2}{k_0^2})(\frac{n^2ab+\pi^2a^2}{n^2b^2+\pi^2a^2})$	$\frac{2R_m}{bZ_0(1-\frac{k_{c,nm}^2}{k_0^2})^{1/2}}\frac{(n^2b^3+m^2a^3)}{(n^2b^2a+m^2a^3)}$

$R_m = [\omega\mu_0/(2\sigma)]^{1/2}$, $\epsilon_{0m} = 1$ for $m = 0$, and $\epsilon_{0m} = 2$ for $m > 0$. The expression for α is not valid for degenerate modes.

Below 5000 MHz, the theoretical formulae for attenuation agree quite well with the experimental values. In general, for higher frequencies, the measured values of $\bar{\alpha}_c$ are considerably

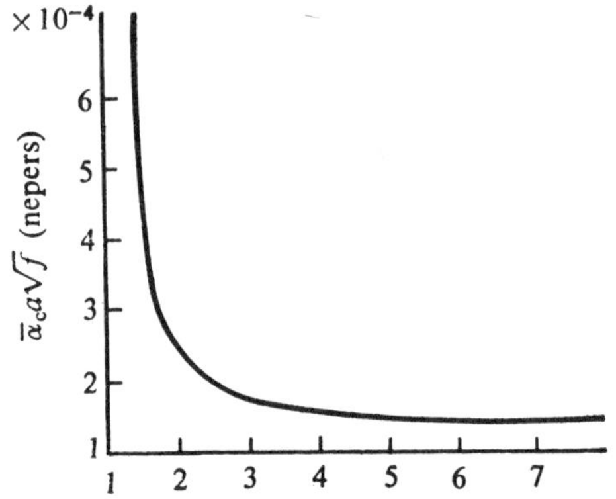

Fig. 5.10 Attenuation of H_{10} mode in rectangular waveguide. ($a = 2b$, f in units of 10^{10} cps.)

greater than the theoretical values, depending on the roughness of the waveguide walls. If surface imperfections are of the order of or greater than the skin depth δ_s, then the effective surface area is very much greater, and hence the loss is higher. The actual losses and the attenuation may be reduced by suitably polishing the surfaces.

5.15 TM Modes in Rectangular Waveguides

For TM modes, $H_z = 0$, but $E_z \neq 0$. e_z plays the role of a potential function from which the remaining field components may be derived, and satisfies the equation

$$\nabla_t^2 e_z + k_c^2 e_z = 0 \tag{5.139}$$

which is the same equation as that satisfied by h_z for TE modes, and hence the solution of Eq. (5.139) is similar to that of h_z for TE modes. The boundary conditions make $e_z = 0$ at $x = 0$, $x = a$ and also at $y = 0$, $y = b$. Hence, the expression for e_z takes the form

$$e_z = A_{nm} \sin\left(\frac{n\pi x}{a}\right) \sin\left(\frac{m\pi y}{b}\right). \tag{5.140}$$

There is a double infinity of TM modes which is similar to that of the TE modes and corresponds to various integral values of n and m. However, if $n = 0$ or $m = 0$, $e_z = 0$, and hence all the other components of $\boldsymbol{H}$ and $\boldsymbol{E}$ are zero. Therefore, the lowest integral value that n or m can take is 1. Hence, the lowest propagating TM mode is the TM_{11} mode, and it has a cut-off wavelength

$$\lambda_{c,11} = \frac{2ab}{(a^2 + b^2)^{1/2}}.$$

On the other hand, the lowest TE mode, which is the TE_{10} mode, has a cut-off wavelength of $\lambda_{c,10} = 2a$. Since $2a$ is greater than $2ab/(a^2 + b^2)^{1/2}$, the TE_{10} mode has a greater cut-off wavelength and a lower cut-off frequency than the TM_{11} mode. This shows that of all TE and TM modes, the TE_{10} (or H_{10}) mode has the lowest cut-off frequency and is therefore the dominant mode. It should be noted that, for the same values of n and m, the TE_{nm} and TM_{nm} modes have the same propagation factor $k_{c,\,nm}$ and also the same cut-off frequency $\lambda_{c,\,nm}$. Hence, these two modes are said to be degenerate. If $a = b$, the four modes, namely, TE_{nm}, TE_{mn}, TM_{nm}, and TM_{mn} are all degenerate. Other degeneracies occur if a is an integral multiple of b or vice versa.

The other components of $\boldsymbol{E}$ and $\boldsymbol{H}$ for the TM modes can be constructed with the help of Eqs. (5.33) and (5.34). A summary of the solutions is given in Table 5.2.

5.16 Dominant TE_{10} Mode in Rectangular Waveguide

The TE_{10} mode is the dominant and most commonly used mode in the rectangular metal waveguide. The field components for this mode (using Table 5.2) are

$$H_z = A \cos\left(\frac{\pi x}{a}\right) \exp(-j\bar{\beta}z), \tag{5.141a}$$

$$H_x = \frac{j\bar{\beta}}{k_c} A \sin\left(\frac{\pi x}{a}\right) \exp(-j\bar{\beta}z), \tag{5.141b}$$

$$E_y = -jZ_h \frac{\bar{\beta}}{k_c} \sin\left(\frac{\pi x}{a}\right) \exp(-j\bar{\beta}z), \tag{5.141c}$$

where $\bar{\beta} = \bar{\beta}_{10}$ and $k_c = k_{c,\,10}$. The quantities k_c, $\bar{\beta}$, Z_h, λ_c, and λ_g are given by

$$k_c = \frac{\pi}{a}, \tag{5.142a}$$

$$\bar{\beta} = \left[k_0^2 - \left(\frac{\pi}{a}\right)^2\right]^{1/2}, \tag{5.142b}$$

$$Z_h = -\frac{E_y}{H_x} = \frac{k_0}{\bar{\beta}} Z_0, \tag{5.142c}$$

$$\lambda_c = 2a, \tag{5.142d}$$

$$\lambda_g = \frac{2\pi}{\bar{\beta}} = \frac{\lambda_0}{[1 - \{\lambda_0/(2a)\}^2]^{1/2}}. \tag{5.142e}$$

The phase velocity v_p and group velocity v_g of the TE_{10} mode are given by

$$v_p = \frac{\lambda_g}{\lambda_0} c, \tag{5.143a}$$

$$v_g = \frac{\lambda_0}{\lambda_g} c. \tag{5.143b}$$

The concepts of phase and group velocities and their relationship to each other are discussed in the latter part of this section.

Figure 5.11 shows the electric and magnetic field lines associated with the TE_{10} mode.

The fields for the TE_{10} mode may be decomposed into the sum of two plane TEM waves

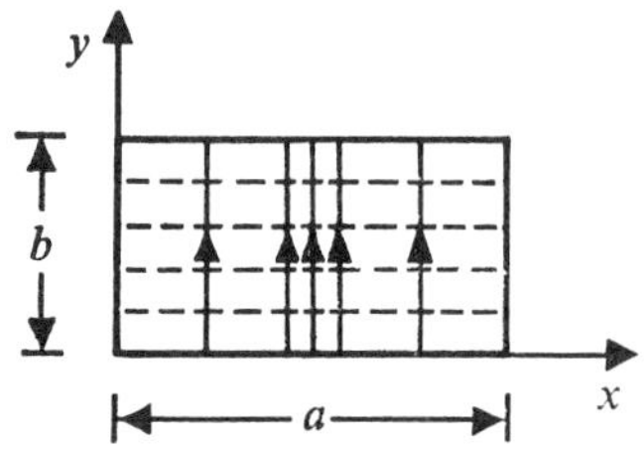

(a) Transverse plane

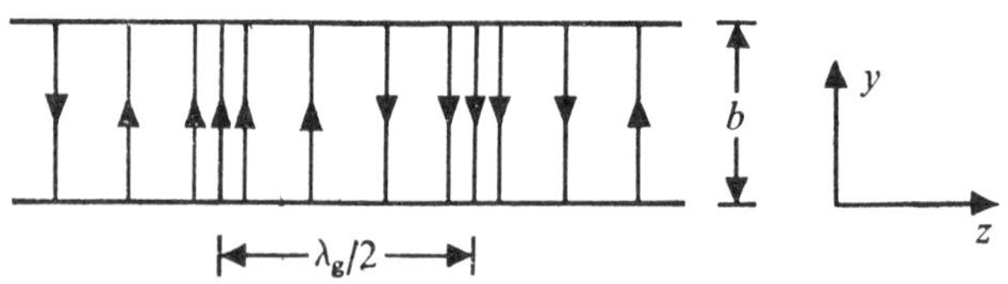

(b) Side view

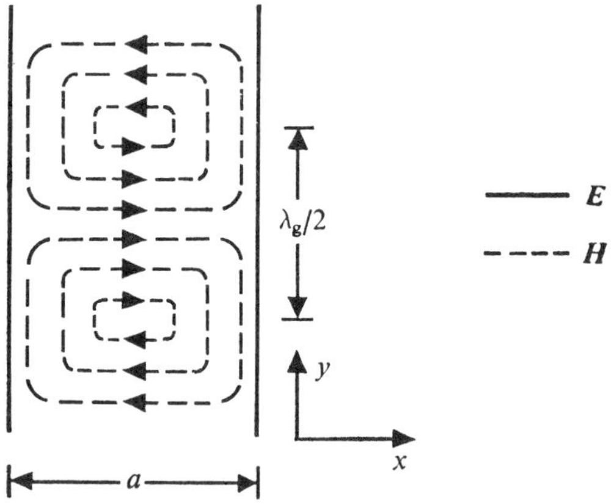

(c) Top view

Fig. 5.11 Electric and magnetic field lines for TE_{10} mode.

propagating along zig-zag paths between the two waveguide walls at $x = 0$ and $x = a$, as shown in Fig. 5.12.

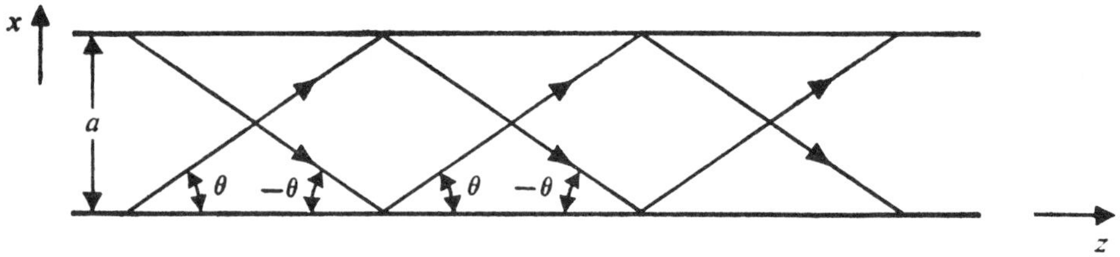

Fig. 5.12 Decomposition of TE_{10} mode into two plane waves.

The electric field for the TE_{10} mode is

$$E_y = -jZ_h\frac{\bar{\beta}}{k_c}\sin\left(\frac{\pi x}{a}\right)\exp(-j\bar{\beta}z) = -\frac{Z_h}{2}\frac{\bar{\beta}}{k_c}\exp\left[j\left(\frac{\pi x}{a} - \bar{\beta}z\right)\right] - \exp\left[-j\left(\frac{\pi x}{a} + \bar{\beta}z\right)\right].$$

If $\pi/a = k_0 \sin\theta$, and $\bar{\beta} = k_0\cos\theta$, the relation $(\pi/a)^2 + \bar{\beta}^2 = k_0^2$ still holds good. Then E_y is given by

$$E_y = \frac{Z_h}{2}\frac{\bar{\beta}}{k_c}[\exp\{-jk_0(x\sin\theta + z\cos\theta)\} - \exp\{-jk_0(-x\sin\theta + z\cos\theta)\}]. \quad (5.144)$$

This equation shows that E_y consists of two plane waves propagating at angles $\pm\theta$, with respect to the z-axis, as shown in Fig. 5.12. The field may be depicted as a plane wave reflecting back and forth between the two guide walls at $x = 0$ and $x = a$. The constant phase planes of these two obliquely propagating plane waves move in the z-direction with the phase velocity $c/\cos\theta = \bar{\beta}c/k_0$, which is greater than c_0. This causes the phase velocity v_p of the TE_{10} mode in the z-direction to exceed the velocity c of light. Since the energy in a TEM wave propagates with the velocity c in the direction in which the plane wave propagates, the energy in the TE_{10} mode in the waveguide propagates in the z-direction, which is along the axis of the waveguide, with a velocity equal to the component of c along the z-axis. This component is $v_z = c\cos\theta = k_0\bar{\beta}/c$, and is the group velocity v_g of the wave. It can now be seen that

$$v_p v_g = c^2. \tag{5.145}$$

When $\theta = \pi/2$, the plane waves reflect back and forth, but do not progress down the guide, and hence the mode is cut-off. Hence, cut-off happens when $\cos\theta = \cos(\pi/2) = \bar{\beta}/k_0 = 0$ or $\bar{\beta} = 0$ or

$$k_0^2 - \left(\frac{\pi}{a}\right)^2 = 0 \quad \text{or} \quad \omega^2\mu_0\epsilon_0 = \frac{\pi^2}{a^2} = \left(\frac{2\pi}{\lambda_0}\right)^2 \quad \text{or} \quad \lambda_0 = 2a = \lambda_c.$$

The last equation gives the cut-off wavelength.

The decomposition of the TE_{10} mode into two plane waves may be extended to the TE_{nm} modes also. When n and m are both different from zero, four plane waves result. This kind of decomposition into plane waves of the different modes of the waveguide gives a physical picture of the properties of the electromagnetic field inside a waveguide.

5.17 Field Configurations of Some Lower-Order Modes in Rectangular Waveguide: Methods of Excitation

The field configuration for the TE_{10} (or dominant) mode has already been shown in Fig. 5.11. Figure 5.13 gives the electric and magnetic field configurations of some of the other lower TE and TM modes in the rectangular metal waveguide.

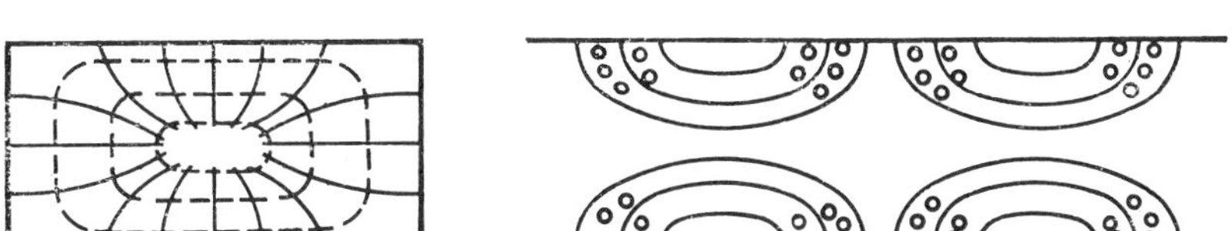

(a) TE_{11} mode

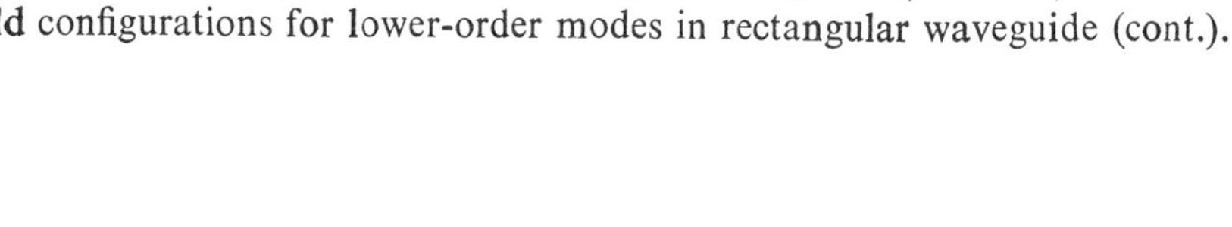

(b) TM_{11} mode

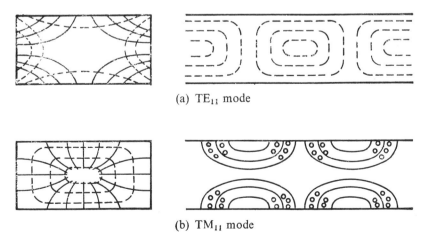

Fig. 5.13 Electric (shown by solid lines) and magnetic (shown by dashed lines) field configurations for lower-order modes in rectangular waveguide (cont.).

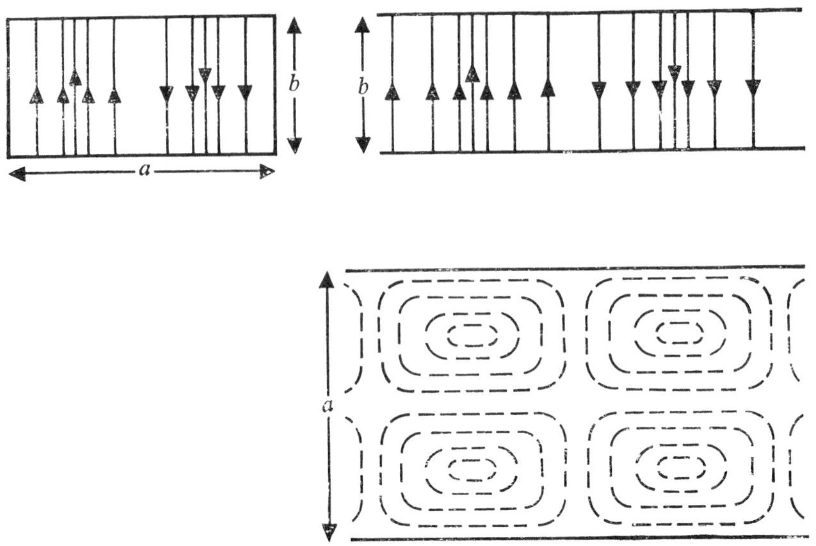

(c) TE_{20} mode

Fig. 5.13 Electric (shown by solid lines) and magnetic (shown by dashed lines) field configurations for lower-order modes in rectangular waveguide.

Methods of excitation of some of the TE and TM modes given in Fig. 5.13 are illustrated in Fig. 5.14. In order to excite a particular mode, a type of probe, which produces lines of ***E*** and ***H***

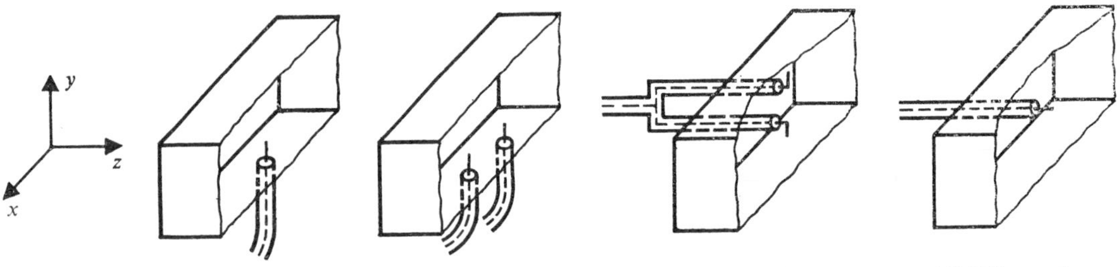

(a) TE_{10} mode (b) TE_{20} mode (c) TE_{11} mode (d) TM_{11} mode

Fig. 5.14 Excitation methods for various modes.

roughly parallel to the lines of ***E*** and ***H*** for that mode, is chosen. Thus, in Fig. 5.14a for the TE_{10} mode, the probe is parallel to the y-axis so that it produces lines of ***E*** in the y-direction and lines of ***H*** which lie in the xz-plane. This gives the correct configuration for the TE_{10} mode. In Fig. 5.14b, the parallel probes fed with opposite phase help to set up the TE_{20} mode. In Fig. 5.14c, the two probes parallel to the y-direction give the correct configuration for the TE_{11} mode. In Fig. 5.14d, the probe parallel to the z-axis produces magnetic field lines in the xy-plane, which are correct for the TM modes.

Several modes can exist simultaneously in a guide if the frequency is above cut-off for these particular modes. However, the guide dimensions are often chosen so that only the dominant mode can exist.

5.18 Wave Impedances and Characteristic Impedance of Rectangular Metal Waveguide

The wave impedances at a point are defined in the rectangular coordinates by the following ratios of electric to magnetic field strengths:

$$Z^+_{xy} = \frac{E_x}{H_y}, \quad Z^+_{yz} = \frac{E_y}{H_z}, \quad Z^+_{zx} = \frac{E_z}{H_x},$$
$$Z^+_{yx} = -\frac{E_y}{H_x}, \quad Z^+_{zy} = -\frac{E_z}{H_y}, \quad Z^+_{xz} = -\frac{E_x}{H_z}, \tag{5.146}$$

$$Z^-_{xy} = -\frac{E_x}{H_y}, \quad Z^-_{yz} = -\frac{E_y}{H_z}, \quad Z^-_{zx} = -\frac{E_z}{H_x},$$
$$Z^-_{yx} = \frac{E_y}{H_x}, \quad Z^-_{zy} = \frac{E_z}{H_y}, \quad Z^-_{xz} = \frac{E_x}{H_z}. \tag{5.147}$$

Here, the wave impedances looking along the positive directions of the coordinates are indicated by the superscript plus signs, and the wave impedances in the opposite (or negative) directions are shown by the superscript minus signs. Similar definitions hold good for any orthogonal coordinate system, for example, the right cylindrical or spherical coordinate system.

For waves guided by transmission lines or waveguides, interest centres on the wave impedance which is seen when looking in the direction of propagation, that is, along the z-axis. An inspection of expressions for the transverse field components of a TM wave (see Table 5.2) in a rectangular waveguide shows that

$$\frac{E_x}{H_y} = -\frac{E_y}{H_x} = \frac{(E_x^2 + E_y^2)^{1/2}}{(H_x^2 + H_y^2)^{1/2}} = \frac{\beta_{nm}}{\omega\epsilon}.$$

Therefore

$$Z_{xy} = Z_{yx} = \frac{\beta_{nm}}{\omega\epsilon} = Z_z = Z_{e,\,nm}. \tag{5.148}$$

This equation shows that the wave impedances looking in the z-direction are equal and may be put equal to Z_z given by

$$Z_z = \frac{|E_t|}{|H_t|} = \frac{(E_x^2 + E_y^2)^{1/2}}{(H_x^2 + H_y^2)^{1/2}} = \frac{\beta_{nm}}{\omega\epsilon} = Z_{e,nm} \tag{5.149}$$

which is the ratio of the total transverse electric field strength to the total magnetic field strength. The expression $Z_z = \beta_{nm}/\omega\epsilon$ is independent of the coordinates x, y, z and is constant at all points inside the waveguide. Recalling that

$$\beta_{mn} = (k^2 - k^2_{c,\,nm})^{1/2} = (\omega^2\mu\epsilon - k^2_{c,\,nm})^{1/2} = \omega\sqrt{\mu\epsilon}\,(1 - \omega_c^2/\omega^2)^{1/2},$$

where ω_c is the cut-off frequency, the impedance looking in the z-direction for TM waves is

$$Z_{e,\,nm} = Z_z(\text{TM}) = [\tfrac{\mu}{\epsilon}(1 - \tfrac{\omega_c^2}{\omega^2})]^{1/2} = Z_0(1 - \tfrac{\omega_c^2}{\omega^2})^{1/2}. \tag{5.150}$$

Similarly, for TE waves it can be seen that (see Table 5.2)

$$Z_{h,\,nm} = Z_z(\text{TE}) = \frac{\omega\mu}{\beta_{nm}} = Z_0[1 - \frac{\omega_c^2}{\omega^2}]^{-1/2} \tag{5.151}$$

so that

$$Z_z(\text{TM})Z_z(\text{TE}) = Z_0^2. \tag{5.152}$$

For ordinary two-conductor transmission lines, the (integrated) characteristic impedance Z_c of the line, which is defined in terms of the voltage-current ratio or in terms of the power transmitted for a given voltage or current, is of major significance. Such a definition has been used for the coaxial line (see Section 5.8) for the TEM mode.

For an infinitely long line, the three definitions for Z that can be used are

$$Z_c = \frac{V}{I}, \quad Z_c = \frac{2W}{II^*}, \quad Z_c = \frac{VV^*}{2W}, \tag{5.153}$$

where V and I are the peak values of the voltage and current. For transmission lines carrying the TEM mode, these three definitions are equivalent and give identical values for Z_c. But, for waveguides, they lead to three different values that depend on the waveguide dimensions in the same way; however, they differ by a constant. For example, for the TE_{10} mode in the rectangular waveguide, the voltage can be taken as the maximum voltage from the lower face of the guide to the upper face. The maximum voltage occurs at $x = a/2$, and is given by

$$V = \int_{y=0}^{b} E_y(\max)\,dy = bE_y(\max) = \frac{-j\omega\mu ba}{\pi}. \tag{5.154}$$

The longitudinal linear current density in the lower face is

$$J_z = -H_x = \frac{j\beta_{10}a}{\pi}\sin\left(\frac{\pi x}{a}\right), \tag{5.155}$$

and hence the total longitudinal current in the lower face is

$$I = \int_{x=0}^{a} J_z\,dx = \frac{-j2a^2\beta_{10}}{\pi^2}. \tag{5.156}$$

The integrated characteristic impedance by the first definition in (5.153) is

$$Z_c(V, I) = \frac{V}{I} = \frac{\pi b\omega\mu}{2a\beta_{10}} = \frac{\pi b}{2a}Z_z = \frac{\pi b Z_0}{2a\sqrt{1-(f_c/f)^2}}, \tag{5.157}$$

where $Z_0 = \sqrt{\mu/\epsilon}$ and f_c is the cut-off frequency.

In terms of the second definition in (5.153), we have

$$W = \tfrac{1}{2}Z_z\int_{x=0}^{a}\int_{y=0}^{b} |H_x|^2\,dx\,dy = \tfrac{1}{2}Z_z\int_{x=0}^{a}\int_{y=0}^{b} \left(\frac{\beta_{10}a}{\pi}\right)^2 \sin^2\left(\frac{\pi x}{a}\right)dx\,dy = \tfrac{1}{4}Z_z\frac{a^3b\beta_{10}^2}{\pi^2},$$

and hence

$$Z_c(W, I) = \frac{Z_z a^3 b\beta_{10}^2}{2\pi^2}\,\frac{\pi^4}{4a^4\beta_{10}^2} = \frac{\pi^2 b}{8a}Z_z = \frac{\pi}{4}Z_c(V, I). \tag{5.158}$$

In terms of the third definition in (5.153), the integrated characteristic impedance is

$$Z_c(W, V) = \frac{VV^*}{2W} = \frac{2b}{a}Z_z = \frac{4}{\pi}Z_c(V, I). \tag{5.159}$$

5.19 Transmission Line Analogy for Waveguides

We can now show that there exists a useful analogy between the electric and magnetic fields of TE and TM waves in waveguides and the voltages and currents on suitably loaded transmission lines. This analogy is very useful in establishing equivalent circuits of waveguides and in making use of the transmission line theory for solving waveguide problems. Transmission line theory is studied in detail in Chapter 6.

For TM waves ($H_z = 0$) in the rectangular coordinates, the field equations are

$$\frac{\partial H_z}{\partial y} - \frac{\partial H_y}{\partial z} = +j\omega\epsilon E_x, \qquad \frac{\partial E_z}{\partial y} - \frac{\partial E_y}{\partial z} = j\omega\mu H_x,$$

$$\frac{\partial H_z}{\partial x} - \frac{\partial H_x}{\partial z} = -j\omega\epsilon E_y, \qquad \frac{\partial E_x}{\partial z} - \frac{\partial E_z}{\partial x} = -j\omega\mu H_y, \tag{5.160}$$

$$\frac{\partial H_y}{\partial x} - \frac{\partial H_x}{\partial y} = j\omega\epsilon E_z, \qquad \frac{\partial E_y}{\partial x} - \frac{\partial E_x}{\partial y} = -j\omega\mu H_z.$$

Since $H_z = 0$ for TM waves, the last equation in the set of equations (5.160) implies that

$$-j\omega\mu H_z = (\nabla \times E)_z = 0 \tag{5.161}$$

which means that the electric field has no curl in the xy-plane. Hence in the xy-plane, $\mathbf{E}$ may be written as the gradient of a scalar potential V (see also Section 5.5) so that

$$E_x = -\frac{\partial V}{\partial x}, \qquad E_y = -\frac{\partial V}{\partial y}. \tag{5.162}$$

From Eqs. (5.160) and (5.162), and remembering that $\partial/\partial z = -j\beta_{nm}$ and that $\beta_{nm}^2 = k^2 - k_{c,nm}^2$, we can see that

$$\frac{\partial}{\partial z}\left(\frac{j\omega\epsilon}{k_{c,nm}^2}\frac{\partial E_z}{\partial x}\right) = -j\omega\epsilon\frac{\partial V}{\partial x}$$

or

$$\frac{\partial}{\partial z}\left(\frac{j\omega\epsilon}{k_{c,nm}^2}\right) = -j\omega\epsilon V. \tag{5.163}$$

Further,

$$\frac{\partial E_x}{\partial z} - \frac{\partial E_z}{\partial x} = -\frac{\omega^2\mu\epsilon}{k_{c,nm}^2}\frac{\partial E_z}{\partial x},$$

and hence

$$\frac{\partial V}{\partial z} = \left(\frac{\omega^2\mu\epsilon}{k_{c,nm}^2} - 1\right)E_z = -\left(j\omega\mu + \frac{k_{c,nm}^2}{j\omega\epsilon}\right)\left(\frac{j\omega\epsilon}{h^2}E_z\right). \tag{5.164}$$

The quantity $j\omega\epsilon E_z$ is the longitudinal displacement current density, and $1/k_{c,nm}^2$ has the dimensions of area, so that $j\omega\epsilon E_z/k_{c,nm}^2$ represents a current in the z-direction and is designated by I_z. Then, Eqs. (5.163) and (5.164) become

$$\frac{\partial I_z}{\partial z} = -j\omega\epsilon V, \tag{5.165a}$$

$$\frac{\partial V}{\partial z} = -(j\omega\mu + \frac{k_{c,nm}^2}{j\omega\epsilon})I_z. \tag{5.165b}$$

Equations (5.165a) and (5.165b) are the differential equations for a lossless transmission line having a series impedance per unit length $Z = j\omega\mu + k_{c,nm}^2/(j\omega\epsilon)$, and a shunt admittance per unit length $Y = j\omega\epsilon$. The equivalent circuit for such a transmission line is shown in Fig. 5.15.

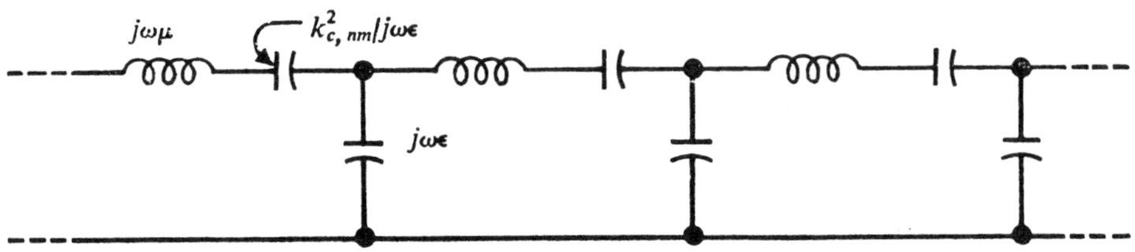

Fig. 5.15 Equivalent transmission-line circuit representation for TM waves in rectangular metal waveguide.

For TE waves, since $E_z = 0$, the two equations of interest in Eqs. (5.160) are

$$\frac{\partial E_y}{\partial z} = j\omega\mu H_x, \qquad \frac{\partial H_z}{\partial y} - \frac{\partial H_y}{\partial z} = j\omega\epsilon E_x.$$

Since $E_z = 0$, we have $(\nabla \times \boldsymbol{H})_z = 0$, and it is possible to define a magnetic scalar potential U in the xy-plane such that

$$H_x = -\frac{\partial U}{\partial x}, \qquad H_y = -\frac{\partial U}{\partial y}. \tag{5.166}$$

Using Eqs. (5.160) in Eqs. (5.166), we get

$$\frac{\partial}{\partial z}(\frac{j\omega\mu}{k_{c,nm}^2}\frac{\partial H_z}{\partial x}) = -j\omega\mu\frac{\partial U}{\partial x}, \qquad \frac{\partial H_z}{\partial y} - \frac{\partial H_y}{\partial z} = \frac{\omega^2\mu\epsilon}{k_{c,nm}^2}\frac{\partial H_z}{\partial y}$$

and therefore

$$\frac{\partial}{\partial z}(\frac{j\omega\mu H_z}{k_{c,nm}^2}) = -j\omega\mu U, \qquad \frac{\partial U}{\partial z} = -(\frac{k_{c,nm}^2}{j\omega\mu} + j\omega\epsilon)(\frac{j\omega\mu H_z}{k_{c,nm}^2}). \tag{5.167}$$

The quantity $j\omega\mu H_z/k_{c,nm}^2$ has the dimensions of voltage and U has the dimensions of current, and hence Eqs. (5.167) may be rewritten as

$$\frac{\partial V_1}{\partial z} = -ZI_1, \qquad \frac{\partial I_1}{\partial z} = -YV_1, \tag{5.168}$$

where

$$V_1 = \frac{j\omega\mu H_z}{k_{c,nm}^2}, \qquad I_1 = U, \qquad Z = j\omega\mu, \qquad Y = j\omega\epsilon + \frac{k_{c,nm}^2}{j\omega\mu}. \tag{5.169}$$

The equivalent circuit for TE waves is shown in Fig. 5.16.

The loaded transmission-line circuits of Figs. 5.15 and 5.16 have high-pass filter characteristics. The cut-off frequency for the line in Fig. 5.15 for TM waves occurs when the series

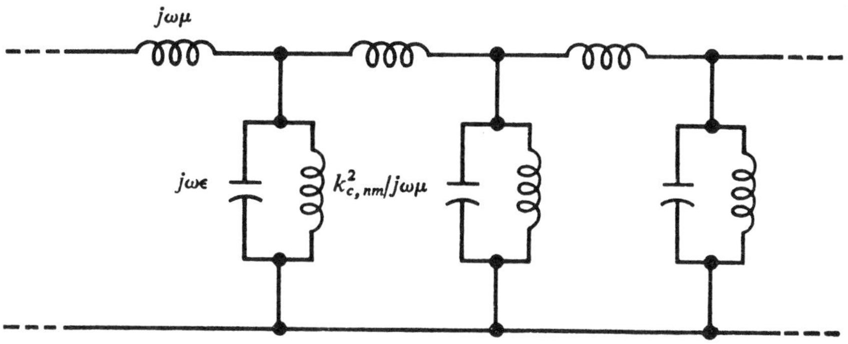

Fig. 5.16 Equivalent transmission-line circuit representation for TE waves in rectangular metal waveguide.

reactance equals zero, whereas for the line in Fig. 5.16 for TE waves, the cut-off frequency occurs when the shunt susceptance is equal to zero. Both these conditions require that

$$k_{c,nm}^2 = \omega_c^2\mu\epsilon \tag{5.170}$$

which was obtained in Sections 5.12 and 5.15. The characteristic impedance for the line in Fig. 5.15 for TM waves is

$$Z_c(\text{TM}) = \sqrt{\frac{Z}{Y}} = [\frac{j\omega\mu + k_{c,nm}^2/(j\omega\epsilon)}{j\omega\epsilon}]^{1/2} = [\frac{\mu}{\epsilon}(1 - \frac{\omega_c^2}{\omega^2})]^{1/2} = Z_z(\text{TM}) = Z_{e,nm} \tag{5.171}$$

and the characteristic impedance of the line in Fig. 5.16 for TE waves is

$$Z_c(\text{TE}) = [\frac{j\omega\mu}{j\omega\epsilon + k_{c,nm}^2/(j\omega\mu)}]^{1/2} = [\frac{\mu}{\epsilon}(\frac{1}{1 - \omega_c^2/\omega^2})]^{1/2} = Z_z(\text{TE}) = Z_{h,nm}. \tag{5.172}$$

It is thus obvious that the characteristic impedances of the equivalent transmission lines are equal to the corresponding wave impedances.

The concept of a waveguide as an equivalent transmission line is a very useful tool in the solution of waveguide problems, because circuit theory and transmission line theory can be used to obtain such a solution. However, it should be remembered that the two transmission-line equivalents can be applied to waveguides carrying a single TM_{nm} or TE_{nm} mode. Waveguides carrying more than one mode must be regarded as combinations of several such transmission lines.

Typical examples of waveguide problems such as discontinuities are shown in Fig. 5.17, and their equivalent circuits are shown in Fig. 5.18. In Figs. 5.17a and 5.18a, the iris with edges perpendicular to the $\boldsymbol{E}$ lines in a rectangular waveguide carrying the dominant TE_{10} mode behaves like a shunt capacitance, whereas in Figs. 5.17b and 5.18b, the iris with edges parallel to the $\boldsymbol{E}$ lines behaves like a shunt inductance. In Figs. 5.17c and 5.18c, an abrupt change in waveguide dimensions is represented by a shunt reactance.

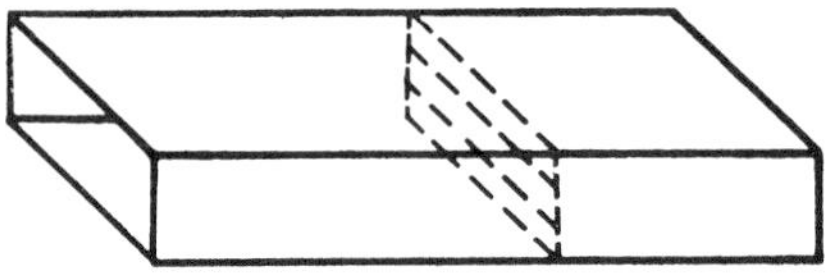

(a) Iris with edges perpendicular to E (TE_{10} mode)

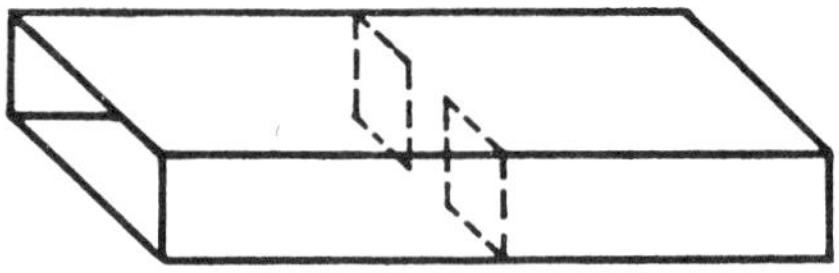

(b) Iris with edges parallel to E (TE_{10} mode)

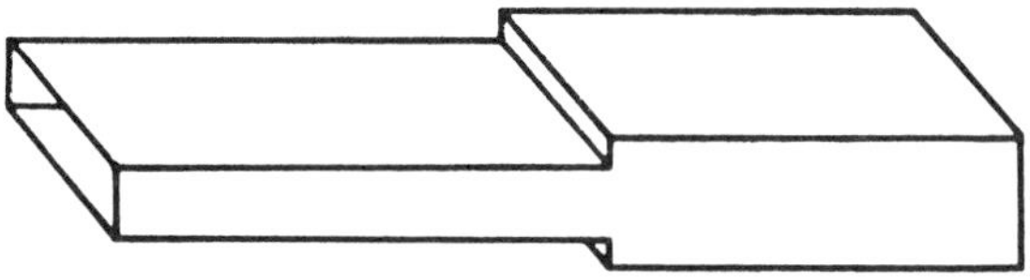

(c) Change of waveguide dimensions

Fig. 5.17 Typical discontinuities in waveguides.

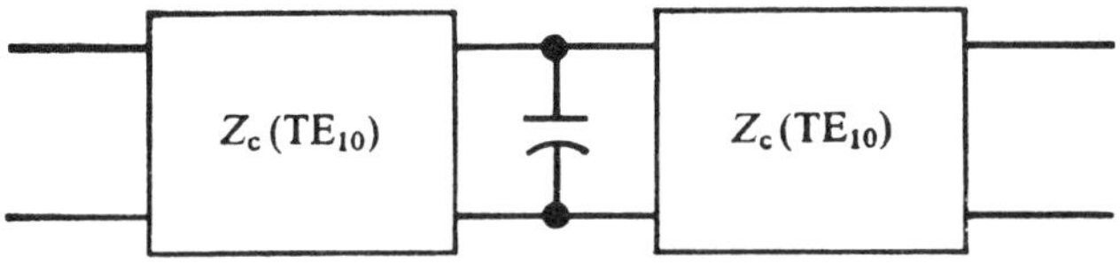

(a) Iris with edges perpendicular to E(TE_{10} mode)

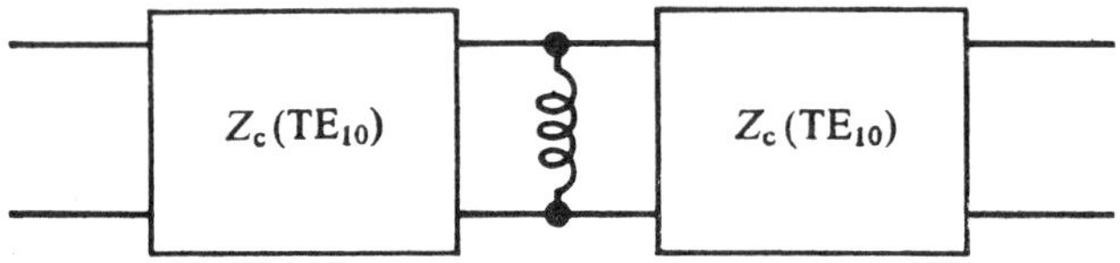

(b) Iris with edges parallel to E(TE_{10} mode)

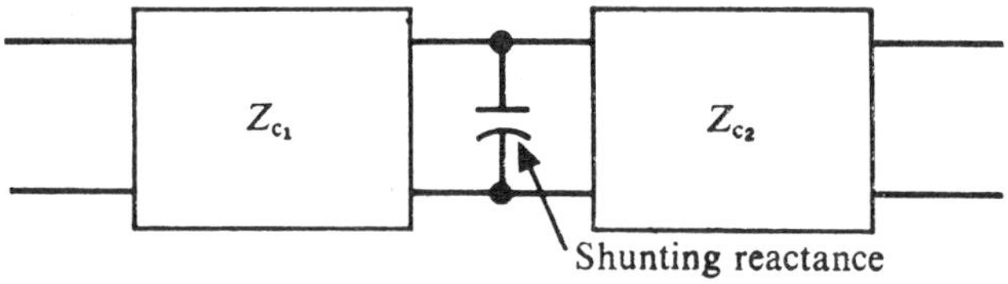

(c) Change of waveguide dimensions

Fig. 5.18 Equivalent circuits for waveguide discontinuities shown in Fig. 5.17.

The calculation of the actual value of shunt reactance or any other circuit equivalent is a field problem that can be solved by matching solutions at the boundary. The field at the junction or discontinuity is represented by the sum of the dominant and higher-order waves, the relative amplitudes of which are obtained by matching the tangential components of $\boldsymbol{E}$ and $\boldsymbol{H}$ at the boundary. The higher-order waves are set up by the discontinuity and are necessary to match the boundary conditions. But, since the higher-order waves have cut-off frequencies which are higher than the frequency of transmission, the waveguide has been designed to carry only the dominant TE_{10} mode, and the waves are attenuated rapidly as they travel away from the junction or discontinuity. The load impedance and the generator are assumed to be sufficiently far removed from the discontinuity so that the higher-order waves do not reach them. For this reason, the waveguide problem can be treated in terms of the effect of the discontinuity on the dominant wave only so that the circuit representation of transmission lines can be used to solve the problem.

The values for equivalent shunting reactances have been calculated by many workers (see Whinnery and Jamieson[1]), and are also found in waveguide handbooks (see Marcuvitz[2]).

5.20 Circular Metal Waveguides

Figure 5.19 shows a circular metal waveguide of radius a. To study the electromagnetic field inside

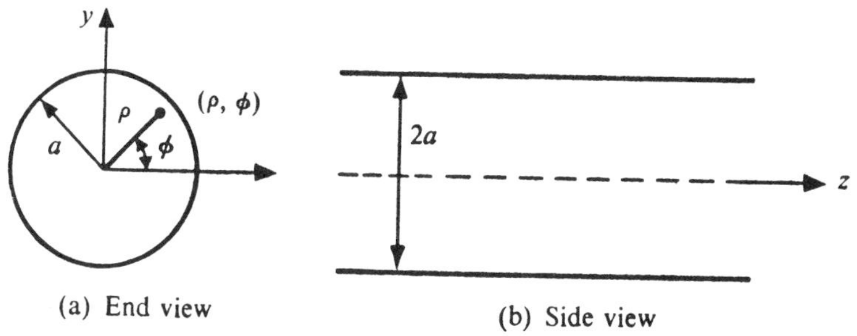

Fig. 5.19 Circular cylindrical waveguide.

a circular waveguide, the cylindrical coordinates (ρ, ϕ, z) are most appropriate.

TM Modes ($H_z = 0$)

For TM modes, a solution of the equation

$$\nabla_t^2 e_z + k_c^2 e_z = 0 \tag{5.173}$$

is required in the cylindrical coordinates such that e_z is zero at $r = a$ on the metal walls of the cylinder. In the cylindrical coordinates, Eq. (5.173) becomes

$$\frac{\partial^2 e_z}{\partial \rho^2} + \frac{1}{\rho}\frac{\partial e_z}{\partial \rho} + \frac{1}{\rho^2}\frac{\partial^2 e_z}{\partial \phi^2} + k_c^2 e_z = 0. \tag{5.174}$$

By using the method of separation of variables (see Section 4.5), the solution of Eq. (5.173) can

[1]J. R. Whinnery and H. W. Jamieson, Equivalent circuits for discontinuities in transmission lines, *Proc. I.R.E.*, **32**, 98, 1944.

[2]N. Marcuvitz (ed.), Waveguide Handbook, MIT Radiation Laboratory Series, Vol. 10, McGraw-Hill, New York, 1951, Ch. 5.

be written in the form

$$e_z(\rho, \phi) = [A_1 \cos(n\phi) + A_2 \sin(n\phi)]J_n(k_c\rho), \tag{5.175}$$

where $n = 0, 1, 2, \ldots$.

The boundary condition on the metal walls is $e_z = 0$ at $\rho = a$. Hence

$$J_n(k_c a) = 0 \tag{5.176}$$

which is the characteristic or eigen equation, with an infinite number of roots for each value of n. If p_{nm} is the m-th root of the equation $J_n(x) = 0$, then

$$k_{c,\,nm} = \frac{p_{nm}}{a}. \tag{5.177}$$

The values of p_{nm} for the first three values of $n = 0, 1, 2$ are given in Table 5.3. Since for each value of n an infinite number of values for p_{nm} exist, there is a double infinity of TM modes.

Table 5.3 Values of p_{nm} for TM Modes

n	p_{n1}	p_{n2}	p_{n3}
0	2.405	5.520	8.654
1	3.832	7.016	10.174
2	5.135	8.417	11.620

Each choice of n and m specifies a particular TM_{nm} mode (eigenfunction). The integer n gives the number of circumferential variations in the field, and since $J_n(x)$ behaves in a quasi-periodic fashion like a damped sinusoidal function, m gives the number of radial variations. The phase constant β_{nm} is given by

$$\beta_{nm} = (k^2 - k^2_{c,\,nm})^{1/2} = \left(k^2 - \frac{p^2_{nm}}{a^2}\right)^{1/2}, \tag{5.178}$$

where $k^2 = \omega^2\mu\epsilon$; the cut-off wavelength is

$$\lambda_{c,\,nm} = \frac{2\pi}{k_{c,\,nm}} = \frac{2\pi a}{p_{nm}}; \tag{5.179}$$

and the wave impedance looking in the z-direction is given by

$$Z_{e,\,nm} = Z_z(\text{TM}) = \frac{\beta_{nm}}{k} Z_0, \tag{5.180}$$

where $Z_0 = \sqrt{\mu/\epsilon}$. For an air-filled waveguide, $k = k_0 = \omega\sqrt{\mu_0\epsilon_0}$. For the lowest TM mode, $\lambda_c = 2\pi a/p_{01} = 2.61a$, which is 30 per cent greater than the waveguide diameter.

Expressions for the other field components may be derived by using the general equations (5.33) and (5.34). Power flow and attenuation may be determined by using methods similar to those used for the rectangular metal waveguide. The results are shown in Table 5.4.

Table 5.4 Properties of Modes in Cylindrical Waveguides

Property	*TE modes*	*TM modes*
H_z	$J_n(\frac{p'_{nm}\rho}{a}) \exp(-j\bar{\beta}_{nm}z)\begin{cases}\cos(n\phi)\\ \sin(n\phi)\end{cases}$	0
E_z	0	$J_n(\frac{p_{nm}\rho}{a}) \exp(-j\bar{\beta}_{nm}z)\begin{cases}\cos(n\phi)\\ \sin(n\phi)\end{cases}$
H_ρ	$\frac{-j\bar{\beta}_{nm}p'_{nm}}{ak^2_{c,nm}}J'_n(\frac{p'_{nm}\rho}{a}) \exp(-j\bar{\beta}_{nm}z)\begin{cases}\cos(n\phi)\\ \sin(n\phi)\end{cases}$	$\frac{-E_\phi}{Z_{e,nm}}$
H_ϕ	$\frac{-jn\bar{\beta}_{nm}}{pk^2_{c,nm}}J_n(\frac{p'_{nm}\rho}{a}) \exp(-j\bar{\beta}_{nm}z)\begin{cases}-\sin(n\phi)\\ \cos(n\phi)\end{cases}$	$\frac{E_\rho}{Z_{e,nm}}$
E_ρ	$Z_{h,nm}H_\phi$	$\frac{-j\bar{\beta}_{nm}p_{nm}}{ak^2_{c,nm}}J'_n(\frac{p_{nm}\rho}{a}) \exp(-j\bar{\beta}_{nm}z)\begin{cases}\cos(n\phi)\\ \sin(n\phi)\end{cases}$
E_ϕ	$-Z_{h,nm}H_\rho$	$\frac{-jn\bar{\beta}_{nm}}{pk^2_{c,nm}}J_n(\frac{p_{nm}\rho}{a}) \exp(-j\bar{\beta}_{nm}z)\begin{cases}-\sin(n\phi)\\ \cos(n\phi)\end{cases}$
$\bar{\beta}_{nm}$	$[k_0^2 - (p_{nm}/a)]^{1/2}$	$[k_0^2 - (p_{nm}/a)^2]^{1/2}$
$Z_{h,nm}$	$(\frac{k_0}{\bar{\beta}_{nm}})Z_0$	
$Z_{e,nm}$		$(\frac{\bar{\beta}_{nm}}{k_0})Z_0$
$k_{c,nm}$	$\frac{p'_{nm}}{a}$	$\frac{p_{nm}}{a}$
$\lambda_{c,nm}$	$\frac{2\pi a}{p'_{nm}}$	$\frac{2\pi a}{p_{nm}}$
Power	$\frac{Z_0k_0\bar{\beta}_{nm}\pi}{4k^4_{c,nm}}(p'^2_{nm} - n^2)J^2_n(p'_{nm}a)$	$\frac{Z_0k_0\bar{\beta}_{nm}\pi}{4k^4_{c,nm}}p^2_{nm}[J'_n(k_{c,nm}a)]^2$
α	$\frac{R_m}{aZ_0}(1 - \frac{k^2_{c,nm}}{k_0^2})^{-1/2}[\frac{k^2_{c,nm}}{k_0^2} + \frac{n^2}{(p'_{nm})^2 - n^2}]$	$\frac{R_m}{aZ_0}(1 - \frac{k^2_{c,nm}}{k_0^2})^{-1/2}$

$Z_0 = \sqrt{\mu_0/\epsilon_0} = 1/Y_0$ for free space
$= \sqrt{\mu/\epsilon}$ for any other dielectric.

TE Modes ($E_z = 0$)

The analysis of TE modes in the circular metal waveguide is similar to that of TM modes. The boundary conditions require that $\partial h_z/\partial\rho$ should vanish at $\rho = a$, because E_ϕ which is proportional to $\partial h_z/\partial\rho$ should vanish at $\rho = a$. An appropriate solution for h_z is

$$h_z(\rho, \phi) = [B_1 \cos(n\phi) + B_2 \sin(n\phi)]J_n(k_c\rho) \quad (5.181)$$

with the boundary condition

$$\frac{d}{d\rho}J_n(k_c\rho) = 0 \quad \text{at } \rho = a. \tag{5.182}$$

If the roots of Eq. (5.181) are designated p'_{nm}, then the eigenvalues $k_{c,nm}$ are given by

$$k_{c,nm} = \frac{p'_{nm}}{a}. \tag{5.183}$$

Table 5.5 gives the values of p'_{nm} for the first few TE modes. It is interesting to note that $p'_{0m} = p_{1m}$ since $dJ_0(x)/dx = -J_1(x)$, and hence the TE_{0m} and TM_{1m} modes are degenerate. The lowest value of p'_{n_1} is $p'_{11} = 1.841$, and hence the TE_{11} mode has the largest cut-off wave-

Table 5.5 Values of p'_{nm} for TE Modes

n	p'_{n1}	p'_{n2}	p'_{n3}
0	3.832	7.016	10.174
1	1.841	5.331	8.536
2	3.054	6.706	9.970

length equal to $\lambda_{c,11} = 3.41a$. Therefore, the TE_{11} mode is the dominant mode in the circular cylindrical waveguide, and is normally the one used in practice. The field configuration for this mode is given in Fig. 5.20.

Comparing Fig. 5.20 with Fig. 5.11 that gives the field lines for the TE_{10} mode (the dominant mode in the rectangular waveguide), we observe that the two field configurations are

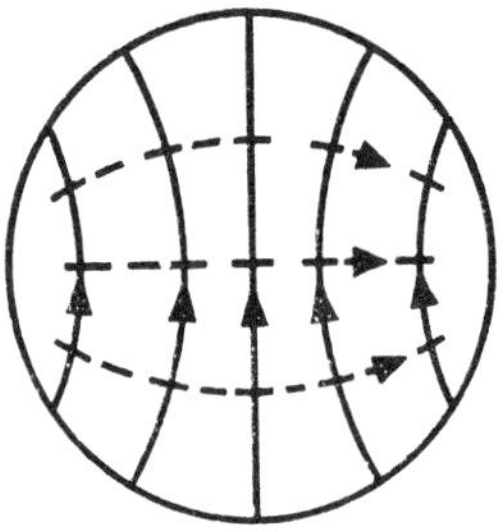

Fig. 5.20 Field lines for the TE_{11} mode in circular waveguide.

similar except for the fact that, in the circular guide, the lines are curved to suit the circular geometry. If a rectangular waveguide that carries only the dominant TE_{10} mode is gradually shaped to become a circular waveguide and is also designed to carry the dominant TE_{11} mode, it is natural for the field lines shown in Fig. 5.11 to gradually become the field lines given in Fig. 5.20.

Figure 5.21 shows the field configurations for the TM_{01} and TE_{01} modes in the cylindrical waveguides.

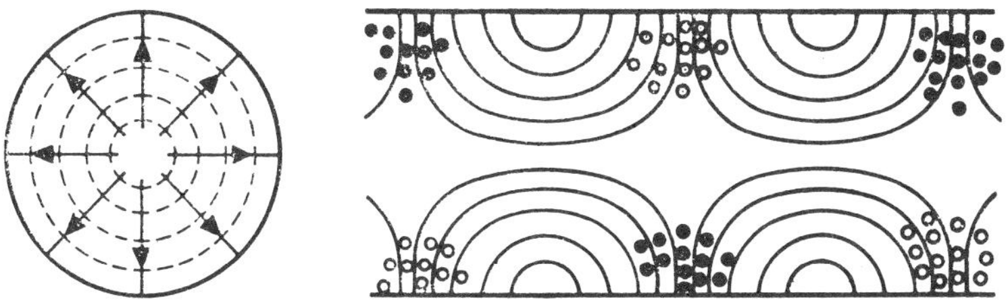

(a) TM_{01} mode

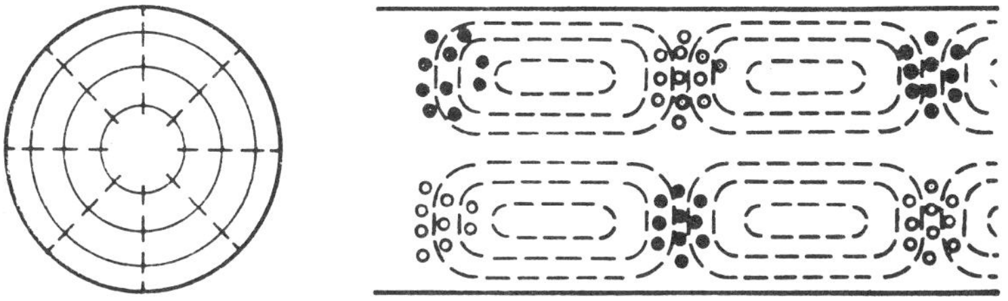

(b) TE_{01} mode

Fig. 5.21 TE_{01} and TM_{01} modes in cylindrical waveguides.

Figure 5.22 illustrates the methods of excitation of the TE_{01}, TE_{11}, and TM_{01} modes in the cylindrical waveguides.

If the expression for the attenuation constant for TE modes in Table 5.4 is examined, it is seen that, for the TE_{0m} modes, the attenuation is

$$\alpha = \frac{R_m}{aZ_0} \frac{f_{c,\,0m}^2}{f(f^2 - f_{c,\,0m}^2)^{1/2}}. \tag{5.184}$$

The attenuation α falls off as $f^{-3/2}$ for high frequencies since R_m increases as $f^{1/2}$. The rapid decrease in attenuation with increasing frequency is a unique property of the TE_{0m} modes in the circular waveguides and is useful in making very long low-wave waveguide communication links (see Miller[1,2]). Figure 5.23 shows the attenuation in decibels for the TE_{0m} modes as a function of frequency for a copper waveguide.

Theoretically, although very low attenuation is obtained for frequencies well above cut-off, in practice, much higher attenuations are obtained. The practical difficulty lies in the fact that, to obtain very low attenuation, the waveguide has to be operated at a frequency very much higher than the cut-off frequency for the dominant TE_{11} mode, and at such a high frequency, many modes propagate. Any small irregularity in the waveguide causes some of the power in the TE_{01} mode to be converted into power in the other modes. Two major effects arise from this mode

[1]S. E. Miller, Waveguide as a communication medium, *B.S.T.J.*, 33, 1209, 1954.

[2]S. E. Miller, Millimeter waves in communications, in J. Fox (ed.) Proc. Symp. Millimeter Waves, Polytechnic Press of Polytechnic Institute of Brooklyn, New York, 1959.

conversion. The first is the loss in power in the TE_{01} mode. The second and more serious one is that, when the power in the TE_{01} mode is converted into power in the other modes, which have

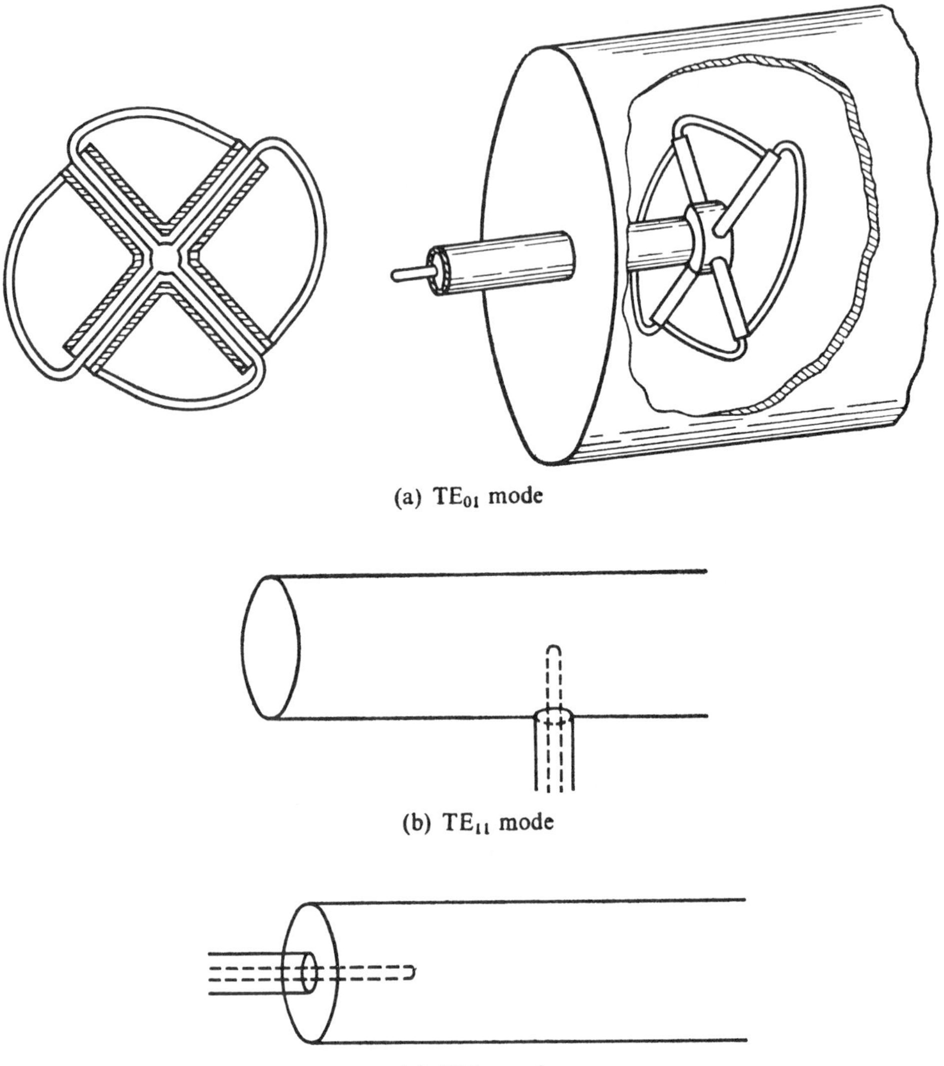

(a) TE_{01} mode

(b) TE_{11} mode

(c) TM_{01} mode

Fig. 5.22 Excitation of TE_{01}, TE_{11}, and TM_{01} modes in cylindrical waveguides.

different propagation constants and hence different phase velocities, the converted power will be reconverted to the TE_{01} mode at a position further along the guide, resulting in signal distortion.

To avoid the signal distortion arising from the mode conversion and reconversion, it is necessary to have a high attenuation for the undesired modes. Since the currents associated with the TE_{0m} modes are in the circumferential direction only, this property may be made use of to construct mode filters that suppress modes having currents directed along the waveguide axis. Special waveguide linings, having high conductivity in the circumferential direction and low conductivity in the axial direction, have been used as mode suppressors. Another technique that

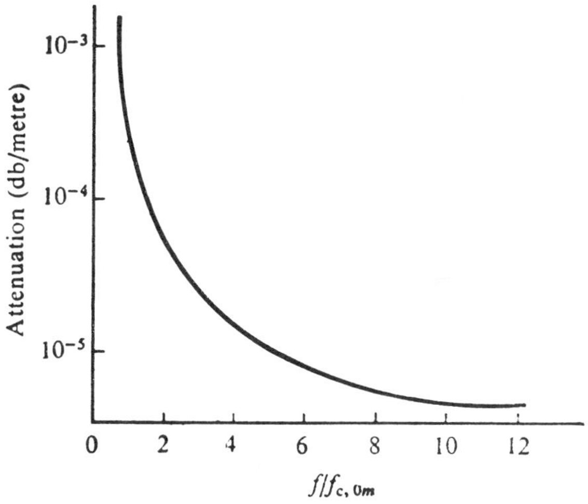

Fig. 5.23 Attenuation of low-loss TE_{0m} modes in a circular copper waveguide. (f in units of 10^{10} cps, $f_{c,10} = 1.83 \times 10^{10}a$ cps, where a = radius in centimetres.)

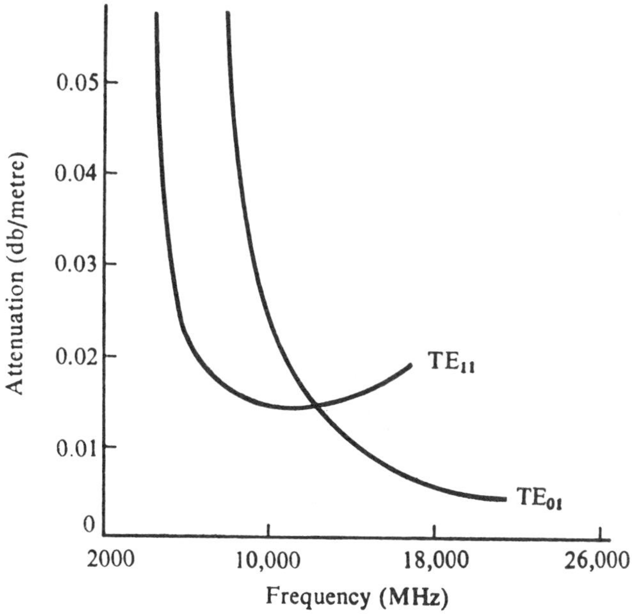

Fig. 5.24 Attenuation due to copper guide wall losses in a circular waveguide of 2 in diameter.

shows considerable promise is the use of a small pitch helical winding with a suitable supporting jacket for the waveguide. The helix guide has a very low conductivity in the axial direction, whereas in the circumferential direction, the conductivity is essentially the same as that for a solid cylindrical waveguide wall (see Morgan and Young[1], Morrison[2], Hosono and Kohno[3], and Luderer and Unger[4]). On the other hand, the attenuation for the dominant TE_{11} mode is higher than that for the TE_{10} mode and is minimum at a certain frequency and increases for higher frequencies, as shown in Fig. 5.24.

5.21 Mode Filters

In Section 5.19, we have seen that a waveguide carrying a single TE or TM mode in a rectangular waveguide can be represented by an equivalent transmission line. This is equally true for a circular cylindrical waveguide. The design of a microwave network employing a single mode waveguide, as a transmission line or as a microwave circuit element (e.g., inductance and capacitance), or as a combination of both, becomes considerably simplified since transmission line analysis can be made use of for the waveguide analysis. On the other hand, when two or more modes simultaneously exist in the waveguide, the analysis for determining the equivalent circuit of the waveguide becomes quite complicated. Also, the existence of more than one mode reduces the efficiency of the waveguide since the power in the unwanted modes cannot be properly utilized. Hence, to suppress the unwanted modes, mode filters or mode absorbers are used.

Mode filters usually consist of suitable metallic screens placed in the transverse plane of the waveguide. The screens are so designed that the undesirable mode (or modes) is reflected without affecting the transmission of the desired mode. If a thin conductor is placed along the transverse electric field lines for a particular waveguide mode, the incident power on the conductor will be scattered and a portion of the incident power corresponding to this mode will be reflected back. However, if the conductor is placed perpendicular to the electric field, then very little power will be reflected back from the conductor, provided the waveguide is matched. Thus, a screen can be designed with thin conducting wires along the direction of the transverse electric fields for the undesirable modes, provided the desired mode has no electric field lines along the same direction. Typical screens to suppress some modes in the cylindrical and rectangular waveguides are shown in Fig. 5.25.

In Fig. 5.25a is shown a filter for a circular waveguide which has to carry the TE_{10} mode. Since the TE_{10} mode in a circular guide has no radial electric field lines (see Fig. 5.21), this filter does not affect the transmission of this mode. The dominant TE_{11} mode and other TE modes in a circular guide, for which $p \neq 0$, are reflected by the screen since they have transverse electric field components along the radial direction. Also, all circular TM modes are reflected by the screen since all of them have radial electric field lines.

Figure 5.25b shows a screen suitable for a rectangular waveguide carrying the dominant TE_{10} mode. This screen reflects all TE_{pq} modes for which $q \neq 0$ as also all TM_{pq} modes for which $p \neq 0$, because all these modes have components of $\boldsymbol{E}$ in the x-direction.

A screen for a circular waveguide carrying the TM_{01} mode is shown in Fig. 5.25c. This screen reflects all TM_{pq} modes for which $p \neq 0$ and all TE_{pq} modes, because all these modes have circular electric field lines.

The single resonant ring of Fig. 5.25d reflects the dominant TE_{11} mode and partially

[1]S. P. Morgan and J. A. Young, Helix waveguide, *B.S.T.J.*, **35**, 1347, 1956.
[2]J. A. Morrison, Heat loss of circular electric waves in helix waveguides, *IRE Trans.*, **MTT-6**, 173, 1958.
[3]T. Hosono and S. Kohno, The transmission of TE_{10} waves in helix waveguides, *IRE Trans.*, **MTT-7**, 370, 1959.
[4]G. W. Luderer and H. G. Unger, Circular electric wave propagation in periodic structures, *B.S.T.J.*, **43**, 755, 1964.

reflects the TE_{01} mode.

It is difficult to achieve both perfect reflection for all undesirable modes and absolutely no

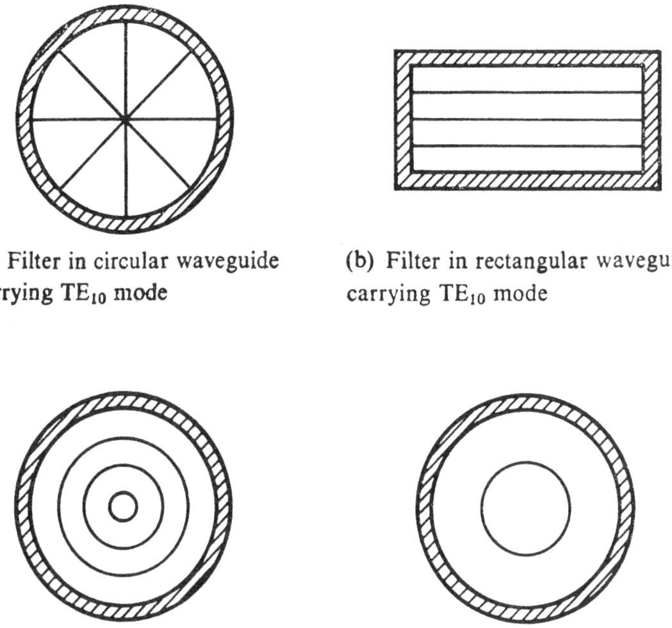

(a) Filter in circular waveguide carrying TE_{10} mode

(b) Filter in rectangular waveguide carrying TE_{10} mode

(c) Filter in circular waveguide carrying TM_{01} mode

(d) Resonant-ring mode filter in circular waveguide carryingTE_{11} mode

Fig. 5.25 Typical mode filters for circular and rectangular waveguides.

reflection for the desired mode. This is not even necessary in most cases.

5.22 Concepts of Voltage and Current in a Bounded Electromagnetic Field as in a Waveguide

The analysis and synthesis of low-frequency circuits and networks are often facilitated by considering voltages and currents at various terminals or sections of the network. In Section 5.18, voltage and current have been defined for the rectangular waveguide carrying the dominant TE_{10} mode. It will be useful to extend the concepts of voltage and current in a waveguide, which is not necessarily carrying a single mode, so that the information available in low-frequency circuit theory can be made use of in analyzing microwave circuits which consist of waveguides and cavities containing bounded electromagnetic fields.

The distribution of electric and magnetic fields inside a waveguide or cavity is such that the familiar concept which postulates that the voltage and currents are line integrals of $\boldsymbol{E}$ and $\boldsymbol{H}$ serves no useful purpose, especially in finding the equivalent low-frequency circuit of a microwave circuit. For instance, in Section 5.18, the voltage in a waveguide carrying the TE_{10} mode was defined as the line integral of the maximum electric field from the lower face of the guide to the upper face, whereas the current was defined as the total longitudinal current in the lower or upper face. This method cannot be easily applied to a waveguide carrying other modes, even if there is only a single mode. However, it is possible to define voltage V and current I in a waveguide such that voltage is proportional to the transverse electric field and current is proportional to the transverse magnetic field in the waveguide. Thus

(i) the voltage V is proportional to the strength of the transverse electric field $\boldsymbol{E}_t = \boldsymbol{e}_t \exp(-j\beta z)$;

(ii) the current I is proportional to the strength of the transverse magnetic field $\boldsymbol{H}_t = \boldsymbol{h}_t \exp(-j\beta z)$; and

(iii) the product of voltage and the conjugate of the current gives a measure of the power flow.

This definition can be used for a waveguide carrying a single mode or more than one mode.

It may be interesting to point out that the foregoing definition is not only different from the definitions used in Section 5.18, but differs also from the concepts of voltage and current defined in Section 5.19, where the voltage V is the scalar potential from which the transverse components of the electric field are derived and the current I is the displacement current in the z-direction, in order to derive the transmission-line equivalent of a waveguide carrying the TM waves. The concepts of voltage and current used for TE waves in Section 5.18 are again different. Thus, it is possible to define voltage and current in a waveguide in several ways. However, the definitions used here are more general.

Thus, if V and I respectively represent the complex voltage and complex current as functions of z, we can put

$$E_u = Vf_u(u, v), \tag{5.185a}$$

$$E_v = Vf_v(u, v), \tag{5.185b}$$

$$H_u = Ig_u(u, v), \tag{5.185c}$$

$$H_v = Ig_v(u, v), \tag{5.185d}$$

where f_u, f_v, g_u, g_v are the real functions of the coordinates u, v in the transverse plane. The coordinates (u, v) may be (x, y) for a rectangular waveguide or (ρ, ϕ) for a cylindrical waveguide. We also assume that f and g are normalized so that

$$\int_S (f_u g_v - f_v g_u)\, dS = -1, \tag{5.186}$$

where S is the cross-sectional area of the waveguide.

The power transmitted along the z-direction is

$$P_z = \tfrac{1}{2} \int_S (\boldsymbol{E}_t \times \boldsymbol{H}_t^*) \cdot \boldsymbol{u}_z\, dS. \tag{5.187}$$

Substituting Eqs. (5.185) in Eq. (5.187) and making use of Eq. (5.186), we obtain

$$P_z = \tfrac{1}{2} VI^* \int_S (f_u g_v - f_v g_u)\, dS = -\tfrac{1}{2} VI^*. \tag{5.188}$$

The minus sign here is used to denote the power flow into the network. The definitions of V and I, as given in Eqs. (5.185), satisfy conditions (i)–(iii). For the low-frequency equivalent circuit of a waveguide, V and I (both independent of the transverse coordinates) should refer to the voltage and current at any value of z.

It will be shown in Section 5.23 that the definitions of voltage and current as given in Eqs. (5.185) are unambiguous or, in other words, there is a one-to-one correspondence between V and I and the $\boldsymbol{E}$ and $\boldsymbol{H}$ fields.

5.23 Uniqueness of Voltage and Current in a Waveguide

We have seen from Eqs. (5.185) that any of the permissible field solutions (E_u, E_v, H_u, H_v) in a waveguide corresponds to a definite voltage or current. To prove the uniqueness of voltage and current, it is necessary to show that any particular value of voltage or current corresponds to a unique field distribution. To do this, consider first Maxwell's equations for a simple periodic field:

$$\nabla \times \boldsymbol{H} = (j\omega\epsilon + \sigma)\boldsymbol{E}, \tag{5.189a}$$

$$\nabla \times \boldsymbol{E} = -j\omega\mu\boldsymbol{H}, \tag{5.189b}$$

$$\nabla \cdot \epsilon\boldsymbol{E} = \rho, \tag{5.189c}$$

$$\nabla \cdot \mu\boldsymbol{H} = 0, \tag{5.189d}$$

where $\boldsymbol{E}$ and $\boldsymbol{H}$ are, in general, complex quantities. Also, consider the identity

$$\nabla \cdot (\boldsymbol{E} \times \boldsymbol{H}^*) = (\boldsymbol{H}^* \cdot \nabla \times \boldsymbol{E}) - (\boldsymbol{E} \cdot \nabla \times \boldsymbol{H}^*) = -j\omega\mu\boldsymbol{H}^* \cdot \boldsymbol{H} - (\sigma + j\omega\epsilon)\boldsymbol{E}^* \cdot \boldsymbol{E}. \tag{5.190}$$

Then, using the divergence theorem, this equation can be rewritten as

$$\int_S \boldsymbol{E} \times \boldsymbol{H}^* \cdot d\boldsymbol{S} = -j\omega \int_V \mu\boldsymbol{H}^* \cdot \boldsymbol{H}\, dV + j\omega \int_V \epsilon\boldsymbol{E}^* \cdot \boldsymbol{E}\, dV - \int_V \sigma\boldsymbol{E}^* \cdot \boldsymbol{E}\, dV. \tag{5.191}$$

From Eqs. (5.187), (5.188), and (5.191), we have

$$VI^* = j\omega\left(\int_V \mu\boldsymbol{H}^* \cdot \boldsymbol{H}^*\, dV - \int_V \epsilon\boldsymbol{E}^* \cdot \boldsymbol{E}\, dV\right) + \int_V \sigma\boldsymbol{E}^* \cdot \boldsymbol{E}\, dV. \tag{5.192}$$

In order to prove the uniqueness of voltage and current and the electromagnetic field, we shall show that given V and I, $\boldsymbol{E}$ and $\boldsymbol{H}$ are uniquely determined and vice versa. To do this, we consider two sets of field components and the corresponding voltages and currents such that

$$\begin{aligned} &\boldsymbol{E}_1, \boldsymbol{H}_1 \text{ correspond to } V_1, I_1, \\ &\boldsymbol{E}_2, \boldsymbol{H}_2 \text{ correspond to } V_2, I_2. \end{aligned} \tag{5.193}$$

Since $\boldsymbol{E}_1$, $\boldsymbol{H}_1$ and $\boldsymbol{E}_2$, $\boldsymbol{H}_2$ satisfy Maxwell's equations and the boundary conditions pertinent to the given waveguide independently, $(\boldsymbol{E}_1 - \boldsymbol{E}_2)$, $(\boldsymbol{H}_1 - \boldsymbol{H}_2)$ and the corresponding voltage and current $(V_1 - V_2)$ and $(I_1 - I_2)$ should also form a solution to these equations and satisfy Eq. (5.192). This is due to the linearity of Maxwell's equations. Therefore

$$\begin{aligned} (V_1 - V_2)(I_1^* - I_2^*) = j\omega\Bigg[&\int_V \mu(\boldsymbol{H}_1^* - \boldsymbol{H}_2^*) \cdot (\boldsymbol{H}_1 - \boldsymbol{H}_2)\, dV \\ &- \int_V \epsilon(\boldsymbol{E}_1^* - \boldsymbol{E}_2^*) \cdot (\boldsymbol{E}_1 - \boldsymbol{E}_2)\, dV\Bigg] \\ &+ \int_V \sigma(\boldsymbol{E}_1^* - \boldsymbol{E}_2^*) \cdot (\boldsymbol{E}_1 - \boldsymbol{E}_2)\, dV \end{aligned} \tag{5.194}$$

or

$$(V_1 - V_2)(I_1^* - I_2^*) = j\omega\left(\int_V \mu|\boldsymbol{H}_1 - \boldsymbol{H}_2|^2\, dV - \int_V \epsilon|\boldsymbol{E}_1 - \boldsymbol{E}_2|^2\, dV\right) + \int_V \sigma|\boldsymbol{E}_1 - \boldsymbol{E}_2|^2\, dV. \tag{5.195}$$

If $V_1 = V_2$ or $I_1 = I_2$, the left-hand side of Eq. (5.195) vanishes, but each volume integral on the right-hand side represents integration of a quantity, non-negative in the entire range of integration. It should be noted that this proof fails in a lossless waveguide, where σ is identically zero. In practical waveguides, σ is never zero, and so we may assume that it is a very small quantity in the so-called lossless waveguides. The only way the right-hand side of Eq. (5.195) can be zero is for each of these integrals to vanish independently. Also, the integral of a non-negative quantity can be zero only when the quantity vanishes identically. Hence

$$\boldsymbol{E}_1 = \boldsymbol{E}_2, \qquad \boldsymbol{H}_1 = \boldsymbol{H}_2. \tag{5.196}$$

Thus, we can see from Eqs. (5.196) that, for a given voltage $V(z)$ and a given current $I(z)$ at any longitudinal position z, as defined by Eqs. (5.185), there cannot exist two $\boldsymbol{E}$ and $\boldsymbol{H}$ fields which are not identically the same. This proves the uniqueness of the definition of voltage and current, as given in Eqs. (5.185). Also, if the waveguide is terminated at any point, the terminal voltage and current and the electromagnetic field inside the waveguide are uniquely related.

5.24 Relationship between Power Flow, Impedance, and Stored and Dissipated Energies

In a low-frequency transmission line, the average power flow is given by $\frac{1}{2}VI^*$, and the real power flow is Re $(\frac{1}{2}VI^*)$. In Section 5.19, the equivalence between a rectangular waveguide carrying the TE or TM modes and a low-frequency transmission line was established. From Eqs. (5.187), (5.188), and (5.191), we may write the expression for the power flow in a waveguide as

$$\tfrac{1}{2}VI^* = -\tfrac{1}{2}\int_S (\boldsymbol{E}\times\boldsymbol{H}^*)_z\, dS = j\omega\left(\int_V \mu\boldsymbol{H}^*\cdot\boldsymbol{H}\, dV - \int_V \epsilon\boldsymbol{E}^*\cdot\boldsymbol{E}\, dV\right) + \tfrac{1}{2}\int_V \sigma\boldsymbol{E}^*\cdot\boldsymbol{E}\, dV, \tag{5.197}$$

the minus sign being introduced to take into account the power flow into the network.

The average power flow can be obtained by integrating the power flow over a period. Although a harmonic time independence $e^{j\omega t}$ of the field is assumed, the electric and magnetic field vectors are the real parts of the respective expressions, that is,

$$\begin{aligned}\boldsymbol{E} &= \text{Re}\,[\boldsymbol{E}(p)e^{j\omega t}],\\ \boldsymbol{H} &= \text{Re}\,[\boldsymbol{H}(p)e^{j\omega t}],\end{aligned} \tag{5.198}$$

where $\boldsymbol{E}(p)$ and $\boldsymbol{H}(p)$ are vectors and functions of position. If $\boldsymbol{E}(p)$ and $\boldsymbol{H}(p)$ are written as

$$\begin{aligned}\boldsymbol{E}(p) &= \boldsymbol{E}_{\text{Re}} + j\boldsymbol{E}_{\text{Im}},\\ \boldsymbol{H}(p) &= \boldsymbol{H}_{\text{Re}} + j\boldsymbol{H}_{\text{Im}},\end{aligned} \tag{5.199}$$

then we obtain

$$\begin{aligned} \boldsymbol{E} &= \boldsymbol{E}_{\text{Re}} \cos \omega t - \boldsymbol{E}_{\text{Im}} \sin \omega t, \\ \boldsymbol{H} &= \boldsymbol{H}_{\text{Re}} \cos \omega t - \boldsymbol{H}_{\text{Im}} \sin \omega t \end{aligned} \tag{5.200}$$

so that, finally,

$$\frac{\epsilon}{2}\boldsymbol{E}^* \cdot \boldsymbol{E} = \frac{\epsilon}{2}|E|^2 = \frac{\epsilon}{2}(E_{\text{Re}}^2 \cos^2 \omega t + E_{\text{Im}}^2 \sin^2 \omega t - 2E_{\text{Re}}E_{\text{Im}} \sin \omega t \cos \omega t), \tag{5.201}$$

$$\frac{\mu}{2}\boldsymbol{H}^* \cdot \boldsymbol{H} = \frac{\mu}{2}|H|^2 = \frac{\mu}{2}(H_{\text{Re}}^2 \cos^2 \omega t + H_{\text{Im}}^2 \sin^2 \omega t - 2H_{\text{Re}}H_{\text{Im}} \sin \omega t \cos \omega t). \tag{5.202}$$

The time averages of Eqs. (5.201) and (5.202) over a period give the average stored energies in the electric and magnetic fields, respectively. If W_{H} and W_{E} respectively denote average stored energies in the magnetic and electric fields in the volume under consideration, the average time integral of Eq. (5.197) gives

$$\frac{1}{2}\frac{1}{T}\int_0^T VI^* \, dt = \frac{j\omega}{2}(4W_{\text{H}} - 4W_{\text{E}}) + P, \tag{5.203}$$

where T is the period of one cycle and P the average dissipated power. The physical interpretation of Eq. (5.203) can be given as follows. The difference between the energy input and energy output of the volume of the waveguide or any other bounded electromagnetic field is equal to the difference of the average stored energies in the electric and magnetic fields, less the dissipated energy. Equation (5.203) may therefore be rewritten as

$$(VI^*)_{\text{av}} = j4\omega(W_{\text{H}} - W_{\text{E}}) + 2P. \tag{5.204}$$

To obtain the equivalence between the impedance of a low-frequency transmission line element and that of a section of a waveguide from Eq. (5.204), IZ is substituted for V, and the equation becomes

$$II^*Z = j4\omega(W_{\text{H}} - W_{\text{E}}) + 2P$$

or

$$Z = \frac{2j\omega(W_{\text{H}} - W_{\text{E}}) + P}{\frac{1}{2}II^*}. \tag{5.205}$$

Similarly, the admittance Y is

$$Y = \frac{2j\omega(W_{\text{E}} - W_{\text{H}}) + P}{\frac{1}{2}VV^*}. \tag{5.206}$$

Equations (5.205) and (5.206) establish two very important and useful concepts in microwave circuit theory: (i) The stored electric and magnetic energies in a microwave circuit determine the reactance, whereas the dissipated energy determines the resistance of the circuit; (ii) if the stored magnetic energy exceeds the stored electric energy, the circuit is inductive, whereas if the stored electric energy exceeds the stored magnetic energy, then the circuit is capacitive.

5.25 Power Orthogonality

We shall now show that, in a hollow cylindrical conducting waveguide with no losses, each possible mode of propagation carries energy independently of all other modes present. This can be proved for any particular waveguide by making use of the actual expressions for the field components. More generally, we can also prove the following.

Let $\boldsymbol{E}_1$, $\boldsymbol{H}_1$ and $\boldsymbol{E}_2$, $\boldsymbol{H}_2$ be two linearly independent solutions (modes) of Maxwell's equations in a given waveguide. These fields may be put in the form

$$\boldsymbol{E}_1 = \boldsymbol{e}_1 \exp(-j\beta_1 z) + \boldsymbol{e}_{z_1} \exp(-j\beta_1 z), \tag{5.207a}$$

$$\boldsymbol{H}_1 = \boldsymbol{h}_1 \exp(-j\beta_1 z) + \boldsymbol{h}_{z_1} \exp(-j\beta_1 z), \tag{5.207b}$$

$$\boldsymbol{E}_2 = \boldsymbol{e}_2 \exp(-j\beta_2 z) + \boldsymbol{e}_{z_2} \exp(-j\beta_2 z), \tag{5.207c}$$

$$\boldsymbol{H}_2 = \boldsymbol{h}_2 \exp(-j\beta_2 z) + \boldsymbol{h}_{z_2} \exp(-j\beta_2 z). \tag{5.207d}$$

In order to show that there is power orthogonality or, in other words, that these modes carry power independently, it is necessary to show that

$$\begin{aligned}
\text{Re} \int_{S_0} (\boldsymbol{E}_1 + \boldsymbol{E}_2) \times (\boldsymbol{H}_1^* + \boldsymbol{H}_2^*) \cdot \boldsymbol{u}_z \, dS \\
= \text{Re} \int_{S_0} (\boldsymbol{E}_1 \times \boldsymbol{H}_1^* + \boldsymbol{E}_2 \times \boldsymbol{H}_2^*) \cdot \boldsymbol{u}_z \, dS \\
= \int_{S_0} [\boldsymbol{e}_1 \times \boldsymbol{h}_1 \exp(-2j\beta_1 z) + \boldsymbol{e}_2 \times \boldsymbol{h}_2 \exp(-2j\beta_2 z)] \cdot \boldsymbol{u}_z \, dS,
\end{aligned} \tag{5.208}$$

where $\boldsymbol{e}_1$, $\boldsymbol{e}_2$, $\boldsymbol{h}_1$, $\boldsymbol{h}_2$ are the transverse components assumed to be real in a loss-free waveguide and S_0 is an arbitrary cross-section of the waveguide.

Now, we have to prove that

$$\begin{aligned}
\int_{S_0} (\boldsymbol{E}_1 \times \boldsymbol{H}_2^* + \boldsymbol{E}_2 \times \boldsymbol{H}_1^*) \cdot \boldsymbol{u}_z \, dS \\
= \int_{S_0} [\boldsymbol{e}_1 \times \boldsymbol{h}_2 \exp(-j\beta_1 z + j\beta_2 z) + \boldsymbol{e}_2 \times \boldsymbol{h}_1 \exp(j\beta_1 z - j\beta_2 z)] \cdot \boldsymbol{u}_z \, dS = 0
\end{aligned} \tag{5.209}$$

so that the rate of energy flow (power) is the sum of the rates due to each mode separately. Figure 5.26 shows a section of a waveguide bounded by planes at z_1 and z_2. The closed surface S bounding the volume V consists of the perfectly conducting walls S_w and the transverse planes S_1

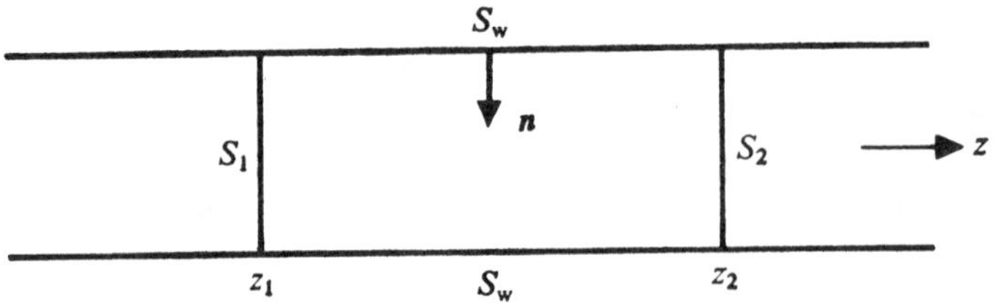

Fig. 5.26 Section of waveguide bounded by surfaces S_1, S_2, and S_w.

and S_2. Since the volume V does not contain any source, the fields in V satisfy the equations

$$\nabla \times \boldsymbol{E}_i = -j\omega\mu \boldsymbol{H}_i, \qquad \nabla \times \boldsymbol{H}_i = j\omega\epsilon \boldsymbol{E}_i,$$

where $i = 1, 2$. Then

$$\nabla \cdot (\boldsymbol{E}_1 \times \boldsymbol{H}_2 - \boldsymbol{E}_2 \times \boldsymbol{H}_1) = j\omega\mu(\boldsymbol{H}_1 \cdot \boldsymbol{H}_2 - \boldsymbol{H}_1 \cdot \boldsymbol{H}_2) + j\omega\epsilon(\boldsymbol{E}_1 \cdot \boldsymbol{E}_2 - \boldsymbol{E}_1 \cdot \boldsymbol{E}_2) = 0.$$

Therefore

$$\oint_S (\boldsymbol{E}_1 \times \boldsymbol{H}_2 - \boldsymbol{E}_2 \times \boldsymbol{H}_1) \cdot \boldsymbol{n}\, dS = 0. \tag{5.210}$$

On S_w,

$$\boldsymbol{E}_1 \times \boldsymbol{H}_2 \cdot \boldsymbol{n} = \boldsymbol{n} \times \boldsymbol{E}_1 \cdot \boldsymbol{H}_2 = 0, \qquad \boldsymbol{E}_2 \times \boldsymbol{H}_1 \cdot \boldsymbol{n} = 0.$$

Therefore, Eq. (5.210) reduces to

$$\int_{S_1} (\boldsymbol{E}_1 \times \boldsymbol{H}_2 - \boldsymbol{E}_2 \times \boldsymbol{H}_1) \cdot \boldsymbol{u}_z\, dS - \int_{S_2} (\boldsymbol{E}_1 \times \boldsymbol{H}_2 - \boldsymbol{E}_2 \times \boldsymbol{H}_1) \cdot \boldsymbol{u}_z\, dS = 0. \tag{5.211}$$

Any term in Eq. (5.211) involves only the transverse fields $\boldsymbol{e}$, $\boldsymbol{h}$ and $\boldsymbol{e} \times \boldsymbol{h}$, the latter field being directed in the z-direction. If Eqs. (5.207) are used in Eq. (5.211), then the integral over S_1 depends on z_1 and the integral over S_2 depends on z_2. As z_1 and z_2 can be arbitrarily chosen, it is necessary for each integral to be zero. However, when $\beta_1 = -\beta_2$, the integrals are independent of z_1 and z_2 and cancel identically. When $\beta_1 \neq \beta_2$,

$$\int_{S_0} (\boldsymbol{E}_1 \times \boldsymbol{H}_2 - \boldsymbol{E}_2 \times \boldsymbol{H}_1) \cdot \boldsymbol{u}_z\, dS = 0, \tag{5.212}$$

where S_0 is any arbitrary transverse plane. Equation (5.212) can be rewritten as

$$\exp(-j\beta_1 z_0 - j\beta_2 z_0) \int_{S_0} (\boldsymbol{e}_1 \times \boldsymbol{h}_2 - \boldsymbol{e}_2 \times \boldsymbol{h}_1) \cdot \boldsymbol{u}_z\, dS = 0$$

or

$$\int_{S_0} (\boldsymbol{e}_1 \times \boldsymbol{h}_2 - \boldsymbol{e}_2 \times \boldsymbol{h}_1) \cdot \boldsymbol{u}_z\, dS = 0. \tag{5.213}$$

In place of $\boldsymbol{E}_2$, $\boldsymbol{H}_2$, we may use the same mode propagating in the $-z$-direction so that

$$\begin{aligned} \boldsymbol{E}_2 &= \boldsymbol{e}_2 \exp(j\beta_2 z) - e_{z_2} \exp(j\beta_2 z), \\ \boldsymbol{H}_2 &= -\boldsymbol{h}_2 \exp(j\beta_2 z) + h_{z_2} \exp(j\beta_2 z). \end{aligned} \tag{5.214}$$

Then, Eq. (5.212) is replaced by

$$\int_{S_0} (-\boldsymbol{e}_1 \times \boldsymbol{h}_2 - \boldsymbol{e}_2 \times \boldsymbol{h}_1) \cdot \boldsymbol{u}_z\, dS = 0. \tag{5.215}$$

Addition and subtraction of Eqs. (5.213) and (5.215) give

$$\int_{S_0} (\boldsymbol{e}_1 \times \boldsymbol{h}_2) \cdot \boldsymbol{u}_z\, dS = \int_{S_0} (\boldsymbol{e}_2 \times \boldsymbol{h}_1) \cdot \boldsymbol{u}_z\, dS = 0 \tag{5.216}$$

which proves Eq. (5.209). Thus, it is shown that the power-interaction terms between two non-degenerate modes ($\beta_1 \neq \beta_2$) vanish.

When $\beta_1 = \beta_2$, the modes are degenerate and the proof is no more applicable because, when the two modes are propagating in the same direction, the two integrals in Eq. (5.211) vanish separately or, if they are propagating in different directions, they become identically the same and cancel each other. When degenerate modes do not satisfy the power-orthogonality relation, it is possible to define new modes, which are linear combinations of the two modes such that power orthogonality is satisfied. For example, if $(\boldsymbol{E}_1, \boldsymbol{H}_1)$ and $(\boldsymbol{E}_2, \boldsymbol{H}_2)$ are two degenerate modes and $(\boldsymbol{E}_1' = \boldsymbol{E}_1 + C\boldsymbol{E}_2, \boldsymbol{H}_1' = \boldsymbol{H}_1 + C\boldsymbol{H}_2)$ is a new mode, then it is possible to choose C so that the interaction terms vanish. If

$$P_{ij} = \tfrac{1}{2} \int_{S_0} (\boldsymbol{E}_i \times \boldsymbol{H}_j^*) \cdot \boldsymbol{u}_z \, dS,$$

then the interaction terms vanish when

$$C = \frac{2P_{11}}{P_{12} + P_{21}}.$$

For lossy waveguides, power orthogonality does not hold, but it is a good approximation to assume that it holds for low-loss guides.

PROBLEMS

1 A hollow metal rectangular waveguide has dimensions of 5 cm × 6 cm. The frequency of the signal is 3000 MHz. What are the different TE and TM modes that the waveguide can support? Calculate, for each one of them, the cut-off wavelength, guide wavelength, phase and group velocities, characteristic wave impedance, and the attenuation constant.

2 A circular hollow metal waveguide is to be operated at a frequency of 6000 MHz, and is to have dimensions such that

$$\frac{\lambda}{\lambda_c} = 0.85$$

for the dominant mode. Calculate the diameter of the guide, phase and group velocities, guide wavelength, characteristic wave impedance, and the attenuation constant.

3 A circular hollow waveguide has a diameter of 8 cm. What are the different TE and TM modes it can support at a frequency of 10 GHz? Calculate the phase velocity, group velocity, guide wavelength, and the attenuation for all the modes at this given frequency.

4 Compare the attenuation of a hollow rectangular waveguide with that of a dielectric filled waveguide at a frequency of 4000 MHz. Both guides are silver plated and have rectangular shape with a height equal to 1/2 that of the width. The guides are operated in the dominant TE_{10} mode and each of them is so designed that

$$\frac{\lambda}{\lambda_c} = 0.9.$$

The dielectric filled guide has a dielectric having the properties

$$\epsilon_r = 2.54, \qquad \frac{\sigma}{\omega\epsilon} = 0.0005.$$

5 A rectangular metal hollow waveguide is designed to carry only the TE_{10} mode at a frequency of 7000 MHz, and has a ratio

$$\frac{\lambda}{\lambda_c} = 0.85.$$

The ratio of the height to width is 0.55. The time-average power flow through the guide is 2 kW. Calculate the maximum values of electric and magnetic intensities in the guide and indicate where these occur in the guide.

6 Show that

$$\nabla \cdot \boldsymbol{E} = 0, \qquad \nabla \cdot \boldsymbol{H} = 0$$

for TE and TM modes in a rectangular waveguide.

7 Show that the Q of a rectangular waveguide carrying the dominant TE_{10} mode approaches the value of b/δ_s, where b is the width of the waveguide and δ_s the skin depth, assuming the metal for the guide to be nonmagnetic.

8 In an infinitely long rectangular metal waveguide, there is a dielectric filling for $z \geqslant 0$, having a permittivity ϵ. A TE_{10} mode is incident from $z < 0$. At $z = 0$, a reflected TE_{10} mode and a transmitted TE_{10} mode are produced because of the discontinuity. Show that the reflection coefficient is given by

$$\frac{Z_2 - Z_1}{Z_2 + Z_1},$$

where Z_1 and Z_2 are the characteristic wave impedances in the empty guide and the dielectric filled guide, respectively. Show also that the ratio of the wave impedances is equal to the ratio of the guide wavelengths.

9 A rectangular guide of dimensions

$$a = 2b = 3 \text{ cm}$$

is operated at a frequency of 9 GHz. A pulse-modulated carrier of this frequency is transmitted through the guide. How much pulse delay time is introduced by a guide 150 m long?

10 Calculate the cut-off wavelength and corresponding frequencies for the first three higher modes in a coaxial line having dimensions

$$a = 1 \text{ cm}, \qquad b = 4.5 \text{ cm}.$$

11 Show that for an air-filled coaxial line, minimum attenuation occurs when

$$x \ln x = 1 + x,$$

where

$$x = \frac{b}{a}.$$

What is the corresponding characteristic impedance?

12 Calculate the characteristic impedance of a coaxial line having

$$b = 3a = 1.2 \text{ cm},$$

$$f = 1200 \text{ MHz},$$

$$\epsilon = (2.56 - j0.001)\epsilon_0.$$

Also, calculate R, G, L, C per metre. Show that

$$\text{Im}(Z_c) \ll \text{Re}(Z_c),$$

$$Z_c \approx \left(\frac{L}{C}\right)^{1/2}.$$

13 Show that for TM modes in a waveguide, the line integral of the transverse electric field between any two points on the boundary is zero, and for TE modes in a waveguide, the line integral of the transverse electric field between two points located on the guide boundary depends on the path of integration.

14 In a rectangular waveguide of dimensions 2.6 cm × 1.3 cm, a signal

$$[1 + \cos(2\pi f_1 t) + \cos(2\pi f_2 t)] \cos(2\pi f_c t)$$

is transmitted through a section 2000 m long. If

$$f_c = 9000 \text{ MHz}, \qquad f_1 = 8 \text{ MHz}, \qquad f_2 = 10 \text{ MHz},$$

and a TE_{10} mode is used, obtain an expression for the exact form of the output signal. Sketch the input and output low-frequency modulating signals to show the distortion.

15 An amplitude-modulated wave $[1 + 0.25 \cos(2\pi f_m t)] \cos(2\pi f_c t)$ is transmitted in a waveguide of dimensions 2.5 cm × 1.25 cm. If

$$f_m = 15 \text{ kHz}, \qquad f_c = 9500 \text{ MHz},$$

how long must the guide be before the upper and lower side-bands can undergo a relative phase shift of 180°?

6 Transmission Line Theory as Applied to Microwave Circuits

6.1 Methods of Analysis of Transmission Lines

Transmission lines may be analyzed by the solution of Maxwell's field equations or by the methods of ordinary circuit analysis. The solution of the field equations involves the determination of the field intensities in three-dimensional space; hence, in addition to the time variable, three-space variables are required. This method may be used to analyze a few systems which have relatively simple geometries such as the coaxial line (see Sections 5.8 and 5.9). The solution of Maxwell's field equations reveals that the energy propagates through the dielectric medium as an electromagnetic wave, the conductors guiding the energy flow. But in many practical problems, the mathematical complications resulting from four independent variables in the solution of Maxwell's equations become very involved.

In the circuit method, the effects of the electric and magnetic fields are taken into consideration. This is done by using the circuit parameters, namely, the series inductance, series resistance, shunt capacitance, and the shunt conductance, per unit length of the transmission line, as defined by Eqs. (5.73)–(5.76). By this procedure, the mathematical analysis is reduced to a problem involving one-space variable in addition to the time variable.

The circuit method can be applied to the guiding structures, e.g., waveguides, which can carry single TE or TM modes, but which may not transmit the TEM mode which is transmitted by the conventional transmission lines such as parallel-wire lines or coaxial lines. However, if the waveguide or any other guiding structure carries more than one mode, then the circuit theory becomes complicated. Also, for other types of guiding structures, for instance, surface-wave structures, field analysis is much more suitable than circuit theory. But for some guiding structures, for example, slow-wave structures, which are used in microwave electron tubes, circuit analysis sometimes gives fast and useful results, whereas field analysis is extremely complicated.

Circuit theory (as applied to a structure which guides electromagnetic waves in the direction of a single coordinate) is usually known as transmission line theory. The single coordinate direction may be the z-direction, which is usually true of all transmission lines or waveguides of cylindrical geometry, or it may be the r-direction for conical transmission lines.

In this chapter, we shall develop the theory for uniform transmission lines.

6.2 Generalized Transmission Line Equations

In Chapter 5, we have studied how the transmission of energy takes place in a transmission line or a waveguide through the electromagnetic field. No restriction on frequency was imposed on such an analysis which is valid even for very low frequencies, including zero frequency. However, for conducting cylindrical waveguides of rectangular or circular cross-section (which have already been studied), we have seen that transmission of energy takes place for only frequencies above the cut-off frequency and, for frequencies below the cut-off frequency, the field is evanescent or decays within a short distance. This is also true for other types of electromagnetic guiding structures such as surface-wave lines and slow-wave structures, which behave like low-pass, high-pass, band-pass, or band-elimination filters.

We are familiar with the concept of energy transmission through voltages and currents at low frequencies. Schelkunoff[1] has shown that there is no contradiction between the circuit concept and field concept and that, in fact, the transmission theory for the low-frequency and d-c systems is merely a special case of the generalized field transmission theory. The concept of equivalent voltage and current for bounded electromagnetic fields in waveguides has been considered in Sections 5.18, 5.19, 5.22, and 5.23. The transmission line equations, which will now be derived, can be used for any of the guiding structures for which equivalent transmission lines and equivalent voltages and currents can be determined.

Consider a two-wire parallel-wire transmission line, as shown in Fig. 6.1. Such a trans-

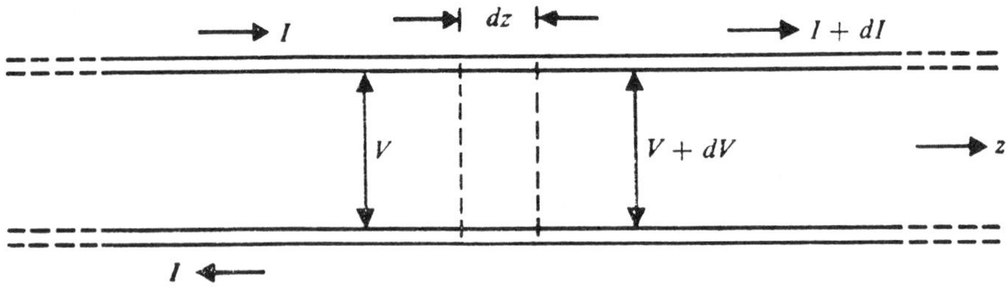

Fig. 6.1 Parallel-wire transmission line.

mission line can be represented by an equivalent distributed circuit (see Fig. 6.2), where *R, L, G,*

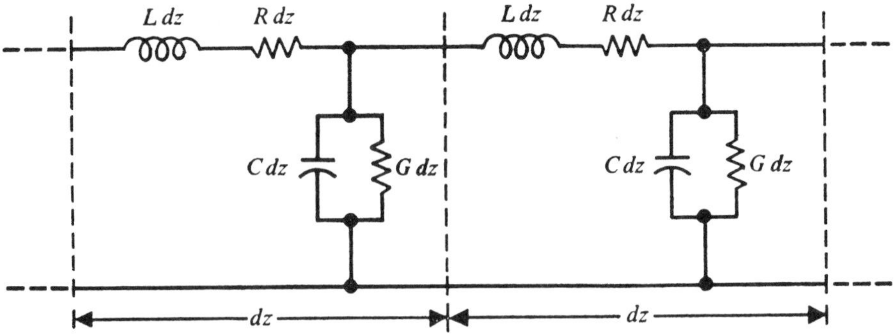

Fig. 6.2 Equivalent distributed circuit representation of transmission line.

and C are the series resistance, series inductance, shunt conductance, and the shunt capacitance, per unit length. Consider also an element of length dz of the transmission line. From elementary

[1]S. A. Schelkunoff, Conversion of Maxwell's equations into generalized telegraphists' equations, *B.S.T.J.*, **34**, 995, 1955.

circuit theory, the voltage drop in the line of length dz is

$$V + dV - V = dV = -(RI + L\frac{dI}{dt})\, dz$$

or

$$\frac{dV}{dz} = -(RI + L\frac{dI}{dt}) \tag{6.1}$$

and the change of current over the length dz is

$$I + dI - I = dI = -(GV + C\frac{dV}{dt})\, dz$$

or

$$\frac{dI}{dz} = -(GV + C\frac{dV}{dt}). \tag{6.2}$$

Equations (6.1) and (6.2) are known as the *generalized transmission line equations* or *generalized telegraphists' equations* which were derived by Lord Kelvin long before Maxwell formulated his field equations. If the voltage V and current I vary harmonically with time as $e^{j\omega t}$, then Eqs. (6.1) and (6.2) become

$$\frac{dV}{dz} = -(R + j\omega L)I = -Z(z)I, \tag{6.3}$$

$$\frac{dI}{dz} = -(G + j\omega C)V = -Y(z)V, \tag{6.4}$$

where $Z(z) = R(z) + j\omega L(z)$ and $Y(z) = G(z) + j\omega C(z)$ may respectively be called the series impedance and shunt admittance per unit length at z. In general, R, L, G, C, Z, and Y are functions of z. For a uniform transmission line, these quantities are constants.

The telegraphists' or transmission line equations hold good for any transmission line or waveguide or any other electromagnetic guiding structure, which can be represented by the equivalent distributed circuit shown in Fig. 6.2.

6.3 Uniform Transmission Lines

A transmission line is said to be uniform if its series impedance Z and shunt admittance Y per unit length are the same throughout the line. A waveguide carrying a single TE or TM mode is a uniform transmission line if Z and Y for that particular mode remain the same throughout the line. This is true also for a coaxial line or parallel-wire line carrying the TEM mode. Parameters of some common transmission lines carrying the TEM mode are shown in Table 6.1.

Since the transmission line equations for a transmission line carrying the TEM mode and those for a waveguide carrying a single TE or TM mode are the same, we shall develop a generalized transmission-line theory which can be applied to both types of guiding structures. For all transmission lines,

$$C = \frac{(\mu_0\epsilon')^{1/2}}{Z_c}, \qquad L = (\mu_0\epsilon')^{1/2}Z_c, \qquad G = \frac{\omega\epsilon'' C}{\epsilon'},$$

$$\alpha_d = \frac{GZ_c}{2}, \qquad \alpha_c = \frac{RY_c}{2}, \qquad R_m = \frac{1}{\sigma\delta_s} = \left(\frac{\omega\mu}{2\sigma}\right)^{1/2},$$

where $\epsilon = \epsilon' + j\epsilon''$ is the complex dielectric constant of the dielectric medium, δ_s is the skin depth

Table 6.1 Parameters of Common Transmission Lines

Type	Z_c	R
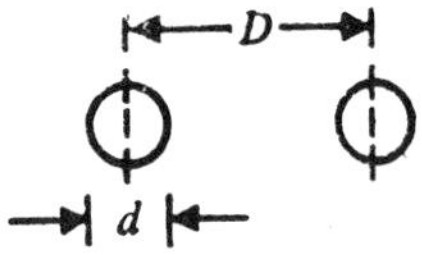	$\frac{1}{\pi}\left(\frac{\mu_0}{\epsilon'}\right)^{1/2} \cosh^{-1} \frac{D}{a}$	$\frac{2R_m}{\pi d} \frac{D/d}{[(D/d)^2 - 1]^{1/2}}$
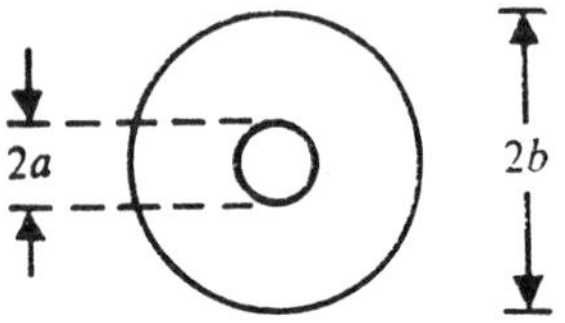	$\frac{1}{2\pi}\left(\frac{\mu_0}{\epsilon'}\right)^{1/2} \ln \left(\frac{b}{a}\right)$	$\frac{R_m}{2\pi}\left(\frac{1}{a} + \frac{1}{b}\right)$
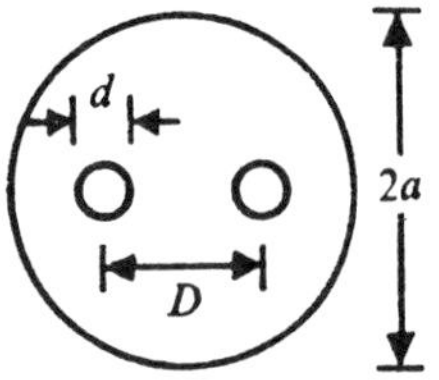 $p = D/d,\ q = D/a$	$\frac{1}{\pi}\left(\frac{\mu_0}{\epsilon'}\right)^{1/2}[\ln \{2p(\frac{1 - q^2}{1 + q^2})\} - \{(\frac{1 + 4p^2}{16p^4})(1 - 4q^2)\}]$	$\frac{2R_m}{\pi d}[1 + (\frac{1 + 2p^2}{4p^4})(1 - 4q^2)] + \frac{8R_m}{\pi a} q^2[(1 + q^2) - (\frac{1 + 4p^2}{8p^4})]$

of conducting walls, σ the conductivity of conducting walls, μ the permeability of conducting walls, α_d the attenuation constant due to loss in the dielectric medium, and α_c is the attenuation constant due to loss in conducting walls.

From Eqs. (6.3) and (6.4), we have

$$\frac{d^2V}{dz^2} - \frac{d}{dz}(\ln Z)\frac{dV}{dz} - YZV = 0, \tag{6.5}$$

$$\frac{d^2I}{dz^2} - \frac{d}{dz}(\ln Y)\frac{dI}{dz} - YZI = 0. \tag{6.6}$$

For a uniform transmission line,

$$\frac{d}{dz}(\ln Z) = \frac{d}{dz}(\ln Y) = 0. \tag{6.7}$$

Hence

$$\frac{d^2V}{dz^2} - YZV = 0, \qquad \frac{d^2I}{dz^2} - YZI = 0. \tag{6.8}$$

If YZ is put equal to $\bar{\gamma}^2$, the solution of Eqs. (6.8) is of the form

$$V = A \exp(-\bar{\gamma}z) + B \exp(\bar{\gamma}z), \qquad I = C \exp(-\bar{\gamma}z) + D \exp(\bar{\gamma}z), \tag{6.9}$$

where A, B, C, D are arbitrary constants. The constant $\bar{\gamma} = \sqrt{YZ}$ is called the propagation constant of the transmission line in the z-direction. In general, $\bar{\gamma}$ is a complex quantity equal to $(\bar{\alpha} + j\bar{\beta})$. If the real part $\bar{\alpha}$ of $\bar{\gamma}$ is positive, then $\exp(\bar{\gamma}z)$ increases indefinitely with increasing z, and hence for an infinitely long line, the constants B and D should be zero for any physically realizable field. Thus, for an infinitely long line,

$$V(z) = A \exp(-\bar{\gamma}z), \qquad I(z) = C \exp(-\bar{\gamma}z). \tag{6.10}$$

Using Eqs. (6.10) in Eqs. (6.3) and (6.4), we obtain

$$-A\bar{\gamma} \exp(-\bar{\gamma}z) = -ZC \exp(-\bar{\gamma}z), \qquad -C\bar{\gamma} \exp(-\bar{\gamma}z) = -YA \exp(-\bar{\gamma}z) \tag{6.11}$$

and therefore

$$Z = \frac{A\bar{\gamma}}{C}, \qquad Y = \frac{C\bar{\gamma}}{A}. \tag{6.12}$$

But A/C is the ratio of the voltage V to the current I of an infinitely long line, and hence by definition of the characteristic impedance of a transmission line, it is the characteristic impedance Z_c of the line. Thus

$$Z_c = \frac{A}{C} = \frac{1}{Y_c}, \tag{6.13}$$

where Y_c is the characteristic admittance of the line. Thus, Eqs. (6.3) and (6.4) become

$$\frac{dV}{dz} = -\bar{\gamma}Z_c I, \qquad \frac{dI}{dz} = -\bar{\gamma}Y_c V. \tag{6.14}$$

Recalling that harmonic variation of time $e^{j\omega t}$ has been assumed, Eqs. (6.9) and (6.10) have to be multiplied by $e^{j\omega t}$, and they become, for a finite length line,

$$\begin{aligned} V(z, t) &= A \exp(j\omega t - \bar{\gamma}z) + B \exp(j\omega t + \bar{\gamma}z), \\ I(z, t) &= C \exp(j\omega t - \bar{\gamma}z) + D \exp(j\omega t + \bar{\gamma}z) \end{aligned} \tag{6.15}$$

and, for an infinitely long line,

$$V(z, t) = A \exp(j\omega t - \bar{\gamma}z), \qquad I(z, t) = C \exp(j\omega t - \bar{\gamma}z). \tag{6.16}$$

Equations (6.15) and (6.16) imply that the voltage and current on a finite length line consist, in general, of a wave travelling in the positive z-direction and a wave travelling in the negative z-direction, whereas for an infinitely long line, there is a single travelling wave in the positive z-direction.

The value of $\bar{\gamma}$ and Z_c may be obtained from Table 6.1 for common transmission lines carrying TE modes. For lossless TEM transmission lines, $\bar{\gamma} = j\omega\sqrt{\mu\epsilon} = k = j\bar{\beta}$, so that there is no attenuation, and the wave travels with a velocity $v = \omega/\bar{\beta} = 1/\sqrt{\mu\epsilon}$ which is the characteristic phase velocity of the dielectric medium. Also, for such transmission lines, all frequencies from zero to very high frequencies can be transmitted or, in other words, there is no cut-off.

For cylindrical metal waveguides of any closed cross-section, there is a cut-off frequency ω_c

and, therefore, propagation is possible only for TE and TM modes, and not for TEM modes. Hence, $\bar{\gamma}^2 = k_c^2 - k^2$, where k_c can take a double infinity of values $k_{c,\,nm}$ and $k^2 = \omega^2\mu\epsilon$, so that

$$\bar{\gamma}^2 = k_{c,\,nm}^2 - k^2. \tag{6.17}$$

Here

$$k_{c,\,nm}^2 = \left(\frac{n\pi}{a}\right)^2 + \left(\frac{m\pi}{b}\right)^2 \quad \text{for TE and TM modes in rectangular waveguide}$$

$$= \frac{p_{nm}^2}{a^2} \quad \text{for TM modes in cylindrical waveguide}$$

$$= \frac{p_{nm}'^2}{a^2} \quad \text{for TE modes in cylindrical waveguide.}$$

μ and ϵ are the permeability and permittivity of the dielectric medium. For a lossy medium, ϵ is replaced by $\epsilon + \sigma/(j\omega) = \epsilon_1$. From Eqs. (5.165), (6.14), and (6.17), the characteristic impedance of the waveguide is obtained as

$$Z_c = \frac{(k_{c,\,nm}^2 - k^2)^{1/2}}{j\omega\epsilon} \quad \text{for TM modes}$$

$$= \frac{j\omega\mu}{(k_{c,\,nm}^2 - k^2)^{1/2}} \quad \text{for TE modes.} \tag{6.18}$$

From Eqs. (6.10) and (6.17), it can be seen that the transmission of a waveguide mode nm along the waveguide of any reasonable length is possible only when the real part of $\bar{\gamma}_{pq}$ is small or negligible. When the dielectric medium inside the waveguide is lossless, $\sigma = 0$, $\epsilon_1 = \epsilon$. Therefore, for a TE_{pq} or TM_{pq} mode to propagate,

$$k_{c,\,nm}^2 < \omega^2\mu\epsilon, \tag{6.19a}$$

$$\bar{\gamma}_{nm} = j\bar{\beta}_{nm}, \tag{6.19b}$$

where

$$\bar{\beta}_{nm} = (\omega^2\mu\epsilon - k_{c,\,nm}^2)^{1/2} = (\beta^2 - k_{c,\,nm}^2)^{1/2} \tag{6.20}$$

with $\beta = \omega\sqrt{\mu\epsilon} = 2\pi/\lambda$ as the intrinsic phase constant in the dielectric medium. For a lossless medium, the characteristic impedance of the waveguide is given by

$$Z_c = \frac{\bar{\beta}_{nm}}{\omega\epsilon} \quad \text{for TM modes}$$

$$= \frac{\omega\mu}{\bar{\beta}_{nm}} \quad \text{for TE modes.} \tag{6.21}$$

Relations (6.18)–(6.21) hold good for rectangular as well as circular cylindrical waveguides, and for cylindrical waveguides of any closed cross-section. Equation (6.21) shows that, for a lossless dielectric medium, the characteristic impedance Z_c is real for a propagating mode.

Relations (6.19) and (6.20) can be used to show some interesting properties of the transmission or propagation of electromagnetic waves inside a waveguide. Since $k_{c,\,nm}$ depend on the dimensions of a waveguide and, in general, increase monotonically with increasing n and m, for any given frequency, only a finite number of modes can propagate freely through any practical waveguide of finite dimensions. Also, the dimensions of the waveguide can be chosen so that only one mode (usually the lowest or dominant mode) can propagate freely. Besides, we have already seen in Sections 5.11 and 5.15 that, above the cut-off frequency,

$$\bar{\beta}_{nm} < \beta = \frac{2\pi}{\lambda}$$

and that

$$\lambda_g = \frac{2\pi}{(\beta^2 - k_{c,\,nm}^2)^{1/2}} = \frac{\lambda}{[1 - (\lambda/\lambda_c)^2]^{1/2}},$$

where

$$\lambda_c = \frac{2\pi}{k_{c,\,nm}} = \text{cut-off wavelength.}$$

The cut-off frequency is

$$f_c = \frac{c}{\lambda_c}.$$

For any propagating mode in a waveguide,

$$\bar{\gamma}^2 = YZ = -\bar{\beta}^2, \tag{6.22}$$

and hence Eqs. (6.8) become

$$\frac{d^2V}{dz^2} + \bar{\beta}^2 V = 0, \qquad \frac{d^2I}{dz^2} + \bar{\beta}^2 I = 0 \tag{6.23}$$

whose solutions may be written as

$$\begin{aligned} V(z) &= A_1 \cos[\bar{\beta}(z - z_0)] + B_1 \sin[\bar{\beta}(z - z_0)], \\ I(z) &= C_1 \cos[\bar{\beta}(z - z_0)] + D_1 \sin[\bar{\beta}(z - z_0)]. \end{aligned} \tag{6.24}$$

If the voltage and current at $z = z_0$ are $V(z_0)$ and $I(z_0)$, then

$$V(z_0) = A_1, \qquad I(z_0) = C_1,$$

$$A_1/C_1 = Z_c = \text{characteristic impedance of the line.}$$

From Eqs. (6.10), remembering that $\bar{\gamma} = j\bar{\beta}$,

$$B_1 = -jA_1, \qquad D_1 = -jC_1$$

so that

$$B_1 = -jV(z_0) = -jZ_0 I(z_0), \qquad D_1 = -jC_1 = -jI(z_0) = -j\frac{V(z_0)}{Z_0}.$$

Equations (6.24) can therefore be rewritten as

$$\begin{aligned} V(z) &= V(z_0) \cos[\bar{\beta}(z - z_0)] - jI(z_0)Z_0 \sin[\bar{\beta}(z - z_0)], \\ I(z) &= I(z_0) \cos[\bar{\beta}(z - z_0)] - j\frac{V(z_0)}{Z_0} \sin[\bar{\beta}(z - z_0)]. \end{aligned} \tag{6.25}$$

Equations (6.25) represent the voltage and current in a waveguide or a two-wire line everywhere along the z-direction in terms of the voltage and current at z_0. The impedance $Z(z)$ at any point $z = z$ on the line, defined by $V(z)/I(z)$, can be determined from Eq. (5.22) in terms of the impedance at z_0. Thus

$$Z(z) = \frac{V(z)}{I(z)} = Z_c \frac{Z_n - j \tan [\beta(z - z_0)]}{1 - jZ_n \tan [\beta(z - z_0)]}, \tag{6.26}$$

where

$$Z_n = \frac{Z(z_0)}{Z_c} = \frac{V(z_0)/I(z_0)}{Z_c}$$

is the impedance at z_0 normalized with respect to Z_c.

Similarly, the admittance $Y(z)$ at z is given in terms of the normalized admittance $Y_n = Y(z_0)/Y_c$, as

$$Y(z) = \frac{Y_n - j \tan [\beta(z - z_0)]}{1 - jY_n \tan [\beta(z - z_0)]}, \tag{6.27}$$

where Y_c, the characteristic admittance of the line, is equal to $1/Z_c$.

6.4 Applications of Uniform Transmission Line Theory to Microwave Circuits

We shall now examine Eqs. (6.25)–(6.27) carefully to bring out the various possible applications of transmission lines and waveguides in microwave circuits. To do this, let us consider a lossless

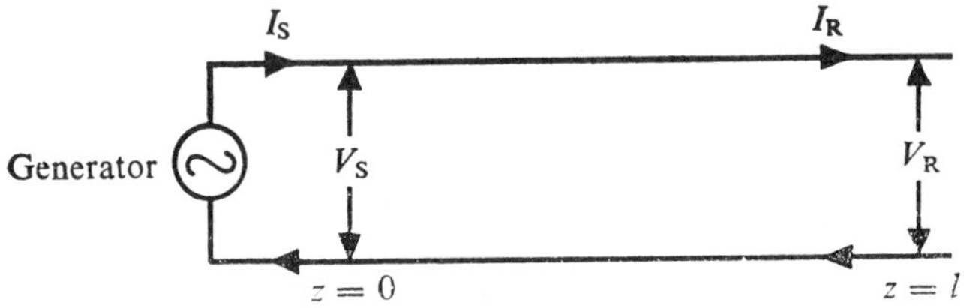

Fig. 6.3 Lossless transmission line or waveguide section.

transmission line or waveguide of length l, as shown in Fig. 6.3. If

$$V(l) = V_R, \quad I(l) = I_R \qquad \text{at } z_0 = l,$$

$$V(0) = V_S, \quad I(0) = I_S \qquad \text{at } z_0 = 0,$$

then Eqs. (6.25) can be used to express the input voltage and input current of the line at $z = 0$, as

$$V_S = V_R \cos \beta l + jI_R Z_c \sin \beta l, \qquad I_S = I_R \cos \beta l + \frac{jV_R}{Z_c} \sin \beta l, \tag{6.28}$$

where V_R and I_R are the voltage and current at the receiving end and V_S and I_S are those at the sending end.

Equations (6.28) show that a lossless transmission line can be used as a current or as a voltage transformer by a proper choice of βl and Z_c. They also show that the magnitudes of V_S and I_S for various values of βl depend on V_R and I_R which are the termination characteristics. For example, if a short circuit is placed at $z = l$, so that $V_R = 0$, as shown in Fig. 6.4a, then

$$|V_S| = |I_R Z_c|\,|\sin \beta l|, \qquad |I_S| = |I_R|\,|\cos \beta l|. \tag{6.29}$$

The variations of $|V_S|$ and $|I_S|$ along the line for various values of l are shown in Fig. 6.4b. Standing waves of voltage and current exist on the line. If, on the other hand, there is an open circuit at $z = l$, such that $I_R = 0$, then

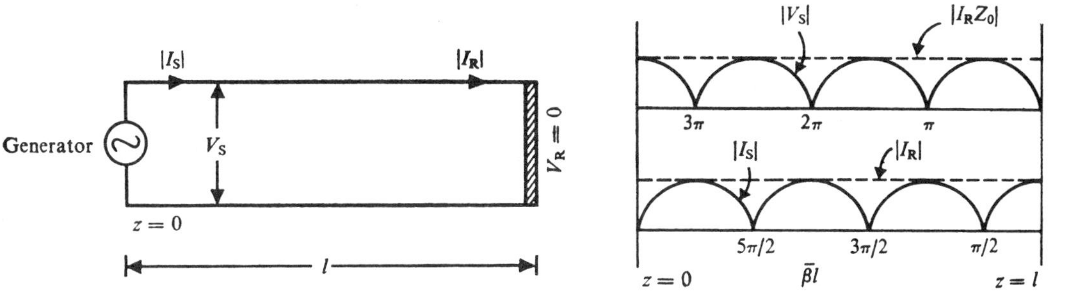

(a) Transmission line shorted at $z = 1$ (b) Variation of V and I on shorted-transmission line

Fig. 6.4 Short-circuited transmission line.

$$|V_S| = |V_R|\,|\cos \beta l|, \qquad |I_S| = \left|\frac{V_R}{Z_c}\right|\,|\sin \beta l|. \tag{6.30}$$

In general, for any terminating impedance $Z_L = R_L + jX_L$, we have

$$|V_S| = |V_R|\left[\left(\cos \beta l + \frac{Z_c X_L}{Z_L^2}\sin \beta l\right)^2 + \left(\frac{Z_c R_L}{Z_L}\right)^2 \sin^2 \beta l\right]^{1/2},$$

$$|I_S| = |I_R|\left[\left(\cos \beta l - \frac{X_L}{Z_c}\sin \beta l\right)^2 + \left(\frac{R_L}{Z_c}\right)^2 \sin^2 \beta l\right]^{1/2}. \tag{6.31}$$

The characteristic impedance Z_c in Eqs. (6.31) is assumed to be real for a lossless transmission line or waveguide.

Standing waves of voltage and current always exist on any arbitrarily terminated line, unless

$$X_L = 0, \qquad Z_L = R_L = Z_c. \tag{6.32}$$

When a terminating impedance Z_L satisfies Eqs. (6.32), the magnitudes of voltage and current remain uniform throughout the line, as seen from Eqs. (6.31); also, there are no standing waves, and the line is said to be matched.

Examining Eqs. (6.26) and (6.27), we can explain the transformer characteristics of a transmission line or waveguide. Consider a transmission line of length L such that $\beta L = \pi/2$, or

$$L = \frac{\pi}{2\beta} = \frac{\pi \lambda_g}{4\pi} = \frac{\lambda_g}{4},$$

where λ_g is the guide wavelength. If the line is terminated at $z = z_0$ by an impedance Z_2, then the input impedance Z_1 at $z = 0$ can be obtained from Eq. (6.26). Thus, $Z_1 = Z_c^2/Z_2$, or

$$Z_1 Z_2 = Z_c^2. \tag{6.33}$$

As an example of Eq. (6.33), we can consider a microwave antenna of impedance Z_2 which has to be fed from a generator of impedance Z_1. To eliminate reflection from the antenna, a matching transformer has to be designed between the antenna and generator. For this a transmission line or waveguide of length equal to $\lambda_g/4$ and characteristic impedance Z_c is selected so that

$$Z_c = (Z_1 Z_2)^{1/2}. \tag{6.34}$$

Such a transmission line or waveguide can be used as the transformer. To have the desired value, Z_c is selected by a proper choice of waveguide dimensions.

Equations (6.26) and (6.27) may be used to investigate the possibility of using transmission lines or waveguides as circuit elements. For example, if a transmission line is terminated at $z = z_0$ by a zero impedance (short circuit), then the impedance $Z(z)$ at any other point is given from Eq. (6.26) by

$$\frac{Z(z)}{Z_c} = j \tan [\bar{\beta}(z_0 - z)]. \tag{6.35}$$

If $\bar{\beta}(z_0 - z) < \pi/2$, then $Z(z)$ is positive imaginary and is equal to a pure inductive reactance, whereas if $\pi/2 < \bar{\beta}(z_0 - z) < \pi$, then $Z(z)$ is negative imaginary and corresponds to a capacitive reactance. Further, a resonance occurs when $\bar{\beta}(z_0 - z)$ is equal to π or $\pi/2$ and when $Z(z)$ approaches a zero or infinite value. Thus, by a proper choice of the length of a short-circuited transmission line, we can obtain practically any value of reactance ranging from $-\infty$ to ∞.

Similarly, it is possible to obtain any value of reactance from an open line or waveguide. In

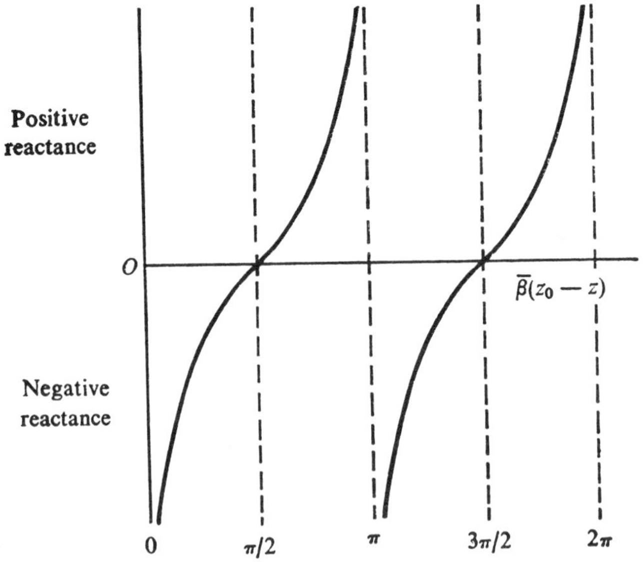

Fig. 6.5 Input impedance of open transmission line or waveguide as function of $\bar{\beta}(z_0 - z)$.

Eq. (6.26), if $Z_n = \infty$ at $z = z_0$, then

$$\frac{Z(z)}{Z_c} = -j \cot [\bar{\beta}(z_0 - z)]. \tag{6.36}$$

As $\cot [\bar{\beta}(z_0 - z)]$ can also assume any value from $-\infty$ to ∞ by a proper choice of $\bar{\beta}(z_0 - z)$, practically any value of reactance can be theoretically obtained from a section of an open waveguide or transmission line. However, it should be remembered that an open-ended waveguide or line also acts like a good radiator. Consequently, $Z(z)$ is not only a pure reactance but also contains a resistive part. The variation of the input impedance of an open transmission line with the length of the line is shown in Fig. 6.5. Here, the effect of radiation is omitted.

6.5 Impedance Matching by Transmission Line Sections

In Section 6.4, we have seen that, if a transmission line is terminated in an impedance other than its characteristic impedance, there are always standing waves of voltage and current along the line. These standing waves are very large if there is considerable mismatch. When a transmission line or waveguide is used for the transfer of energy from one point to another, the presence of standing waves is usually undesirable because it increases the effective line losses and thus reduces the efficiency of transmission. Also, the well known power transfer theorem states that, if a variable load impedance is connected to a constant voltage generator, maximum power transfer occurs when the load impedance is equal to the complex conjugate of the generator impedance. When this condition prevails, the impedances are said to be matched. It is sometimes helpful to determine the power sacrificed as a result of not having matched impedances.

Let $Z_g = R_g + jX_g$ be the generator impedance and $Z_L = R_L + jX_L$ represent the load impedance (see Fig. 6.6). The scalar value of the current I is

$$|I| = \frac{V_g}{[(R_g + R_L)^2 + (X_g + X_L)^2]^{1/2}}, \tag{6.37}$$

and the power consumed in the load is

$$P = |I|^2 R_L = \frac{V_g^2 R_L}{(R_g + R_L)^2 + (X_g + X_L)^2}. \tag{6.38}$$

If the load impedance is the only variable, the power is a maximum when $R_L = R_g$, $X_L = -X_g$,

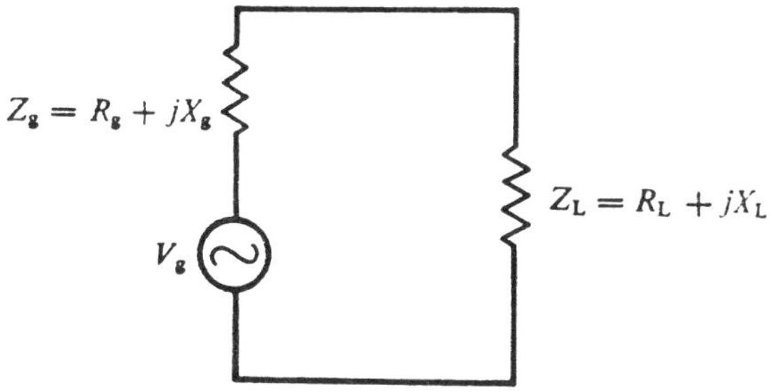

Fig. 6.6 Generator and load.

that is, when the generator impedance and load impedance are complex conjugates. Then, the power is

$$P_{max} = \frac{V_g^2}{4R_g}. \tag{6.39}$$

The ratio of the power for any load impedance to the maximum power is found by dividing Eq. (6.38) by Eq. (6.39), and is given by

$$\frac{P}{P_{max}} = \frac{4R_gR_L}{(R_g + R_L)^2 + (X_g + X_L)^2}. \tag{6.40}$$

If R' is put equal to R_L/R_g and $X' = (X_g + X_L)/R_g$, Eq. (6.40) becomes

$$\frac{P}{P_{max}} = \frac{4R'}{(1 + R')^2 + (X')^2}. \tag{6.41}$$

If R' and X' are the coordinate axes in a rectangular coordinate system, the loci representing the constant values of P/P_{max} are circles, as shown in Fig. 6.7. This figure enables us to evaluate the

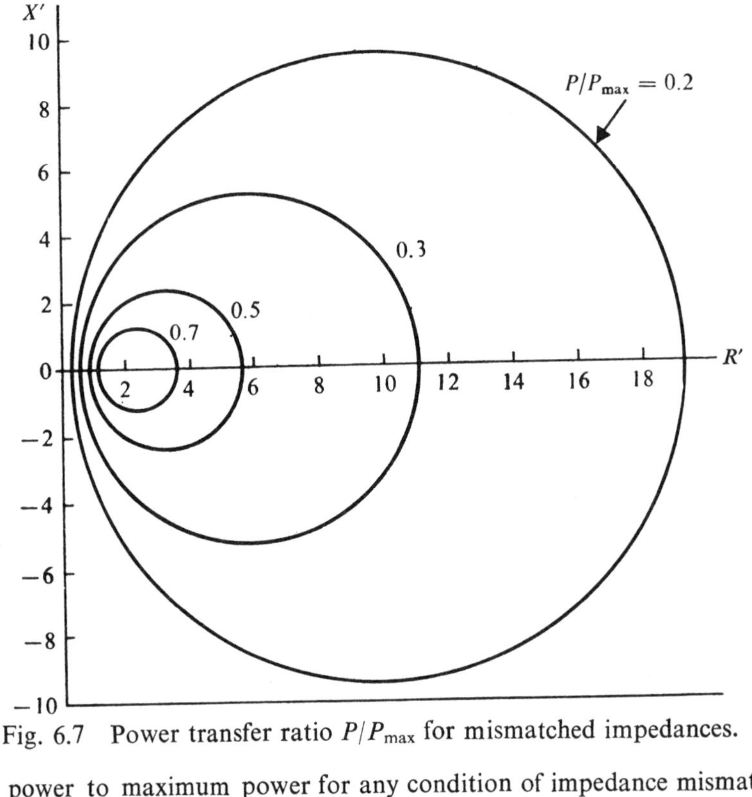

Fig. 6.7 Power transfer ratio P/P_{max} for mismatched impedances.

ratio of actual power to maximum power for any condition of impedance mismatch. If the load consists of a transmission line with a load impedance at the distant end, then the impedance Z_L is the input impedance of the line.

There is also another useful power transfer theorem which states that, if a generator is connected through one or more pure reactance networks to a load, as shown in Fig. 6.8, and the conditions are such that there is a conjugate impedance match at one pair of terminals, then there will be a conjugate impedance match at every pair of terminals and maximum power will be transferred from the generator to the load.

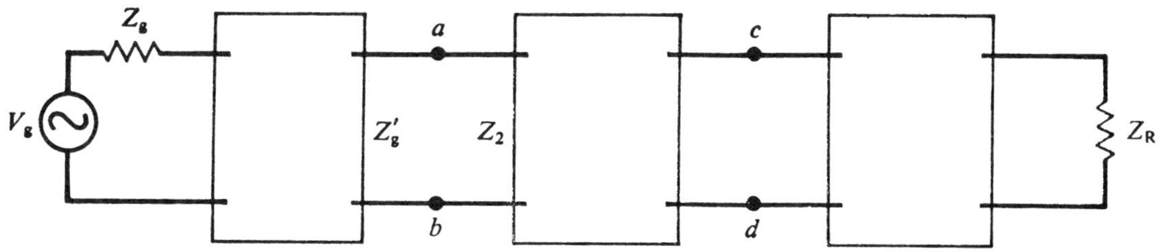

(a) Network consisting of generator connected to load impedance through pure reactance networks

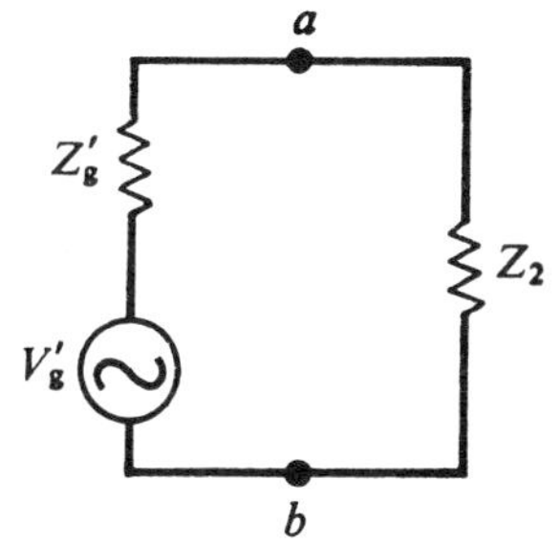

(b) Equivalent circuit obtained from Thevenin's theorem

Fig. 6.8 Impedance matching of network.

This theorem may be verified by using Thevenin's theorem. If we break into the network at any junction, such as *ab* in Fig. 6.8b, Thevenin's theorem permits us to replace the network to the left of *ab* by a generator. The generator impedance Z'_g in the equivalent network is equal to the impedance looking to the left at terminals *ab*, and the generator voltage V'_g is the open-circuit voltage at *ab*. The network to the right of *ab* in Fig. 6.8a is replaced by its equivalent impedance Z_2 in Fig. 6.8b. Now, assume that there is a conjugate match of impedances at *ab* in both these figures. A conjugate impedance match in the circuit of Fig. 6.8b signifies that there is maximum power transfer past the junction *ab*. If there is maximum power transfer in the equivalent circuit, there must also be maximum power transfer in the original circuit. Since there is maximum power flow past the junction *ab* in Fig. 6.8a and there is no power lost in the reactive networks, it follows that there is maximum power flow at every junction and thus maximum power transfer to the load. Consequently, there must be a conjugate impedance match at every junction since, if there were not a conjugate impedance match at any junction, there could not be maximum power flow past the junction.

For most transmission lines operating at microwave frequencies, we have $\omega L \gg R$ and $\omega C \gg G$; hence, the lines may be treated as pure reactance networks. The foregoing power transfer theorem makes it possible to match impedance at any point on the line between the generator and the load and ensures a conjugate impedance match throughout the entire system, resulting in maximum power transfer to the load.

The use of the quarter-wavelength line as an impedance transformer, which was considered in Section 6.4, is restricted largely to the matching of resistive impedances, where the frequency and impedances are constant. Stub-impedance matching systems are versatile in that they may be used to match complex impedances and are readily adjustable. A stub consists of an open-circuited or short-circuited line which is shunted across the transmission line between the generator and load. One or more stubs may be used for impedance matching.

Consider now a microwave antenna of resistance R and a waveguide (with a single mode) of characteristic impedance $Z_c = R_0$, as shown in Fig. 6.9. At a point distant $\bar{l}$ from the horn to

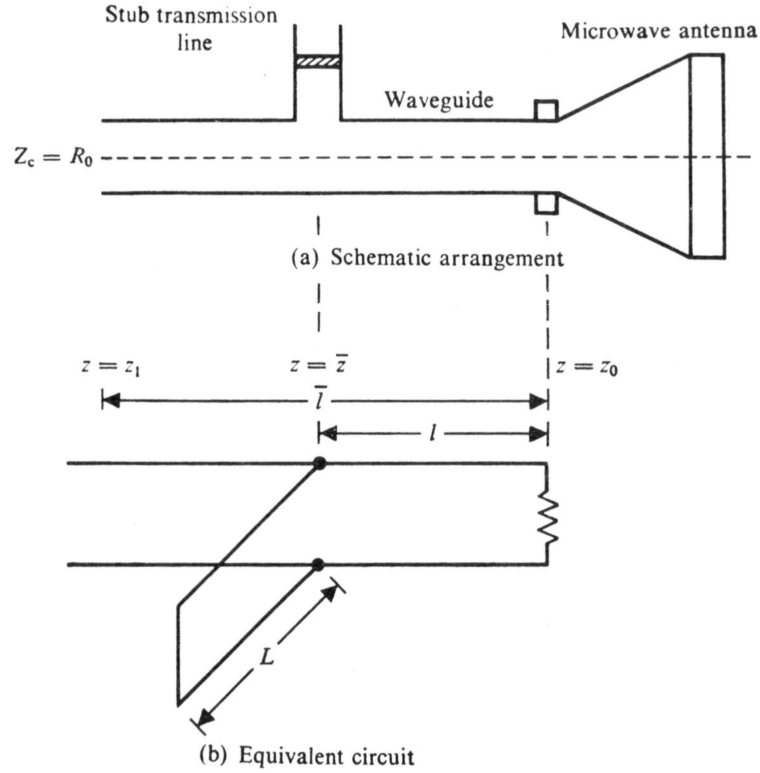

Fig. 6.9 Microwave antenna and matching stub line.

waveguide junction such that $\bar{l} = z_0 - z_1 = \pi/(2\bar{\beta})$, the input impedance is a pure resistance of value $R(\bar{l}) = R_0^2/R$. If R is less than R_0, then $R(\bar{l})$ is greater than R_0, whereas if R is greater than R_0, then $R(\bar{l})$ is less than R_0. It is obvious then that, at some z between z_0 and z_1, the resistive component of the input impedance will equal R_0. There will also be a reactive component at such a point, as given by Eq. (6.26). Making use of the power transfer theorem, this reactive component can be tuned out with the help of an equal and opposite reactance that can be obtained from a section of a transmission line. Then, $R(l) = R_0$, where $l = z_0 - \bar{z}$, and the line coming up to this point is properly terminated. From Eq. (6.26),

$$Z(l) = R(l) + jX(l) = R_0 \frac{R \cos \bar{\beta}l + jR_0 \sin \bar{\beta}l}{R_0 \cos \bar{\beta}l + jR \sin \bar{\beta}l}. \tag{6.42}$$

From Eq. (6.27), the input admittance at this point is

$$Y(l) = G(l) + jB(l) = \frac{R_0 \cos \beta l + jR \sin \beta l}{R_0(R \cos \beta l + jR_0 \sin \beta l)}$$

$$= \frac{RR_0}{R_0(R^2 \cos^2 \beta l + R_0^2 \sin^2 \beta l)} + j[\frac{(R^2 - R_0^2) \sin \beta l \cos \beta l}{R_0(R^2 \cos^2 \beta l + R_0^2 \sin^2 \beta l)}]. \tag{6.43}$$

The matching condition requires that

$$G(l) = \frac{1}{R_0}. \tag{6.44}$$

From Eqs. (6.43) and (6.44),

$$RR_0 = R^2 \cos^2 \beta l + R_0^2 \sin^2 \beta l,$$

$$\cos^2 \beta l = \frac{RR_0 - R_0^2}{R^2 - R_0^2}, \qquad \sin^2 \beta l = \frac{R^2 - RR_0}{R^2 - R_0^2}, \qquad \tan^2 \beta l = \frac{R}{R_0}$$

so that

$$l = \frac{1}{\beta} \tan^{-1} (\frac{R}{R_0})^{1/2}. \tag{6.45}$$

Equation (6.45) gives the proper location of the section of transmission line (also called stub line) that can be used for the matching purpose. The length L of the stub line can be calculated by making its reactance equal and opposite to that at l. If the characteristic impedance of the short-circuited stub line is the same as that of the main line, the stub line will have a reactance

$$X = jR_0 \tan \beta L. \tag{6.46}$$

Equating this reactance to $-1/jB(l)$, we have

$$L = \frac{1}{\beta} \tan^{-1} \frac{(R/R_0)^{1/2}}{(R/R_0 - 1)}. \tag{6.47}$$

If the characteristic impedance and phase constant of the short-circuited stub line are different from those of the main line and are equal to R_0' and β', then

$$X' = jR_0' \tan \beta L. \tag{6.48}$$

The reactance X' given by Eq. (6.48) has to be equated to $-1/jB(l)$ to obtain the value of L. Equations (6.46)–(6.48) give the stub criteria to match a line to a resistance load. Similar equations can be derived for matching a complex load impedance Z_L by merely replacing R in Eq. (6.41) by Z_L and by evaluating the corresponding values of l and L. Thus, if the main line and the stub line have the same characteristic impedance and phase constant, then

$$l = \frac{1}{\beta} \tan^{-1} [\frac{X_n}{R_n - 1} + \{\frac{X_n^2}{(R_n - 1)^2} - \frac{R_n - R_n^2 - X_n^2}{R_n - 1}\}^{1/2}], \tag{6.49}$$

$$L = \frac{1}{\beta} \tan^{-1} [\frac{R_n^2 + (X_n + A)^2}{A(R_n^2 + X_n^2) + X_n(A^2 - 1) - A}], \tag{6.50}$$

where $A = \tan \beta l$ and R_n and X_n are respectively the resistance and reactance of the load normalized with respect to the characteristic impedance of the line.

Sometimes, in practice, it is necessary to vary l and L slightly about their theoretical values to obtain the best matching results. This is so because small junction impedances are introduced at the load and at the junction of the stub and main line.

Since it is difficult to vary l and L in a waveguide or a coaxial line, two stub lines (Fig. 6.10) are often used for matching purposes. In addition to the ease in adjusting the matching parameters mechanically, the double-stub tuner provides facilities for matching a load at two different frequencies, particularly when the load is resistive. Thus, a load whose resistance changes significantly at two frequencies in the pass-band can be matched simultaneously with the double-stub tuner. Usually, the distance l between the two stubs is made approximately equal to a quarter wavelength. If Y_c is the characteristic admittance of the main line, to obtain maximum power transfer, it is necessary to adjust the lengths of the stubs so that the admittance looking to the right at ab is equal to Y_c. Stub 2 at cd serves to make the conductance part of the admittance at ab equal to Y_c, and stub 1 is then adjusted to cancel the susceptance portion of the admittance at ab. The range of admittances which can be matched by this method is slightly increased by spacing the stubs a little less than a quarter wavelength apart. Triple-stub impedance transformers are sometimes used when accurate impedance matching is required.

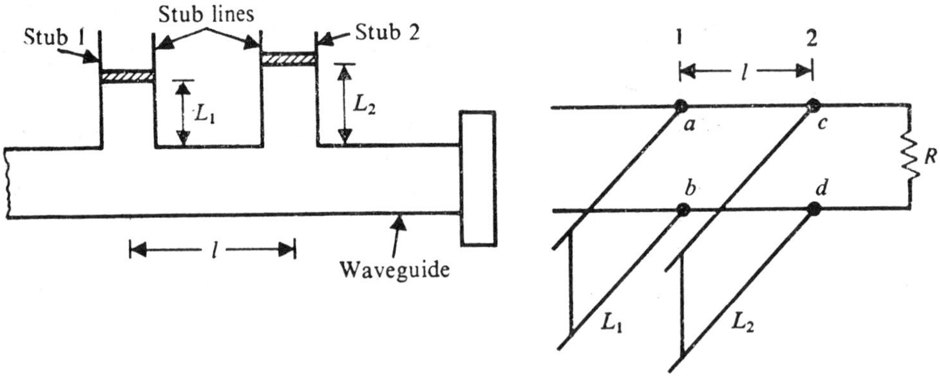

(a) Schematic arrangement of double-stub tuner (b) Transmission-line equivalent circuit of double-stub tuner

Fig. 6.10 Double-stub tuner.

6.6 Graphical Solution of Transmission Line Equations

Transmission lines and waveguides are not only used for guiding electromagnetic energy at microwave frequencies, but also sections of them are used as, for example, resonators, circuit elements, and matching devices. The theory of the design of a transmission line for any of these purposes is based primarily on the transmission line equations. Most of the equations associated with the design of a transmission line involve transcendental functions such as trigonometric and hyperbolic functions, and often the numerical computation for the design parameters is fairly complicated. However, a few useful graphical methods are available for the solution of transmission line problems. We shall consider two such methods, both called impedance diagrams. They are (i) the rectangular impedance diagram and (ii) the polar impedance diagram or the Smith chart. While examining these diagrams, we should consider two other transmission line parameters, namely, the *reflection coefficient* and the *standing-wave ratio*.

We have seen that, if a uniform transmission line is not properly matched, that is, if the line is not terminated in its characteristic impedance, there will be standing waves of the voltage and current along the line. These standing waves result from interference between two sets of travelling waves, one travelling in the positive z-direction and the other travelling in the negative z-direction [see Eqs. (6.9)]. If the line is infinitely long or terminated in its characteristic impedance, then B and D in Eqs. (6.9) are zero. But, for a finite line or a mismatched line, there is some reflection from the termination, and B and D are not zero. The term exp $(\bar{\gamma}z)$ in Eqs. (6.9) results from reflection. The terms involving exp $(-\bar{\gamma}z)$ are called the outgoing or forward waves; the terms involving exp $(+\bar{\gamma}z)$ are called the incoming or reflected waves.

If the transmission line is lossless, $\bar{\gamma} = j\bar{\beta}$. Then, from Eqs. (6.25),

$$\begin{aligned} V(z) &= V(z_0) \cos [\bar{\beta}(z - z_0)] - jI(z_0)Z_0 \sin [\bar{\beta}(z - z_0)] \\ &= \tfrac{1}{2}[V(z_0) + I(z_0)Z_0] \exp [-j\bar{\beta}(z - z_0)] + \tfrac{1}{2}[V(z_0) - I(z_0)Z_0] \exp [j\bar{\beta}(z - z_0)] \\ &= V_1 \exp [-j\bar{\beta}(z - z_0)] + V_2 \exp [j\bar{\beta}(z - z_0)], \end{aligned} \tag{6.51}$$

where V_1 and V_2 are respectively the amplitudes of the forward and reflected voltage travelling waves. Here, a time factor of $e^{j\omega t}$ is assumed, and z_0 is a reference point along z. Also, from Eqs. (6.25),

$$I(z) = \left(\frac{V_1}{Z_0}\right) \exp [-j\bar{\beta}(z - z_0)] - \left(\frac{V_2}{Z_0}\right) \exp [j\bar{\beta}(z - z_0)]. \tag{6.52}$$

The impedance at z can also be expressed in terms of the travelling waves. Hence, from Eqs. (6.51) and (6.52),

$$Z(z) = Z_c \frac{(1 + V_2/V_1) \exp [j2\bar{\beta}(z - z_0)]}{(1 - V_2/V_1) \exp [j2\bar{\beta}(z - z_0)]}. \tag{6.53}$$

If $Z(z)/Z_c$ is the normalized impedance, and if the transmission line is terminated at an impedance Z_L at $z = z_0$, from Eq. (6.53), the normalized impedance becomes

$$\frac{Z_L}{Z_c} = \frac{1 + V_2/V_1}{1 - V_2/V_1}. \tag{6.54}$$

It is seen from Eq. (6.53) that the ratio of V_2 to V_1 is determined by the terminating impedance Z_L. The ratio of the amplitude of the forward and reflected waves at any value of z is known as the reflection coefficient, and is designated as $\Gamma(z)$. The reflection coefficient $\Gamma(z_0)$ of the termination at $z = z_0$ is V_2/V_1. Similarly, the reflection coefficient $\Gamma(z)$ at any other z is given by

$$\Gamma(z) = \frac{V_2}{V_1} \exp [j2\bar{\beta}(z - z_0)] = \Gamma(z_0) \exp [2j\bar{\beta}(z - z_0)]. \tag{6.55}$$

From Eqs. (6.53) and (6.55), we have

$$\Gamma(z) = \frac{Z(z) - Z_c}{Z(z) + Z_c} = \frac{Z(z)/Z_c - 1}{Z(z)/Z_c + 1}, \tag{6.56}$$

$$\frac{Z(z)}{Z_c} = \frac{1 + \Gamma(z)}{1 - \Gamma(z)}. \tag{6.57}$$

The impedance diagrams are based primarily on Eqs. (6.56) and (6.57).

The *voltage standing-wave ratio* (VSWR) is defined as the ratio of the maximum to the minimum voltage in the standing waves on a transmission line. Symbolically,

$$\text{VSWR} = \frac{|V|_{\text{max}}}{|V|_{\text{min}}} = r. \tag{6.58}$$

From Eq. (6.51), it is seen that the standing-wave voltage becomes maximum when $V_1 \exp[-j\beta(z - z_0)]$ and $V_2 \exp[j\beta(z - z_0)]$ are 180° out of phase. Thus

$$\begin{aligned} |V|_{\text{max}} &= |V_1| + |V_2|, \\ |V|_{\text{min}} &= |V_1| - |V_2|. \end{aligned} \tag{6.59}$$

From Eqs. (6.58) and (6.59),

$$|\Gamma| = \frac{r - 1}{r + 1}$$

or

$$r = \frac{1 + |\Gamma|}{1 - |\Gamma|}. \tag{6.60}$$

In general, Γ and Z are complex, but r is always real. Let

$$Z(z) = R(z) + jX(x) = (R^2 + X^2)^{1/2} e^{j\phi},$$

where $\phi = \tan^{-1}(X/R)$.

If $\Gamma = |\Gamma| e^{j\psi}$, we have, from Eqs. (6.56),

$$|\Gamma|^2 = \frac{(R/Z_c - 1)^2 + (X/Z_c)^2}{(R/Z_c + 1)^2 + (X/Z_c)^2}, \tag{6.61}$$

$$\tan \psi = \frac{2(X/Z_c)}{(R/Z_c)^2 + (X/Z_c)^2 - 1}. \tag{6.62}$$

Equations (6.61) and (6.62) may also be written as

$$\left(\frac{R}{Z_c} - \frac{1 + |\Gamma|^2}{1 - |\Gamma|^2}\right)^2 + \left(\frac{X}{Z_c}\right)^2 = \left(\frac{2|\Gamma|}{1 - |\Gamma|}\right)^2, \tag{6.63}$$

$$\left(\frac{R}{Z_c}\right)^2 + \left(\frac{X}{Z_c} - \cot \psi\right)^2 = \operatorname{cosec}^2 \psi. \tag{6.64}$$

Rectangular Impedance Diagram

Examining Eq. (6.63), we see that, if $|\Gamma|$ is kept constant in a transmission line, this equation describes a circle of radius $\rho = 2|\Gamma|/(1 - |\Gamma|^2)$, with its centre on the real axis, at a distance $(1 + |\Gamma|^2)/(1 - |\Gamma|^2)$ from the origin. Since $|\Gamma|$ is uniquely related to the VSWR through Eq. (6.60), we find that the contours of R/Z_0 and X/Z_0, for any constant r, lie on a circle of radius

$\rho = (r^2 - 1)/(2r)$, the centre of the circle being at $(r^2 + 1)/(2r)$. This circle intersects the real axis at points r and $1/r$, which correspond to the values of the impedance when it is real. On the rectangular impedance chart, these two points correspond respectively to the points (values of z) on the line at which the voltage maxima and voltage minima occur since, from Eqs. (6.59), the impedance of the line is real at these points. The family of circles for constant values of $|\Gamma|$ or r is shown in Fig. 6.11. The contours of R/Z_c and X/Z_c also lie on a circle when ψ is a constant, as

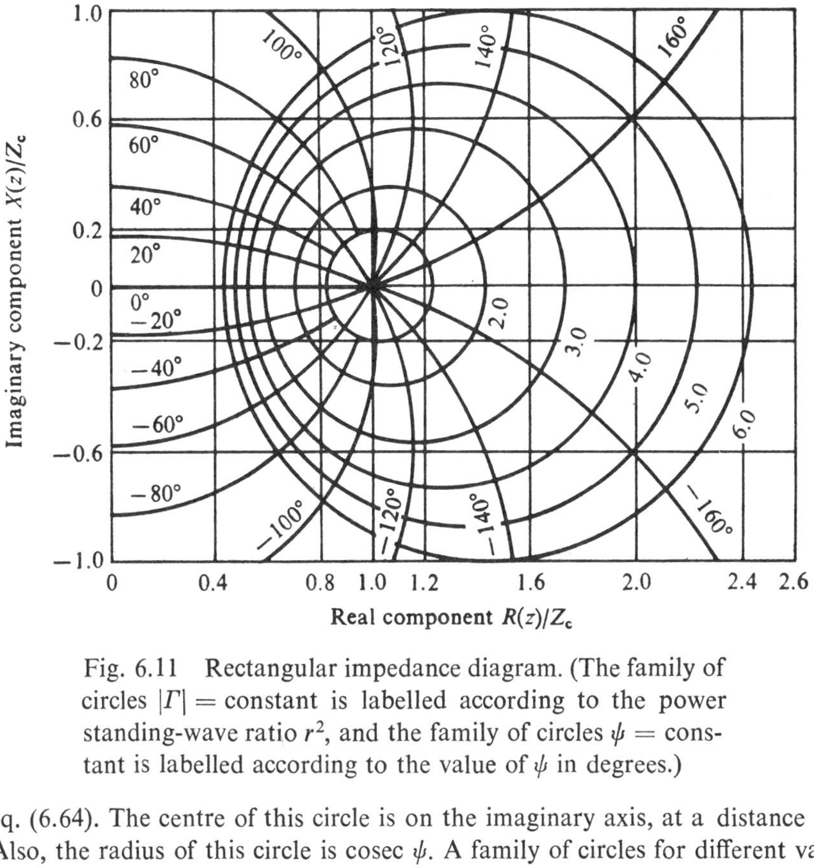

Fig. 6.11 Rectangular impedance diagram. (The family of circles $|\Gamma|$ = constant is labelled according to the power standing-wave ratio r^2, and the family of circles ψ = constant is labelled according to the value of ψ in degrees.)

seen from Eq. (6.64). The centre of this circle is on the imaginary axis, at a distance cot ψ from the origin. Also, the radius of this circle is cosec ψ. A family of circles for different values of ψ is shown in Fig. 6.11. This family of circles is orthogonal to the first family of circles for which r is equal to a constant. The two families of orthogonal circles constitute the rectangular impedance chart. The family of circles ψ = constant intersects at the point (1, 0) on the impedance chart.

We now give an example of how the rectangular impedance chart is used. The input impedance of a uniform lossless transmission line which is terminated by an impedance Z_L at one end is to be determined. For a lossless line, the voltage standing-wave ratio, as also $|\Gamma|$, does not change along the line. But the phase angle of Γ changes by an amount equal to $2\bar{\beta}l$, where $\bar{\beta}$ is the propagation constant and l is the length of the line. Thus, to determine the input impedance normalized with respect to Z_c, adopt the following procedure.

Step 1 Plot Z_L/Z_c on Fig. 6.11.

Step 2 Read the values of r and ψ corresponding to this point.

Step 3 Add $2\bar{\beta}l$ to ψ.

Step 4 Obtain on the chart the point corresponding to r and $\psi + 2\beta l$.

The horizontal and vertical coordinates of this point give the normalized input resistance and reactance of the line, respectively. To obtain the actual value of the input resistance and reactance in ohms, we have to multiply the value by Z_c.

Smith Chart

The Smith chart[1] is a polar impedance diagram, which is used quite frequently to study the impedance or admittance transformation due to a transmission line or waveguide.

Since Γ and Z are, in general, complex, we may write

$$\Gamma = \Gamma_{\text{Re}} + j\Gamma_{\text{Im}}, \tag{6.65}$$

$$\frac{Z}{Z_c} = \frac{R}{Z_c} + \frac{jX}{Z_c}. \tag{6.66}$$

Substituting Eqs. (6.65) and (6.66) in Eq. (6.57), we obtain

$$\frac{R}{Z_c} = \frac{1 - (\Gamma_{\text{Re}}^2 + \Gamma_{\text{Im}}^2)}{(1 - \Gamma_{\text{Re}}^2) + \Gamma_{\text{Im}}^2}, \tag{6.67}$$

$$\frac{X}{Z_c} = \frac{2\Gamma_{\text{Im}}}{(1 - \Gamma_{\text{Re}})^2 + \Gamma_{\text{Im}}^2} \tag{6.68}$$

which may also be written as

$$(\Gamma_{\text{Re}} - \frac{R/Z_c}{1 + R/Z_c})^2 + \Gamma_{\text{Im}}^2 = \frac{1}{(1 + R/Z_c)^2}, \tag{6.69}$$

$$(\Gamma_{\text{Re}} - 1)^2 + (\Gamma_{\text{Im}} - \frac{Z_c}{X})^2 = (\frac{Z_c}{X})^2. \tag{6.70}$$

Equation (6.69) means that the locus of R/Z_c = constant is a circle in the Γ-plane, with its centre on the positive real axis, at a distance $(R/Z_c)/(1 + R/Z_c)$ from the origin. The radius of this circle is $(1 + R/Z_c)^{-1}$. Similarly, Eq. (6.70) implies that the locus of X/Z_c = constant is also a circle with its centre at $(1, Z_c/X)$, and its radius is equal to $|Z_c/X|$. The two families of circles corresponding to R/Z_c = constant and X/Z_c = constant are mutually orthogonal. They form the Smith chart shown in Figs. 6.12 and 6.20. Sometimes, the chart also includes circles of constant standing-wave ratio P. The centre of the family of circles $|\Gamma|$ = constant is at (1, 0) on the impedance plane.

In determining the input impedance of a lossless transmission line or a waveguide of length l terminated by an impedance Z_L, the following steps are involved.

Step 1 Compute Z_L/Z_c and l/λ.

Step 2 Enter the normalized impedance on the Smith chart.

Step 3 Draw a line that passes through the points (1, 0) and Z_L/Z_c and extend the line up to the periphery.

Step 4 Read the equivalent length l'/λ of a shorted-transmission line corresponding to Z_L/Z_c from the chart and add l/λ to it.

Step 5 Draw a line from the periphery to the point (1, 0) so that it passes through $l'/\lambda + l/\lambda$.

[1] P. H. Smith, Transmission line calculator, *Electronics*, **12**, 29, January 1939.

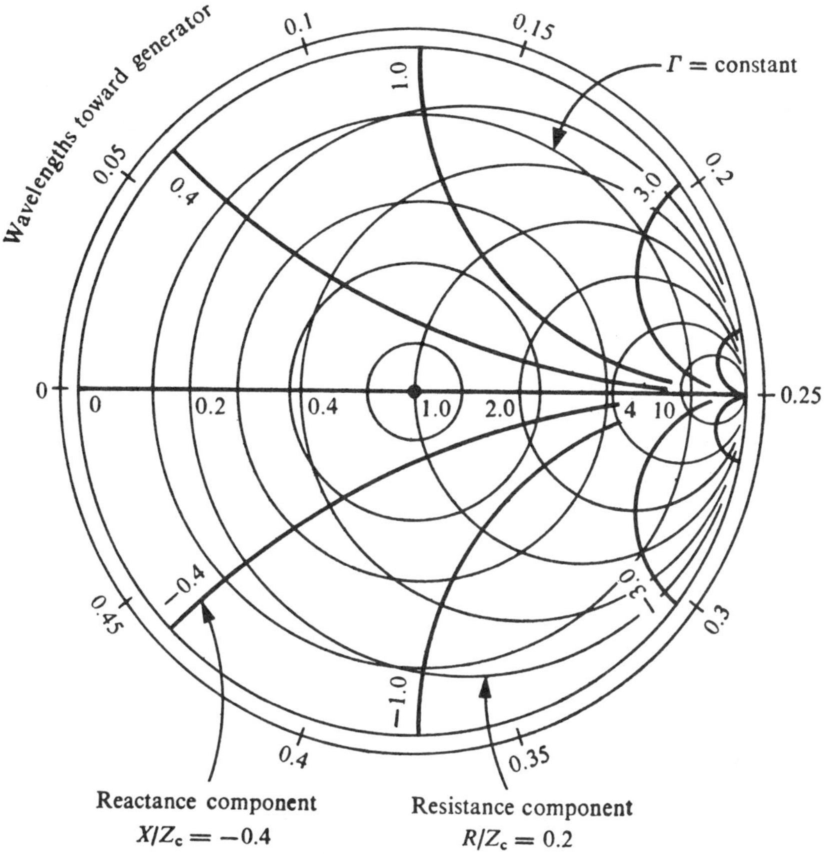

Fig. 6.12 Polar impedance diagram or the Smith chart.

Step 6 Draw an arc with (1, 0) as the centre that passes through Z_L/Z_c and intersects the line corresponding to the length $l'/\lambda + l/\lambda$. [The point of intersection gives the required input impedance (resistance and reactance) (see Problems 6 and 7 at the end of this chapter).]

There are many uses of the Smith chart in microwave engineering, some of which are (i) to design stubs for matching purposes, (ii) to design microwave filters, and (iii) to determine the effect of a number of small discontinuities in a waveguide or a transmission line.

6.7 Transmission Lines with Losses

In Sections 5.8 and 5.14, we have studied the losses due to the finite conductivity of the metal walls and the losses in the dielectric medium and obtained expressions for the attenuation constant $\bar{\alpha}$ in coaxial lines and metal waveguides. Due to these losses, the propagation constant $\bar{\gamma}$ is a complex quantity $\bar{\alpha} + j\bar{\beta}$, where $\bar{\alpha}$ is, in general, equal to $\bar{\alpha}_d + \bar{\alpha}_c$, $\bar{\alpha}_d$ being the attenuation due to losses in the dielectric medium and $\bar{\alpha}_c$ the attenuation due to the finite conductivity of the metal walls. This $\bar{\alpha}$ is the attenuation constant for propagating modes. In waveguides, for frequencies below the cut-off frequency, there is no propagation but only attenuation; the wave gets attenuated very rapidly, and the field is called an evanescent field.

The transmission line equations describe the propagating waves in a waveguide or transmission line. Hence, $\bar{\beta}$ should be replaced by $\bar{\gamma} = \bar{\alpha} + j\bar{\beta}$ in all the equations discussed earlier in this chapter. For example, the expressions for $V(z)$ and $I(z)$ corresponding to Eqs. (6.25) become

$$V(z) = V(z_0) \cosh [\bar{\gamma}(z - z_0)] - I(z_0)Z_c \sinh [\bar{\gamma}(z - z_0)],$$
$$I(z) = I(z_0) \cosh [\bar{\gamma}(z - z_0)] - [V(z_0)/Z_c] \sinh [\bar{\gamma}(z - z_0)]. \tag{6.71}$$

The input impedance $Z(z)$ corresponding to Eq. (6.26) becomes

$$Z(z) = \frac{V(z)}{I(z)} = Z_c \frac{Z_n - \tanh [\bar{\gamma}(z - z_0)]}{1 - Z_n \tanh [\bar{\gamma}(z - z_0)]}, \tag{6.72}$$

where Z_n is the normalized impedance at z_0 and is equal to

$$Z(z_0)/Z_c = \frac{V(z_0)/I(z_0)}{Z_c}.$$

In terms of the travelling waves, the voltage, current, and impedance at any z in a dissipative waveguide can be expressed as

$$V(z) = V_1 \exp [-\bar{\gamma}(z - z_0)] + V_2 \exp [\bar{\gamma}(z - z_0)],$$
$$I(z) = \frac{V_1}{Z_c} \exp [-\bar{\gamma}(z - z_0)] - \frac{V_2}{Z_c} \exp [\bar{\gamma}(z - z_0)], \tag{6.73}$$

$$Z(z) = Z_c \frac{(1 + V_2/V_1) \exp [2\bar{\gamma}(z - z_0)]}{(1 - V_2/V_1) \exp [2\bar{\gamma}(z - z_0)]}, \tag{6.74}$$

$$\frac{Z(z)}{Z_c} = \frac{1 + \Gamma(z)}{1 - \Gamma(z)}, \tag{6.75}$$

where $\Gamma(z)$ is the voltage reflection coefficient. And the equation corresponding to Eq. (6.55) is

$$\Gamma(z) = \frac{V_2}{V_1} \exp [2j\bar{\beta}(z - z_0)] = \Gamma(z_0) \exp [j2\bar{\beta}(z - z_0)] \tag{6.76a}$$

or

$$|\Gamma(z)| = |\Gamma(z_0)| \exp [2\bar{\alpha}(z - z_0)] \tag{6.76b}$$

or

$$r = \frac{1 + |\Gamma|}{1 - |\Gamma|} = -\coth [\tfrac{1}{2} \ln |\Gamma(z_0)| + \bar{\alpha}(z - z_0)], \tag{6.77}$$

where r is the VSWR. We see from Eq. (6.77) that the VSWR in a dissipative line is not a constant, as in the lossless line, but changes in the z-direction. We shall also see that the period of the standing wave (distance between two consecutive voltage maxima or voltage minima) in a dissipative line also changes along the z-direction, as shown in Figs. 6.13 and 6.14.

Consider a uniform transmission line with losses, shown in Fig. 6.15, which is connected to an arbitrary load of impedance Z_R at $z = z_1$. If we put $z_0 - z = l$ in Eqs. (6.71), the generalized transmission line equations can be written as

$$V_S = V_R \cosh \bar{\gamma} l + Z_c I_R \sinh \bar{\gamma} l,$$
$$I_S = I_R \cosh \bar{\gamma} l + \frac{V_R}{Z_c} \sinh \bar{\gamma} l. \tag{6.78}$$

Here, V_S and I_S represent the voltage and current at the sending end at z; V_R and I_R denote the corresponding voltage and current at the receiving end at z_1; and $\bar{\gamma}$ is the complex propagation constant equal to $\bar{\alpha} + j\bar{\beta}$. In general, V_S, I_S, V_R, I_R, and Z_R are complex quantities. We shall

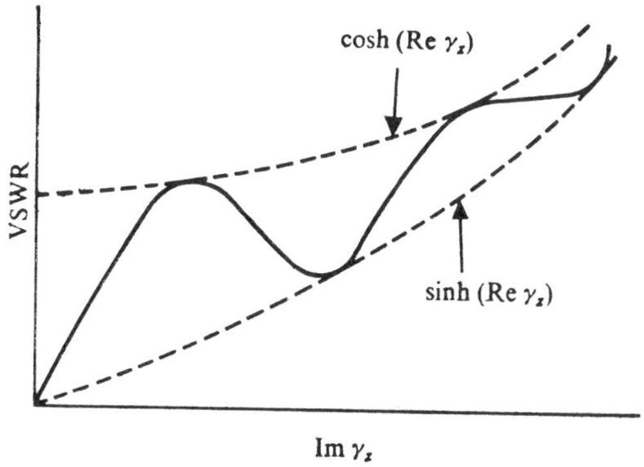

Fig. 6.13 Voltage standing-wave ratio for a short-circuited transmission line with losses.

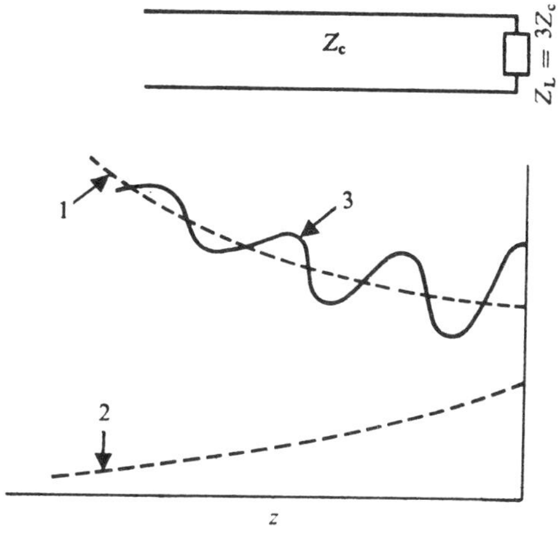

Fig. 6.14 Voltage standing-wave pattern on a lossy transmission line. (1, envelope of incident wave amplitude; 2, envelope of reflected wave amplitude; 3, standing-wave pattern.)

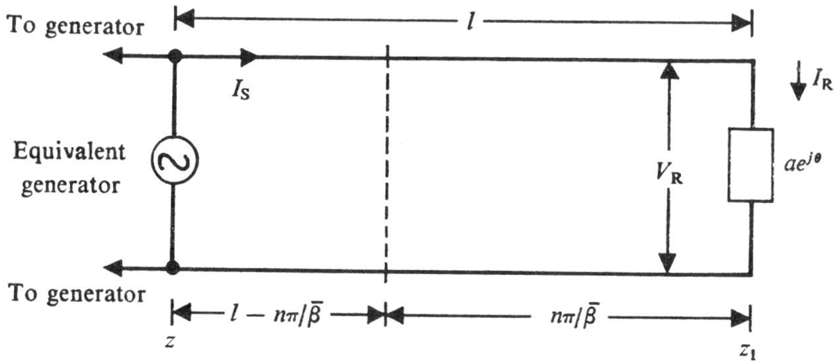

Fig. 6.15 Lossy transmission line.

assume, with no loss of generality, that the phase angle of V_R is zero. The other complex quantities can then be expressed with respect to the reference zero phase angle.

We now define

$$\frac{I_R Z_c}{V_R} = \frac{Z_c}{Z_R} = ae^{j\theta}, \tag{6.79}$$

where a and θ are arbitrary real quantities.

Substituting $\bar{\gamma} = \bar{\alpha} + j\bar{\beta}$ in Eq. (6.78), we obtain the equation

$$\begin{aligned}|V_S'|^2 &= (\cosh \bar{\alpha}l \cos \bar{\beta}l + a \cos \theta \sinh \bar{\alpha}l \cos \bar{\beta}l - a \sin \theta \cosh \bar{\alpha}l \sin \bar{\beta}l)^2 \\ &\quad + (\sinh \bar{\alpha}l \sin \bar{\beta}l + a \sin \theta \sinh \bar{\alpha}l \cos \bar{\beta}l + a \cos \theta \cosh \bar{\alpha}l \sin \bar{\beta}l)^2,\end{aligned} \tag{6.80}$$

where $V_S' = V_S/V_R$. From the nature of the variation of the standing wave in the line, it can be seen that $|V_S|$ is maximum or minimum when $|V_S'|^2$ is maximum or minimum, respectively. From Eq. (6.80),

$$\begin{aligned}|V_S'|^2 &= a^2 \sinh^2 \bar{\alpha}l \cos^2 \bar{\beta}l + a^2 \cosh^2 \bar{\alpha}l \sin^2 \bar{\beta}l + \sinh^2 \bar{\alpha}l \sin^2 \bar{\beta}l \\ &\quad + \cosh^2 \bar{\alpha}l \cos^2 \bar{\beta}l + 2a \cos \theta \sinh \bar{\alpha}l \cosh \bar{\alpha}l - 2a \sin \theta \sin \bar{\beta}l \cos \bar{\beta}l.\end{aligned} \tag{6.81}$$

Differentiating Eq. (6.81) with respect to l and setting the expression for the derivative equal to zero, we obtain

$$(a^2 - 1) \sin (2\bar{\beta}l) = 2a \sin \theta \cos (2\bar{\beta}l) - (a^2 + 1) \sinh (2\bar{\alpha}l) - 2a \cos \theta \cosh (2\bar{\alpha}l). \tag{6.82}$$

If l_1 is the position of a voltage maximum, then the other voltage maxima are given by $l_2 = l_1 + n\pi/\bar{\beta}$, where n is a positive or negative integer.

From Eq. (6.82), we find that the condition for a maximum or a minimum (as the case may be) necessitates

$$\begin{aligned}\frac{\cosh [2\bar{\alpha}(l_1 + \pi/\bar{\beta})] - \cosh (2\bar{\alpha}l_1)}{\sinh [2\bar{\alpha}(l_1 + \pi/\bar{\beta})] - \sinh (2\bar{\alpha}l_1)} &= \text{constant} \\ &= \frac{\cosh [2\bar{\alpha}(l_1 + 2\pi/\bar{\beta})] - \cosh (2\bar{\alpha}l_1)}{\sinh [2\bar{\alpha}(l_1 + 2\pi/\bar{\beta})] - \sinh (2\bar{\alpha}l_1)}.\end{aligned} \tag{6.83}$$

Equation (6.83) can be true only if $\bar{\alpha} = 0$ or $\bar{\beta} = \infty$, or if $\bar{\alpha} = 0$ and $\bar{\beta} = \infty$. Hence, it can be concluded that, in a lossy transmission line, two consecutive voltage maxima or minima are not, in general, separated by π or 180 electrical degrees. However, for some particular choice of a and θ, any two voltage maxima or minima can be exactly 180 electrical degrees apart, even on a lossy line. The same conclusion is true also for current maxima and minima.

If the constant in Eq. (6.83) is equal to $-(a^2 + 1)/\cos \theta$, then it is obvious that it has to be positive for $\bar{\alpha} > 0$. This necessitates some value of θ corresponding to a negative conductance of the load. Physically, this is realizable by connecting a generator as the load.

In many problems, it is desirable to determine the phase deviation of the voltage due to nonzero $\bar{\alpha}$ at $\bar{\beta}l = n\pi$, where n is an integer. For a lossless line, the phase angle is zero. For a lossy line, we have, at $\bar{\beta}l = n\pi$,

$$|V_S'|^2 = (\cosh \bar{\alpha}l + a \cos \theta \sinh \bar{\alpha}l)^2 + a^2 \sin^2 \theta \sinh^2 \bar{\alpha}l \tag{6.84}$$

and the phase angle of V_S' is

$$\bar{\theta} = \tan^{-1} \left(\frac{a \sin \theta \sinh \bar{\alpha}l}{\cosh \bar{\alpha}l + a \cos \theta \sinh \bar{\alpha}l}\right). \tag{6.85}$$

The phase angle $\bar{\theta}$ can be determined from a simple graph shown in Fig. 6.16. In order to

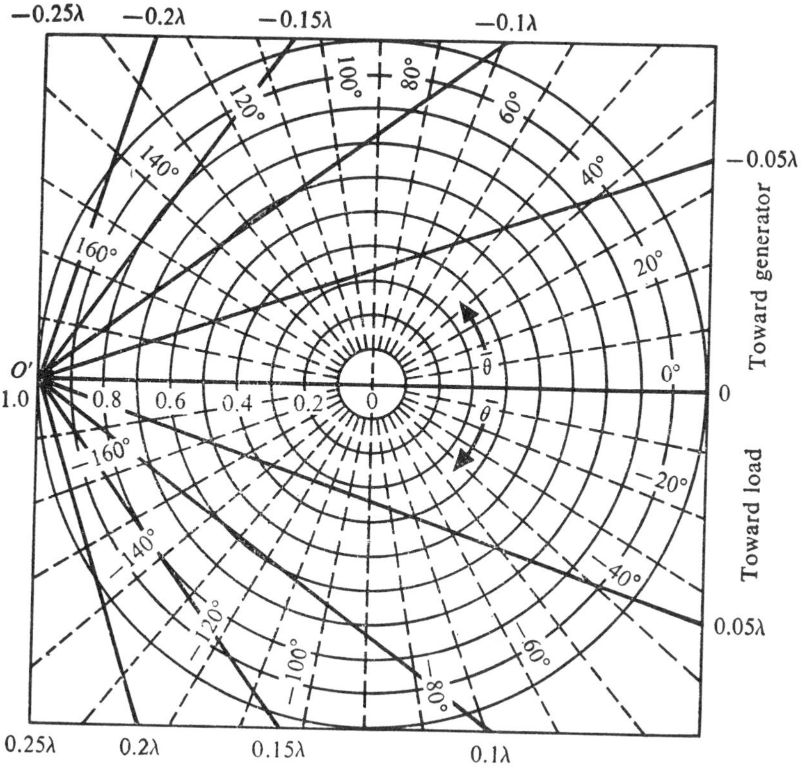

Fig. 6.16 Nomograph for determining $\bar{\theta}$ in transmission line with losses. (Radius of the concentric circle $R = a \tanh \bar{\alpha}l$; $\bar{\theta}$ = phase angle of load.)

obtain $\bar{\theta}$, adopt the following procedure.

Step 1 Compute $\bar{\alpha}l$ and $\tanh \bar{\alpha}l$ when $\bar{\beta}l = n\pi$.

Step 2 Draw a line, making an angle θ with the horizontal from 0.

Step 3 Join the point of intersection of the line and the circle of radius $R = a \tanh \bar{\alpha}l$ and O'.

The angle subtended at O' by the line represents the phase angle of the voltage at $\bar{\beta}l = n\pi$.

6.8 Rectangular Impedance Diagram for Lossy Transmission Lines

The rectangular impedance diagram (see Section 6.6) can be used for the solution of lossy transmission line problems also. From Eq. (6.57),

$$\frac{Z(z)}{Z_c} = \frac{1 + \Gamma(z)}{1 - \Gamma(z)}. \tag{6.86}$$

For a transmission line with losses,

$$\Gamma(z) = \Gamma(z_0) \exp\left[-2\gamma(z - z_0)\right], \tag{6.87}$$

where $\gamma = \bar{\alpha} + j\bar{\beta}$ is the propagation constant of the line, $\Gamma(z_0)$ the reflection coefficient at the termination $z = z_0$, and $\Gamma(z)$ is the reflection coefficient at $z = z$.

Put

$$\Gamma(z_0) = \exp\left[-2(t_0 + ju_0)\right], \tag{6.88}$$

$\exp(-2t_0)$ being the magnitude and $-2u_0$ the angle of the reflection coefficient. Now, let

$$t = t_0 + \bar{\alpha}z, \qquad u = u_0 + \bar{\beta}z. \tag{6.89}$$

Then, using Eqs. (6.86)–(6.89) in Eq. (6.85), we obtain

$$\frac{Z(z)}{Z_c} = \frac{1 + \exp\left[-2(t + ju)\right]}{1 - \exp\left[-2(t + ju)\right]}. \tag{6.90}$$

At the receiving end of the line, we have

$$Z(z) = Z_R, \qquad \bar{\alpha}_z = 0, \qquad \bar{\beta}_z = 0,$$

and therefore Eq. (6.90) becomes

$$\frac{Z_R}{Z_c} = \frac{1 + \exp\left[-2(t_0 + ju_0)\right]}{1 - \exp\left[-2(t_0 + ju_0)\right]}. \tag{6.91}$$

Equation (6.90) is the basic relationship used in the construction of impedance diagrams. The impedance diagram is a plot of Eq. (6.90) which enables us to obtain values of Z/Z_c if the values of t and u are known or, conversely, to obtain values of t and u if Z/Z_c is known. When impedances are expressed as ratio, e.g., Z/Z_c, they are known as normalized impedances.

For the sake of convenience, the real and imaginary parts of Z/Z_c are separated in Eq. (6.90). Thus

$$\frac{Z}{Z_c} = r + jx = \frac{1 + \exp\left[-2(t + ju)\right]}{1 - \exp\left[-2(t + ju)\right]}. \tag{6.92}$$

The rectangular impedance diagram is a plot of Eq. (6.92) in the form of constant t- and constant u-loci which are two sets of orthogonal circles, as shown in Fig. 6.17. Note that all the circles $\bar{\beta}l =$ constant pass through the point (1, 0).

To evaluate the input impedance of a transmission line terminated in a known impedance Z_R, use the following steps (assuming that Z_R, Z_c, $\bar{\alpha}$, $\bar{\beta}$, and the length l of the line are known).

Step 1 Compute the values of Z_R/Z_c, $\bar{\alpha}l$, and $\bar{\beta}l$.

Step 2 Enter the chart at the known value of Z_R/Z_c, and observe the values of t_0 and u_0 (on the $\bar{\alpha}l$- and $\bar{\beta}l$-circles).

Step 3 Compute the values of $t = t_0 + \bar{\alpha}l$ and $u = u_0 + \bar{\beta}l$. These are the values of t and u corresponding to the sending end of the line.

Step 4 Re-enter the chart at the new values of t and u (on the $\bar{\alpha}l$- and $\bar{\beta}l$-circles) and then read the corresponding impedance Z/Z_c. This is the normalized input impedance.

As an example, assume that $Z_R/Z_c = 2 + j0$, $\bar{\alpha}l = 0.2$ neper, and $\bar{\beta}l = 0.6$ radian. Entering the impedance diagram of Fig. 6.17 at $Z_R/Z_c = 2 + j0$, we obtain the values of $t_0 = 0.534$ and $u_0 = 90°$ or 1.57 radians. Step 3 yields $t = 0.734$ and $u = 124°$ or 2.17 radians. Re-entering the

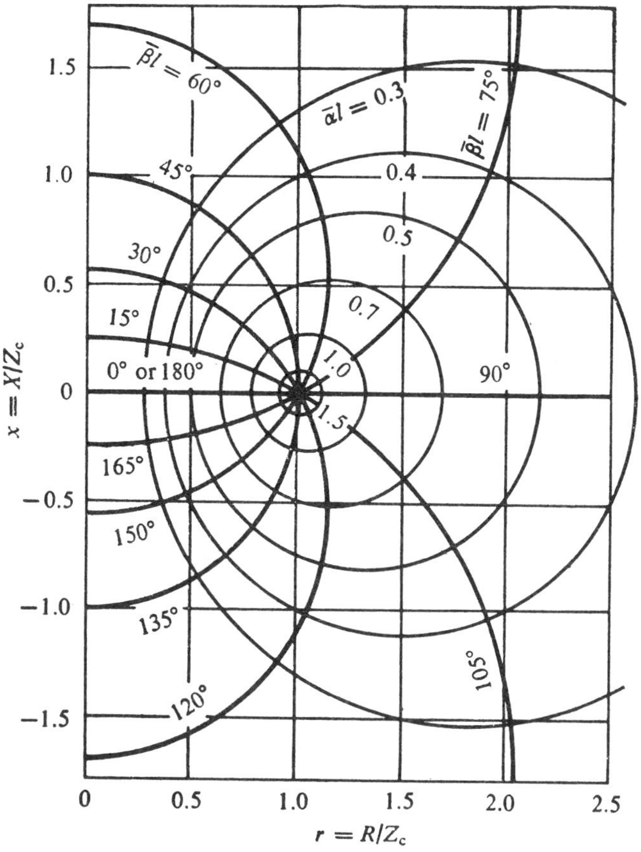

Fig. 6.17 Rectangular impedance diagram.

impedance diagram at these values of t and u, the normalized input impedance is found to be $Z/Z_c = 1.08 - j0.5$.

6.9 Smith Chart or Polar Impedance Diagram for Lossy Transmission Lines

In the rectangular impedance diagram (Fig. 6.17), the circle $\bar{\alpha}l = 0$ has an infinitely large radius. Therefore, an infinitely large diagram is required to solve all possible problems. When dealing with low-loss lines which are open circuited, short circuited, or terminated in a pure reactance, the solution is found to be beyond the limits of any practical diagram. In the polar impedance diagram introduced by Smith[1, 2], the entire impedance diagram is contained within a circle of any desired radius.

We shall now introduce a new variable $(p + jq)$ such that

$$(p + jq) = e^{-2(t+ju)}. \tag{6.93}$$

[1]P. H. Smith, Transmission line calculator, *Electronics*, **12**, 29, January 1939.
[2]P. H. Smith, An improved transmission line calculator, *Electronics*, **17**, 130, January 1944.

Hence, Eq. (6.92) becomes

$$\frac{Z}{Z_c} = r + jx = \frac{1 + (p + jq)}{1 - (p + jq)}. \tag{6.94}$$

If p and q represent the rectangular coordinate axes, the constant r- and constant x-loci are two sets of orthogonal circles, as shown in Fig. 6.18.

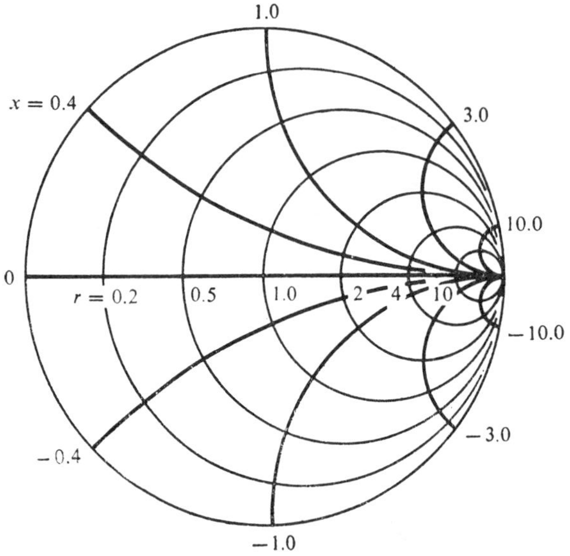

Fig. 6.18 Polar impedance diagram or Smith chart: constant r- and constant x-loci.

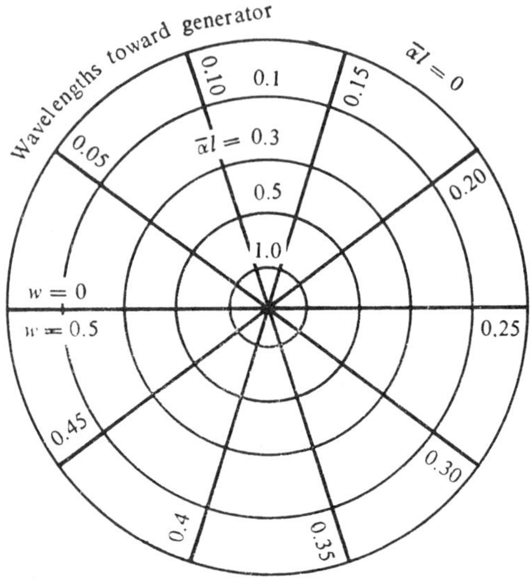

Fig. 6.19 Polar impedance diagram: constant $\bar{\alpha}l$- and constant w-circles.

The constant t- and constant w-loci, where $w = w_0 + z/\lambda = u/(2\pi)$ and $w_0 = u_0/(2\pi)$ and $\bar{\beta} = 2\pi/\lambda$, are again two sets of orthogonal curves, as shown in Fig. 6.19. w is a measure of the length of the line in wavelengths.

The Smith chart or polar impedance diagram given in Fig. 6.20 is a superposition of Figs. 6.18 and 6.19. Sometimes, instead of $\bar{\alpha}l$, the decibel or standing-wave ratio circles are included. The $\bar{\alpha}l$-circles facilitate the solution of problems where the line has losses.

It should be observed that both the rectangular and polar impedance diagrams may be used to solve problems in terms of admittances and impedances. The normalized admittance at any point on the line is the reciprocal of the normalized impedance, and hence

$$\frac{Y}{Y_c} = \frac{1 - \exp[-2(t + ju)]}{1 + \exp[-2(t + ju)]}, \tag{6.95}$$

where $Y_c = 1/Z_c$ is the characteristic admittance of the line. The constant r- and constant x-circles become the constant-conductance and constant-susceptance circles.

In using the polar impedance diagram, the following steps are involved.

Step 1 Compute Z_R/Z_c, $\bar{\alpha}l$, and l/λ.

Step 2 Enter the impedance diagram at the known value of Z_R/Z_c and read the corresponding values of t_0 on the $\bar{\alpha}l$-circles and w_0 on the "wavelengths toward generator" scale.

Step 3 Compute the values of $t = t_0 + \overline{\alpha}l$ and $w = w_0 + l/\lambda$.

Step 4 Re-enter the diagram at the new values of t and w and read the corresponding normalized sending-end impedance.

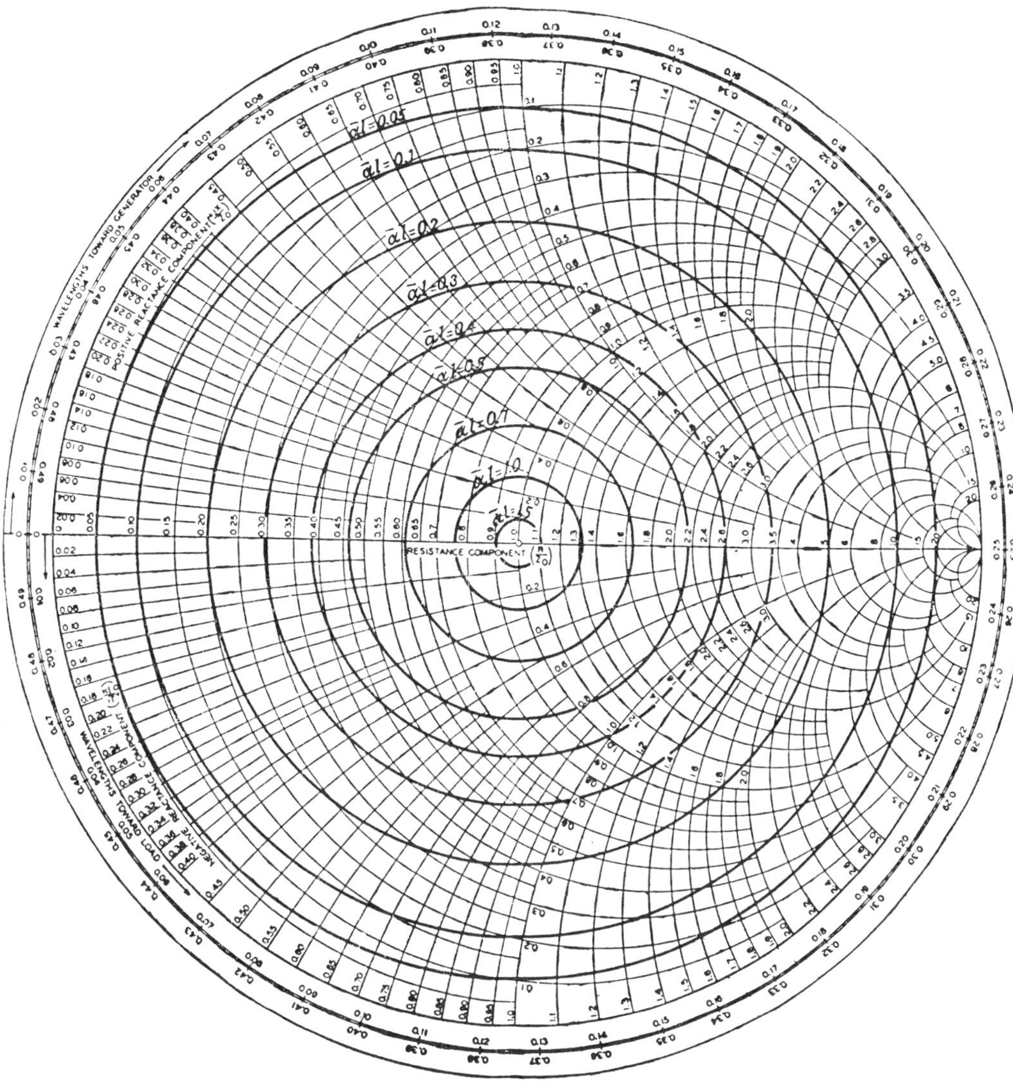

Fig. 6.20 Smith chart or polar impedance diagram.

Now, assume that point P (see Fig. 6.21) corresponds to the terminal impedance Z_R/Z_c and that the line is lossless. As we move toward the generator on the transmission line, the impedance point moves in the clockwise direction on the constant $\overline{\alpha}l$-circle along the path PQ_1. However, if the line has losses, then the impedance point spirals inward, as indicated by the path PQ_2. As the length of the line increases, the impedance point continues to spiral inward, eventually winding up on the point $Z/Z_c = 1$. If we move toward the generator on the transmission line, the impedance point moves in the clockwise direction in the impedance diagram, and the values of w are read on the "wavelengths toward generator" scale. If we move toward the load, the impedance

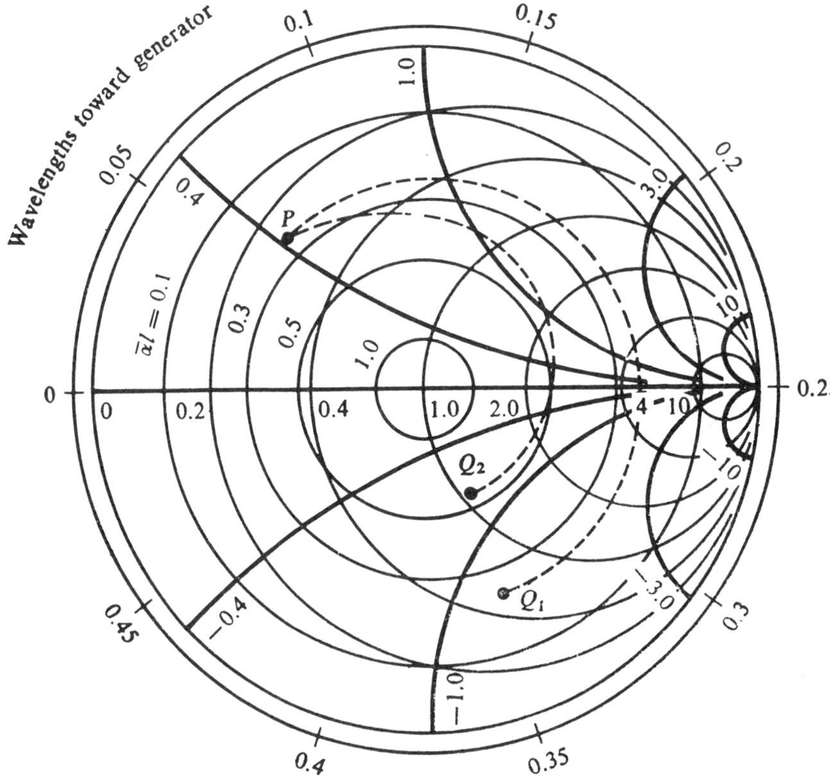

Fig. 6.21 Use of polar impedance diagram.

point in the diagram moves in the counter-clockwise direction, and the values of w are read on the "wavelengths toward load" scale. One complete revolution on the impedance diagram corresponds to a half-wavelength line.

The three examples we now give illustrate the uses of the Smith chart.

Examples (i) Input impedance of line. Let us determine the input impedance of a line having the characteristics

$$Z_c = 75 \text{ ohms}, \qquad Z_R = 150 + j100 \text{ ohms}, \qquad l/\lambda = 0.2, \qquad \bar{\alpha}l = 0.15 \text{ neper.}$$

We first compute $Z_R/Z_c = 2 + j1.333$. Entering the polar impedance diagram at this value of normalized impedance, we obtain $t_0 = 0.35$ and $w_0 = 0.210$ (on the "wavelengths toward generator scale"). Now, compute $t = t_0 + \bar{\alpha}l = 0.50$ and $w = w_0 + l/\lambda = 0.410$. These are the values of t and l at the sending end of the line. Re-entering the impedance diagram at these values of t and w, we obtain the normalized input impedance $Z/Z_c = 0.60 - j0.46$, or $Z = 45 - j34.5$ ohms.

(ii) Single-stub matching. Here, a generator having an internal impedance $Z_g = 75$ ohms is connected to a load impedance $Z_R = 250$ ohms by means of a transmission line having a characteristic impedance of 75 ohms. The wavelength of the impressed signal is $\lambda = 20$ cm. Find the position and length of a short-circuited stub which will yield maximum power transfer to the load. We enter the polar impedance diagram at the normalized load admittance $Y_R/Y_c = 0.30 + j0$ (point P_0 in Fig. 6.22) and observe that $t_0 = 0.30$ (on the $\bar{\alpha}l$-circles) and $w_0 = 0$ (on the "wavelengths toward generator" scale). As we move back toward the generator, the admittance point

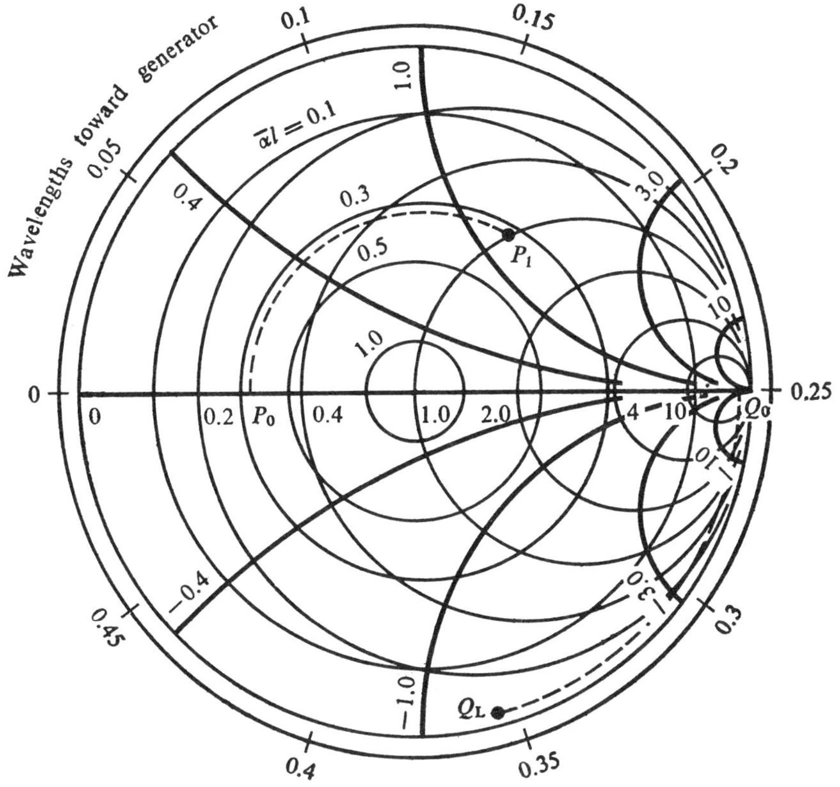

Fig. 6.22 Graphical solution of single-stub impedance matching.

moves in a clockwise direction on the constant $\bar{\alpha}l$-circle. At P_1 we intersect the unity-conductance circle, and at the corresponding point on the transmission line, we place the stub. At this point the admittance is

$$\frac{Y_1}{Y_c} = 1 + j1.26,$$

and we have $w = 0.17$. The distance l_1 from the load to the stub is determined by

$$\frac{l_1}{\lambda} = w_1 - w_0 = 0.17.$$

The stub must provide a normalized susceptance of -1.26 mho. Let us now determine the length of the stub. At the short-circuited end of the stub, we have $Y/Y_c = \infty$. Entering the diagram at this point (Q_0 in Fig. 6.22), we observe that $w_0' = 0.25$, and as we move in the clockwise direction around the constant $\bar{\alpha}l$-circle, we stop at the susceptance $Y_S/Y_c = -j1.26$. The corresponding value of w' is $w' = 0.356$. The stub length is obtained from

$$\frac{l_2}{\lambda} = w' - w_0' = 0.106.$$

To compare the graphical solution with the analytical solution, use Eqs. (6.45) and (6.47) and put

$$\beta = \frac{2\pi}{20/100} = 31.4,$$

$$R = 250 \text{ ohms},$$

$$R_0 = Z_c = 75 \text{ ohms}.$$

Then, we obtain

$$l_1 = \frac{1}{\beta} \tan^{-1} \left(\frac{R}{R_0}\right)^{1/2} \approx 3.38 \text{ cm},$$

$$l_2 = L = \frac{1}{\beta} \tan^{-1} \left[\frac{(R/R_0)^{1/2}}{(R/R_0) - 1}\right] \approx 2.13 \text{ cm}.$$

The values obtained by the two methods are in agreement.

(iii) Double-stub impedance matching. This consists of two or more adjustable stubs spaced approximately a quarter wavelength apart, as shown in Fig. 6.23. To obtain maximum

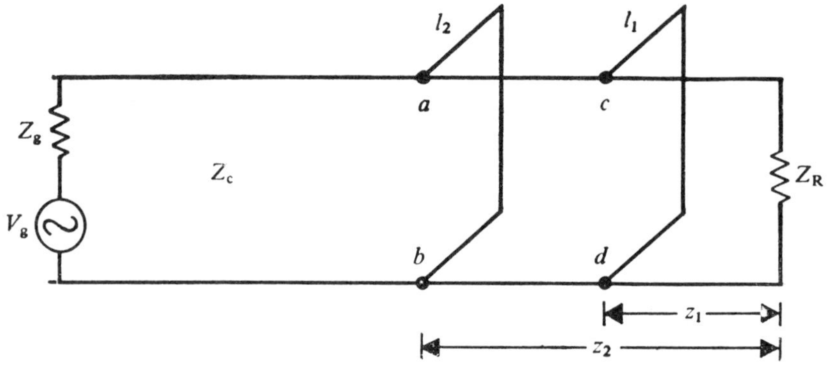

Fig. 6.23 Double-stub impedance matching.

power transfer, it is necessary to adjust the lengths of the stubs so that the admittance looking to the right at *ab* is equal to Y_c. Stub 1 serves to make the conductance part of the admittance at *ab* equal to Y_c, and stub 2 is then adjusted to cancel the susceptance portion of the admittance at *ab*. Assume that the admittance Y_{cd} at points *cd* in Fig. 6.23 (with both stubs disconnected) corresponds to point P_0 in Fig. 6.24. Connecting of stub 1 adds a pure susceptance, causing the admittance point to move on the constant-conductance circle to a new position P_1 which is determined by the length of the stub. If the stubs are a quarter wavelength apart, the admittance Y_{ab} (with stub 2 disconnected) is at point P_2 in Fig. 6.24, which is half way around the diagram

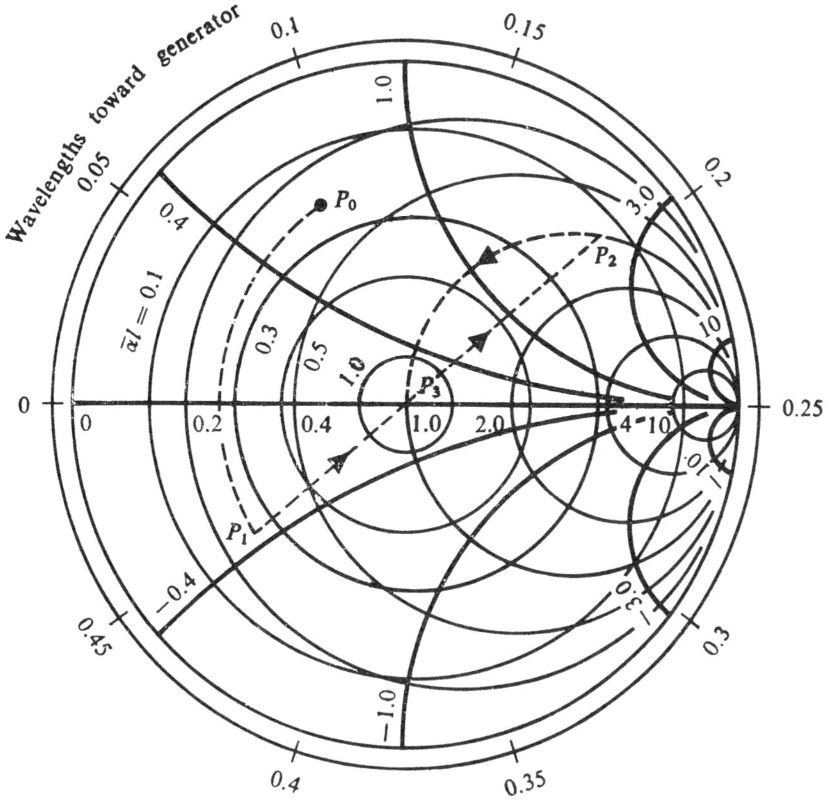

Fig. 6.24 Graphical solution of double-stub impedance matching.

from P_1 on the same $\bar{\alpha}l$-circle. In order to obtain the conditions necessary for matched admittances, the length of stub 1 should be such that point P_2 falls on the unity-conductance circle. With stub 2 connected and adjusted to cancel the susceptance at P_2, the admittance point moves from P_2 to P_3. Point P_3 corresponds to

$$\frac{Y}{Y_c} = 1 + j0,$$

which is the requirement for matched admittances.

PROBLEMS

1 A lossless transmission line is terminated in a pure resistance which is not equal to the characteristic impedance of the line. Prove that the standing wave has its maximum and minimum values at the receiving end and at integral multiples of quarter-wavelength distances from the receiving end. Derive an expression for the standing-wave ratio for this case. Will the voltage be a maximum or minimum at the receiving end?

2 A silver-plated coaxial line has inner and outer conductors of diameter 0.2 cm and 0.4 cm, respectively, and is to be used at a frequency of 800 MHz. Assume that the dielectric is lossless. Calculate

(i) the attenuation constant and Q of the line,

(ii) the input impedance of a quarter-wavelength short-circuited section of line and a quarter-wavelength open-circuited section of line, and

(iii) the magnitude of input impedance as a function of frequency in the vicinity of the antiresonant impedance for the short-circuited line, and plot a graph.

3 A transmission line having a characteristic impedance of 75 ohms is terminated in an unknown impedance which is to be measured. The maximum and minimum voltages on the line are 130 volts and 30 volts, respectively, and the distance of the maximum voltage from the terminal impedance is 32 cm. The frequency is 350 MHz. Calculate

(i) the value of the terminal impedance and verify it by using the impedance diagram, and

(ii) the length and position of a single stub that will match the impedance to the line.

4 An oscillator operates at 550 MHz and has to deliver an appreciable amount of third harmonic in its output. A short-circuited coaxial line having a characteristic impedance of 75 ohms is used as the tuned circuit, and is shunted at its input end by the grid-plate capacitance of 1.0×10^{-12} F of a tube. Calculate the length of the line and additional shunt capacitance which should be added to the input end of the line in order to make it simultaneously antiresonant at 550 MHz and 1600 MHz.

5 If $|V_{max}|$ and $|V_{min}|$ are the maximum and minimum voltages on the transmission line supplying the load, show that the power in the load impedance is given by

$$P = \frac{|V_{max}|\,|V_{min}|}{Z_0},$$

regardless of the magnitude or phase angle of the load impedance.

6 A coaxial line of characteristic impedance 50 ohms and $5\lambda/8$ long is terminated in an impedance of $(70 + j50)$ ohms. A condenser of capacitance 2×10^{-12} F is connected in series with the centre conductor $\lambda/8$ from the receiving end. Using the impedance diagram, find the input impedance of the line at a frequency of 900 MHz.

7 Find the input impedance of a line consisting of two coaxial lines in series, the first being 0.55λ long and having characteristic impedance of 50 ohms and $\bar{\alpha} = 0.3$ neper per metre, and the second being 0.45λ long and having characteristic impedance of 75 ohms and $\bar{\alpha} = 0.4$ neper per metre. The load impedance is $(40 - j60)$ ohms, and the wavelength is 1.5 metres.

8 If a generator with an impedance Z_g equal to Z_0 is connected to a lossless transmission line having characteristic impedance Z_0, show that the voltage at the receiving end of the line is always equal to the internal voltage of the generator, regardless of the length of the line.

9 A line terminated in a normalized load impedance $\bar{Z}_L$ has a standing-wave ratio equal to 2.5 and a voltage minimum occurring at a distance of $\lambda/4$ from the load. Find $\bar{Z}_L$, and find the position and length of a single shunt stub that will match the load to the line.

10 A double stub with spacing 0.25λ is used to match a normalized load impedance of $(0.4 + j1.1)$. Find the required stub impedances.

11 A line having an attenuation constant of $\bar{\alpha} = 0.02$ neper/metre is used as a short-circuited stub. Find the maximum value of normalized susceptance the stub can give. If $\bar{\alpha}$ is increased to 0.03, what is the maximum value of stub susceptance that can be obtained?

12 Find the length and impedance of a quarter-wave transformer which will match a 120-ohm load to a 50-ohm line at a frequency of 9000 MHz. Over what frequency band will the reflection coefficient remain less than 0.15?

7 Resonators

7.1 Fundamental Properties of Resonant Circuits

Resonant circuits are used quite extensively at all usable frequencies. They are used for oscillator circuits, tuned amplifiers, filter networks, and wavemeters. Resonant circuits used at all frequencies have some common properties, and it is useful to review some of these by using a lumped-parameter *RLC* series or parallel circuit as an example. Figure 7.1 shows low-frequency lumped-parameter series and parallel resonant circuits. The resistance R is an equivalent resistance which accounts for the power loss in the inductor L and capacitor C and sometimes also the power extracted from the resonant system by some external load coupled to the resonant system.

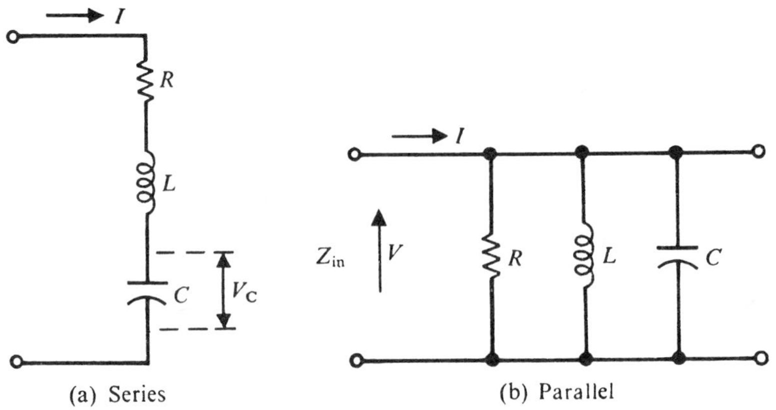

Fig. 7.1 Lumped-parameter series and parallel resonant circuits.

From Eq. (5.205),

$$Z_{\text{in}} = \frac{2j\omega(W_{\text{H}} - W_{\text{E}}) + P}{\frac{1}{2}II^*} = R + jX. \tag{7.1}$$

At resonance, $W_{\text{H}} = W_{\text{E}}$, and the input impedance is a pure real quantity and equal to R.

The time-average energy stored in the electric field in the capacitor is

$$W_{\text{E}} = \frac{C}{2}\frac{1}{T}\int_0^T V_{\text{C}}^2(t)\, dt, \tag{7.2}$$

where T is the time period for one cycle and $V_{\text{C}}(t)$ is the voltage across C. For a sinusoidally time-varying impressed voltage and current in the series circuit,

$$V_{\text{C}}(t) = \frac{I_0}{\omega} \cos \omega t, \tag{7.3}$$

where I_0 is the maximum amplitude of the series current. From Eqs. (7.2) and (7.3),

$$W_E = \tfrac{1}{2}\frac{I_0^2}{\omega^2 C}\frac{\omega}{2\pi}\int_0^{2\pi/\omega} \cos^2 \omega t \, dt = \tfrac{1}{4}\frac{I_0^2}{\omega^2 C}. \tag{7.4}$$

The average stored magnetic energy W_H in the circuit is the energy stored in the inductance and is given by

$$W_H = \tfrac{1}{2}L\frac{1}{T}\int_0^T I^2(t)\, dt = \tfrac{1}{4}LI_0^2. \tag{7.5}$$

At resonance, $\omega = \omega_0$, $\omega_0 L = 1/(\omega_0 C)$, so that $\omega_0^2 = 1/(LC)$, and

$$\tfrac{1}{4}\frac{I_0^2}{\omega_0^2 C} = \tfrac{1}{4}LI_0^2.$$

Also,

$$W_E = W_H. \tag{7.6}$$

Equation (7.6) holds good also for the parallel resonant circuit. Thus, we may say that the criterion for resonance in any electromagnetic circuit is that, at the resonant frequency, the average stored electric and magnetic energies should be equal.

If we plot $U = |W_H - W_E|$ as a function of frequency, U will be minimum and equal to zero at the resonant frequency, as shown in Fig. 7.2. Conversely, the frequency at which U is

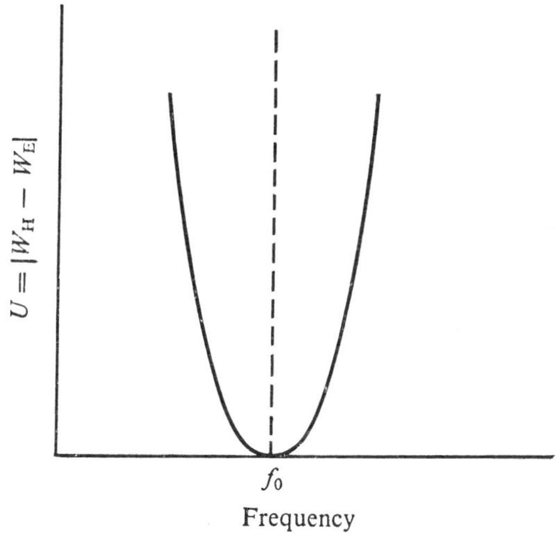

Fig. 7.2 Variation of U with frequency.

minimum may be called the resonant frequency. This does not imply the uniqueness of the resonant frequency since, for some resonators such as microwave cavity resonators, there may be an infinite number of frequencies at which U is minimum and equal to zero, and hence there may be an infinite number of resonant frequencies. This is the fundamental difference between a low-frequency lumped-parameter resonant circuit, which has a single resonant frequency, and a microwave cavity resonator, which has an infinite number of resonant frequencies. We shall discuss the reasons for an infinite number of resonant frequencies in a cavity resonator in the later sections of this chapter.

In a low-frequency resonant circuit, for sinusoidal input, the phases of $V_C(t)$ and $I(t)$ differ always by $\pi/2$ at any instant of time, t. Hence, $V_C(t)$ is maximum when $I(t)$ is zero, and vice versa. Therefore, from Eqs. (7.2) and (7.5), we see that, when the instantaneous stored energy in C is maximum, the stored energy in L is zero, and vice versa. Also, the maximum stored energies in L and C are the same. Besides, in a resonant circuit, the stored energy is continuously transferred from L to C, and vice versa. Such transfer of energy is identically the same in a cavity resonator also. It should be noted that, since L and C are interpreted only in terms of the stored energies and are not constants in a cavity at all frequencies, they have no unique values for the cavity. In fact, the stored energies in the electric and magnetic fields of a cavity change considerably, and in a complex manner, with the frequency.

For series and parallel resonant circuits, the input impedance Z_{in} of the resonant circuit given by Eq. (7.1) varies with frequency, as shown in Fig. 7.3.

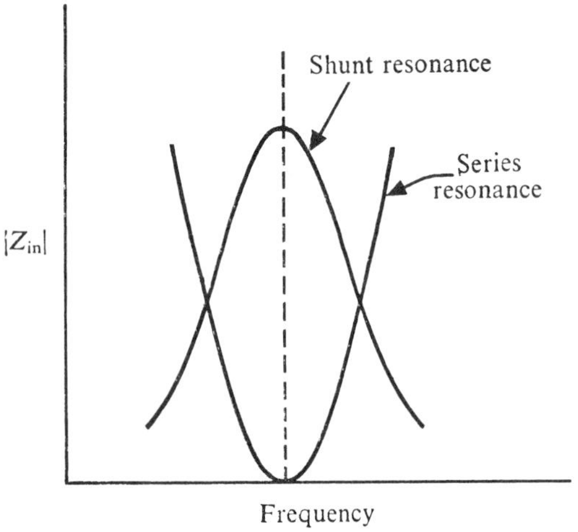

Fig. 7.3 Input impedance of resonant circuit.

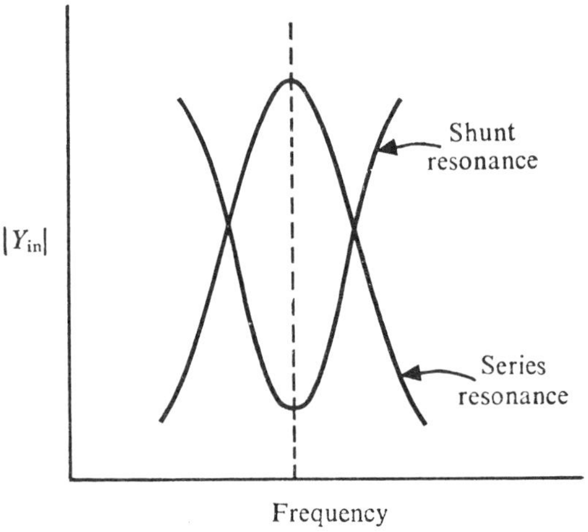

Fig. 7.4 Input admittance of resonant circuit.

The input admittance of a resonant circuit is given by

$$Y_{in} = \frac{2j\omega(W_E - W_H) + P}{\frac{1}{2}VV^*} = G + jB \tag{7.7}$$

and varies with frequency, as shown in Fig. 7.4.

An important parameter specifying the frequency selectivity and performance, in general, of a resonant circuit is the *quality factor* Q. A very general definition of Q applicable to all resonant systems is

$$Q = \frac{\omega \text{ (time-average energy stored in system)}}{\text{energy loss per second in system}}. \tag{7.8}$$

At resonance, $W_H = W_E$, and the peak value of energy stored in the capacitor is $2W_E$ which occurs when the energy stored in the inductor is zero, and vice versa. The average energy stored in the circuit is

$$W = W_H + W_E = 2W_H = 2W_E = \tfrac{1}{2}CVV^*. \tag{7.9}$$

The power loss is $\frac{1}{2}GV^*V^* = \frac{1}{2}(VV^*/R)$, and is the energy loss per second. Hence, for the parallel resonant circuit of Fig. 7.1,

$$Q = \frac{\omega C}{G} = \omega RC = \frac{R}{\omega L} \tag{7.10}$$

since $\omega^2 LC = 1$ at resonance.

The input impedance in the vicinity of resonance, say, at $\omega = \omega_0 + \Delta\omega$, can be expressed in a relatively simple form. For example, for the parallel resonant circuit,

$$Z_{in} = (\frac{1}{R} + \frac{1}{j\omega L} + j\omega C)^{-1} = (\frac{1}{R} + j\omega_0 C + j\Delta\omega C + \frac{1 - \Delta\omega/\omega_0}{j\omega_0 L})^{-1}, \tag{7.11}$$

where the approximation $1/(\omega_0 + \Delta\omega) \approx (1 - \Delta\omega/\omega_0)(1/\omega_0)$ has been used. Since $\omega_0^2 LC = 1$, we obtain

$$Z_{in} = \frac{\omega_0^2 RL}{\omega_0^2 L + j2R\Delta\omega} = \frac{R}{1 + j2Q(\Delta\omega/\omega_0)}. \tag{7.12}$$

A plot of Z_{in} as a function of $\Delta\omega/\omega_0$ is given in Fig. 7.5. When $|Z_{in}|$ has fallen to 0.707 of its

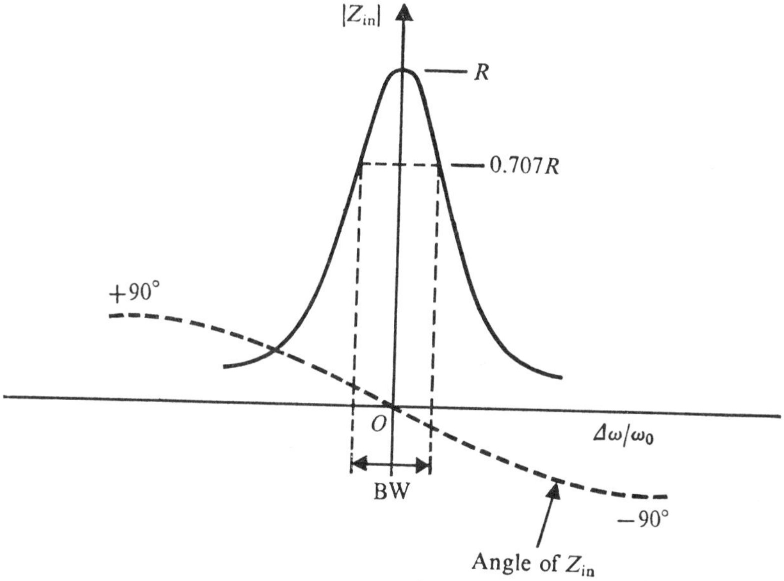

Fig. 7.5 Z_{in} for parallel resonant circuit.

maximum value, its phase angle is 45° if $\omega < \omega_0$ and −45° if $\omega > \omega_0$. From Eq. (7.12), the corresponding value of $\Delta\omega$ is found to be given by

$$2Q\frac{\Delta\omega}{\omega_0} = 1$$

or

$$\Delta\omega = \frac{\omega_0}{2Q}. \tag{7.13}$$

The fractional bandwidth (BW) between the $0.707R$ or half-power points is twice this; hence

$$Q = \frac{\omega_0}{2\Delta\omega} = \frac{1}{\text{BW}}. \tag{7.14}$$

Relation (7.14) gives an alternative definition of Q, that is, Q is equal to the fractional bandwidth between the points, where $|Z_{in}|$ is equal to 0.707 of its maximum value. (For a series resonant circuit, this applies to $|Y_{in}|$.)

If the resistor R in Fig. 7.1 represents the losses in the resonant circuit only, then the Q given by Eq. (7.10) is called the unloaded Q. If the resonant circuit is coupled to an external load that absorbs a certain amount of power, this loading effect can be represented by an additional resistor R_L in parallel with R for the parallel resonant circuit. The total resistance is now less and, consequently, the Q is also smaller.

The Q, called the loaded Q and denoted by Q_L, is

$$Q_L = \frac{RR_L/(R + R_L)}{\omega L}. \tag{7.15}$$

The external Q, denoted by Q_e, is defined as the Q that would result if the resonant circuit were loss free and only the loading by the external load were present. Thus

$$Q_e = R_L/\omega L. \tag{7.16}$$

From Eqs. (7.10), (7.15), and (7.16), we obtain

$$\frac{1}{Q_L} = \frac{1}{Q_e} + \frac{1}{Q}. \tag{7.17}$$

For a series resonant circuit, Q is defined as

$$Q = \frac{\omega L}{R} = \frac{1}{\omega CR}, \tag{7.18}$$

and the loading effect of the external load can be represented by a resistance R_L in series with R so that the loaded Q becomes smaller than the unloaded Q.

Another parameter of importance for a resonant circuit is the *damping factor* δ. This parameter measures the rate at which the oscillations would decay if the driving force were removed. For a high Q circuit, δ may be evaluated in terms of Q by using the perturbation technique. Due to the losses present, the energy stored in the resonant circuit decays at a rate proportional to the average energy present at any time (since P is proportional to VV^* and W also is proportional to VV^*, we have P proportional to W) so that

$$\frac{dW}{dt} = -2\delta W \tag{7.19}$$

or

$$W = W_0 e^{-2\delta t}, \tag{7.20}$$

where W_0 is the average energy present at $t = 0$. But the rate of decrease in W must equal the power loss so that

$$\frac{-dW}{dt} = 2\delta W = P. \tag{7.21}$$

Consequently, using the definition of Q given by Eq. (7.8), we obtain

$$\delta = \frac{P}{2W} = \frac{\omega P}{2\omega W} = \frac{\omega}{2Q}. \tag{7.22}$$

The damping factor δ is inversely proportional to Q, and Eq. (7.20) becomes

$$W = W_0 e^{-\omega t/Q}. \tag{7.23}$$

In this equation, Q is replaced by Q_L if an external load is coupled to the circuit. The damping factor is also a measure of the speed with which the amplitude of oscillations in the resonant circuit can build up when a driving force is applied.

In microwave systems, sections of transmission lines or metallic enclosures, called cavities, are used as resonators in place of the lumped-parameter circuits, because these circuits are not only very small in size but also have very high losses (both conductor and radiation losses) at microwave frequencies. Recently, dielectric and ferrite spheres, for example, yttrium iron garnet (YIG) spheres, have also been used as resonant circuits at microwave frequencies. These YIG spheres are very small in size, have low loss, and can be tuned over a broad band by changing the applied d-c magnetic field.

In calculating the impedance of a microwave cavity, it is sometimes convenient to assume that there are no losses. Q can be evaluated separately and, in terms of this parameter, the impedance can be modified to account for losses by replacing the resonant frequency ω_0 by an equivalent complex resonant frequency $\omega_0[1 + j/(2Q)] = \omega_0 + j\delta$. Then, Eq. (7.12) becomes

$$Z_{\text{in}} = \frac{\omega_0 R/(2Q)}{j[\omega - \omega_0\{1 + j/(2Q)\}]} \tag{7.24}$$

which shows that, when losses are present, it is equivalent to having a complex resonant frequency $\omega_0[1 + j/(2Q)]$. Equation (7.24) neglects the small change in resonant frequency that occurs when small losses are present.

7.2 Transmission-Line Resonant Circuits

Short-circuited Line, Series Resonance

At frequencies in the range 100 to 1000 MHz, short-circuited or open-circuited sections of transmission lines are commonly used to replace the usual lumped-LC resonant circuit. If air filled lines are used, the only losses are those due to the series resistance R of the line, which is the result of the losses in the conducting walls. If dielectric filled lines are used, the shunt conductance will also have to be considered, resulting in a lower Q. Therefore, in our study, only air filled lines will be considered.

Let us consider a short-circuited line of length l, with parameters R, L, C per unit length, as shown in Fig. 7.6. Let $l = \lambda_0/2$ at $f = f_0$, i.e., at $\omega = \omega_0$. For f near f_0, say, $f = f_0 + \Delta f$,

$\bar{\beta}l = 2\pi fl/c = \pi\omega/\omega_0 = \pi + \pi\Delta\omega/\omega_0$, since at $\omega = \omega_0$, $\bar{\beta}l = \pi$. Putting $Z_n = 0$ and $z_0 - z = l$ in

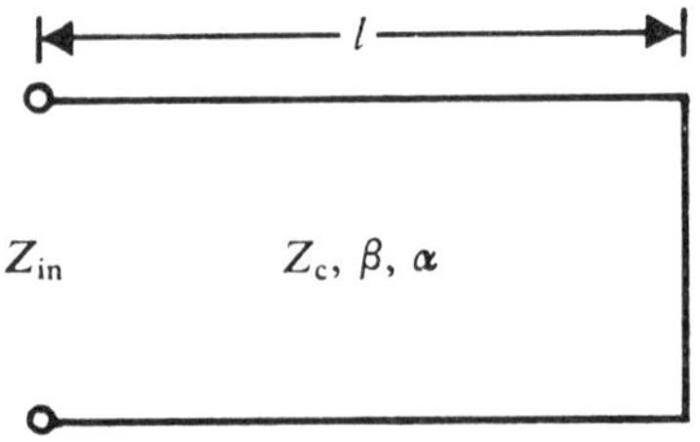

Fig. 7.6 Short-circuited transmission line.

Eq. (6.72), the input impedance is given by

$$Z_{in} = Z_c \tanh (j\bar{\beta}l + \bar{\alpha}l) = Z_c \tanh \bar{\gamma}l = Z_c \frac{\tanh \bar{\alpha}l + j \tan \bar{\beta}l}{1 + j \tan \bar{\beta}l \tanh \bar{\alpha}l}. \tag{7.25}$$

If the losses are small, then $\tanh \bar{\alpha}l \approx \bar{\alpha}l$, as $\bar{\alpha}l \ll 1$. Also,

$$\tan \bar{\beta}l = \tan (\pi + \pi\Delta\omega/\omega_0) = \tan (\pi\Delta\omega/\omega_0) \approx \pi\Delta\omega/\omega_0$$

since $\Delta\omega/\omega_0$ is small. Hence

$$Z_{in} = Z_c \frac{\bar{\alpha}l + j\pi\Delta\omega/\omega_0}{1 + j\bar{\alpha}l\pi\Delta\omega/\omega_0} = Z_c(\bar{\alpha}l + j\pi\Delta\omega/\omega_0), \tag{7.26}$$

where

$$Z_c = (L/C)^{1/2}, \qquad \bar{\alpha} = \tfrac{1}{2}RY_c = \frac{R}{2}(C/L)^{1/2}, \qquad \bar{\beta}l = (LC)^{1/2}l = \pi,$$

so that $\pi/\omega_0 = l(LC)^{1/2}$, and the expression for Z_{in} becomes

$$Z_{in} = (L/C)^{1/2}[\tfrac{1}{2}R(C/L)^{1/2} + j\Delta\omega l(LC)^{1/2}] = \tfrac{1}{2}Rl + jlL\Delta\omega. \tag{7.27}$$

Comparing Eq. (7.27) with the series $R_0L_0C_0$ circuit shown in Fig. 7.7, we obtain, for the

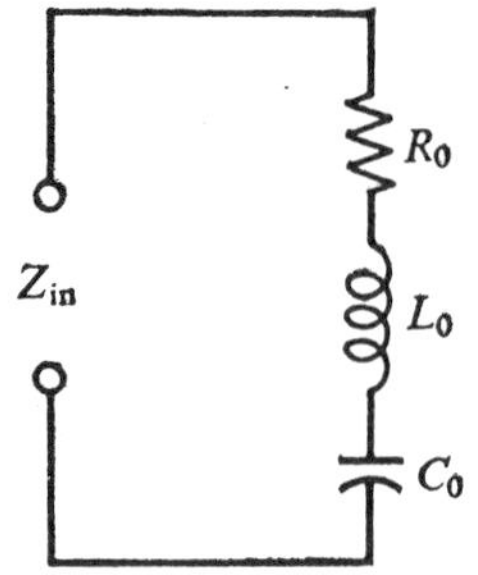

Fig. 7.7 Series resonant circuit.

series resonant circuit,

$$Z_{in} = R_0 + j\omega L_0[1 - 1/(\omega^2 L_0 C_0)]. \tag{7.28}$$

If $\omega_0^2 = 1/(L_0C_0)$, then for the series resonant circuit,

$$Z_{\text{in}} = (R_0 + j\omega L_0)\frac{\omega^2 - \omega_0^2}{\omega^2}. \tag{7.29}$$

If $\omega - \omega_0 = \Delta\omega$ is small,

$$\omega^2 - \omega_0^2 = (\omega - \omega_0)(\omega + \omega_0) \approx \Delta\omega 2\omega.$$

Then

$$Z_{\text{in}} = (R_0 + j\omega L_0)\frac{2\omega\Delta\omega}{\omega^2} = R_0 + 2jL_0\Delta\omega. \tag{7.30}$$

Comparing Eqs. (7.27) and (7.30), we see that, in the vicinity of the frequency for which $l = \lambda_0/2$, the short-circuited line behaves as a series resonant circuit with resistance $R_0 = \frac{1}{2}Rl$ and inductance $L_0 = \frac{1}{2}Ll$. We note that Rl and Ll are the total resistance and inductance of the line. The factors $\frac{1}{2}$ arise because the current on the short-circuited line is a half sinusoid, and hence the effective circuit parameters R_0 and L_0 are only one-half of the total line quantities.

The Q of the short-circuited line may be defined as

$$Q = \frac{\omega_0 L_0}{R_0} = \frac{\omega_0 L}{R} = \frac{\bar{\beta}}{2\bar{\alpha}}. \tag{7.31}$$

As an alternative to Eq. (7.31), the general definition of Eq. (7.8) for Q may be used. The Q of the short-circuited line may be evaluated by using this definition by means of a method, called the perturbation method, which is valid for high Q (low loss) systems.

The perturbation method is very useful for many microwave devices. For small losses, the energy stored in the system is to first order, the same as if there were no losses. For a loss-free short-circuited line, the current on the line is a pure standing wave, i.e.,

$$I = I_0 \cos(\bar{\beta}z)e^{j\omega t}, \tag{7.32}$$

where z is the distance measured from the short towards the input end. In a length dz, the energy stored in the magnetic field is

$$dW_{\text{H}} = \tfrac{1}{4}II^*\,dz = \tfrac{1}{4}I_0^2 L\cos^2(\bar{\beta}z)\,dz. \tag{7.33}$$

The total time-average stored energy in the magnetic field is

$$W_{\text{H}} = \tfrac{1}{4}I_0^2 L\int_0^{\lambda/2}\cos^2(\bar{\beta}z)\,dz = \frac{\lambda_0}{16}I_0^2 L. \tag{7.34}$$

The energy W_{E} stored in the electric field, i.e., in the line capacitance, is equal to W_{H} at resonance, and hence the total time-average energy stored in the system is

$$W = W_{\text{E}} + W_{\text{H}} = 2W_{\text{H}} = \frac{\lambda_0}{8}I_0^2 L. \tag{7.35}$$

To a first approximation, the power losses do not modify the current distribution along the line. Hence, the power loss is given by

$$P_{\text{L}} = \tfrac{1}{2}\int_0^{\lambda_0/2} RII^*\,dz = \frac{R}{2}I_0^2\int_0^{\lambda_0/2}\cos^2(\bar{\beta}z)\,dz = \frac{\lambda_0}{8}RI_0^2. \tag{7.36}$$

Thus at $\omega = \omega_0$,

$$Q = \frac{\omega_0 W}{P_L} = \frac{\omega_0 \lambda_0 L I_0^2/8}{R I_0^2 \lambda_0/8} = \frac{\omega_0 L}{R} \tag{7.37}$$

which conforms to the earlier result.

Typical values of Q range from several hundreds up to about 10,000. In contrast with low-frequency lumped-parameter circuits, the practical values of Q are very much higher for microwave resonators. Though, in the analysis, the losses in the short circuit have been neglected, it introduces very little error if the length of the line is considerably greater than the spacing between the conductors.

Open-circuited Line

We now show that an open-circuited transmission line is equivalent to a series resonant circuit in the vicinity of the frequency for which it is an odd multiple of a quarter wavelength long. The

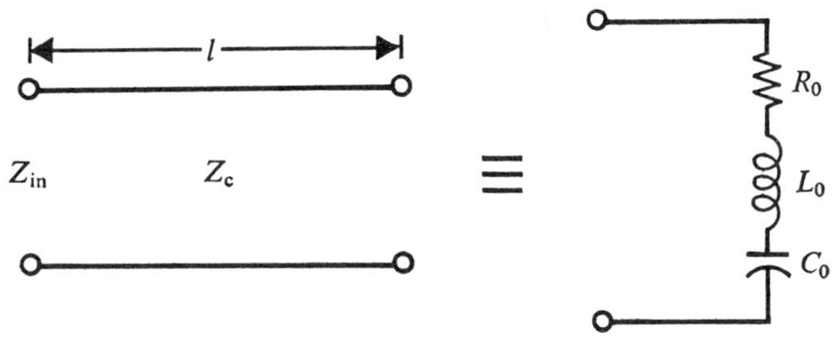

Fig. 7.8 Open-circuited transmission-line resonator.

equivalent relations are (see Fig. 7.8)

$$l = \frac{\lambda_0}{4} \quad \text{at } \omega_0, \tag{7.38a}$$

$$Z_{in} \approx (\bar{\alpha} l + j\frac{\Delta\omega}{\omega_0}\frac{\pi}{2})Z_c = \tfrac{1}{2}Rl + j\Delta\omega Ll, \tag{7.38b}$$

$$R_0 = \tfrac{1}{2}Rl, \tag{7.38c}$$

$$L_0 = \tfrac{1}{2}Ll, \tag{7.38d}$$

$$\omega_0^2 = (LC)^{-1}. \tag{7.38e}$$

Short-circuited and Open-circuited Antiresonant Lines

If the length of a short-circuited transmission line is close to an odd multiple of a quarter wavelength, and if the length of an open-circuited transmission line is close to an even multiple of a quarter wavelength, then they behave like parallel resonant circuits or as antiresonant circuits. Figure 7.9 shows a short-circuited transmission line of length l equal to $\lambda_0/4$ at $\omega = \omega_0$, and its equivalent parallel resonant circuit. $\bar{\beta} l$ is

$$\bar{\beta} l = \omega_0\sqrt{LC}\,l + \Delta\omega\sqrt{LC}\,l \quad \text{at } \omega. \tag{7.39}$$

The input admittance is given by

$$Y_{\text{in}} = Y_c \coth (\bar{\alpha} + j\bar{\beta})l = Y_c\left(\frac{1 + j \tan \bar{\beta}l \tanh \bar{\alpha}l}{\tanh \bar{\alpha}l + j \tan \bar{\beta}l}\right) \approx Y_c\frac{1 - j\bar{\alpha}l/(\Delta\omega l\sqrt{LC})}{\bar{\alpha}l - j/(\Delta\omega l\sqrt{LC})} \tag{7.40}$$

since

$$\tan (\omega_0\sqrt{LC}\,l + \Delta\omega\sqrt{LC}\,l) = \tan \left(\frac{\pi}{2} + \Delta\omega\sqrt{LC}\,l\right) \approx -(\Delta\omega\sqrt{LC}\,l)^{-1},$$

$$\tanh \bar{\alpha}l \approx \bar{\alpha}l.$$

Finally,

$$Y_{\text{in}} \approx Y_c\frac{j\Delta\omega\sqrt{LC}\,l + \bar{\alpha}l}{1 + j\Delta\omega\bar{\alpha}l^2\sqrt{LC}} \approx Y_c(j\Delta\omega\sqrt{LC}\,l + \bar{\alpha}l) = \frac{RC}{2L}l + j\Delta\omega Cl \tag{7.41}$$

after replacing Y_c by $\sqrt{C/L}$ and $\bar{\alpha}$ by $Y_cR/2$.

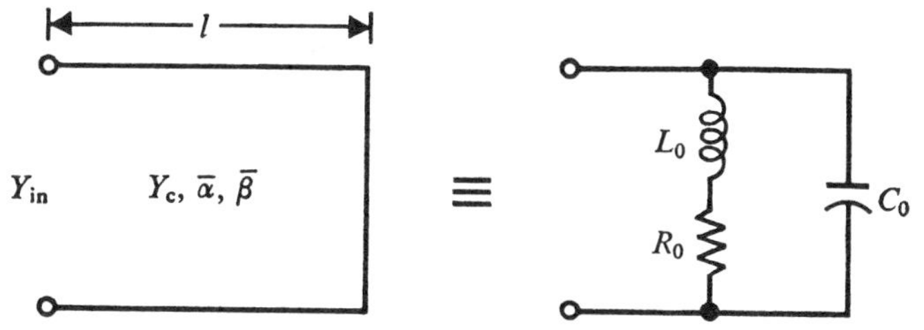

Fig. 7.9 Short-circuited antiresonant transmission line and its equivalent circuit.

For the parallel resonant circuit of Fig. 7.9, we have

$$Y_{\text{in}} = \frac{1}{R_0 + j\omega L_0} + j\omega C_0 = \frac{j\omega C_0(R_0 + j\omega L_0) + 1}{R_0 + j\omega L_0} \approx \frac{j\omega C_0 R_0 - \omega^2 L_0 C_0 + 1}{j\omega L_0} \tag{7.42}$$

since we assume $R_0 \ll \omega L_0$. If $\omega_0^2 L_0 C_0 = 1$, then

$$Y_{\text{in}} = \frac{C_0 R_0}{L_0} + \frac{\omega_0^2 L_0 C_0 - \omega^2 L_0 C_0}{j\omega L_0}$$

$$= R_0\frac{C_0}{L_0} - jL_0C_0\frac{(\omega_0 - \omega)(\omega_0 + \omega)}{\omega L_0} \approx R_0\frac{C_0}{L_0} + jC_0 2\Delta\omega. \tag{7.43}$$

Comparing Eqs. (7.41) and (7.43), we notice that the short-circuited line in the vicinity of a quarter wavelength long is equivalent to a parallel resonant circuit with

$$\frac{C_0R_0}{L_0} = \frac{RC}{2L}l, \qquad Cl = 2C_0, \qquad \frac{R_0}{L_0} = \frac{R}{L}.$$

The Q of the circuit is given by

$$Q = \frac{\omega_0 L_0}{R_0} = \frac{\omega L}{R} = \frac{\bar{\beta}}{2\bar{\alpha}}. \tag{7.44}$$

Although particular lengths of transmission lines behave as simple lumped-parameter resonant circuits in the vicinity of a particular resonant frequency, they have an infinite number of resonant and antiresonant frequencies. The resonant frequencies (series resonance) occur approximately when the short-circuited line is a multiple of a half wavelength long, that is, $f_n = nc/(2l)$, and the antiresonant frequencies (parallel resonance) occur when the line is an odd multiple of a quarter wavelength long, that is, $f_n = (2n + 1)c/(4l)$, where n is an integer and c is the velocity of transmission. Thus, the exact equivalent circuit consists of an infinite number of resonant circuits coupled together. However, in practice, the frequency range of interest is normally such that a simple single resonant-frequency circuit represents the transmission-line resonator with adequate accuracy.

7.3 Microwave Cavity Resonators

A microwave cavity resonator consists of a bounded electromagnetic field enclosed almost entirely by a conducting (metallic) wall, except perhaps at the regions of excitation and coupling. At frequencies above 1000 MHz, transmission-line resonators have relatively low values of Q, and it is preferable to use cavity resonators. The electric and magnetic energies are stored in the volume of the cavity. The finite conducting walls give rise to power loss and are therefore equivalent to some effective resistance. The fields in the cavity may be excited, or coupled to an external circuit, by means of small coaxial-line probes or loops. Alternatively, the cavity may be coupled to a waveguide by means of a small aperture in a common wall. The coupling methods are illustrated in Fig. 7.10.

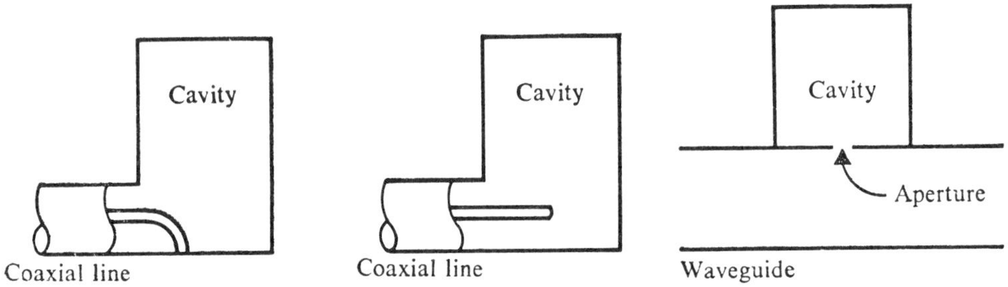

(a) Loop coupling (b) Probe coupling (c) Aperture coupling

Fig. 7.10 Cavity-coupling methods.

Theoretically, a cavity resonator can be of any arbitrary shape. In practice, however, simple geometrical shapes such as rectangular, cylindrical, spherical, or toroidal shapes are commonly used.

7.4 Rectangular Cavity

A rectangular cavity may be considered as a section of a rectangular waveguide terminated in a short circuit at $z = 0$, $z = d$, as shown in Fig. 7.11. If d equals a multiple of a half guide wavelength at the frequency f, the resultant standing-wave pattern is such that the x- and y-component of electric field are zero at $z = 0$, $z = d$. The field solution may be obtained directly from the corresponding waveguide solutions. For the nm-th TE or TM mode, the propagation constant is given by

$$\beta_{nm}^2 = k_0^2 - (\frac{n\pi}{a})^2 - (\frac{m\pi}{b})^2, \tag{7.45}$$

where

$$k_0 = \frac{2\pi}{\lambda_0} = \frac{2\pi f_0}{c} = \omega\sqrt{\mu_0\epsilon_0}.$$

If the dielectric inside the waveguide is not air, then in Eq. (7.45), k_0 is replaced by $k = \omega\sqrt{\mu_0\epsilon}$ if it is some other dielectric. For the cavity to be a multiple of a half guide wavelength long, it is

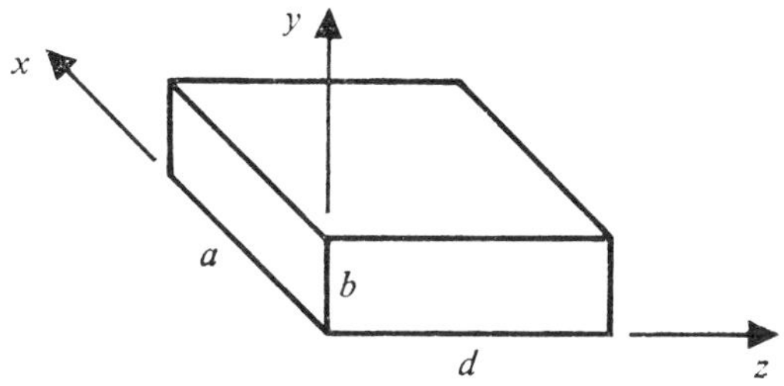

Fig. 7.11 Rectangular cavity.

necessary that $\beta l = l\pi$, where l is an integer. Thus, when d is specified, β_{mn} is given by

$$\beta_{mn} = \frac{l\pi}{d} \qquad (l = 1, 2, 3, \ldots). \tag{7.46}$$

From Eqs. (7.45) and (7.46), k_0 can have only certain discrete values k_{nml} given by

$$k_0^2 = k_{rml}^2 = (\frac{l\pi}{d})^2 + (\frac{m\pi}{a})^2 + (\frac{n\pi}{b})^2. \tag{7.47}$$

These discrete values of k_0 given by k_{nml} give the resonant frequencies of the cavity, i.e.,

$$f_{nml} = \frac{ck_{nml}}{2\pi} = c[(\frac{l}{2d})^2 + (\frac{m}{2a})^2 + (\frac{n}{2b})^2]^{1/2}, \tag{7.48}$$

where c, the velocity of light, is equal to $1/\sqrt{\mu_0\epsilon_0}$.

Equation (7.48) shows that there is a triply infinite number of resonant frequencies corresponding to different field configurations. This equation holds good for both TE and TM modes, and hence there is more than one field solution for a given resonant frequency. Such modes are called degenerate modes, e.g., TE_{nm} and TM_{nm} modes.

Since the TE_{10} mode is the dominant mode in the rectangular waveguide, the TE_{101} mode is treated in detail. If $b < a < d$, this is the mode with the lowest resonant frequency and corresponds to the dominant TE_{10} mode in the rectangular waveguide. The mode subscripts m, n, l represent the number of half-sinusoid variations in the standing-wave pattern along the x-, y-, and z-axis, respectively. Using Eqs. (5.141), the field solution for a TE_{10} mode in a rectangular cavity is

$$H_z = [A^+ \exp(-j\beta_{10}z) + A^- \exp(-j\beta_{10}z)] \cos(\frac{\pi x}{a}), \tag{7.49a}$$

$$H_x = \frac{j\beta_{10}a}{\pi}[A^+ \exp(-j\beta_{10}z) - A^- \exp(j\beta_{10}z)] \sin(\frac{\pi x}{a}), \tag{7.49b}$$

$$E_y = \frac{-jk_0Z_0a}{\pi}[A^+ \exp(-j\beta_{10}z) + A^- \exp(j\beta_{10}z)] \sin(\frac{\pi x}{a}), \tag{7.49c}$$

$$H_y = E_x = E_z = 0, \tag{7.49d}$$

where A^+ and A^- are the amplitude constants for the modes propagating in the $+z$- and $-z$-direction, respectively. The boundary condition is $E_y = 0$ at $z = 0$, $z = d$. Hence, $A^- = A^+$, so that

$$A^+ \exp(-j\beta_{10}z) + A^- \exp(-j\beta_{10}z) = -2jA^+ \sin \beta_{10}z$$

and $\beta_{10} = \pi/d$. The corresponding value of k_0 is

$$k_0 = k_{101} = [(\frac{\pi}{a})^2 + (\frac{\pi}{d})^2]^{1/2} = [(\frac{\pi}{a})^2 + \beta_{10}^2] \tag{7.50}$$

which determines the resonant frequency. The expressions for the field components may now be written as

$$E_y = \frac{-2A^+k_{101}Z_0a}{\pi} \sin(\frac{\pi x}{a}) \sin(\frac{\pi z}{d}), \tag{7.51a}$$

$$H_x = \frac{2jA^+a}{d} \sin(\frac{\pi x}{a}) \cos(\frac{\pi z}{d}), \tag{7.51b}$$

$$H_z = -2jA^+ \cos(\frac{\pi x}{a}) \sin(\frac{\pi z}{d}). \tag{7.51c}$$

Equations (7.51) show that the magnetic field is $\pm 90°$ out of phase relative to the electric field. This is always true for a lossless cavity and corresponds to the voltage and current which are $\pm 90°$ out of phase with each other in a lossless *LC* circuit.

At resonance the time-average stored electric and magnetic energies in the cavity are equal. The average stored electric energy is given by

$$W_E = \frac{\epsilon_0}{4}\int_0^a\int_0^b\int_0^d E_yE_y^* \, dx\, dy\, dz = \frac{\epsilon_0}{4\pi^2}a^3bdk_{101}^2Z_0^2|A^+|^2 \tag{7.52}$$

and the average stored magnetic energy by

$$W_H = \frac{\mu_0}{4}\int_0^a\int_0^b\int_0^d (H_xH_x^* + H_zH_z^*) \, dx\, dy\, dz. \tag{7.53}$$

W_H is found to be equal to W_E.

The losses due to the finite conductivity of the walls have to be evaluated to determine the Q of the cavity. For small losses, the surface currents are assumed to be the same as those under loss-free conditions. Hence, the surface current density is given by

$$\boldsymbol{J}_S = \boldsymbol{n} \times \boldsymbol{H},$$

where $\boldsymbol{n}$ is the unit outward normal to the surface. Hence, the power loss in the walls is

$$P_L = \frac{R_m}{2} \int_{\text{walls}} \boldsymbol{J}_S \cdot \boldsymbol{J}_S^* \, dS = \frac{R_m}{2} \int_{\text{walls}} |H_t|^2 \, dS. \tag{7.54}$$

Here, $R_m = 1/(\sigma\delta_s)$ is the resistive part of the surface impedance of the conducting wall having a conductivity σ and skin depth $\delta_s = [2/(\omega\mu\sigma)]^{1/2}$, and H_t is the tangential magnetic field at the surface of the walls. Finally, substituting for H_t from Eqs. (7.51), we obtain

$$P_L = |A^+|^2 R_m \frac{2a^3b + 2d^3b + ad^3 + da^3}{d^2}. \tag{7.55}$$

Hence, by definition, the Q of the cavity is

$$Q = \frac{\omega(W_E + W_H)}{P_L} = \frac{2\omega W_E}{P_L} = \frac{\omega k_{101}^2 Z_0^2 a^3 d^3 b \epsilon_0}{2\pi^2 R_m(2a^3b + 2d^3b + a^3d + d^3a)}$$

$$= \frac{(k_{101}ad)^3 b Z_0}{2\pi^2 R_m(2a^3b + 2d^3b + a^3d + d^3a)} \tag{7.56}$$

by replacing $\omega Z_0 \epsilon_0 = \omega\sqrt{\mu_0\epsilon_0}$ by k_{101}. For example, for a copper cavity ($\sigma = 5.8 \times 10^7$ mhos/metre), $a = b = d = 3$ cm. The resonant frequency for the TE_{101} mode is 7070 MHz. The surface resistance at this frequency is $R_m = 0.022$ ohm, and Q comes out to be 12,700. The damping factor $\delta = \omega/(2Q)$ equals 1.74×10^6 nepers/sec or about 2.5×10^{-4} neper/cycle of oscillation. Because of the high value of Q, 4000 cycles of free oscillation can occur before the amplitude has decreased by a factor of e^{-1}.

If the cavity is filled with a lossy dielectric material with permittivity $\epsilon = \epsilon' - j\epsilon''$, the time-average stored electric energy in the cavity volume is given by

$$W_E = \frac{\epsilon'}{4} \int_V |E|^2 \, dV. \tag{7.57}$$

The lossy dielectric has an effective conductivity $\omega\epsilon''$, and hence the power loss in the dielectric is

$$P_{Ld} = \tfrac{1}{2} \int_V \boldsymbol{J} \cdot \boldsymbol{E}^* \, dV = \frac{\omega\epsilon''}{2} \int_V |E|^2 \, dV. \tag{7.58}$$

If Q_d is the Q when a lossy dielectric is present and the walls are perfectly conducting, then

$$Q_d = \frac{2\omega W_E}{P_{Ld}} = \frac{\epsilon'}{\epsilon''}. \tag{7.59}$$

When wall losses are also present, the resulting net Q is Q', i.e.,

$$Q' = \left(\frac{1}{Q_d} + \frac{1}{Q}\right)^{-1}, \tag{7.60}$$

where Q is the quality factor when lossy walls are present and $\epsilon'' = 0$ and is given by Eq. (7.56), with ϵ_0 replaced by ϵ'.

The resonant frequency of a cavity filled with dielectric for the TE_{nml} or TM_{nml} mode is

$$f_{nml} = \sqrt{\frac{\epsilon_0}{\epsilon'}}\frac{c}{2\pi}k_{nml}. \tag{7.61}$$

7.5 Cylindrical Cavity

The cylindrical cavity is a section of a circular metal waveguide of length d and radius a, short circuited at both ends, as shown in Fig. 7.12.

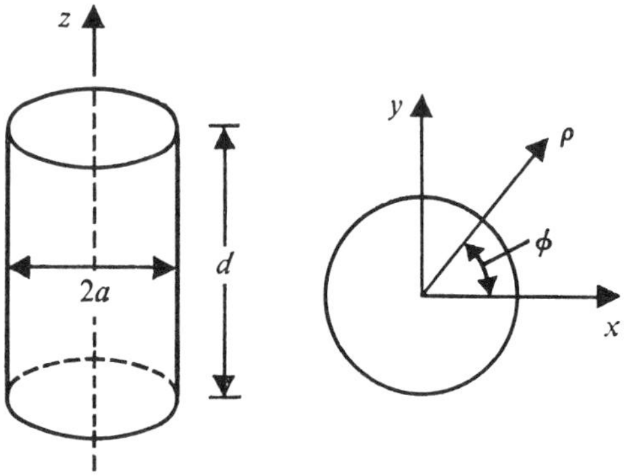

Fig. 7.12 Cylindrical cavity.

Cylindrical cavities are very commonly used for wavemeters to measure frequency because of the high Q and wide frequency range of operation they provide. A high Q is necessary for a wavemeter so that a high degree of resolution is obtained in the measurement of an unknown frequency. When the cavity is tuned to the frequency of the unknown source, it absorbs maximum power from the input line. A crystal detector connected to the input line can be used to indicate this dip in power at resonance.

The fields in the cylindrical cavity may be determined from the corresponding waveguide solutions. The lowest resonant mode is the TE_{111} mode corresponding to the dominant TE_{11} mode in the cylindrical waveguide. Using Table 5.4 and combining a forward and a backward propagating TE_{11} mode, we obtain the field components for the lowest TE mode as

$$H_z = J_1\left(\frac{p'_{11}\rho}{a}\right)\cos\phi\,[A^+\exp(-j\beta_{11}z) + A^-\exp(+j\beta_{11}z)], \tag{7.62a}$$

$$H_\rho = \frac{-j\beta_{11}a}{p'_{11}}J'_1\left(\frac{p'_{11}\rho}{a}\right)\cos\phi\,[A^+\exp(-j\beta_{11}z) - A^-\exp(+j\beta_{11}z)], \tag{7.62b}$$

$$H_\phi = \frac{j\beta_{11}a^2}{(p'_{11})^2\rho}J_1\left(\frac{p'_{11}\rho}{a}\right)\sin\phi\,[A^+\exp(-j\beta_{11}z) - A^-\exp(+j\beta_{11}z)], \tag{7.62c}$$

$$E_\rho = \frac{jk_0Z_0a^2}{(p'_{11})^2\rho}J_1\left(\frac{p'_{11}\rho}{a}\right)\sin\phi\,[A^+\exp(-j\beta_{11}z) + A^-\exp(j\beta_{11}z)], \tag{7.62d}$$

$$E_\phi = \frac{jk_0Z_0a}{p'_{11}}J'_1\left(\frac{p'_{11}\rho}{a}\right)\cos\phi\,[A^+\exp(-j\beta_{11}z) + A^-\exp(-j\beta_{11}z)], \tag{7.62e}$$

$$E_z = 0, \tag{7.62f}$$

where $p'_{11} = 1.841$. To make E_ρ and E_ϕ vanish at $z = 0$, $z = d$, we must choose $A^- = A^+$ so that the factor $[A^+ \exp(-j\bar{\beta}_{11}z) + A^- \exp(-j\bar{\beta}_{11}z)]$ becomes $-2jA^+ \sin(\bar{\beta}_{11}z)$. To make $\sin(\bar{\beta}_{11}d)$ vanish at $z = 0$, $z = d$, $\bar{\beta}_{11}$ must be chosen equal to $l\pi/d$. For the lowest TE_{111} mode, $l = 1$ so that $\bar{\beta}_{11} = \pi/d$. The resonant frequency is given by

$$k_0 = \omega_{111}(\mu_0\epsilon_0)^{1/2} = \frac{2\pi f_{111}}{c} = [\bar{\beta}_{11}^2 + (\frac{p'_{11}}{a})^2]^{1/2} = [(\frac{\pi}{d}) + (\frac{p'_{11}}{a})^2]^{1/2}. \tag{7.63}$$

To find the Q of the cavity for the TE_{111} mode, the time-average energy W stored in the cavity and the power loss P_L in the walls must be calculated. The general expressions for W and P_L are

$$W = 2W_E = \frac{\epsilon_0}{2}\int_{\rho=0}^{a}\int_{\phi=0}^{2\pi}\int_{z=0}^{d}(|E_\rho|^2 + |E_\phi|^2)\rho\, d\phi\, d\rho\, dz, \tag{7.64}$$

$$P_L = \frac{R_m}{2}\int_{\text{walls}} |H_t|^2\, dS. \tag{7.65}$$

By evaluating W and P_L, the expression for Q for the TE_{nml} mode may be obtained as

$$Q = \frac{\omega W}{P_L} = \frac{\lambda_0}{\delta_s}\frac{[1 - (n/p'_{nm})^2][(p'_{nm})^2 + (l\pi a/d)^2]^{3/2}}{2\pi[(p'_{nm})^2 + (2a/d)(l\pi a/d)^2 + (1 - 2a/d)(nl\pi a/p'_{nm}d)^2]}, \tag{7.66}$$

where δ_s is the skin depth. For the TE_{111} mode, $l = n = m = 1$. Since all other terms in Eq. (7.66) are independent of frequency, Q varies as λ_0/δ_s for any cavity and thus decreases as $f^{-1/2}$.

The TE_{011} mode in the cylindrical cavity is of great advantage for wavemeters, because its Q is two to three times that of the TE_{111} mode. Another advantage of this mode is that $H_\phi = 0$, and hence there are no axial currents. This means that the end plate of the cavity can be free to move to adjust the cavity length d for tuning purposes, without introducing any significant loss since no axial currents flow across the gap between the circular end plate and the cylindrical walls. However, the TE_{011} mode is not the dominant mode, and so care must be taken to choose a coupling scheme that does not excite the other possible modes which could resonate within the frequency tuning range of the cavity.

A mode chart is useful for determining the type of mode that can resonate for a given value of $2a/d$. For any given TE_{nml} or TM_{nml} mode,

$$f_{nml} = \frac{k_{nml}}{2\pi}c = [(\frac{x_{nm}}{a})^2 + (\frac{l\pi}{d})^2]^{1/2}\frac{c}{2\pi}$$

or

$$(2af_{nml})^2 = (\frac{cx_{nm}}{\pi})^2 + (\frac{cl}{2})^2(\frac{2a}{d})^2, \tag{7.67}$$

where $x_{nm} = p'_{nm}$ for TE modes and p_{nm} for TM modes. Figure 7.13 gives a plot of $(2af_{nml})^2$ against $(2a/d)^2$ for several modes. It constitutes a mode chart. By examining this chart, we can determine the range of frequency and $(2a/d)$ over which a single mode can resonate. For degenerate modes, two modes resonate at the same frequency. For example, for $(2a/d)^2$ between 2 and 3, only the TE_{011} and TM_{111} modes can resonate in the frequency range corresponding to $(2af)^2$ between

16.3×10^8 and 20.4×10^8. If the TM_{111} mode is not excited, this frequency range can be tuned so that spurious resonances from other modes do not occur.

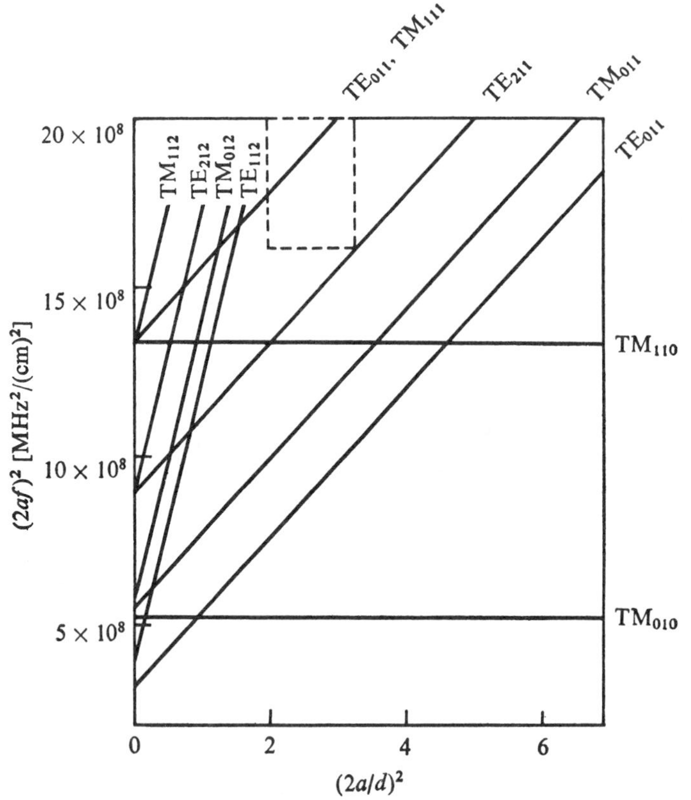

Fig. 7.13 Mode chart for circular cylindrical cavity.

For the TM modes, the Q of the cavity can be shown as

$$Q\frac{\delta_s}{\lambda_0} = \begin{cases} \dfrac{[p_{nm}^2 + (l\pi a/d)^2]^{1/2}}{2\pi(1 + 2a/d)} & \text{for } l > 0 \\[2ex] \dfrac{p_{nm}}{2\pi(1 + a/d)} & \text{for } l = 0 \end{cases} \tag{7.68}$$

Figure 7.14 gives a plot of $Q\delta_s/\lambda_0$ against $2a/d$ for several modes. It can be noticed that the Q for the TE_{011} mode is higher than that for the TE_{111} mode. Optimum Q occurs for $d \approx 2a$. At $\lambda_0 = 3$ cm, $\delta_s/\lambda_0 = 2.2 \times 10^{-5}$, and hence from Fig. 7.14, it is apparent that the typical values of Q range from 10,000 to 40,000 or more. At $\lambda_0 = 10$ cm, the corresponding values of Q would be greater by $\sqrt{10/3}$.

7.6 Equivalent Circuits for Cavities

We shall consider the equivalent circuits of cavities coupled to transmission lines and waveguides, with two examples, namely, the aperture-coupled cavity and the loop-coupled cavity.

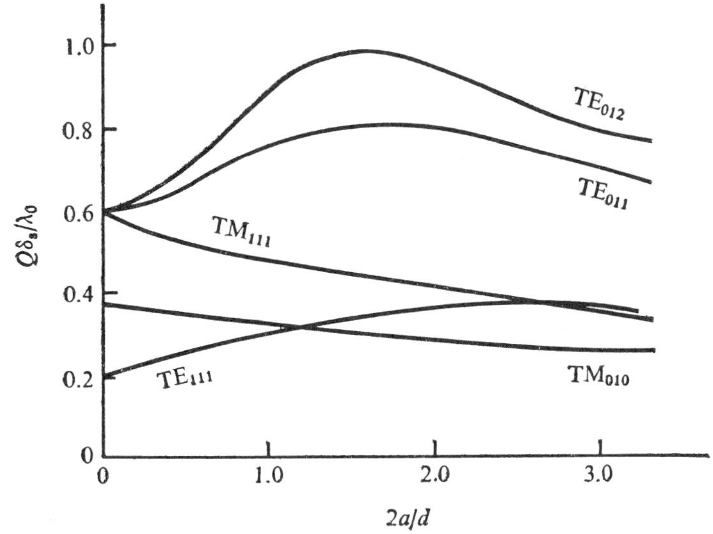

Fig. 7.14 Q of cylindrical cavity.

Aperture-coupled Cavity

Consider a rectangular cavity coupled to a rectangular waveguide by means of a small circular hole centred in the transverse wall at $z = 0$, as shown in Fig. 7.15. It can be shown that a small

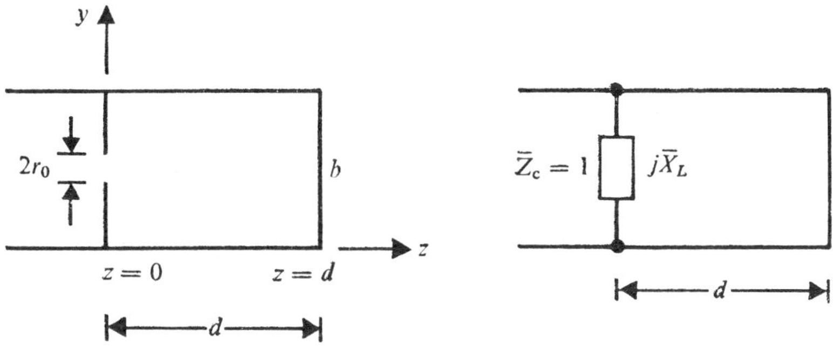

(a) Aperture-coupled rectangular cavity (b) Equivalent transmission-line circuit

Fig. 7.15 Coupled cavity.

circular aperture in a transverse wall behaves as a shunt inductive susceptance with a normalized value given by

$$\bar{B}_L = \frac{3ab}{8\bar{\beta}r_0^3}, \tag{7.69}$$

where a is the guide width, b the guide height, r_0 the aperture radius, and $\bar{\beta} = [k_0^2 - (\pi/a)^2]^{1/2}$ is the propagation constant for the TE_{10} mode. The equivalent circuit of the aperture-coupled cavity is thus a short-circuited transmission line of length d, shunted by a normalized susceptance $\bar{B}_L$.

We shall first assume that there are no losses in the cavity and then show that the cavity will exhibit an infinite number of resonances and the input impedance will have an infinite number of zeros interlaced with an infinite number of poles. This type of behaviour is true for any distributed-parameter one-port microwave network. The poles correspond to parallel resonant circuits and the zeros correspond to series resonant circuits.

The input impedance consists of $j\bar{X}_L$ and $j \tan(\bar{\beta}d)$, and is given by

$$\bar{Z}_{\text{in}} = \frac{-\bar{X}_L \tan(\bar{\beta}d)}{j\bar{X}_L + j \tan(\bar{\beta}d)}, \tag{7.70}$$

where $j\bar{X}_L = (-j\bar{B}_L)^{-1}$. The antiresonances occur when the denominator of Eq. (7.70) vanishes at the poles of $\bar{Z}_{\text{in}}$, or when

$$\bar{X}_L = -\tan(\bar{\beta}d) = \frac{8r_0^3\bar{\beta}d}{3abd}. \tag{7.71}$$

It is convenient to solve Eq. (7.71) by graphical methods. Figure 7.16 shows the two sides of Eq. (7.71) plotted against $\bar{\beta}d$. The points of intersection of $\tan(\bar{\beta}d)$ and $-\bar{X}_L$ give the resonant

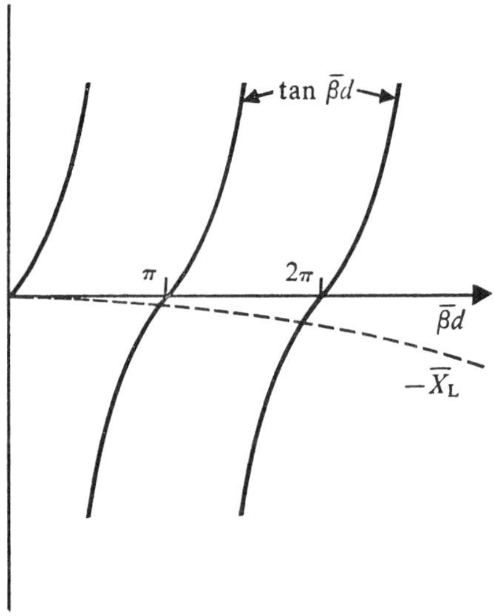

Fig. 7.16 Illustration of graphical solution for resonant frequency of aperture-coupled cavity.

frequencies. Knowing the value of $\bar{\beta}$ at the points of intersection, we get the resonant frequency

$$\frac{\omega}{2\pi} = f = \frac{c}{2\pi}[\bar{\beta}^2 + (\frac{\pi}{a})^2]^{1/2}. \tag{7.72}$$

It should be observed that there are an infinite number of solutions. Usually, $\bar{X}_L$ is very small so that the value $\bar{\beta}d$ for the fundamental mode is approximately equal to π. The higher-order modes occur for $\bar{\beta}d = \bar{\beta}ld \approx (l - \frac{1}{2})\pi$ when n is very large, l being an integer. The resonant frequencies for the first mode and the lower-order modes are very nearly the same as those for the cavity with no aperture.

If the value of $\bar{\beta}$ for the first mode is denoted by $\bar{\beta}_1$ and the corresponding value of ω by ω_1, an infinite number of equivalent lumped-parameter networks can be used to represent $\bar{Z}_{\text{in}}$ in the vicinity of ω_1. Usually, the simplest possible network is used. Such an equivalent network must be chosen so that its input impedance $\bar{Z}$ equals $\bar{Z}_{in}$ at ω_1, and also for small variations $\Delta\omega$ about ω_1. This is achieved by expanding the two impedance functions in the two power series in $\omega - \omega_1 = \Delta\omega$ about ω_1 and by equating the two series term by term. Since $\bar{Z}_{in}$ has a pole at $\omega = \omega_1$, the Taylor expansion cannot be applied directly to $\bar{Z}_{in}$. On the other hand, it is applied to $(\omega - \omega_1)\bar{Z}_{in}$ to give

$$(\omega - \omega_1)\bar{Z}_{in} = \lim_{\omega\to\omega_1} (\omega - \omega_1)\bar{Z}_{in}(\omega) + \frac{d}{d\omega}(\omega - \omega_1)\bar{Z}_{in}\Big|_{\omega_1} (\omega - \omega_1)$$
$$+ \tfrac{1}{2}\frac{d^2}{d\omega^2}(\omega - \omega_1)\bar{Z}_{in}\Big|_{\omega_1} (\omega - \omega_1)^2 + \ldots$$

so that

$$\bar{Z}_{in}(\omega) = \frac{\lim_{\omega\to\omega_1} (\omega - \omega_1)\bar{Z}_{in}(\omega)}{\omega - \omega_1} + \frac{d}{d\omega}(\omega - \omega_1)\bar{Z}_{in}\Big|_{\omega_1} + \ldots . \tag{7.73}$$

Similarly, $\bar{Z}(\omega)$ can be expanded as

$$\bar{Z}(\omega) = \frac{\lim_{\omega\to\omega_1} (\omega - \omega_1)\bar{Z}(\omega)}{\omega - \omega_1} + \frac{d}{d\omega}(\omega - \omega_1)\bar{Z}\Big|_{\omega_1} + \ldots . \tag{7.74}$$

Equations (7.73) and (7.74) are called *Laurent series expansions*, and the coefficient of the $(\omega - \omega_1)^{-1}$ term is referred to as the residue at the pole ω_1. These two series must be equated term by term up to the highest power in $\omega - \omega_1 = \Delta\omega$ which is required to represent $\bar{Z}_{in}$ with enough accuracy in the frequency range of interest. For a microwave cavity whose Q is very high, the frequency range $\Delta\omega/\omega_1$ of interest is approximately the range between the two points, where $|\bar{Z}_{in}|$ equals 0.707 of its maximum or the half-power points. This fractional frequency of $\Delta\omega/\omega_1$ is very small so that, normally, only the first term in the series expansion (7.73) is required to represent $\bar{Z}_{in}$ with sufficient accuracy in the vicinity of ω_1. In such a case, a simple parallel LC circuit is sufficient to represent $\bar{Z}_{in}(\omega)$ around ω.

To specify the parallel LC circuit, the first terms on the right-hand side of Eqs. (7.73) and (7.74) have to be evaluated. For this circuit,

$$\bar{Z} = \frac{j\omega L}{1 - \omega^2 LC}, \qquad \lim_{\omega\to\omega_1} (\omega - \omega_1)\bar{Z} = \frac{-j\omega\omega_1^2 L}{\omega + \omega_1}\Big|_{\omega_1} = \frac{-j\omega_1^2 L}{2}$$

so that

$$\bar{Z}(\omega) = \frac{-j\omega_1^2 L}{2(\omega - \omega_1)} \tag{7.75}$$

for ω near ω_1.

To evaluate $\bar{Z}_{in}$ near ω_1, ω is made equal to ω_1 in the numerator of Eq. (7.70). The denominator of Eq. (7.70) is expanded into a Taylor series in $\bar{\beta}$ about $\bar{\beta}_1$ so that

$$\bar{X}_L(\bar{\beta}) + \tan(\bar{\beta}d) \approx \bar{X}_L(\bar{\beta}_1) + \tan(\bar{\beta}_1 d) + [\frac{d\bar{X}_L}{d\bar{\beta}} + \frac{d\{\tan(\bar{\beta}d)\}}{d\bar{\beta}}]\Big|_{\bar{\beta}_1} (\bar{\beta} - \bar{\beta}_1)$$
$$= [\frac{\bar{X}_{L1}}{\bar{\beta}_1} + d\sec^2(\bar{\beta}_1 d)](\bar{\beta} - \bar{\beta}_1),$$

where

$$\bar{X}_L(\beta_1) = \bar{X}_{L_1} = -\tan(\beta_1 d), \qquad \frac{d\bar{X}_L}{d\beta} = \frac{1}{\beta}\bar{X}_L.$$

Further, β is expanded in terms of ω about ω_1 so that

$$\beta \approx \beta_1 + \left.\frac{d\beta}{d\omega}\right|_{\omega_1} (\omega - \omega_1).$$

If $d\beta/d\omega$ at ω_1 is denoted by β_1', we see that $\bar{Z}_{\text{in}}$ is given by

$$\bar{Z}_{\text{in}} = \frac{j\bar{X}_{L_1} \tan(\beta_1 d)}{[\bar{X}_{L_1} + \beta_1 d\{1 + \tan^2(\beta_1 d)\}][(\beta_1'/\beta_1)(\omega - \omega_1)]}$$

by replacing $\sec^2(\beta_1 d)$ by $[1 + \tan^2(\beta_1 d)]$. Now, replacing $\tan(\beta_1 d)$ by $-\bar{X}_{L_1}$, we obtain

$$\bar{Z}_{\text{in}} = -j\frac{\bar{X}^2_{L_1}}{[\bar{X}_{L_1} + \beta_1 d(1 + \bar{X}^2_{L_1})][(\beta_1'/\beta_1)(\omega - \omega_1)]}. \tag{7.76}$$

Usually, $\bar{X}_{L_1} \ll 1$, and $\beta_1 d \approx \pi$. $\bar{Z}_{\text{in}}$ may therefore be approximated to

$$\bar{Z}_{\text{in}} = -j\frac{\bar{X}^2_{L_1}}{\beta_1' d(\omega - \omega_1)}. \tag{7.77}$$

Equating relations (7.75) and (7.77), we get

$$\frac{\omega_1^2 L}{2} = \frac{\bar{X}^2_{L_1}}{\beta_1' d}$$

or

$$L = \frac{2\bar{X}^2_{L_1}}{\omega_1^2 \beta_1' d}. \tag{7.78}$$

The capacitance C of the equivalent parallel LC circuit is given by $\omega_1^2 LC = 1$.

If we do not neglect the losses in the cavity, we may compensate for these losses in a high Q cavity by replacing the resonant frequency ω_1 by a complex resonant frequency $\omega_1[1 + j/(2Q)]$, as shown in Section 7.1. Then, the natural response of a lossy cavity is proportional to $\exp(-\delta t + j\omega_1 t)$, instead of $\exp(j\omega_1 t)$, where $\delta = \omega_1/(2Q)$. This is the same as having a complex resonant frequency $\omega_1[1 + j/(2Q)]$. Let the field in the cavity be approximately the TE_{101} mode. Then, there will be some local fringing due to the presence of the aperture. The Q of the cavity is given by Eq. (7.56).

For the lossy cavity, we have

$$\bar{Z}_{\text{in}} = -j\frac{\bar{X}^2_{L_1}}{\beta_1' d[\omega - \omega_1 - j\omega_1/(2Q)]}. \tag{7.79}$$

At resonance, $\omega = \omega_1$, and we obtain a pure resistive impedance given by

$$\bar{Z}_{\text{in}} = \bar{R}_{\text{in}} = \frac{2\bar{X}^2_{L_1} Q}{\omega_1 \beta_1' d}. \tag{7.80}$$

If the cavity is to be matched to the waveguide at resonance, the aperture reactance $\bar{X}_{L_1}$ is chosen so that $\bar{R}_{in} = 1$, because the characteristic impedance $\bar{Z}_c$ of the waveguide is normalized and is equal to 1 (see Fig. 7.15). This makes

$$\bar{X}_{L_1} = \left(\frac{\omega_1 \bar{\beta}_1' d}{2Q}\right)^{1/2}. \tag{7.81}$$

The matched condition given by Eq. (7.81) is called *critical coupling*. If $\bar{R}_{in}$ is greater than the characteristic impedance of the waveguide, that is, greater than 1, the cavity is said to be *over-coupled*, whereas if $\bar{R}_{in}$ is less than 1, then it is termed *undercoupled*. If $\bar{R}_{in}$ is the normalized input resistance at resonance for a parallel resonant cavity, then $\bar{R}_{in}$ is defined as the coupling parameter K. For series resonance, the coupling parameter equals the input normalized conductance at resonance. For example, for a rectangular copper cavity whose dimensions are $a = b = d = 3$ cm, the first resonant frequency is $f_1 = 7070$ MHz and $Q = 12{,}700$. For this cavity,

$$\bar{\beta}_1' = \frac{\omega_1}{\bar{\beta}_1 c^2} = 4.7 \times 10^{-11} \text{ sec/cm}$$

so that $\bar{X}_{L_1}$ given by Eq. (7.81) is equal to 0.0157 for critical coupling. The aperture radius r_0 given by Eq. (7.69) is 0.37 cm. It is observed that $\bar{X}_{L_1} \ll 1$. Also, the first resonant frequency should be found by solving Eq. (7.71). However, for a high Q cavity, the first resonant frequency (approximate) is obtained by the relation $\bar{\beta} d = \pi$, which is true for a lossless cavity.

For the lossy cavity, the equivalent circuit must include a resistance $\bar{R}_{in}$ in parallel with L and C, as shown in Fig. 7.17a. In this case, the input impedance $\bar{Z}$ for ω near ω_1 becomes equal to

$$\bar{Z} = -j\frac{\omega_1^2 L}{2[\omega - \omega_1 - j\omega_1/(2Q)]}, \tag{7.82}$$

where $Q = \bar{R}_{in}/(\omega_1 L)$.

The cavity is coupled to an input waveguide, and hence the cavity terminals are loaded by an impedance equal to the impedance seen looking toward the generator from the aperture plane.

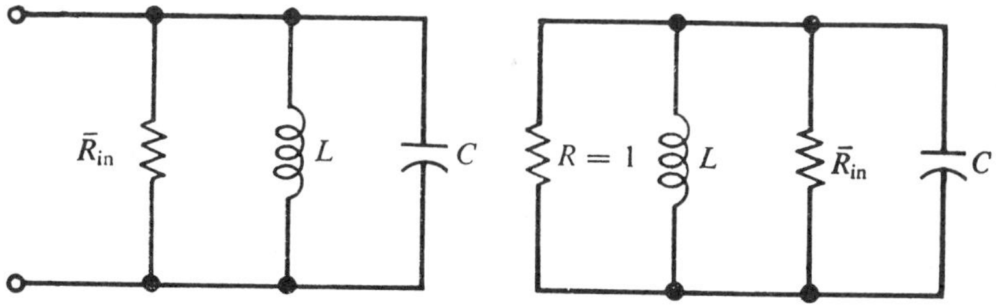

(a) Lossy cavity (b) Matched-coupled cavity

Fig. 7.17 Equivalent circuit for aperture-coupled cavity.

If the generator is matched to the waveguide, a normalized resistance of unity is connected across the cavity terminals, as illustrated in Fig. 7.17b. The external Q, denoted by Q_e, is

$$Q_e = \frac{R}{\omega_1 L} = \frac{1}{\omega_1 L} \tag{7.83a}$$

and the loaded Q, denoted by Q_L, is

$$Q_L = \left(\frac{1}{Q} + \frac{1}{Q_e}\right)^{-1} = \frac{(1 + \bar{R}_{in})\omega_1 L}{\bar{R}_{in}} = \omega_1 L + \frac{\omega_1 L}{\bar{R}_{in}} = \frac{1}{Q}(1 + K) \tag{7.83b}$$

or

$$Q = (1 + K)Q_L, \tag{7.84}$$

where K is the coupling parameter defined as

$$K = \frac{Q}{Q_e}. \tag{7.85}$$

For a parallel resonant circuit,

$$K = \frac{\bar{R}_{in}\omega_1 L}{\omega_1 L} = \bar{R}_{in}. \tag{7.86}$$

This conforms to the original definition.

For a series resonant circuit with normalized input conductance $\bar{G}_{in}$, the unloaded and external Q's are given by

$$Q = \omega_1 L \bar{G}_{in}, \tag{7.87a}$$

$$Q_e = \omega_1 L \tag{7.87b}$$

so that

$$K = \bar{G}_{in} = Q/Q_e. \tag{7.87c}$$

The coupling parameter K is a measure of the degree of coupling between the cavity and the input waveguide or transmission line.

The external quality factor, Q_e, is sometimes called the radiation Q. This is so because the cavity may radiate power through the aperture into the input waveguide. This power loss by radiation through the aperture is equal to the power lost in the normalized unit resistance connected across the resonator terminals in the equivalent circuit shown in Fig. 7.17b.

For the rectangular cavity, the next higher resonance frequency corresponds to a series-type resonance at which $|\bar{Z}_{in}|$ is a minimum. In the loss-free cavity, $\bar{Z}_{in} = 0$. Hence, from Eq. (7.70), this happens when $\beta d = 0$ or $\beta d = \pi$, which is very nearly equal to the condition for parallel resonance for which $\bar{X}_L = -\tan(\beta d)$ is a very small quantity. Therefore, the first series resonance frequency is a little greater than the first shunt resonance frequency. At series resonance, $\bar{Y}_{in}$ has a pole, and hence an analysis similar to that used for $\bar{Z}_{in}$ for shunt resonance is applied to $\bar{Y}_{in}$. It is then found that $\omega = \omega_2$, where ω_2 is obtained from $\beta d = \pi$, so that

$$\bar{Y}_{in} = -j\frac{1}{\beta_2'[\omega - \omega_2 - j\omega_2/(2Q)]}, \tag{7.88}$$

where

$$\beta_2' = \left.\frac{d\beta}{d\omega}\right|_{\omega_2}.$$

For the first series resonance, the aperture has no effect, because the standing-wave pattern in the cavity is such that the transverse electric field is zero at the aperture plane and $\bar{\beta}_2$ and ω_2 do not depend on $\bar{X}_L$. The input admittance near ω_2 is the same as that of a short-circuited guide of length equal to nearly a half guide wavelength. This resonance is not of much practical interest since it corresponds to a very loosely coupled cavity.

Loop-coupled Cavity

Figure 7.18 shows a cavity coupled to a coaxial line by means of a small loop. If the loop is very small, the current in the loop may be assumed to be constant. Any mode in the cavity having a

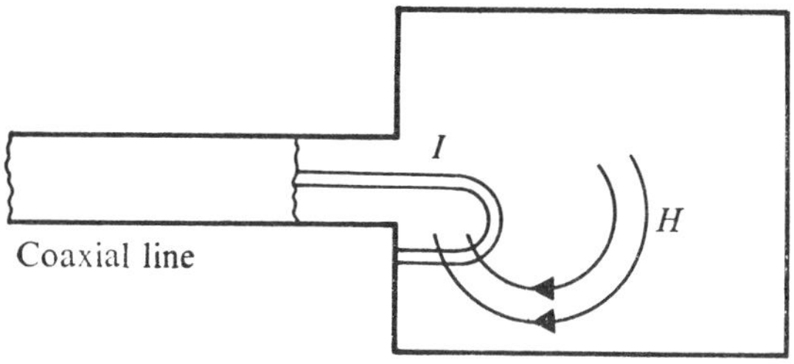

Fig. 7.18 Loop-coupled cavity.

magnetic field with flux lines that thread through the loop can be coupled by the loop. However, at any particular frequency ω, only that mode which is resonant at that frequency will be excited with an appreciable magnitude. The fields excited in the cavity by the current I in the loop can be found by solving for the magnetic vector potential which satisfies the proper boundary conditions. By knowing the magnetic vector potential, the magnetic field intensity, as also the flux passing through the loop, may be found. For a unit current, if the magnetic flux of the n-th mode that threads through the loop is ψ_n, then ψ_n is equal to the mutual inductance M_n between the coupling loop and the n-th mode. Each mode presents an impedance equivalent to that of a series LCR circuit presented to the coupling loop. Therefore, a suitable equivalent circuit for the loop-coupled cavity is an infinite number of series LCR circuits coupled by mutual inductance to the input coaxial line, as shown in Fig. 7.19. The input impedance is thus of the form

$$\bar{Z}_{\text{in}} = j\omega L_0 + j \sum_{n=1}^{\infty} \frac{\omega^3 M_n^2 C_n}{(1 - \omega^2 L_n C_n + j\omega C_n R_n)}, \tag{7.89}$$

where L_0 is the self-inductance of the coupling loop. If we define the resonant frequencies ω_n by $\omega_n^2 L_n C_n = 1$ and the unloaded Q of the n-th mode by $Q_n = \omega_n L_n / R_n$, we can rewrite Eq. (7.89) as

$$\bar{Z}_{\text{in}} = j\omega L_0 + j \sum_{n=1}^{\infty} \frac{\omega^3 M_n^2}{L_n(\omega_n^2 - \omega^2 + j\omega\omega_n/Q_n)}. \tag{7.90}$$

If $\omega \approx \omega_n$, then all the terms in the series in Eq. (7.90), except the n-th term, are small. Hence, in the vicinity of the n-th resonance,

$$\bar{Z}_{\text{in}} = j\omega L_0 + \frac{j\omega^3 M_n^2}{L_n(\omega_n^2 - \omega^2 + j\omega\omega_n/Q_n)} \approx j\omega L_0 - j\frac{\omega_n^2 M_n^2}{2L_n[\omega - \omega_n - j\omega_n/(2Q_n)]}. \tag{7.91}$$

The equivalent circuit now reduces to a single LCR circuit mutually coupled to the input line. For efficient excitation of a particular mode, the loop should be located at a point where the mode provides a maximum flux linkage.

The foregoing analysis gives only a formal general solution to the loop-coupled cavity. To find the actual equivalent circuit parameters for a particular cavity excited by a given loop, the

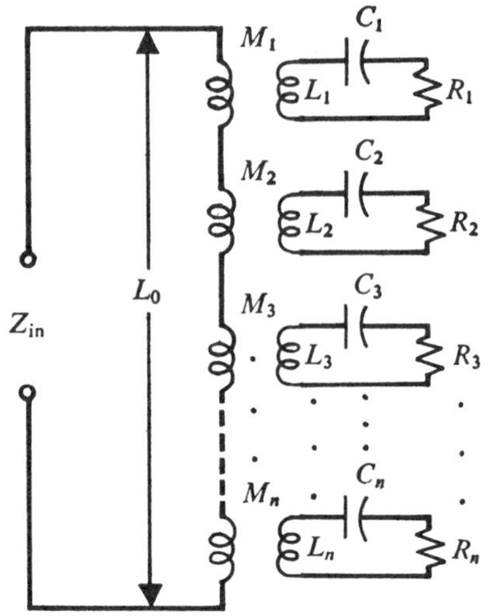

Fig. 7.19 Equivalent circuit of loop-coupled cavity.

boundary value problem for the fields excited in the cavity by the given loop must be solved. Also, the Q's of the cavity for different modes must be determined. For simple cavity shapes, this can be done fairly accurately.

7.7 Spherical Cavity Resonator

The sphere has the highest ratio of volume-to-surface area and therefore it offers attractive possibilities as a high Q resonator.

By using the results given in Section 4.10 for the solution of the vector wave equation in the spherical polar coordinates, the components of the electric and magnetic fields are obtained as follows:

For TM modes

$$E_r = -\mu n(n+1) Y_{mn} \frac{z_n(kr)}{kr} e^{j\omega t}, \tag{7.92a}$$

$$E_\theta = -\mu \frac{\partial Y_{mn}}{\partial \theta} \frac{1}{kr} [krz_n(kr)]' e^{j\omega t}, \tag{7.92b}$$

$$E_\phi = \frac{-\mu}{\sin\theta} \frac{\partial Y_{mn}}{\partial \phi} \frac{1}{kr} [krz_n(kr)]' e^{j\omega t}, \tag{7.92c}$$

$$H_r = 0, \tag{7.92d}$$

$$H_\theta = \frac{k}{j\omega} \frac{1}{\sin\theta} \frac{\partial Y_{mn}}{\partial \phi} z_n(kr) e^{j\omega t}, \tag{7.92e}$$

$$H_\phi = \frac{-k}{j\omega}\frac{\partial Y_{mn}}{\partial \theta} z_n(kr)e^{j\omega t}. \tag{7.92f}$$

For TE modes

$$E_r = 0, \tag{7.93a}$$

$$E_\theta = \frac{-\mu}{\sin\theta}\frac{\partial Y_{mn}}{\partial \phi} z_n(kr)e^{j\omega t}, \tag{7.93b}$$

$$E_\phi = \frac{\partial Y_{mn}}{\partial \theta} z_n(kr)e^{j\omega t}, \tag{7.93c}$$

$$H_r = \frac{n(n+1)}{j\omega} Y_{mn}\frac{1}{r} z_n(kr)e^{j\omega t}, \tag{7.93d}$$

$$H_\phi = \frac{1}{j\omega}\frac{1}{\sin\theta}\frac{\partial Y_{mn}}{\partial \phi}\frac{1}{r}[krz_n(kr)]'e^{j\omega t}, \tag{7.93e}$$

$$H_\theta = \frac{1}{j\omega}\frac{\partial Y_{mn}}{\partial \theta}\frac{1}{r}[krz_n(kr)]'e^{j\omega t}, \tag{7.93f}$$

where

$$Y_{mn} = [A_{mn}\cos(m\phi) + B_{mn}\sin(m\phi)]P_n^m(\cos\theta)$$

and $z_n(kr)$ is one of the spherical Bessel functions $j_n(kr)$, $n_n(kr)$, $h_n^{(1)}(kr)$, and $h_n^{(2)}(kr)$.

Now, considering a spherical metal cavity of radius a, as shown in Fig. 7.20, if the metal has very low losses, there is no electromagnetic field outside the cavity, and inside the cavity, $z_n(kr) = j_n(kr)$. If it is an air filled cavity, $k = k_0 = \omega\sqrt{\mu_0\epsilon_0} = 2\pi/\lambda_0$.

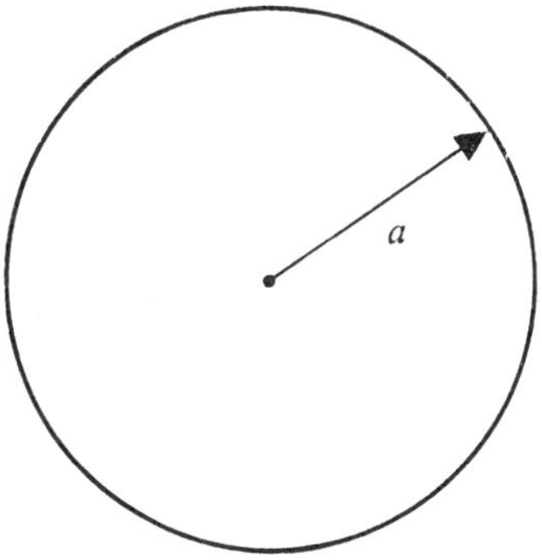

Fig. 7.20 Spherical cavity resonator.

The boundary condition to be satisfied at the metal walls is $E_r = E_\phi = 0$ at $r = a$. This requires that

$$j_n(ka) = 0 \qquad \text{for TE modes,} \tag{7.94}$$

$$\frac{\partial}{\partial(kr)}[(kr)j_n(kr)]'_{r=a} = 0 \qquad \text{for TM modes.} \tag{7.95}$$

The roots of Eqs. (7.94) and (7.95) determine the values of the resonant frequencies f_r given by

$$f_r = \frac{\omega_r}{2\pi} = \frac{k}{2\pi\sqrt{\mu_0\epsilon_0}} = \frac{(ka)c}{2\pi a}, \tag{7.96}$$

where the values of (ka) are determined from Eqs. (7.94) and (7.95).

The modes of the spherical resonator are designated TE_{mnp} and TM_{mnp}, where m is the periodicity of the ϕ-function, n is the order of the spherical Bessel function, and p is an integer denoting the rank of the roots of Eqs. (7.94) and (7.95). From Eqs. (7.94)–(7.96), it is seen that the resonant frequencies are independent of the integer m. However, for the field to exist, it is necessary that $m \leqslant n$ since $P_n^m(\cos\theta)$ vanishes when m is greater than n. Hence, for a given value of n, the integer m may have values from 0 to n (inclusive), each corresponding to a separate mode, but all having identically the same resonant frequency. Furthermore, the intensity distribution in the ϕ-direction may be of the form $\sin(m\phi)$, $\cos(m\phi)$, or any linear combination of the two. A spherical cavity resonator may therefore oscillate in a number of different modes having the same resonant frequency.

The field vanishes if $n = 0$. This does not, however, correspond to an allowed mode. The lowest resonant frequency for the TE (or TM) mode occurs when $n = p = 1$. The corresponding roots of Eqs. (7.94) and (7.95) are

$$(ka)_{1,1} = 4.49 \quad \text{for TE modes,}$$

$$(ka)_{1,1} = 2.75 \quad \text{for TM modes.}$$

Hence, the resonant wavelengths for these modes are

$$\lambda_r = 1.40a \quad \text{for } TE_{m11},$$

$$\lambda_r = 2.29a \quad \text{for } TM_{m11}.$$

If $m = 0$, the field has no variation in the ϕ-direction and is circularly symmetrical, and the function $P_n^m(\cos\theta)$ becomes the Legendre function $P_n(\cos\theta)$. Then, the electric and magnetic field intensities given by Eqs. (7.92) and (7.93) become:

For TM modes

$$E_\phi = H_r = H_\theta = 0, \tag{7.97a}$$

$$E_r = -\mu n(n+1)P_n(\cos\theta)\frac{j_n(kr)}{kr}e^{j\omega t}, \tag{7.97b}$$

$$E_\theta = -\mu\frac{dP_n(\cos\theta)}{d\theta}\frac{1}{kr}[krj_n(kr)]'e^{j\omega t}, \tag{7.97c}$$

$$H_\phi = -\frac{k}{j\omega}\frac{dP_n(\cos\theta)}{d\theta}j_n(kr)e^{j\omega t}. \tag{7.97d}$$

For TE modes

$$E_r = E_\theta = H_\phi = 0, \tag{7.98a}$$

$$E_\phi = \frac{dP_n(\cos\theta)}{d\theta}j_n(kr)e^{j\omega t}, \tag{7.98b}$$

$$H_r = \frac{n(n+1)}{j\omega}P_n(\cos\theta)\frac{1}{r}j_n(kr)e^{j\omega t}, \tag{7.98c}$$

$$H_\theta = \frac{1}{j\omega} \frac{dP_n(\cos \theta)}{d\theta} \frac{1}{r}[krj_n(kr)]' e^{j\omega t}. \tag{7.98d}$$

Let us consider the TM_{011} mode and evaluate its Q. Using the expression for $j_1(kr)$, namely,

$$j_1(kr) = \frac{1}{kr}[\frac{\sin (kr)}{kr} - \cos (kr)],$$

and putting $P_1(\cos \theta) = \sin \theta$, we can evaluate the stored energy as

$$\begin{aligned} W_S &= \frac{\mu}{2} \int_{\theta=0}^{\pi} \int_{\phi=0}^{2\pi} \int_{r=0}^{a} |H_\phi|^2 \, dV \\ &= \frac{\mu}{2} \int_{\theta=0}^{\pi} \int_{\phi=0}^{2\pi} \int_{r=0}^{a} [Aj_1(kr) \sin \theta]^2 r^2 \sin \theta \, d\theta \, d\phi \\ &= \frac{4\pi\mu A^2}{3k^3}[\frac{(ka)^3}{2}\{j_1^2(ka) - j_0(ka)j_2(ka)\}], \end{aligned} \tag{7.99}$$

where H_ϕ is put equal to $Aj_1(kr) \sin \theta$. The time-average power loss P_L is given by

$$P_L = \frac{R_m A^2}{2} j_1^2(ka) \int_{\theta=0}^{\pi} 2\pi a^2 \sin^3 \theta \, d\theta = \frac{4\pi}{3} a^2 A^2 R_m j_1^2(ka), \tag{7.100}$$

where R_m is the surface resistance of the metal walls. The Q of the resonator for the TM_{011} mode is

$$Q = \frac{\omega_r W_S}{P_L} = \frac{1.37\eta}{R_m}[1 - \frac{j_0(ka)j_2(ka)}{j_1^2(ka)}], \tag{7.101}$$

where $\eta = \sqrt{\mu/\epsilon}$. For this mode, $(ka)_{11} = 2.75$, $j_0(ka) = 0.139$, $j_1(ka) = 0.386$, and $j_2(ka) = 0.282$, so that

$$Q = \frac{1.01\eta}{R_m}.$$

For a silver-plated spherical resonator operating in the TM_{011} mode with a resonant frequency of 3×10^9 cycles per second, $R_m = \sqrt{\omega\mu_2/(2\sigma_2)} = 0.0139$, where μ_2, σ_2 are the constants for silver. The Q for this resonator is found to be 27,400. Figure 7.21 gives the electric and magnetic field distributions for the TE_{011} and TM_{011} modes in a spherical resonator. The electric and magnetic intensities for either mode are in time quadrature.

7.8 Fabry-Perot Resonators

We have seen that, in a cylindrical cavity, Q decreases as $f^{-1/2}$ when the frequency is increased. Therefore, for very high frequencies, say, in the millimeter or submillimeter range, Q becomes too small for the usual type of cavity to be of any practical use. Another difficulty arises from the very small size of the cavity at very short wavelengths. In microwave tubes which use these cavities, the volume available for interaction of electron beams and the electromagnetic field becomes too small for sustaining oscillations. To overcome these difficulties at very short wavelengths, including the infra-red region, the optical Fabry-Perot interferometer has been adopted as a resonator.

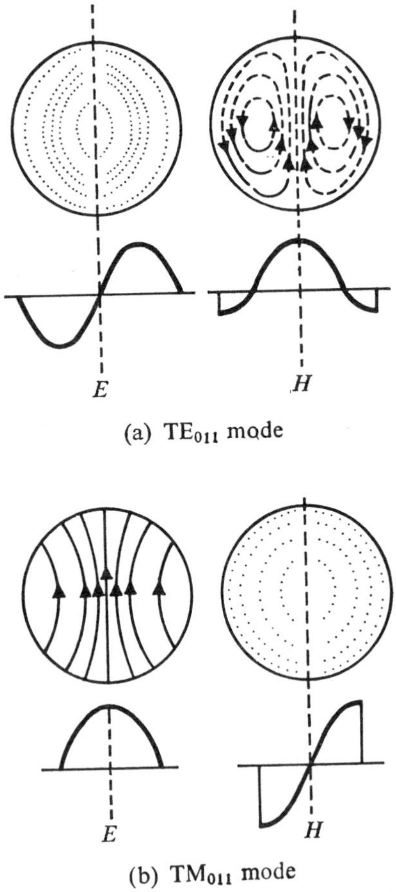

Fig. 7.21 Field patterns in spherical resonator.

A simple form of an ideal Fabry-Perot resonator is shown in Fig. 7.22, where we have two infinite parallel plates with a spacing d. The field solutions in such a resonator are the TEM standing waves. The solutions can be expressed as

$$E_x = E_0 \sin (k_0 z), \tag{7.102}$$

$$H_y = jY_0 E_0 \cos (k_0 z), \tag{7.103}$$

where E_x and H_y are related by Maxwell's equations. The boundary conditions are $E_x = 0$ at $z = 0$, $z = d$. This gives $k_0 d = l\pi$ or

$$f = \frac{c}{2\pi} k_0 = \frac{cl}{2d} \qquad (l = 1, 2, 3, \ldots). \tag{7.104}$$

Equation (7.104) determines the resonant frequencies. To evaluate the Q of the resonator, the energy stored and the power loss in the plates must be evaluated. In a rectangular parallel plate resonator of length d in the z-direction and of unit width and height in the x- and y-direction, the time-average stored electric energy is

$$W_S = W_E = \frac{\epsilon_0}{4}\int_{z=0}^{d} |E_x|^2\, dz = \frac{\epsilon_0|E_0|^2}{4}\int_{z=0}^{d} \sin^2\left(\frac{l\pi z}{d}\right) dz$$

$$= \frac{\epsilon_0|E_0|^2 d}{8} = W_H = \frac{\mu_0}{4}\int_0^d |H_y|^2\, dz. \qquad (7.105)$$

The current on each plate is equal to H_y, and hence the power loss per unit area for the two plates is

$$P_L = R_m|Y_0E_0|^2.$$

Therefore, the Q of the resonator for the l-th mode is obtained as

$$Q = \frac{\omega W}{P_L} = \frac{2\omega W_E}{P_L} = \frac{\omega\epsilon_0 Z_0^2 d}{4R_m} = \frac{\pi\epsilon_0 c l Z_0^2}{4R_m} = \frac{\pi l Z_0}{4R_m} \qquad (7.106)$$

by putting $c = (\mu_0\epsilon_0)^{-1/2}$. By making l very large, Q too can be made very large. For a given spacing d, only one mode of oscillation is possible. (In fact, two modes are possible if the other polarization with the electric field along y is counted as a separate mode.) In a totally enclosed cavity, such as a rectangular cavity, Q can be made very large also by using a large cavity oscillating in a higher-order mode. But in a pure mode, it is difficult to excite such a totally enclosed cavity because of the close spacing between the resonant frequencies of the higher-order modes.

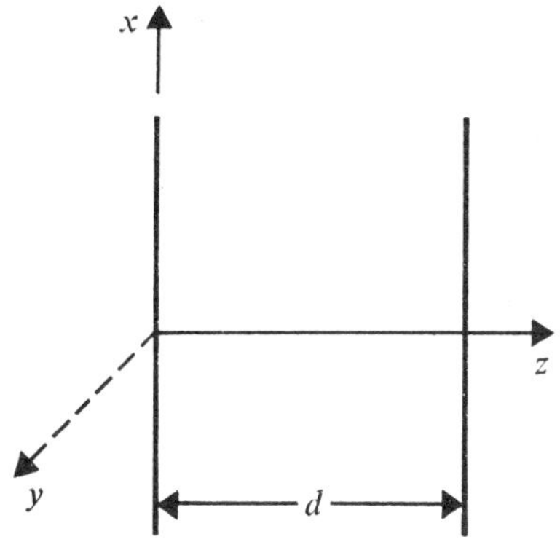

Fig. 7.22 Fabry-Perot resonator.

As a numerical example, consider a Fabry-Perot resonator with copper plates spaced 5 cm apart and operating at $\lambda_0 = 0.1$ mm. Then, the mode index is

$$l = \frac{2d}{\lambda_0} = 1000.$$

The surface resistance for copper is $R_m = (\sigma\delta_s)^{-1} = 0.45$ ohm. Hence, Eq. (7.106) gives a Q of 650,000, which is very high.

A practical Fabry-Perot resonator cannot have infinite plates and must therefore have some means of coupling power into the resonator. One such method is shown in Fig. 7.23, where the resonator is excited by cutting a square array of small circular holes in the end plates. One end plate is illuminated with electromagnetic radiation from a horn with a lens in the opening for collimating the beam. A similar horn is used as a receiving antenna on the other side. The field received by the receiving horn is proportional to the amplitude of the field in the resonator. It will be shown that this field is large only at certain discrete resonant frequencies.

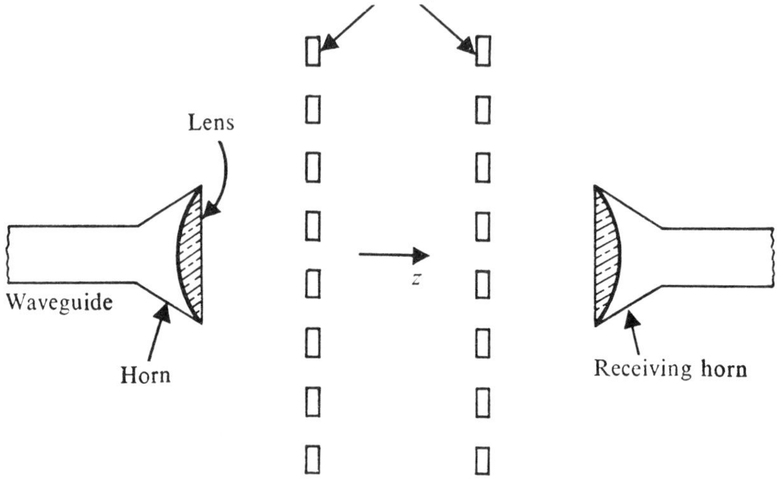

Fig. 7.23 Method of exciting a Fabry-Perot resonator.

A thin plate, with thickness very much less than λ_0, and perforated by a regular array of small circular holes of radius r_0 and spacing a, as shown in Fig. 7.24, has a normalized shunt inductive susceptance $-j\bar{B}_L$ when illuminated by a normally incident TEM wave. This behaviour

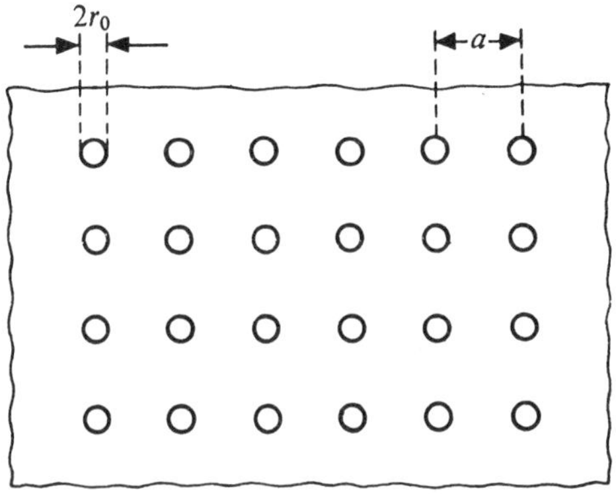

Fig. 7.24 Array of circular holes in end plate of Fabry-Perot resonator.

is similar to that of a circular aperture in a transverse wall in a waveguide, and hence the cavity is essentially aperture-coupled. The normalized susceptance is given by

$$\bar{B}_L = \frac{3a^2}{4k_0 r_0^3} = \frac{3a^2 Z_0}{4\omega\mu_0 r_0^3}. \tag{7.107}$$

The reflection coefficient for the infinite plate is given by

$$\Gamma = \frac{1 - \bar{Y}_{in}}{1 + \bar{Y}_{in}} = \frac{1 - 1 + j\bar{B}_L}{2 - j\bar{B}_L} = \frac{j\bar{B}_L}{2 - j\bar{B}_L} \tag{7.108}$$

because $\bar{Y}_{in} = 1 + j\bar{B}_L$, and the conductive part of $\bar{Y}_{in}$ is taken as the normalized intrinsic admittance of free space, which is equal to 1. For finite-size plates, there will always be some diffraction effects, resulting in some loss of power past the plates at each reflection, and hence not all the energy coupled into the resonator through one plate will be reflected from the other plate. If the plates are large as compared to a wavelength, and if the horns illuminate the centre portion of the plates, then the diffraction losses can be minimized.

In the analysis of the aperture-coupled resonator of finite size, an approach based on multiple reflections is used. The reflection coefficient Γ is replaced by an equivalent reflection coefficient Γ_e which takes care of the diffraction losses. If there are no losses, let the plane wave coupled into the resonator by the transmitting horn have an electric field with an amplitude E_0 at $z = 0$. This wave arrives at $z = d$ with an amplitude $E_0 \exp(-jk_0d)$. Upon reflection at $z = d$, it arrives back at $z = 0$ with an amplitude $\Gamma E_0 \exp(-2jk_0d)$ and is re-reflected. The total field in the resonator is obtained by summing up all these multiple reflected waves, as illustrated in Fig. 7.25.

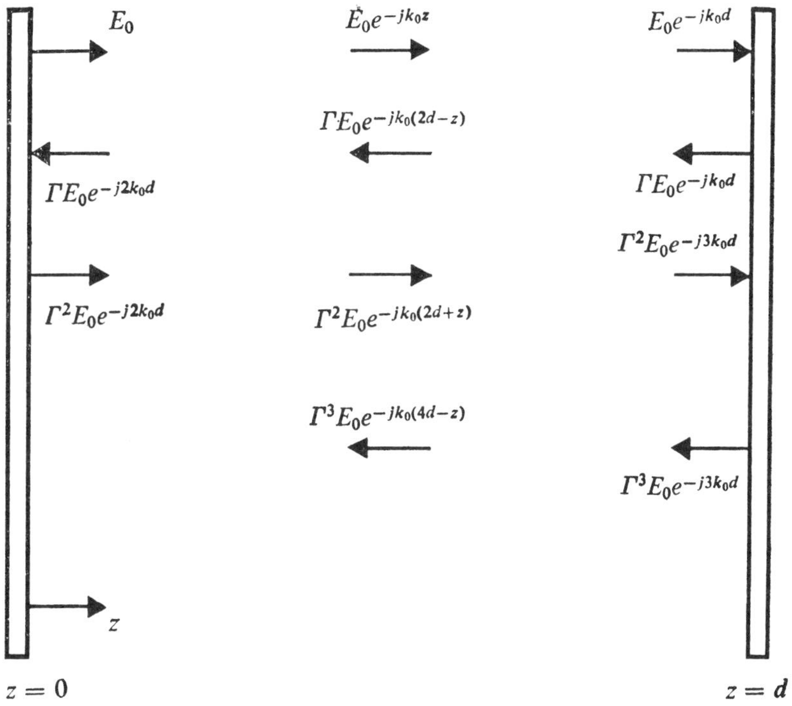

Fig. 7.25 Multiple reflections in Fabry-Perot resonator.

Therefore, for any z,

$$E_x = E_0 \exp(-jk_0z) + \Gamma E_0 \exp[-jk_0(2d - z)] + \Gamma^2 \exp[-jk_0(2d + z)] + \ldots$$

$$= E_0 \exp(-jk_0z) \sum_{n=0}^{\infty} \Gamma^{2n} \exp(-jk_0 2nd) + \Gamma E_0 \exp[-jk_0(2d - z)]$$

$$\times \sum_{n=0}^{\infty} \Gamma^{2n} \exp(-jk_0 2nd) = \frac{E_0 [\exp(-jk_0z) + \Gamma \exp\{-jk_0(2d - z)\}]}{1 - \Gamma^2 \exp(-j2k_0d)}. \tag{7.109}$$

To obtain a high external Q, the susceptance $\bar{B}_L$ must be made very large so that $\Gamma \approx -[1 + 2/(j\bar{B}_L)] \approx 1$. Then

$$E_x = \frac{2jE_0 \exp(-jk_0d) \sin[k_0(d - z)]}{1 - \Gamma^2 \exp(-j2k_0d)}. \tag{7.110}$$

E_x is maximum when

$$|1 - \Gamma^2 \exp(-j2k_0d)| = D = |1 - (\frac{j\bar{B}_L}{2 - j\bar{B}_L})^2 \exp(-j2k_0d)|$$

is a minimum. If $\Gamma = \rho e^{j\theta}$, where

$$\rho = \frac{\bar{B}_L}{(4 + \bar{B}_L^2)^{1/2}}, \qquad \theta = \frac{\pi}{2} + \tan^{-1} \frac{\bar{B}_L}{2}.$$

D is minimum when $|1 - \rho^2 \exp[-j2(k_0d - \theta)]|$ is a minimum, which occurs when $2(k_0d - \theta) = 2(l - 1)\pi$, provided we neglect the small variation in ρ with k_0. Substituting for θ, we obtain

$$2k_0d = 2(l - 1)\pi + 2\theta = (2l - 1)\pi + 2\tan^{-1}(\bar{B}_L/2). \tag{7.111}$$

Since $\bar{B}_L$ is large, $\theta \approx \pi$; so, the resonant frequency is given by

$$k_0d = l\pi \quad \text{or} \quad f = \frac{lc}{2d}$$

which is the same as that for the uncoupled resonator.

The minimum value of the denominator in Eq. (7.110) is $(1 - \rho^2)$. For a value of k_0, which is slightly removed from the resonance value of $l\pi/d$, the denominator has a magnitude $\sqrt{2}(1 - \rho^2)$. If the deviation in k_0 is Δk_0, then the external Q of the resonator is

$$Q_e = \frac{k_0}{2\Delta k_0}.$$

To find Δk_0, we replace $k_0d - \theta$ by

$$(l - 1)\pi + [d - \Delta\theta/(\Delta k_0)]\Delta k_0$$

in the expression for D^2 and equate it to $2(1 - \rho^2)^2$. Thus

$$[1 - \rho^2 \cos 2(d - \frac{\Delta\theta}{\Delta k_0})\Delta k_0]^2 + \rho^4 \sin^2 2(d - \frac{\Delta\theta}{\Delta k_0})\Delta k_0$$

$$= 2(1 - \rho^2)^2 = 1 - 2\rho^2 + \rho^4 + 2\rho^2[1 - \cos 2(d - \frac{\Delta\theta}{\Delta k_0})\Delta k_0].$$

As Δk_0 is small, we may use the approximation

$$\cos 2[d - \Delta\theta/(\Delta k_0)]\Delta k_0 \approx 1 - 2(\Delta k_0)^2[d - \Delta\theta/(\Delta k_0)]^2$$

to obtain

$$\Delta k_0 = \frac{1-\rho^2}{2\rho[d-\Delta\theta/(\Delta k_0)]} \approx \frac{1-\rho^2}{2[d-\Delta\theta/(\Delta k_0)]}$$

since ρ is very nearly equal to unity for a high Q resonator. And

$$\frac{\Delta\theta}{\Delta k_0} = \frac{\Delta[\tan^{-1}(\bar{B}_L/2)]}{\Delta k_0} = \frac{4}{(\bar{B}_L^2+4)}\frac{\Delta(\bar{B}_L/2)}{\Delta k_0} = -\frac{2\bar{B}_L}{(4+\bar{B}_L^2)k_0} = \frac{-2\rho^2}{k_0\bar{B}_L} \approx \frac{-2}{k_0\bar{B}_L}. \tag{7.112}$$

Therefore

$$Q_e = \frac{k_0}{2\Delta k_0} = \frac{k_0[d-\Delta\theta/(\Delta k_0)]}{1-\rho^2} = \frac{k_0[d+2/(k_0\bar{B}_L)]}{1-\rho^2}. \tag{7.113}$$

In a typical resonator, $k_0 d = l\pi$ is very large as compared to $2/\bar{B}_L$, and hence

$$Q_e \approx \frac{k_0 d}{1-\rho^2} = \frac{l\pi}{1-\rho^2}. \tag{7.114}$$

If the losses in the end plates are considered, we may take the losses per unit area to be approximately the same as those for the unperforated plates, and hence the loaded Q of the resonator for the l-th resonant mode is

$$Q_L = \left(\frac{1}{Q_e}+\frac{1}{Q}\right)^{-1} = \frac{\pi l Z_0}{4R_m + Z_0(1-\rho^2)}, \tag{7.115}$$

where

$$Q = \frac{\pi l Z_0}{4R_m}.$$

The external Q arises from the power radiated by the fields in the cavity through the perforated plates into the external region. Further, if power is lost by diffraction around the edges of the plates, then the effective value of ρ is decreased, and this will reduce both Q_e and Q_L.

As the frequencies increase up to the optical frequencies, to keep the diffraction losses low, it is necessary that the plates be flat and parallel to each other to within a fraction of a wavelength. Such tolerances are very difficult to achieve. However, by using spherical or paraboloidal mirrors instead of flat mirrors, these difficulties can be avoided, because spherical and paraboloidal mirrors have focussing effects.

A TEM standing wave in a Fabry-Perot resonator exists only if the plates are infinitely extending. For finite plates, the field must decrease to a negligible value at the edge of the mirrors. If the problem is analyzed rigorously, it can be shown that there are many field distributions satisfying this requirement. Hence, many modes of oscillation are possible. If the plates are very large, all these modes are approximately like the TEM modes.

PROBLEMS

1 The dimensions of a shielded room are 4 m $\times$ 5 m $\times$ 2.5 m. Calculate the six lowest resonant frequencies of the room. If the walls are of copper, calculate the Q at the lowest frequency.

2 A rectangular waveguide carrying the TE_{10} mode at a frequency of 10,000 MHz is shorted at both ends to form a resonator. If the width of the waveguide is 12 per cent greater than the minimum width necessary to support the TE_{10} mode, calculate the dimensions of the cavity. If the inner surfaces of the cavity are plated by silver, calculate the Q at the lowest resonant frequency.

3 Calculate the dimensions of a cylindrical resonator designed to resonate in the TM_{111} mode at a frequency of 6000 MHz if it is (i) air filled, and (ii) filled with dielectric having $\epsilon_r = 4$.

4 A cylindrical waveguide designed to carry the TE_{01} mode is to be converted into a resonator by placing shorts at both ends. If the frequency is 8000 MHz, calculate the minimum cavity dimensions. Evaluate the Q at the lowest resonant frequency.

5 Calculate the inner radius of a spherical resonator to operate in the TE_{111} mode at a frequency of 3000 MHz. If the inner surface is silver plated, calculate the Q. Where should an exciting loop be placed in the cavity to excite this mode?

6 A two-wire copper coaxial line is short circuited. The radii of the inner and outer conductors are 2 mm and 4 mm, respectively, and the length of the line is 20 cm. Find the antiresonant frequency and Q.

7 Design a rectangular cavity from a rectangular waveguide of dimensions 2.5 cm × 1.2 cm to resonate at a frequency of 9500 MHz. The cavity is critically coupled to a rectangular waveguide of the same dimensions. Calculate the radius of the centred circular aperture. Determine the unloaded and loaded Q's of the cavity if it is made of copper.

8 In a Fabry-Perot resonator, the copper plates are perforated by a square array of small circular holes. The spacing between apertures is 5 mm, and the resonator is operated at a wavelength of 10 mm in the fortieth resonant mode. Find the aperture radius required for critical coupling and the unloaded and loaded Q's.

8 Microwave Passive Components

8.1 General Considerations

At low frequencies, circuit elements such as resistances, inductances, and capacitances are combined in various ways to form circuits or networks for different purposes, and the electrical characteristics of such networks can be calculated fairly accurately. In practice, each of these circuit elements is made up of different simpler parts. For example, an inductor may be a coil of wire, and a capacitor may be made of two or more conducting plates of proper shape and size, spaced in a definite manner. The practical circuit element is only an approximate equivalent of a resistance, capacitance, or inductance. For example, a coil of wire, which is used as an inductance, usually has also a low resistance and a low capacitance.

The types of circuit elements just mentioned are not the only ones that are used. In Chapter 6, we have shown that an open or short-circuited low-loss transmission line can behave like an inductance or a capacitance by varying its length. Also, a lossy transmission line (open or short) can behave like an impedance having both a resistive and a reactive part. A waveguide or coaxial line, or any other type of transmission line used at microwave frequencies, may be used in the same manner to replace conventional forms of circuit elements. Another method of obtaining circuit elements is by introducing discontinuities along transmission lines or waveguides. These discontinuities could be introduced in the form of certain shapes of conductors or dielectrics or magnetic materials, and they produce reflections and may change the propagation characteristics of the electromagnetic wave. Some of these configurations may enhance the resultant magnetic field at a point in the waveguide, producing a discontinuity similar to that of an inductance coil connected across a transmission line. Such a discontinuity may be called an *inductive discontinuity*. Similarly, another type of configuration may enhance the electric field intensity, producing a *capacitive discontinuity*. These localized discontinuities may be called *waveguide elements* and their useful combinations, *waveguide components*.

The techniques used at low frequencies with transmission lines consisting of conducting wires and the techniques used at microwave frequencies with waveguides differ from each other in some respects, though they are similar in many respects. One of these differences is due to the fact that, in a waveguide, the phenomena associated with reflection take place in a space comparable with the waveguide itself, whereas in the low-frequency wire line, they are distinctly localized. For this reason, waveguide elements are much more complicated than their low-frequency counterparts, and it is more difficult to evaluate their characteristics. For certain simple types of elements, the characteristics may be calculated, but for others, it is necessary to evaluate the characteristics by experimental methods.

In this chapter, we shall deal with (i) waveguide elements and components which are passive invariant elements that do not include any active source of power, and (ii) nonreciprocal elements and components using anisotropic materials such as ferrites. (It should be observed that there is no sharp dividing line between waveguide elements and waveguide components.)

Some examples of common waveguide elements are:

(i) A straight piece of open or short waveguide of uniform section whose length may be varied by means of a tightly fitting piston

(ii) A bend or a branch in a waveguide

(iii) A thin septum made of conducting or lossy material placed at right angles to the axis of transmission, which may be provided with an opening

(iv) A transverse rod or port made of a conducting or lossy or dielectric material.

Some of these obstacles placed inside the waveguide may act as inductances, capacitances, resistors, or resonant elements.

8.2 Waveguide Components

Properly proportioned discontinuities, when placed in a section of waveguide, perform many useful functions. The simplest parts of these devices may be arbitrarily called waveguide elements, whereas the composite structures, built from these elements, may be called waveguide components. Many of these components have low-frequency counterparts to which they are quite analogous. Some examples of waveguide components are transformers, filters, and hybrids. There are, however, other arrangements which may not have low-frequency counterparts and therefore have new names which are more or less descriptive of the function performed, for instance, directional couplers, isolators, and circulators. In the sections that follow, we shall deal with some of these passive waveguide components. In this chapter, we shall also discuss ferrite microwave passive devices such as isolators and circulators.

8.3 Waveguide Terminations

The matched load and variable short circuit which produce an adjustable reactive load are two common types of waveguide and transmission-line terminations. These are used extensively in the laboratory for measuring the impedance and scattering parameters of a microwave circuit element. The matched load is a termination that absorbs all the incident power and is therefore equivalent to terminating the line in its characteristic impedance. On the other hand, the variable short circuit reflects all the incident power, and the phase of the reflected wave is varied by changing the position of the short circuit, which is equivalent to changing the reactance of the termination.

Matched Load

A tapered edge or slab of lossy material inserted into the guide is a usual form of matched load

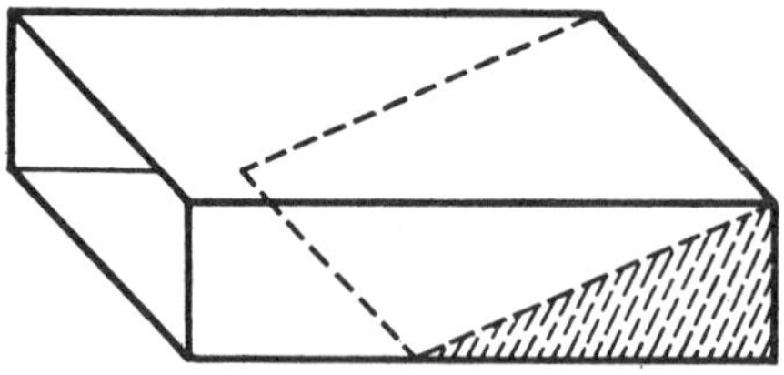

(a) Lossy wedge

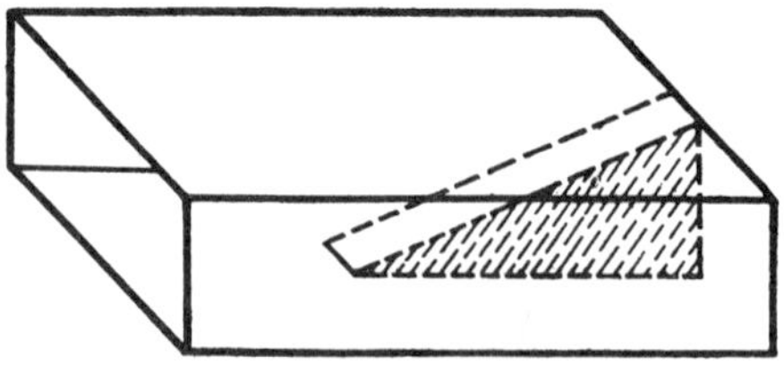

(b) Tapered resistive card

Fig. 8.1 Matched loads for waveguide.

(see Fig. 8.1). The incident power is absorbed by the lossy material, and reflections are prevented by tapering the lossy material into a wedge. An overall length of one or more wavelengths is usually sufficient to provide a matched load with an input standing-wave ratio of 1.01 or smaller.

Variable Short Circuit

A sliding block of copper or of some other good conducting metal that makes a snug fit in the guide, as shown in Fig. 8.2, is a simple form of adjustable short circuit. A micrometer drive is used to vary the position of the block. The performance of this simple type of structure is not

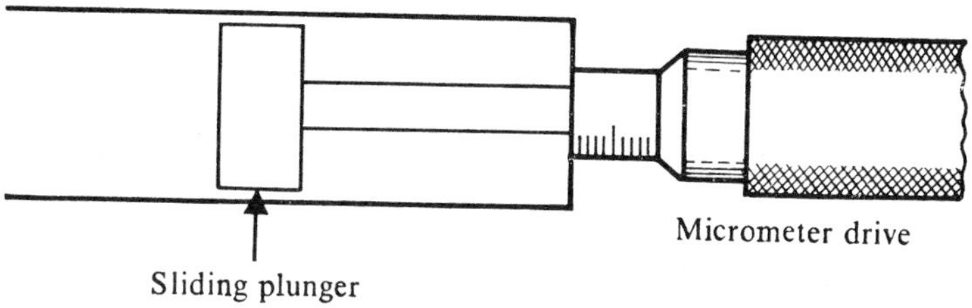

Fig. 8.2 Simple variable short circuit for waveguide.

very satisfactory electrically. The contact between the sliding block and the waveguide walls is erratic and causes the equivalent electrical short-circuit position to deviate in a random fashion from the physical short-circuit position which is the front face of the sliding block. Also, there may be power leakage past the block, which results in a reflection coefficient less than unity. These drawbacks can be overcome by using a choke-type plunger.

In the choke-type plunger, the transformation properties of a quarter-wave transformer are used. Figure 8.3 shows a two-section quarter-wave transformer. A load impedance Z_s, which is

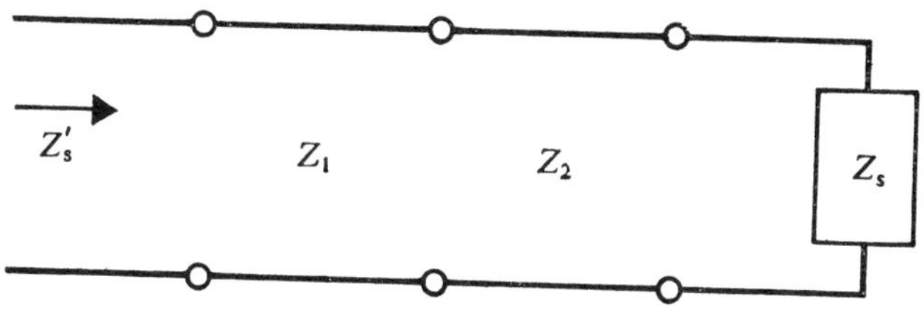

Fig. 8.3 Two-section quarter-wave transformer.

approximately zero, is viewed through this two-section quarter-wave transformer, and the impedance seen at the input is

$$Z_s' = \left(\frac{Z_1}{Z_2}\right)^2 Z_s. \tag{8.1}$$

If $Z_2 \gg Z_1$, the new impedance Z_s' is approximately a short circuit, and is better than Z_s by a factor $(Z_1/Z_2)^2$. This is the principle used in choke-type plungers. Such plungers are, however, frequency sensitive, but, by proper design, may be made to work satisfactorily over a bandwidth of 10 per cent or more. More than two sections may be used for very critical applications. As an example of a choke-type plunger, let us take one such plunger for use in a rectangular waveguide carrying the dominant TE_{10} mode. For such a waveguide, the surface currents on the interior wall flow up and down along the side walls and both across and in the axial direction on the broad walls. This erratic behaviour occurs only due to the axial currents flowing on the upper and lower walls. To avoid such behaviour, the plunger may be made in the form shown in Fig. 8.4. The width of the plunger is uniform and is slightly less than the interior guide width, but the

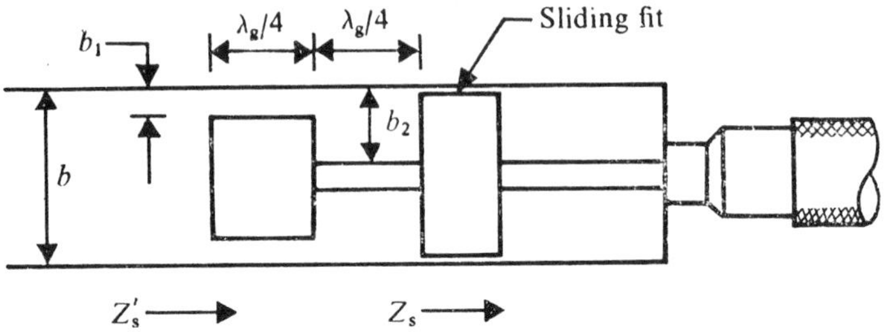

Fig. 8.4 Choke-type adjustable short circuit.

height of the plunger is made nonuniform. The quarter-wave sections have equivalent characteristic impedances proportional to $2b_1/b$ and $2b_2/b$ relative to that of the input guide. Thus, from Eq. (8.1),

$$Z'_s = \left(\frac{b_1}{b_2}\right)^2 Z_s$$

for the normalized input impedance. By making b_2 about $10b_1$, an improvement in performance by a factor of 100 over non-choke-type short circuit can be achieved.

A design different from that in Fig. 8.4 shown in Fig. 8.5 is also used. Here, a two-section folded quarter-wave transformer is used. The inner line transforms the short-circuit

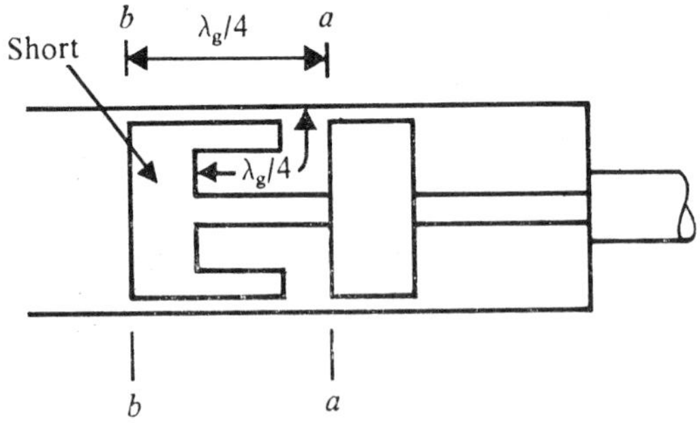

Fig. 8.5 Choke-type plunger.

impedance into an ideal open circuit at the plane *aa*, where an infinite impedance exists and the axial current is zero. Hence, current does not flow across the gap between the waveguide wall and the plunger at the contact point *aa*. The outer quarter-wave transformer transforms the open-circuit impedance at *aa* into a short-circuit impedance at *bb*. Such plungers give very satisfactory performance.

The quarter-wave transformers are also used in the construction of choke joints for joining two waveguide sections together, in rotary joints, as also in plungers used to tune cavity resonators.

8.4 Waveguide Attenuators

There are fixed as well as variable attenuators. Fixed attenuators are used only if a fixed amount of attenuation is to be introduced in a system. Variable attenuators, on the other hand, have many more uses.

A simple form of variable attenuator consists of a thin tapered resistive card, whose depth of penetration into the waveguide is adjustable. The card is inserted into the waveguide through a longitudinal slot cut in the centre of the broad wall of a rectangular waveguide, as shown in Fig. 8.6. Such an attenuator has a rather complicated attenuation variation with depth of insertion and frequency.

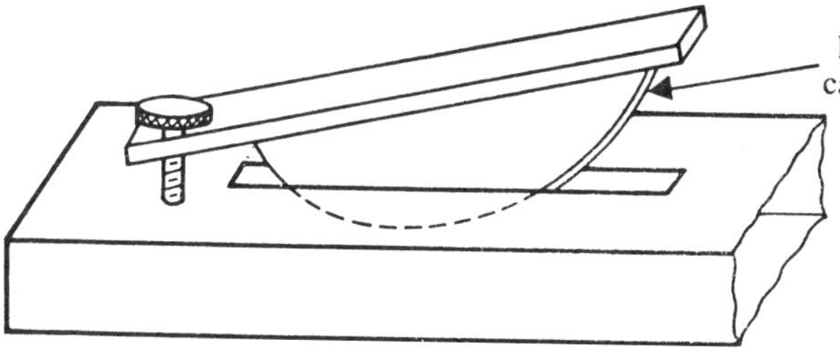

Fig. 8.6 Waveguide flap attenuator.

A better precision-type attenuator employs an adjustable length of waveguide operated below its cut-off frequency. Such an attenuator, however, has a disadvantage in that the output is attenuated by reducing the coupling between the input and output guides, and not by absorption of the incident power. This results in a high degree of attenuation corresponding to a reflection coefficient near unity in the input guide, which is rather undesirable.

A more satisfactory precision attenuator is the rotary attenuator. It consists of two rectangular-to-circular waveguide tapered transitions, together with an intermediate section of circular waveguide which is free to rotate, as shown in Fig. 8.7. A very thin tapered resistive

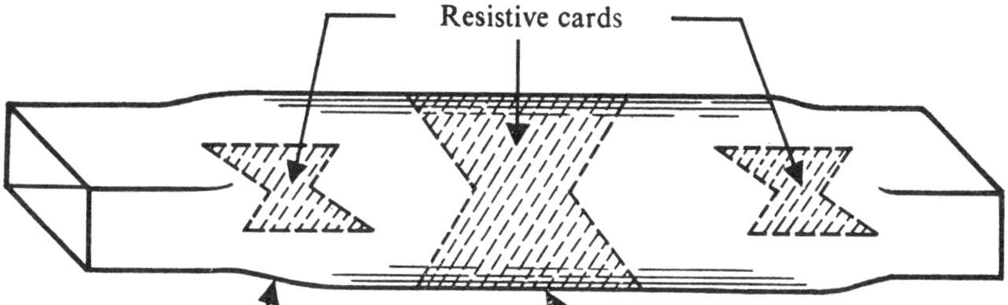

Fig. 8.7 Rotary attenuator.

card is placed at the output end of each transition section and is oriented parallel to the broad walls of the rectangular guide. A similar resistive card is located in the intermediate circular guide section. The incoming TE_{10} mode in the rectangular guide is transformed into the TE_{11} mode in the circular guide, with negligible reflection, by means of the tapered transition. The polarization of the TE_{11} mode is such that the electric field is perpendicular to the thin resistive card in the transition section. This resistive card has a negligible effect on the TE_{11} mode. The resistive card in the centre section can be rotated, and its orientation relative to the electric field of the incoming TE_{11} mode can be varied so that the amount by which this mode is attenuated is adjustable.

Let the centre resistive card be oriented at an angle θ to the direction of the electric field polarization of the TE_{11} mode (see Fig. 8.8). The TE_{11} mode polarized in the x-direction may be

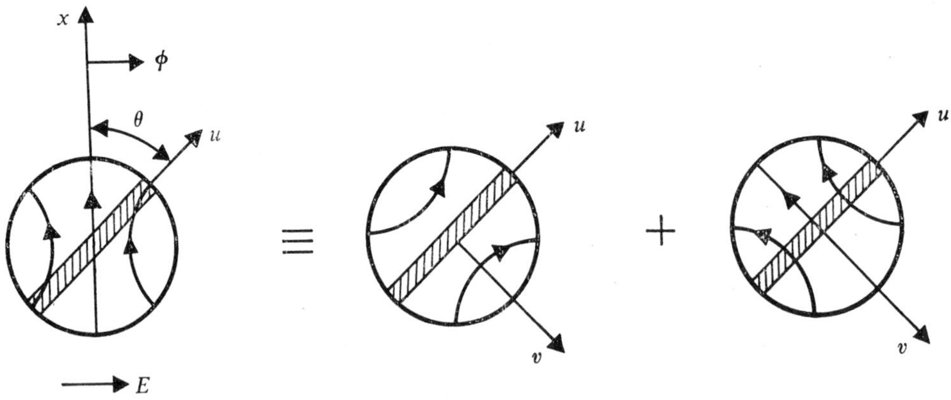

Fig. 8.8 Decomposition of TE_{11} mode into two orthogonally-polarized modes.

decomposed into two TE_{11} modes polarized along the u- and v-direction. The mode polarized parallel to the resistive slab will be absorbed, whereas the mode polarized perpendicular to the slab will be transmitted. This transmitted mode is not perpendicular to the resistive card in the transition section, and hence some additional attenuation occurs.

We shall now derive an expression for attenuation as a function of the rotation angle θ. For polarization in the x-direction, omitting the propagation factor $\exp(-j\beta z)$, we obtain the electric field for the TE_{11} mode as

$$E = \frac{J_1(p'_{11}\rho/a)}{\rho}\mathbf{u}_\rho \cos\phi = \frac{p'_{11}}{a}J'_1\frac{p'_{11}\rho}{a}\mathbf{u}_\phi \sin\phi. \tag{8.2}$$

However,

$$\sin\phi = \sin(\phi - \theta + \theta) = \cos\theta \sin(\phi - \theta) + \sin\theta \cos(\phi - \theta),$$

$$\cos\phi = \cos(\phi - \theta + \theta) = \cos\theta \cos(\phi - \theta) - \sin\theta \sin(\phi - \theta),$$

and hence

$$E = \cos\theta\,[\frac{J_1}{\rho}\mathbf{u}_\rho \cos(\phi - \theta) - \frac{p'_{11}}{a}J'_1\mathbf{u}_\phi \sin(\phi - \theta)]$$
$$- \sin\theta\,[\frac{J_1}{\rho}\mathbf{u}_\rho \sin(\phi - \theta) + \frac{p'_{11}}{a}J'_1\mathbf{u}_\phi \cos(\phi - \theta)]. \tag{8.3}$$

In Eq. (8.3), the first term is a TE_{11} mode polarized with the electric field along the u-direction in Fig. 8.8, and the second term is a TE_{11} mode polarized along the v-direction. The first part is completely absorbed by the resistive card, and only the portion multiplied by $\sin\theta$ is transmitted into the output transition section. If the resistive card in the output transition section is along the y-axis, then only the component of the transmitted field, which is polarized along the x-direction, is transmitted. Therefore, the electric field at the input to the transition section is given by

$$\mathbf{E} = -\sin\theta\,[\frac{J_1}{\rho}\mathbf{u}_\rho \sin(\phi-\theta) + \frac{p'_{11}}{a}J'_1\mathbf{u}_\phi \cos(\phi-\theta)]$$

$$= \sin^2\theta\,(\frac{J_1}{\rho}\mathbf{u}_\rho\cos\phi - \frac{p'_{11}}{a}J'\mathbf{u}_\phi\sin\phi) - \sin\theta\cos\theta\,(\frac{J_1}{\rho}\mathbf{u}_\rho\sin\phi + \frac{p'_{11}}{a}J'_1\mathbf{u}_\phi\cos\phi). \tag{8.4}$$

In Eq. (8.4), the first part is a TE_{11} mode polarized along the x-direction and is transmitted. Comparing Eq. (8.4) with Eq. (8.2), we can see that the transmitted field is reduced by a factor $\sin^2\theta$ from the amplitude of the incident field. Hence, the attenuation produced is given (in decibels) by

$$\alpha = -20\log(\sin^2\theta) = -40\log(\sin\theta). \tag{8.5}$$

It should be noted that the attenuation depends only on the angle θ.

8.5 Phase Changers

A phase changer is an instrument that produces an adjustable change in the phase angle of the wave transmitted through it. An ideal phase changer should be perfectly matched to the input lines as well as to the output lines and should produce zero attenuation. In practical phase changers, these requirements can be met within a degree of approximation. There are different types of phase changers. However, we shall confine our discussion to the linear phase changer and rotary phase changer.

Linear Phase Changer

A linear phase changer which consists of three dielectric slabs placed in a rectangular waveguide is shown in Fig. 8.9. The centre slab is free to move longitudinally, and is moved by a suitable drive mechanism to which it is keyed by means of a dielectric key that protrudes through a long centred slot cut in one broad face of the guide. Each end of the dielectric slab is cut in a step-wise fashion to provide a broad-band multisection quarter-wave transformer to match the partially filled guide to the empty and completely filled guide. If the centre slab is $0.3a$ wide, it is found that the sum of the two propagation constants β_1 and β_3 is greater than $\beta_2 + \beta_4$. If the centre slab is displaced a distance x to the right, the effect is to lengthen lines 1 and 3 by an amount x and to shorten lines 2 and 4 by the same amount x. Therefore, the phase change undergone by a wave propagating through the structure is

$$\Delta\phi = [(\beta_1 + \beta_3) - (\beta_2 + \beta_4)]x. \tag{8.6}$$

The phase change $\Delta\phi$ is proportional to the displacement x. The amount of phase change increases if the dielectric constant of the slab is increased. If a material of ϵ_r equal to 2.56 is used (e.g., perspex) in a 3-cm waveguide of dimensions $a = 2.25$ cm, the phase change obtained is about 0.4 rad/cm of displacement. About 16 cm of displacement gives a phase change of more than

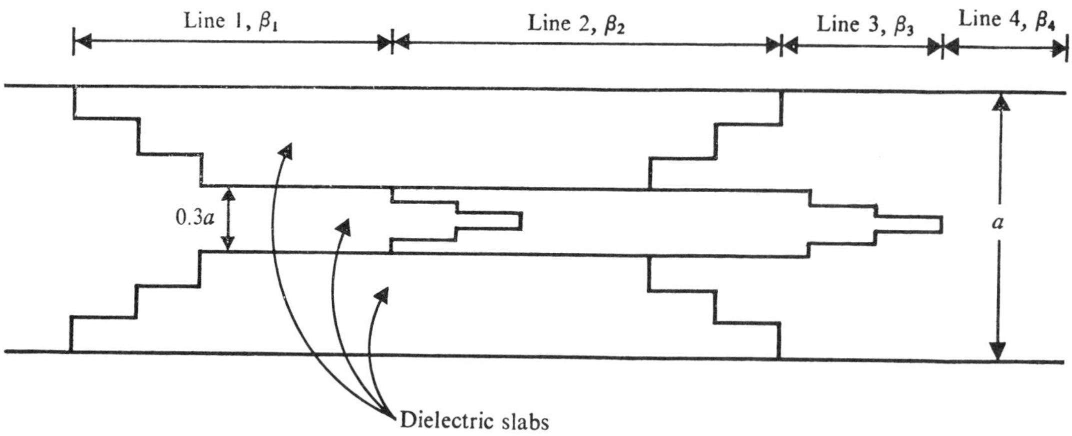

(a) Top view

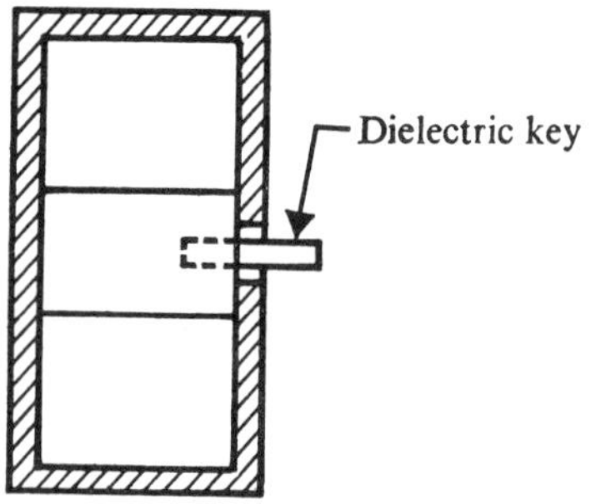

(b) End view

Fig. 8.9 Linear phase changer.

360°. Very accurate setting and reading are possible because of the differential action. For further details, see Collin[1].

Rotary Phase Changer

The rotary phase changer is a far better precision instrument than the linear phase changer. Its construction is similar to that of the rotary attenuator, except that the centre resistive card is replaced by a half-wave plate and the two outer resistive cards are replaced by quarter-wave plates (see Fig. 8.10). The half-wave plate produces a phase shift equal to twice the angle θ through which it is rotated. The quarter-wave plates convert a linearly-polarized TE_{11} mode into a circularly-polarized mode, and vice versa.

A quarter-wave plate is a device that produces a circularly-polarized wave when a linearly-polarized wave is incident on it. This circularly-polarized wave should have the x- and y-component of electric field, which are equal in magnitude but 90° apart in time phase. A quarter-wave plate may be constructed from a slab of dielectric material, as shown in Fig. 8.10b. When the TE_{11} mode is polarized parallel to the slab, the propagation constant $\bar{\beta}_1$ is greater than for the

[1]R. E. Collin, Waveguide phase changer, *Wireless Engineer*, **32**, 82, 1955.

case where the mode is polarized perpendicular to the slab, that is, $\bar{\beta}_1 > \bar{\beta}_2$, where $\bar{\beta}_2$ is the propagation constant for perpendicular polarization. The length l of a quarter-wave plate is chosen to obtain a differential phase change $(\bar{\beta}_1 - \bar{\beta}_2)l$ equal to 90°. The ends of the dielectric slab are tapered to reduce reflections to a negligible value. The half-wave plate is similar in construction, except that its length is increased to produce a differential phase change of 180°.

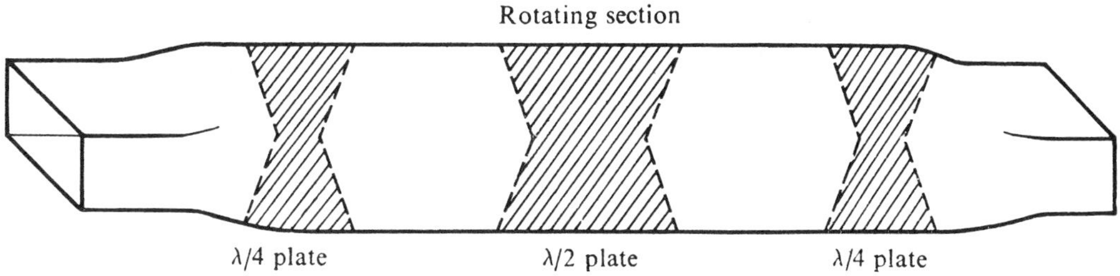

(a) Rotary phase changer

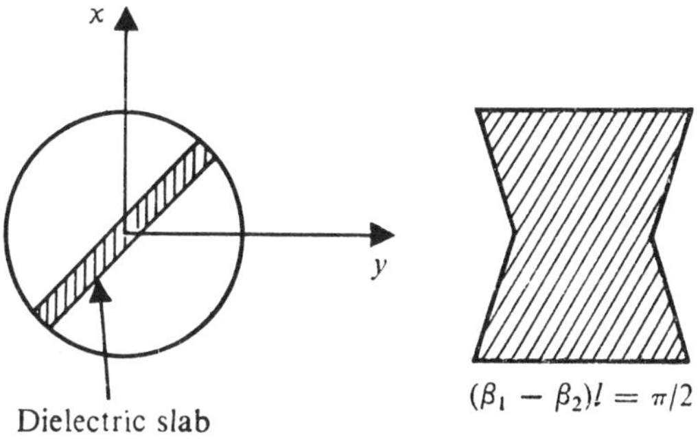

(b) Quarter-wave plate

Fig. 8.10 Phase changer.

In the rotary phase changer, the quarter-wave plates are oriented at a 45° angle in relation to the broad wall of the rectangular waveguide. The incoming TE_{11} mode will be decomposed into two modes polarized parallel and perpendicular to the quarter-wave plate, as shown in Fig. 8.11.

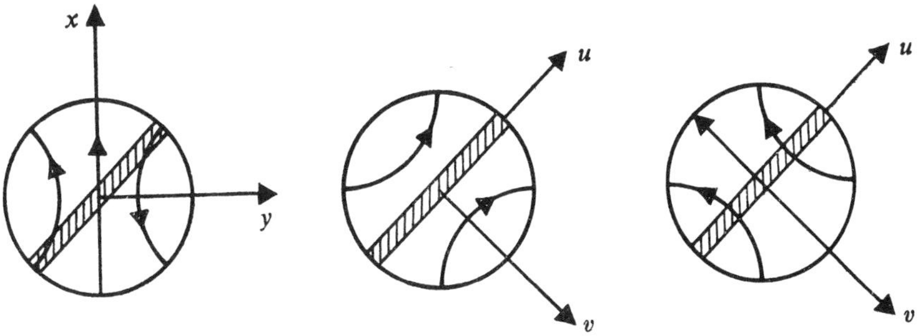

Fig. 8.11 Decomposition of incident TE_{11} mode.

The incident mode may be assumed to have an electric field given by

$$\boldsymbol{E} = \frac{J_1}{\rho}\boldsymbol{u}_\rho \cos\phi - \frac{p'_{11}}{a}J'_1\boldsymbol{u}_\phi \sin\phi. \tag{8.7}$$

Here, replace $\cos\phi$ by

$$\cos\left(\phi - \frac{\pi}{4} + \frac{\pi}{4}\right) = \frac{\sqrt{2}}{2}\left[\cos\left(\phi - \frac{\pi}{4}\right) - \sin\left(\phi - \frac{\pi}{4}\right)\right]$$

and $\sin\phi$ by

$$\frac{\sqrt{2}}{2}\left[\cos\left(\phi - \frac{\pi}{4}\right) + \sin\left(\phi - \frac{\pi}{4}\right)\right].$$

Then, $\boldsymbol{E}$ can be obtained as

$$\boldsymbol{E} = \boldsymbol{E}_1 + \boldsymbol{E}_2, \tag{8.8}$$

where

$$\boldsymbol{E}_1 = \frac{\sqrt{2}}{2}\left[\frac{J_1}{\rho}\boldsymbol{u}_\rho \cos\left(\phi - \frac{\pi}{4}\right) - \frac{p'_{11}}{a}J'_1\boldsymbol{u}_\phi \sin\left(\phi - \frac{\pi}{4}\right)\right], \tag{8.9}$$

$$\boldsymbol{E}_2 = -\frac{\sqrt{2}}{2}\left[\frac{J_1}{\rho}\boldsymbol{u}_\rho \sin\left(\phi - \frac{\pi}{4}\right) + \frac{p'_{11}}{a}J'_1\boldsymbol{u}_\phi \cos\left(\phi - \frac{\pi}{4}\right)\right]. \tag{8.10}$$

The field $\boldsymbol{E}_1$ is polarized parallel to the slab and $\boldsymbol{E}_2$ is polarized perpendicular to the slab. After propagation through the quarter-wave plate, these fields become

$$\boldsymbol{E}'_1 = \boldsymbol{E}_1 \exp(-j\bar{\beta}_1 l), \tag{8.11}$$

$$\boldsymbol{E}'_2 = \boldsymbol{E}_2 \exp(-j\bar{\beta}_2 l) = \boldsymbol{E}_2 \exp(-j\bar{\beta}_1 l) \exp[-j(\bar{\beta}_2 - \bar{\beta}_1)l] = j\boldsymbol{E}_2 \exp(-j\bar{\beta}_1 l) \tag{8.12}$$

since $(\bar{\beta}_2 - \bar{\beta}_1)l = -\pi/2$. Equations (8.11) and (8.12) show that $\boldsymbol{E}'_1$ and $\boldsymbol{E}'_2$ are two orthogonally-polarized TE_{11} modes of equal amplitude and 90° out of time phase, and hence constitute a circularly-polarized field.

We shall now consider the effect of the half-wave plate on the circularly-polarized field. Let the half-wave plate be rotated by an angle θ with respect to the quarter-wave plate, as shown in Fig. 8.12. The fields $\boldsymbol{E}'_1 + \boldsymbol{E}'_2$ may be expressed in terms of TE_{11} modes polarized parallel and perpendicular to the half-wave plate by changing the origin of the variable angle ϕ to $\pi/4 + \theta$; then we have

$$\boldsymbol{E}'_1 = \frac{\sqrt{2}}{2}\exp(-j\bar{\beta}_1 l)\left[\cos\theta\left\{\frac{J_1}{\rho}\boldsymbol{u}_\rho \cos\left(\phi - \theta - \frac{\pi}{4}\right) - \frac{p'_{11}}{a}J'_1\boldsymbol{u}_\phi \sin\left(\phi - \theta - \frac{\pi}{4}\right)\right\}\right.$$
$$\left.- \sin\theta\left\{\frac{J_1}{\rho}\boldsymbol{u}_\rho \sin\left(\phi - \theta - \frac{\pi}{4}\right) + \frac{p'_{11}}{a}J'_1\boldsymbol{u}_\phi \cos\left(\phi - \theta - \frac{\pi}{4}\right)\right\}\right], \tag{8.13}$$

$$\boldsymbol{E}'_2 = -j\frac{\sqrt{2}}{2}\exp(-j\bar{\beta}_1 l)\left[\cos\theta\left\{\frac{J_1}{\rho}\boldsymbol{u}_\rho \sin\left(\phi - \theta - \frac{\pi}{4}\right) + \frac{p'_{11}}{a}J'_1\boldsymbol{u}_\phi \cos\left(\phi - \theta - \frac{\pi}{4}\right)\right\}\right.$$
$$\left.+ \sin\theta\left\{\frac{J_1}{\rho}\boldsymbol{u}_\rho \cos\left(\phi - \theta - \frac{\pi}{4}\right) - \frac{p'_{11}}{a}J'_1\boldsymbol{u}_\phi \sin\left(\phi - \theta - \frac{\pi}{4}\right)\right\}\right] \tag{8.14}$$

by replacing $\cos(\phi - \pi/4)$ by $\cos(\phi - \pi/4 - \theta + \theta)$ and $\sin(\phi - \pi/4)$ by $\sin(\phi - \pi/4 - \theta + \theta)$.

The field polarized parallel to the half-wave plate has a ρ-component varying as $\cos(\phi - \theta - \pi/4)$, whereas the perpendicularly-polarized mode has a ρ-component of electric

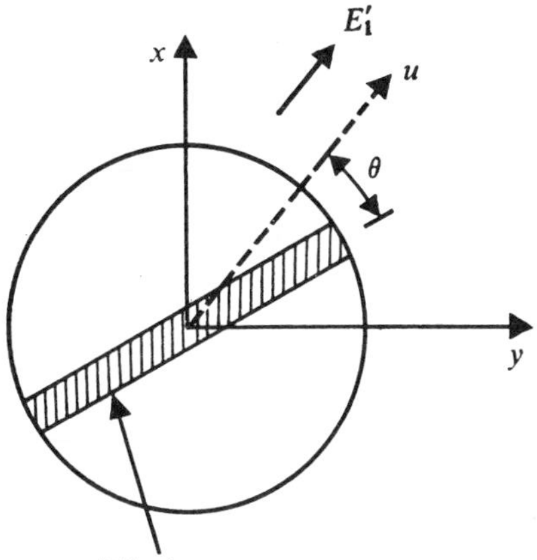

Fig. 8.12 Orientation of half-wave plate.

field varying as $\sin(\phi - \theta - \pi/4)$. Hence, from Eqs. (8.13) and (8.14), we obtain

$$\boldsymbol{E}_1' + \boldsymbol{E}_2' = \boldsymbol{E}_1'' + \boldsymbol{E}_2'',$$

where

$$\boldsymbol{E}_1'' = \frac{\sqrt{2}}{2} \exp(-j\bar{\beta}_1 l - j\theta)\left[\frac{J_1}{\rho}\boldsymbol{u}_\rho \cos\left(\phi - \theta - \frac{\pi}{4}\right) - \frac{p_{11}'}{a} J_1' \boldsymbol{u}_\phi \sin\left(\phi - \theta - \frac{\pi}{4}\right)\right], \tag{8.15}$$

$$\boldsymbol{E}_2'' = -j\frac{\sqrt{2}}{2} \exp(-j\bar{\beta}_1 l - j\theta)\left[\frac{J_1}{\rho}\boldsymbol{u}_\rho \sin\left(\phi - \theta - \frac{\pi}{4}\right) + \frac{p_{11}'}{a} J_1' \boldsymbol{u}_\phi \cos\left(\phi - \theta - \frac{\pi}{4}\right)\right]. \tag{8.16}$$

After propagating through the half-wave plate of length $2l$, this field becomes

$$\boldsymbol{E}_3 = \boldsymbol{E}_1'' \exp(-2j\bar{\beta}_1 l), \tag{8.17}$$

$$\boldsymbol{E}_4 = \boldsymbol{E}_2'' \exp(-2j\bar{\beta}_2 l) = \boldsymbol{E}_2'' \exp[-2j(\bar{\beta}_2 - \bar{\beta}_1)l - 2j\bar{\beta}_1 l] = -\boldsymbol{E}_2'' \exp(-2j\bar{\beta}_1 l) \tag{8.18}$$

since $2(\bar{\beta}_2 - \bar{\beta}_1)l = \pi$.

The new field thus obtained may now be decomposed once again into two TE_{11} modes polarized parallel and perpendicular to the quarter-wave plate in the output guide. If this plate is parallel to the input quarter-wave plate, we can obtain the required decomposition by referring the variable angle ϕ to $\pi/4$ as the origin. Then

$$\boldsymbol{E}_3 + \boldsymbol{E}_4 = \boldsymbol{E}_3' + \boldsymbol{E}_4', \tag{8.19}$$

where

$$\boldsymbol{E}_3' = \frac{\sqrt{2}}{2} \exp(-3j\bar{\beta}_1 l - 2j\theta)\left[\frac{J_1}{\rho}\boldsymbol{u}_\rho \cos\left(\phi - \frac{\pi}{4}\right) - \frac{p_{11}'}{a} J_1' \boldsymbol{u}_\phi \sin\left(\phi - \frac{\pi}{4}\right)\right], \tag{8.20}$$

$$\boldsymbol{E}_4' = j\frac{\sqrt{2}}{2} \exp(-3j\bar{\beta}_1 l - 2j\theta)\left[\frac{J_1}{\rho}\boldsymbol{u}_\rho \sin\left(\phi - \frac{\pi}{4}\right) + \frac{p_{11}'}{a} J_1' \boldsymbol{u}_\phi \cos\left(\phi - \frac{\pi}{4}\right)\right]. \tag{8.21}$$

After propagating through the second quarter-wave plate, the output field becomes

$$\boldsymbol{E}_0 = \boldsymbol{E}_3'' + \boldsymbol{E}_4'', \tag{8.22}$$

where

$$\boldsymbol{E}_3'' = \boldsymbol{E}_3' \exp(-j\beta_1 l), \tag{8.23}$$

$$\boldsymbol{E}_4'' = \boldsymbol{E}_4' \exp(-j\beta_2 l) = j\boldsymbol{E}_4' \exp(-j\beta_1 l). \tag{8.24}$$

Hence

$$\boldsymbol{E}_0 = \exp(-4j\beta_1 l - 2j\theta)[\frac{J_1}{\rho}\boldsymbol{u}_\rho \cos\phi - \frac{p'_{11}}{a}J'_1\boldsymbol{u}_\phi \sin\phi] \tag{8.25}$$

which is a linearly-polarized TE_{11} mode having the same direction of polarization as the incident field given by Eqs. (8.8)–(8.10). Note that the phase has been changed by an amount $4\beta_1 l + 2\theta$. Thus, rotation of the half-wave plate through an angle θ changes the phase of the transmitted wave by an amount 2θ. This simple dependence of the phase change or a mechanical rotation is the chief advantage of the rotary phase changer.

8.6 Microwave Hybrid Junction (Magic-Tee or Twin-Tee)

Examples of four-port microwave junctions used extensively at microwave frequencies are microwave hybrid junctions and directional couplers. In this section, we shall consider the hybrid junction, which is also called the magic-tee, and in the next section, we shall deal with the directional coupler.

The equivalent circuit of a four-port microwave junction should have 10 independent circuit parameters, because the impedance matrix has 16 elements of which only 10 are independent due to the symmetry property. An equivalent circuit which is normally used is shown in Fig. 8.13. This circuit is in the form of a Wheatstone bridge. Such a circuit has many important properties. It is possible to decouple ports 1 and 3 by properly balancing the bridge by altering the impedances connected to ports 2 and 4. These impedances may consist of matched terminations and reflecting irises or probes to tune out the reactance. Also, it is necessary to connect proper irises or probes in ports 1 and 3 so that no power is coupled between these two ports. If irises are placed in such a way that reflection does not occur when waves are incident on ports 1 and 2, then the four-port junction becomes a directional coupler.

A microwave rectangular waveguide hybrid or magic-tee is shown in Fig. 8.14. By considering an imaginary plane of symmetry which divides the magic-tee symmetrically, we can determine the couplings among the different arms. If a wave in the dominant TE_{10} mode is incident on the junction in arm 4, then the magic-tee is symmetrical about the plane of symmetry with respect to this wave, and hence the powers transmitted to arms 1 and 3 will be the same. If E_m^n represents the transmitted electric field intensity to the n-th arm when the incident wave is in the m-th arm, then

$$E_4^1 = E_4^3. \tag{8.26}$$

Besides, it can be seen that no power is transmitted to arm 2 from arm 4. On the other hand, if a TE_{10} wave is incident in arm 2, it has an odd symmetry about the plane of symmetry and therefore excites fields having odd symmetries in arms 1 and 3. Hence

$$E_2^1 = -E_2^3. \tag{8.27}$$

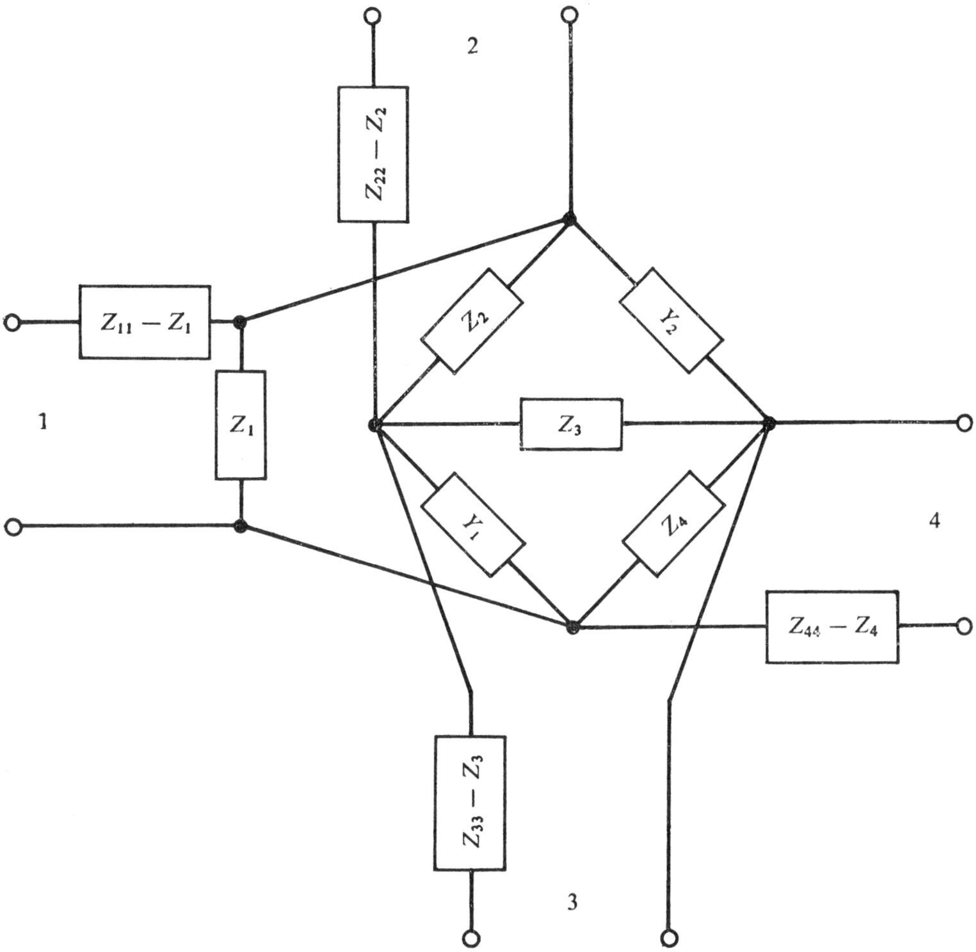

Fig. 8.13 Equivalent circuit of four-port junction.

Also, power incident in arm 2 cannot be transmitted to arm 4. Therefore

$$E_4^2 = E_2^4 = 0. \tag{8.28}$$

Even though $E_4^1 = E_4^3$ and $E_2^1 = -E_2^3$, the coupling coefficient may not be exactly one-half since arms 2 and 4 may have reflections. To ensure a coupling coefficient of exactly one-half, which is desirable, the junction has to be tuned by irises or probes.

8.7 Directional Couplers

Figure 8.15 shows a typical bridge which is an equivalent of a four-port microwave junction. Such a bridge can be balanced by adjusting the impedances connected to ports 2 and 4. Now, if ports 1 and 3 are connected to matched loads, and if suitable irises are used to tune out the junction reactance, no power will be coupled between these two ports. Also, it is possible to place irises so that ports 1 and 2 are matched to incident waves such that there is no coupling

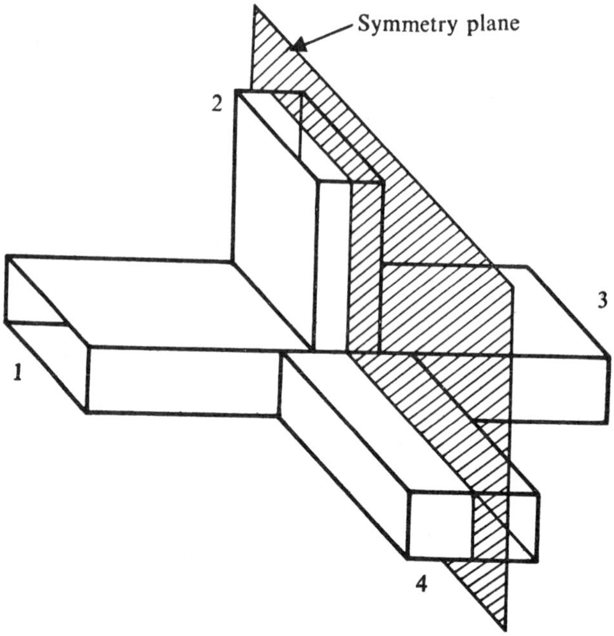

Fig. 8.14 Microwave hybrid or magic-tee.

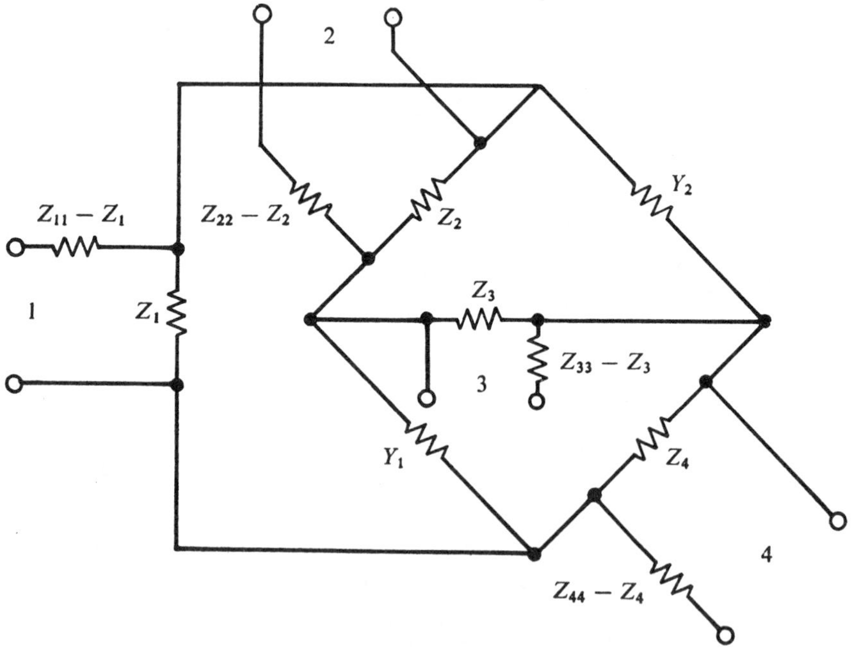

Fig. 8.15 Bridge equivalent of four-port junction.

between ports 2 and 4. Under these conditions, the four-port junction becomes a directional coupler.

A directional coupler can be defined as a four-port junction such that, when all the four ports shown in Fig. 8.16 are terminated by their characteristic impedance or, in other words, matched, there is no reflection in either port 1 or port 2, and there is no coupling between ports 1 and 3 and between ports 2 and 4. Under these circumstances, a wave incident on port 1 propagates through ports 2 and 4, and a wave incident on port 2 leaves by ports 3 and 1. Such a device can be used to monitor power flow in two directions through ports 1 and 2. The powers absorbed by the matched loads in ports 3 and 4 indicate the power flow in the direction of ports

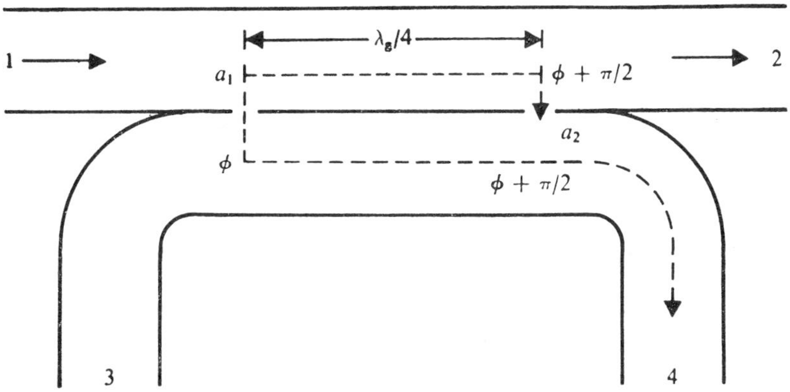

Fig. 8.16 Waveguide directional coupler.

2 and 1, respectively. If the ratios of power flow between ports 1 and 4 and between ports 2 and 3 are known, then the relative amplitudes of the incident and reflected waves in a line connected to any load can be determined by inserting the directional coupler into the line. These two ratios of power flow are known as the *coupling coefficients* of the directional coupler.

A portion of the energy of the wave incident on the junction from port 1 escapes into the bent waveguide forming ports 3 and 4 through hole a_1. The energy that remains travels to hole a_2 in the main waveguide (1, 2), and a portion of it again escapes into the bent waveguide. If the magnitudes of the escaping energy in holes a_1 and a_2 are the same, and if the distance between a_1 and a_2 is $\lambda_g 4$, where λ_g is the guide wavelength, then the two escaped energies are reinforced at port 4, because they arrive with equal phase. On the other hand, the energies of the two waves that escape through holes a_1 and a_2 cancel each other at port 3, because they are 180° out of phase with each other. Similarly, a wave incident on the junction from port 2 travels to ports 1 and 3, but energy does not travel to port 4. In general, the leakage of energy through holes a_1 and a_2 is kept quite small. Hence, a directional coupler can be regarded as a hybrid junction in which the division of power between adjacent ports is unequal. This is the principle of operation of a directional coupler.

By placing matched receivers in ports 3 and 4, it is possible to obtain the value of the reflection coefficient as well as the transmission coefficient of any load connected to port 2 if the incident power is from port 1. However, this method of measurement of the reflection coefficient is not very accurate unless port 4 is accurately matched.

Figure 8.17 shows an alternative arrangement for measuring the transmission and reflection coefficients. Such an arrangement is called a double-directional coupler. In this arrangement, ports 4 and 5 are matched while detectors are placed in ports 3 and 6. The detector in port 3 monitors the reflected power, whereas the detector in port 6 monitors the transmitted power if the incident power is from port 1. In the double-directional coupler, it is possible to select different coupling coefficients for the transmitted and reflected powers.

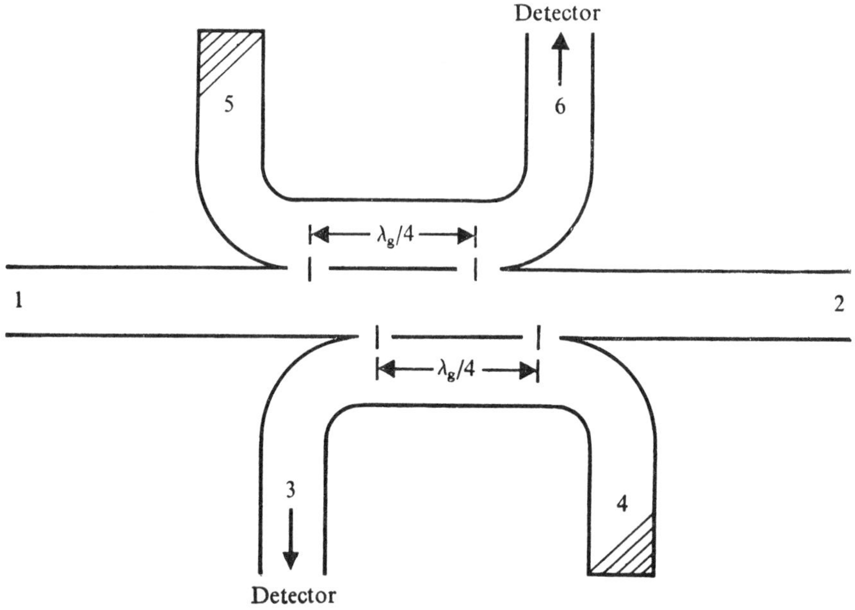

Fig. 8.17 Double-directional coupler.

A directional coupler can also be used as a standing-wave detector, but only the amplitude, and not the phase, can be measured.

Some typical waveguide directional couplers are shown in Fig. 8.18.

In the *reverse-coupling directional coupler* (see Fig. 8.18a), the auxiliary waveguide is coupled to the main waveguide by two slots so that the wave in the auxiliary guide travels in a direction opposite to that in the main waveguide. The coupling slots are situated on opposite sides of the centre line of the broad side of one waveguide and that of the narrow side of the other, as shown in the figure, and these slots are excited by two oppositely directed longitudinal magnetic fields.

In the *two-hole directional coupler* shown in Fig. 8.18b, the holes are smaller and are excited by the longitudinal magnetic fields which excite equivalent magnetic dipoles within the holes. The radiations from the holes into the auxiliary guide are equal in magnitude but 90° out of phase. They are therefore reinforced inside the auxiliary guide in the direction of the original wave and cancelled in the opposite direction.

In the *Bethe-hole directional coupler* illustrated in Fig. 8.18c, the coupling hole is located at the centre of the broad face of the main waveguide and is therefore excited by both the normal electric field and normal transverse magnetic field for the dominant TE_{10} mode. The relative magnitudes of the electric and magnetic couplings are adjusted by rotating the auxiliary guide with respect to the main guide.

Figure 8.18d shows a *multihole directional coupler*.

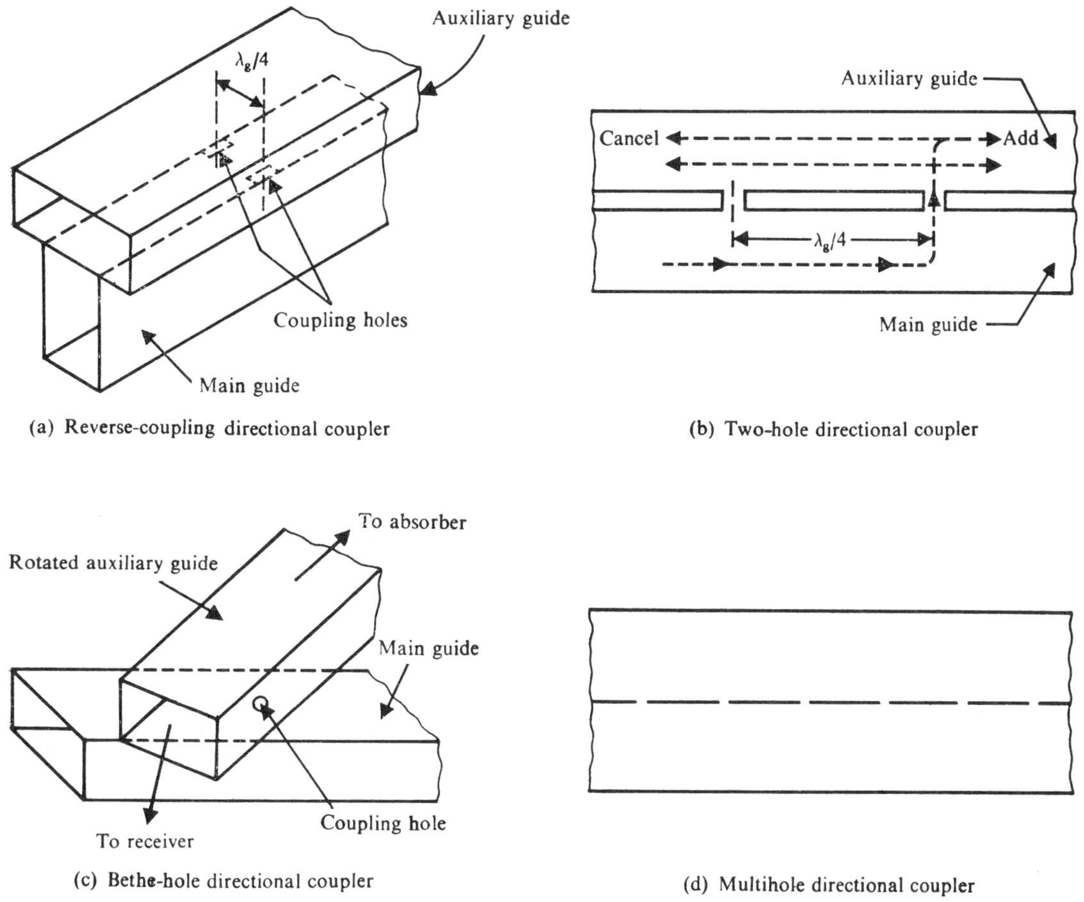

Fig. 8.18 Typical waveguide directional couplers.

8.8 Concept of Nonreciprocity

In the early days of electrical engineering, linear passive networks which obeyed the theorem of reciprocity were in vogue. The reciprocity theorem states that, when a network is terminated by a receiver at one end and a generator at the other end, as shown in Fig. 8.19, the signal received at the receiver remains unchanged when the positions of the receiver and generator are interchanged, if it is assumed that the amplitude and phase of the input signal do not change when the positions of the receiver and generator are interchanged. Such a network is called a *reciprocal network*. A network which does not obey the reciprocity theorem may be defined as a *nonreciprocal network*. Today, we find various types of nonreciprocal networks. They are not only used in various electrical communication systems, but also have numerous other uses. They operate over certain restricted frequency ranges.

The nonreciprocal networks help communication engineers and others to perform system functions which are impossible with passive reciprocal networks. One of these functions, for example, is to build a one-way transmission system at microwave frequencies. Such a microwave transmission system allows an electromagnetic wave to propagate in one direction with negligible

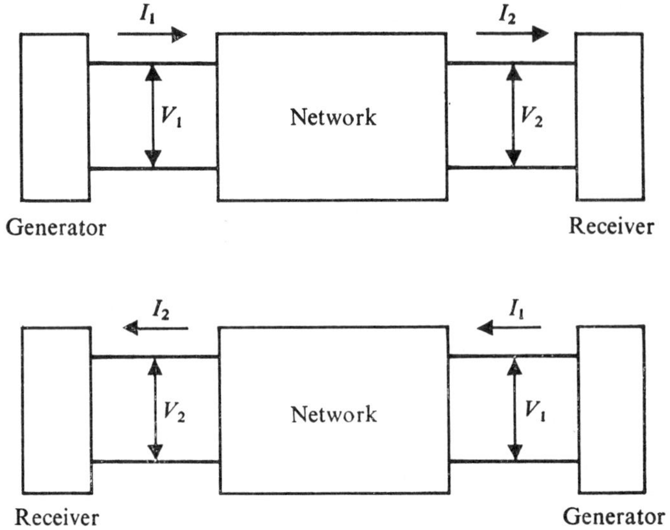

Fig. 8.19 Illustration of reciprocity theorem.

attenuation, but absorbs almost completely the energy contained in the wave which is travelling in the opposite direction through the system. An example is a microwave system having different phase-shift characteristics for two oppositely directed electromagnetic waves. Another function of the nonreciprocal networks is to perform a duplexing operation in which the same antenna can be used for transmitting as well as receiving microwave signals at the same time and at the same frequency. Also, directional couplers having peculiar characteristics, not possessed by ordinary directional couplers, can be designed by using nonreciprocal networks.

8.9 Microwave Properties of Ferrites

Most microwave circuit elements are reciprocal, and so it is necessary to investigate some elements which must be incorporated in a microwave network to make it nonreciprocal.

A nonreciprocal element should have some discriminatory property such that the transmission characteristics through the element along different directions differ in amplitude, phase, or polarization, or in any combination of these properties. Hence, any material whose physical properties (e.g., permeability or permittivity) are not the same in all directions can be used as a reciprocal element. For example, an ionized medium such as plasma in the presence of a magnetic field has a tensor permittivity, and hence it can be used as a nonreciprocal element. Similarly, a ferrite material in the presence of a magnetic field has a tensor permeability, and can be used for this purpose. Ferrites are more convenient to handle and are therefore used extensively in the design of microwave nonreciprocal networks.

Ferrites are ceramic-like magnetic materials with resistivities of the order of 10^{12} times larger than the resistivity of iron and with dielectric constants of about 10 to 15. This low-loss property of ferrites is very useful in microwave circuits. An electromagnetic wave at microwave frequencies can propagate through a ferrite medium with very little attenuation. However, its transmission through a medium such as iron or other magnetic materials is almost impossible. The relative permeability of ferrites is several thousands. Commercially, ferrites are derived from loadstone or magnetite ($Fe^{2+}O \cdot Fe^{3+}O_3$) by substituting, in place of the divalent iron Fe^{2+}, other metal atoms, e.g., those of manganese, magnesium, nickel, copper, cobalt, zinc, and cadmium. If the substituting atoms are divalent and are approximately of the same diameter as the replaced iron atoms, the basic spinel-type crystal structure of the magnetite remains unchanged. But the new material now possesses a greatly increased value of permeability and resistivity. Figure 8.20 shows the spinel-type structure of a typical ferrite crystal.

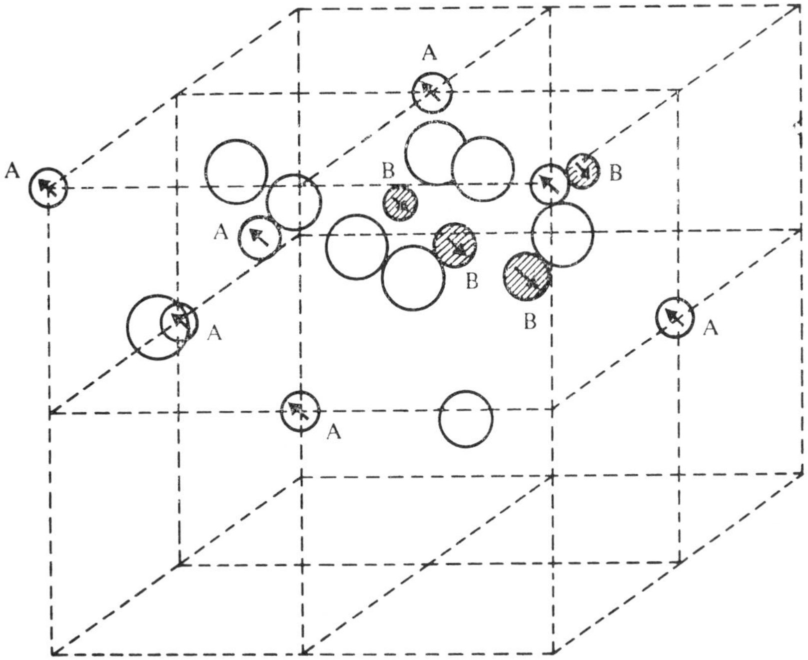

Fig. 8.20 Spinel structure of ferrites. [Positions of ions are different in two octants sharing a cube face and are shown for only two such octants. A and B are metallic ions surrounded by four and six oxygen ions (large circles), respectively.]

A distinguishing property of ferrites is that, under the influence of an external magnetic field, they exhibit different permeability constants in different directions. Hence, permeability has to be expressed as a tensor quantity.

8.10 Microwave Propagation in Ferrites

The magnetic properties of ferrites arise mainly from the magnetic dipole moment associated with the electron spin. By treating the spinning electron as a gyroscopic top, a classical picture of the magnetization process and, in particular, the anisotropic properties may be obtained.

The charge of the electron is $-q_e = -1.602 \times 10^{-19}$ coulomb, its mass $m_e = 9.107 \times 10^{-31}$ kg, its angular momentum $P = \frac{1}{2}\hbar = 0.527 \times 10^{-34}$ joule-m, where $\hbar$ = Planck's constant divided by 2π, and its magnetic dipole moment $m = q_e\hbar/(2m_e) = 9.27 \times 10^{-24}$ amp-m^2. For the electron, the angular momentum $\boldsymbol{P}$ and the magnetic dipole moment $\boldsymbol{m}$ are antiparallel. The ratio of the magnetic moment to the angular momentum is called the gyromagnetic ratio γ, i.e.,

$$\gamma = m/P. \tag{8.29}$$

If an electron is located in a uniform static magnetic field $\boldsymbol{B}_0$, then a torque $\boldsymbol{T}$ given by

$$\boldsymbol{T} = \boldsymbol{m} \times \boldsymbol{B}_0 = -\gamma \boldsymbol{P} \times \boldsymbol{B}_0 \tag{8.30}$$

is exerted on the dipole moment. This gives the equation of motion. The torque causes the dipole axis to precess about an axis parallel to $\boldsymbol{B}_0$, as shown in Fig. 8.21. The rate of change of angular

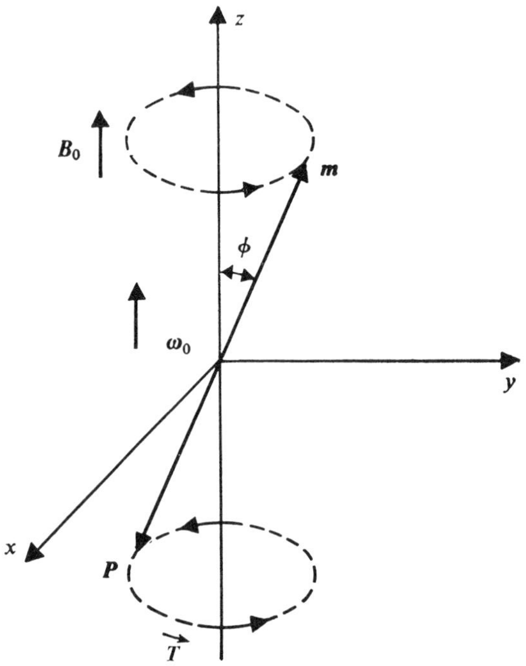

Fig. 8.21 Free precession of spinning electron.

momentum is equal to the torque, and hence the equation of motion is

$$\frac{d\boldsymbol{P}}{dt} = \boldsymbol{T} = -\gamma \boldsymbol{P} \times \boldsymbol{B} = \boldsymbol{\omega}_0 \times \boldsymbol{P} \tag{8.31a}$$

or

$$\gamma P B_0 \sin \phi = \omega_0 P \sin \phi = m B_0 \sin \phi, \tag{8.31b}$$

where $\boldsymbol{\omega}_0$ is the vector-precession angular velocity directed along $\boldsymbol{B}_0$ and ϕ is the angle between $\boldsymbol{m}$ and $\boldsymbol{B}_0$. For free precession, the angular velocity ω_0 is given by $\omega_0 = \gamma B_0$ and is independent of the angle ϕ. The angular velocity ω_0 is called the *Larmor frequency*.

If a small a-c magnetic field is superimposed on the static field $\boldsymbol{B}_0$, the magnetic dipole moment undergoes a forced precession. The case of an a-c magnetic field circularly polarized in the plane perpendicular to $\boldsymbol{B}_0$ is of particular interest. Let this circularly-polarized a-c magnetic field be of the form

$$\boldsymbol{B}_1^- = B_1(\boldsymbol{u}_x + j\boldsymbol{u}_y)e^{j\omega t}. \tag{8.32a}$$

If B_1 is real, the physical field is given by

$$\boldsymbol{B}_1^- = B_1[\text{Re}\,(\boldsymbol{u}_x + j\boldsymbol{u}_y)e^{j\omega t}] = B_1(\boldsymbol{u}_x \cos \omega t - \boldsymbol{u}_y \sin \omega t). \tag{8.32b}$$

The resultant field has a constant magnitude B_1, but the orientation of the field in space rotates with time. The resultant field vector makes an angle

$$\tan^{-1}(B_y/B_x) = -\tan^{-1}(\tan \omega t) = -\omega t$$

at time t with the x-axis, and hence rotates at the rate ω, as shown in Fig. 8.22a. Looking in the z-direction of propagation, if the rotation is clockwise, it is called right circularly polarized

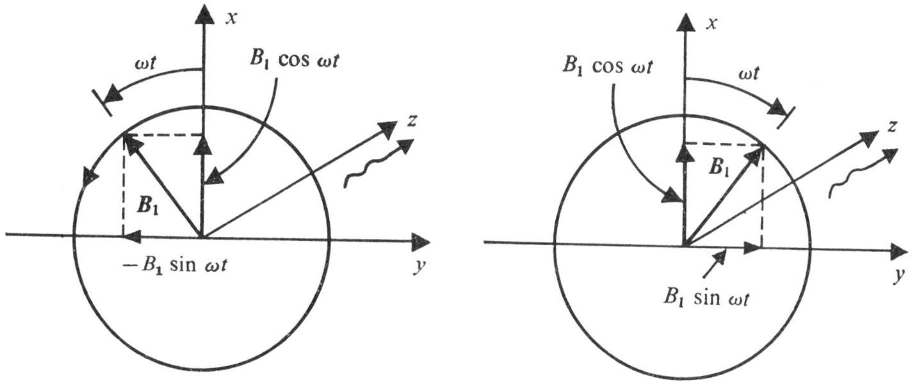

(a) Left or negative circular polarization (b) Right or positive circular polarization

Fig. 8.22 Magnetic field of circularly-polarized wave.

(Fig. 8.22b); and if it is anticlockwise, it is called left circularly polarized (Fig. 8.22a). The magnetic field represented by Eqs. (8.32) is left circularly polarized. A right circularly-polarized field is represented by

$$\boldsymbol{B}_1^+ = B_1(\boldsymbol{u}_x - j\boldsymbol{u}_y)e^{j\omega t} \tag{8.33a}$$

or by

$$\boldsymbol{B}_1^+ = B_1(\boldsymbol{u}_x \cos \omega t + \boldsymbol{u}_y \sin \omega t). \tag{8.33b}$$

If a left circularly-polarized a-c magnetic field is superimposed on the static field $\boldsymbol{B}_0 = B_0\boldsymbol{u}_z$, the resultant total field $\boldsymbol{B}_t$ is inclined at an angle $\theta = \tan^{-1}(B_1/B_0)$ with the z-axis and rotates at a rate ω about the z-axis, as illustrated in Fig. 8.23a. When steady-state conditions exist, the magnetic dipole axes are forced to precess about the z-axis at the same rate. Thus, to obtain a torque to cause precession in a counter-clockwise direction, the precession angle ϕ must be less than θ. Then, the equation of motion is

$$\boldsymbol{T} = \boldsymbol{m} \times \boldsymbol{B}_t = -\gamma \boldsymbol{P} \times \boldsymbol{B}_t = d\boldsymbol{P}/dt = -\omega \boldsymbol{u}_z \times \boldsymbol{P} \tag{8.34a}$$

or

$$-\gamma P B_t \sin(\theta - \phi) = -\omega P \sin\phi. \tag{8.34b}$$

Replacing $B_t \sin\theta$ by B_1 and $B_t \cos\theta$ by B_0, and solving for $\tan\phi$, we obtain

$$\tan\phi = \frac{\gamma B_1}{\gamma B_0 + \omega} = \frac{\gamma B_1}{\omega_0 + \omega}. \tag{8.35}$$

The component of $\boldsymbol{m}$, which rotates in synchronism with $\boldsymbol{B}_1^-$ in the xy-plane for the left circularly-polarized a-c field, is $m^- = m \sin\phi = m_0 \tan\phi$, where $m_0 = m\cos\phi$ is the z-directed component of $\boldsymbol{m}$. If B_1 is very small as compared to B_0, the angle ϕ is small so that m_0 is approximately equal

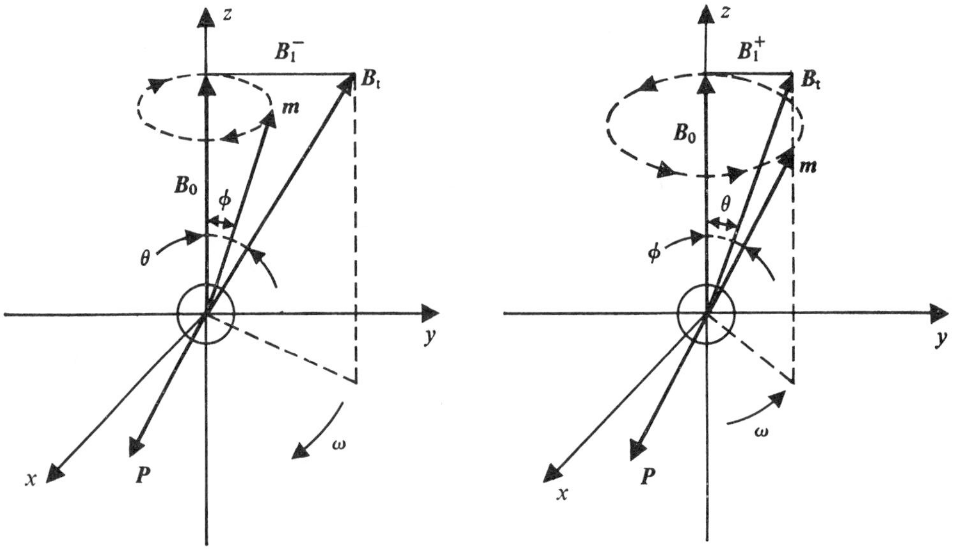
z
B_1^-
B_t
m
B_0
φ
θ
y
x
P
ω
(a) Left circularly-polarized field
z
B_1^+
B_t
B_0
m
θ
φ
y
x
P
ω
(b) Right circularly-polarized field

Fig. 8.23 Forced precession of spinning electron.

to $\boldsymbol{m}$. Thus, from Eq. (8.35), we obtain

$$m^- = m_0 \tan\phi = \frac{\gamma m_0 B_1}{\omega_0 + \omega}. \tag{8.36}$$

On the other hand, if there is a positive or right circularly-polarized a-c field superimposed on the static field $\boldsymbol{B}_0$, the forced precession is in a clockwise sense about the z-axis. Only if the angle ϕ is greater than the angle θ, then a torque that gives precession in this direction is obtained, as shown in Fig. 8.23b. In this case, the equation of motion (8.31) gives

$$\gamma B_t \sin(\phi - \theta) = \omega \sin\phi,$$

and hence

$$\tan\phi = \frac{\gamma B_1}{\omega_0 - \omega}. \tag{8.37}$$

The component of magnetization in the xy-plane rotating in synchronism with the positive circularly-polarized a-c field is

$$m^{+} = m_0 \tan \phi = \frac{\gamma m_0 B_1}{\omega_0 - \omega}. \tag{8.38}$$

The foregoing discussion has brought out the main features of the motion of a single spinning electron in a magnetic field which is the superimposition of a static field along the z-axis and a small circularly-polarized a-c field in the xy-plane.

A ferrite material may be thought of as consisting of N effective spinning electrons per unit volume. The distance between the electrons is of atomic dimensions, and hence the density of magnetic dipoles per unit volume may be considered as a smeared-out continuous distribution from a macroscopic viewpoint. The total magnetic dipole moment per unit volume is $\boldsymbol{M} = N\boldsymbol{m}$. If the static field $\boldsymbol{B}_0$ is large enough to cause saturation of the magnetization, then $\boldsymbol{M} = \boldsymbol{M}_s$. In such a saturated ferrite, all the spins are very tightly coupled so that the whole sample acts essentially as a large single magnetic dipole. The total internal magnetic field $\boldsymbol{B}$ is related to the magnetization $\boldsymbol{M}_s$ as

$$\boldsymbol{B} = \mu_0(\boldsymbol{H}_0 + \boldsymbol{M}_s).$$

The torque acting on $\boldsymbol{M}_s$ is due only to the field $\boldsymbol{B}_0 = \mu_0 \boldsymbol{H}_0$, because $\mu_0 \boldsymbol{M}_s \times \boldsymbol{M}_s = 0$. Therefore, if $\boldsymbol{H}_t$ is the total magnetic field intensity, which is the sum of the static and a-c fields, then $\mu_0 \boldsymbol{H}_t$ is the field producing the torque in the equation of motion. Thus

$$d\boldsymbol{M}/dt = -\gamma(\boldsymbol{M} \times \boldsymbol{B}) = -\gamma\mu_0 \boldsymbol{M} \times (\boldsymbol{H} + \boldsymbol{M}) = -\gamma\mu_0 \boldsymbol{M} \times \boldsymbol{H}. \tag{8.39}$$

If $\boldsymbol{H}_1^{\pm}$ is a circularly-polarized a-c field, and if the magnetic intensity in the ferrite is $\boldsymbol{H}_0 + \boldsymbol{H}_1^{\pm}$, then the resultant a-c magnetization will be given by expressions similar to Eqs. (8.36) and (8.38), with m_0 replaced by $M_0 = Nm_0$. The total a-c magnetic field in the xy-plane is the field $\mu_0 \boldsymbol{H}_1^{\pm} = \boldsymbol{B}_1^{\pm}$, together with the contribution from a-c magnetization. Therefore, the total a-c fields for positive and negative circular polarization are

$$\boldsymbol{B}^{+} = \mu_0 \boldsymbol{M}^{+} + \boldsymbol{B}_1^{+} = \mu_0(N\boldsymbol{m} + \boldsymbol{H}_1^{+}) = \mu_0\left(1 + \frac{\mu_0 \gamma M_0}{\omega_0 - \omega}\right)\boldsymbol{H}_1^{+}, \tag{8.40a}$$

$$\boldsymbol{B}^{-} = \mu_0 \boldsymbol{M}^{-} + \boldsymbol{B}_1^{-} = \mu_0\left(1 + \frac{\mu_0 \gamma M_0}{\omega_0 + \omega}\right)\boldsymbol{H}_1^{-}, \tag{8.40b}$$

where

$$M_0 = M \cos \phi, \qquad B_1 = \mu_0 H_1, \qquad \boldsymbol{H}_1^{+} = H_1(\boldsymbol{u}_x - j\boldsymbol{u}_y), \qquad \boldsymbol{H}_1^{-} = H_1(\boldsymbol{u}_x + j\boldsymbol{u}_y).$$

The quantity M may be replaced by the saturation magnetization M_s in the ferrite since the static field B_0 is usually large enough to cause saturation.

If $B_1 \ll B_0$, so that $M_0 \approx M = M_s$, then the effective permeabilities for positive and negative circularly-polarized a-c fields are

$$\mu_{+} = \mu_0\left(1 + \frac{\gamma\mu_0 M_s}{\omega_0 - \omega}\right), \tag{8.41a}$$

$$\mu_{-} = \mu_0\left(1 + \frac{\gamma\mu_0 M_s}{\omega_0 + \omega}\right). \tag{8.41b}$$

Therefore, plane positive circularly, and plane negative circularly-polarized TEM waves travelling in the direction of the static field $\boldsymbol{B}_0$ will have the propagation constants

$$\beta_+ = \omega\sqrt{\epsilon\mu_+}, \tag{8.42a}$$

$$\beta_- = \omega\sqrt{\epsilon\mu_-}, \tag{8.42b}$$

where ϵ is the dielectric constant of the ferrite.

If it cannot be assumed that $B_1 \ll B_0$, then M_0 cannot be replaced by M_s. Instead of Eqs. (8.35) and (8.37), we may solve for sin ϕ and obtain

$$\sin\phi = \frac{\tan\phi}{(1+\tan^2\phi)^{1/2}} = \frac{\gamma B_1}{[(\gamma B_1)^2 + (\omega_0+\omega)^2]^{1/2}}, \tag{8.43a}$$

$$\sin\phi = \frac{\gamma B_1}{[(\gamma B_1)^2 + (\omega_0-\omega)^2]^{1/2}}. \tag{8.43b}$$

The magnetizations M^+ and M^- are given by

$$M^+ = M_s \sin\phi = \frac{\gamma\mu_0 M_s H_1}{[(\gamma\mu_0 H_1)^2 + (\omega_0-\omega)^2]^{1/2}}, \tag{8.44a}$$

$$M^- = \frac{\gamma\mu_0 M_s H_1}{[(\gamma\mu_0 H_1)^2 + (\omega_0+\omega)^2]^{1/2}}. \tag{8.44b}$$

Thus, it can be seen that a-c magnetization depends nonlinearly on the a-c field strength H_1 and, therefore, under large-signal conditions, μ_+ and μ_- become functions of the applied a-c field strength. This nonlinear behaviour of ferrites under large-signal conditions results in the generation of harmonics of the fundamental frequency ω. Hence, this property may be used in harmonic generators.

If $B_1 \ll B_0$, then $\gamma B_1 \ll \gamma B_0 = \omega_0$, and Eq. (8.44b) approximately becomes

$$M^- = \frac{\gamma\mu_0 M_s H_1}{\omega_0+\omega}. \tag{8.45a}$$

If ω is not very close to the resonant frequency ω_0, then Eq. (8.44a) approximately becomes

$$M^+ = \frac{\gamma\mu_0 M_s H_1}{\omega_0-\omega}. \tag{8.45b}$$

These values of M^- and M^+ lead to the expressions for μ_+ and μ_- given by Eqs. (8.41).

In any available ferrite, damping effects are always present so that M^+ remains finite and small as compared with M_s even when $\omega = \omega_0$. Hence, for small-signal conditions, it may be assumed that $M_0 \approx M_s$ in an actual ferrite medium.

8.11 Permeability and Susceptibility Tensors in Ferrite Medium

Let a static magnetic field $\boldsymbol{B}_0 = \mu_0\boldsymbol{H}_0 = B_0\boldsymbol{u}_z$ be applied in an infinite unbounded ferrite medium. Let $\boldsymbol{M}_s$ be the magnetization per unit volume in the ferrite when no time-varying magnetic field is applied. Let $\boldsymbol{M}$ be a time-varying component of magnetization when a time-varying magnetic field $\mu_0\boldsymbol{H}$ is applied. The equation of motion for the total magnetization per unit volume is similar to that for a single electron, and hence

$$\frac{d(\boldsymbol{M}_s + \mathcal{M})}{dt} = \frac{d\mathcal{M}}{dt} = -\gamma[\boldsymbol{M}_s + \mathcal{M}) \times (\boldsymbol{B}_0 + \mu_0 \mathcal{H})]$$

$$= -\gamma\mu_0[\boldsymbol{M}_s \times \boldsymbol{H}_0 + \boldsymbol{M}_s \times \mathcal{H} + \mathcal{M} \times \boldsymbol{H}_0 + \mathcal{M} \times \mathcal{H}]. \tag{8.46}$$

Under small-signal conditions,

$$|\mathcal{M}| \ll |\boldsymbol{M}_s|, \qquad |\mathcal{H}| \ll |\boldsymbol{H}_0|$$

and the nonlinear term $\mathcal{M} \times \mathcal{H}$ in Eq. (8.46) is negligible. Then, the equation of motion becomes

$$d\mathcal{M}/dt = -\gamma(\mu_0 \boldsymbol{M}_s \times \mathcal{H} + \mathcal{M} \times \boldsymbol{B}_0) \tag{8.47}$$

which is a linearized equation. $\boldsymbol{M}_s \times \boldsymbol{B}_0 = 0$, because the saturation magnetization is in the same direction as that of the applied static field.

Let the time dependence be $e^{j\omega t}$, and let $\mathcal{M}$ and $\mathcal{H}$ be represented by the phasors $\boldsymbol{M}$ and $\boldsymbol{H}$. From Eq. (8.47), we then obtain

$$j\omega \boldsymbol{M} + \gamma \boldsymbol{M} \times \boldsymbol{B}_0 = j\omega \boldsymbol{M} + \omega_0 \boldsymbol{M} \times \boldsymbol{\mu}_z = -\gamma\mu_0 \boldsymbol{M}_s \times \boldsymbol{H}, \tag{8.48}$$

where $\omega_0 = \gamma\mu_0 H_0 = \gamma B_0$. The components of Eq. (8.48) in the rectangular coordinates are

$$j\omega M_x + \omega_0 M_y = \gamma M_s \mu_0 H_y, \tag{8.49a}$$

$$j\omega M_y - \omega_0 M_x = -\gamma M_s \mu_0 H_x, \tag{8.49b}$$

$$j\omega M_z = 0. \tag{8.49c}$$

The solutions of Eqs. (8.49) are

$$M_x = \frac{\omega_0 \gamma\mu_0 M_s H_x + j\omega\gamma\mu_0 M_s H_y}{\omega_0^2 - \omega^2}, \tag{8.50a}$$

$$M_y = \frac{\omega_0 \gamma\mu_0 M_s H_y - j\omega\gamma\mu_0 M_s H_x}{\omega_0^2 - \omega^2}, \tag{8.50b}$$

$$M_z = 0. \tag{8.50c}$$

In applying Maxwell's equations, it is more convenient to introduce susceptibility and permeability than to use the magnetization vector. Hence, the relations

$$M = \chi_m H, \qquad B = \mu_0(M + H) = \mu_0(1 + \chi_m)H = \mu H$$

will be used. For a ferrite, χ_m and μ will not be scalar quantities, but will be represented by matrices. Thus, we use the matrix relation

$$\begin{pmatrix} M_x \\ M_y \\ M_z \end{pmatrix} = \begin{pmatrix} \chi_{xx} & \chi_{xy} & 0 \\ \chi_{yx} & \chi_{yy} & 0 \\ 0 & 0 & 0 \end{pmatrix} \begin{pmatrix} H_x \\ H_y \\ H_z \end{pmatrix}, \tag{8.51a}$$

where

$$\chi_{xx} = \chi_{yy} = \frac{\omega_0 \gamma\mu_0 M_s}{\omega_0^2 - \omega^2}, \tag{8.51b}$$

$$\chi_{xy} = -\chi_{yx} = \frac{j\omega\gamma\mu_0 M_s}{\omega_0^2 - \omega^2}. \tag{8.51c}$$

The matrix

$$\begin{pmatrix} \chi_{xx} & \chi_{xy} & 0 \\ \chi_{yx} & \chi_{yy} & 0 \\ 0 & 0 & 0 \end{pmatrix}$$

represents the susceptibility tensor of the ferrite. $\boldsymbol{B}$ and $\boldsymbol{H}$ are related by

$$\begin{pmatrix} B_x \\ B_y \\ B_z \end{pmatrix} = \mu_0 \begin{pmatrix} 1 + \chi_{xx} & \chi_{xy} & 0 \\ \chi_{yx} & 1 + \chi_{yy} & 0 \\ 0 & 0 & 1 \end{pmatrix} \begin{pmatrix} H_x \\ H_y \\ H_z \end{pmatrix}. \tag{8.52}$$

The matrix

$$\mu_0 \begin{pmatrix} 1 + \chi_{xx} & \chi_{xy} & 0 \\ \chi_{yx} & 1 + \chi_{yy} & 0 \\ 0 & 0 & 1 \end{pmatrix}$$

is the permeability tensor of the ferrite and is denoted by $\bar{\mu}$. Hence

$$\boldsymbol{B} = \bar{\mu}\boldsymbol{H}.$$

Losses in a ferrite may be accounted for by introducing into the equation of motion a damping term that will produce a torque tending to reduce the precession angle ϕ. Hence, the following modified form of the equation of motion is used:

$$\frac{d\mathcal{M}}{dt} = -\gamma\mu_0(\boldsymbol{M}_s + \mathcal{M}) \times (\boldsymbol{H}_0 + \mathcal{H}) + \frac{\alpha}{\mathcal{M}}\mathcal{M} \times \frac{d\mathcal{M}}{dt}, \tag{8.53}$$

where α is a dimensionless damping constant. Using small-signal analysis, the elements of the susceptibility matrix are found to be:

$$\chi_{xx} = \chi_{yy} = \chi' - j\chi'' = \chi, \tag{8.54a}$$

$$\chi_{xy} = -\chi_{yx} = j(K' - jK'') = jK, \tag{8.54b}$$

where

$$\chi' = \frac{\omega_0\omega_m(\omega_0^2 - \omega^2) + \omega_m\omega_0\omega^2\alpha^2}{[\omega_0^2 - \omega^2(1 + \alpha^2)]^2 + 4\omega_0^2\omega^2\alpha^2}, \tag{8.55a}$$

$$\chi'' = \frac{\omega\omega_m\alpha[\omega_0^2 + \omega^2(1 + \alpha^2)]}{[\omega_0^2 - \omega^2(1 + \alpha^2)]^2 + 4\omega_0^2\omega^2\alpha^2}, \tag{8.55b}$$

$$K' = \frac{\omega\omega_m[\omega_0^2 - \omega^2(1 + \alpha^2)]}{[\omega_0^2 - \omega^2(1 + \alpha^2)]^2 + 4\omega_0^2\omega^2\alpha^2}, \tag{8.55c}$$

$$K'' = \frac{2\omega^2\omega_0\omega_m\alpha}{[\omega_{02} - \omega^2(1 + \alpha^2)]^2 + 4\omega_0^2\omega^2\alpha^2}, \tag{8.55d}$$

$$\omega_m = \mu_0 \gamma M_s. \tag{8.56}$$

8.12 Plane Wave Propagation in Unbounded Ferrite Medium

To illustrate the general technique of linearization to be applied in the small-signal analysis of propagation through an unbounded ferrite medium, it is useful to study the propagation of a plane wave in such a medium.

We shall find solutions for Maxwell's equations in a ferrite medium, which are of the form

$$\nabla \times \boldsymbol{E} = -j\omega \boldsymbol{B} = -j\omega \bar{\mu} \bullet \boldsymbol{H}, \tag{8.57a}$$

$$\nabla \times \boldsymbol{H} = j\omega\epsilon \boldsymbol{E}, \tag{8.57b}$$

$$\nabla \bullet \boldsymbol{B} = \nabla \bullet \boldsymbol{E} = 0. \tag{8.57c}$$

We shall find solutions for a TEM wave propagating in the z-direction, i.e., in the direction of $\boldsymbol{B}_0$. Let $\boldsymbol{E}$ be of the form

$$\boldsymbol{E} = \boldsymbol{E}_0 e^{-j\beta z},$$

where $\boldsymbol{E}_0$ is a constant vector in the xy-plane. Then, from Eq. (8.57a),

$$\nabla \times \boldsymbol{E} = -\boldsymbol{E}_0 \times \nabla e^{-j\beta z} = j\beta \boldsymbol{E}_0 \times \boldsymbol{u}_z e^{-j\beta z} = -j\omega \bar{\mu} \bullet \boldsymbol{H}. \tag{8.58}$$

Let the solution for $\boldsymbol{H}$ be $\boldsymbol{H}_0 e^{-j\beta z}$, where $\boldsymbol{H}_0$ is also a constant vector in the xy-plane. Then, from Eq. (8.57b), we have

$$j\beta \boldsymbol{H}_0 \times \boldsymbol{u}_z = j\omega\epsilon \boldsymbol{E}_0. \tag{8.59}$$

Substituting Eq. (8.59) in Eq. (8.58) and eliminating $\boldsymbol{E}_0$, we obtain

$$j\beta \frac{j\beta}{j\omega\epsilon} (\boldsymbol{H}_0 \times \boldsymbol{u}_z) \times \boldsymbol{u}_z = -j\omega \bar{\mu} \boldsymbol{H}_0. \tag{8.60}$$

In matrix form, Eq. (8.60) becomes

$$\begin{pmatrix} \beta^2 - \omega^2\epsilon\mu_0(1+X) & -j\omega^2\epsilon\mu_0 K \\ j\omega^2\epsilon\mu_0 K & \beta^2 - \omega^2\epsilon\mu_0(1+X) \end{pmatrix} \begin{pmatrix} H_{0x} \\ H_{0y} \end{pmatrix} = 0. \tag{8.61}$$

If H_0 is to have a nontrivial solution, the determinant of the square matrix in Eq. (8.61) should vanish. This gives the eigenvalue equation for the propagation constant β as

$$[\beta^2 - \omega^2\epsilon\mu_0(1+X)]^2 - \omega^4\epsilon^2\mu_0^2 K^2 = 0$$

or

$$\beta^2 = \omega^2\epsilon_0\mu_0(1+X) \pm \omega^2\epsilon\mu_0 K. \tag{8.62}$$

Assuming a lossless ferrite so that $X'' = K'' = 0$, and substituting for X and K, we obtain

$$\beta^2 = \beta_+^2 = \omega^2\epsilon\mu_+, \tag{8.63a}$$

$$\beta^2 = \beta_-^2 = \omega^2\epsilon\mu_-, \tag{8.63b}$$

where μ_+ and μ_- are given by Eqs. (8.41).

For each eigenvalue or solution for β^2, the ratio of H_{0x} to H_{0y} is determined. For $\beta^2 = \beta_+^2$, the first equation in the pair of equations given by Eq. (8.61) becomes

$$[\beta_+^2 - \omega^2\epsilon\mu_0(1 + X)]H_{0x} - j\omega^2\epsilon\mu_0 K H_{0y} = 0$$

or

$$H_{0x}/H_{0y} = j \tag{8.64}$$

which means that $\boldsymbol{H}_0 = H_0(+\boldsymbol{u}_x - j\boldsymbol{u}_y)$, or that the wave is positive circularly polarized. Similarly, the solution $\beta^2 = \beta_-^2$ gives

$$H_{0x}/H_{0y} = -j. \tag{8.65}$$

This implies that the wave is negative circularly polarized.

The foregoing analysis shows that the natural modes of propagation along the direction of the static field in a ferrite are circularly-polarized TEM waves. If the wave propagates along a direction different from that of $\boldsymbol{B}_0$, then also there are two modes of propagation, but they are no longer circularly-polarized TEM waves.

8.13 Faraday Rotation

Let a plane TEM wave, linearly polarized along the x-axis at $z = 0$, be propagating in the z-direction in an infinite lossless ferrite medium with a static field $\boldsymbol{B}_0$ applied along the z-direction (see Fig. 8.24). It can be shown that the plane of polarization of this wave rotates as it propagates.

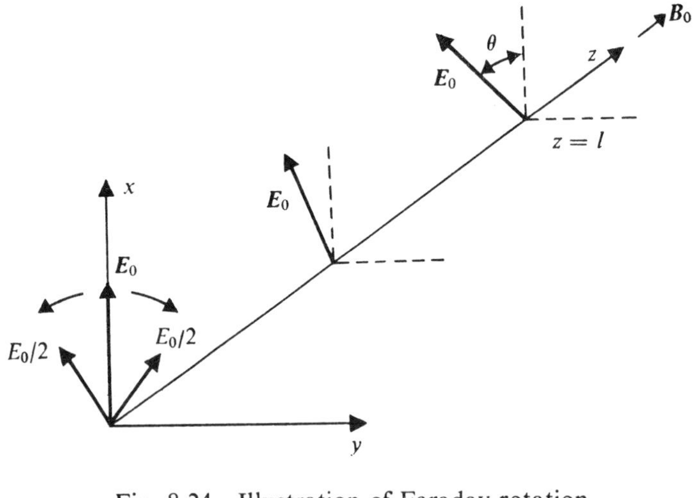

Fig. 8.24 Illustration of Faraday rotation.

This phenomenon is a nonreciprocal one, and is called the *Faraday rotation*.

The linearly-polarized wave may be decomposed into the sum of a left and right circularly-polarized wave as follows:

$$\boldsymbol{E} = \boldsymbol{u}_x E_0 = (\boldsymbol{u}_x + j\boldsymbol{u}_y)\frac{E_0}{2} + (\boldsymbol{u}_x - j\boldsymbol{u}_y)\frac{E_0}{2} \quad \text{at } z = 0. \tag{8.66}$$

The two circularly-polarized waves propagate with different phase constants β_+ and β_-, and hence at $z = l$, the wave becomes

$$\begin{aligned} E &= (u_x + ju_y)\frac{E_0}{2} \exp(-j\beta_- l) + (u_x - ju_y)\frac{E_0}{2} \exp(-j\beta_+ l) \\ &= u_x\frac{E_0}{2} [\exp(-j\beta_- l) + \exp(-j\beta_+ l)] + ju_y\frac{E_0}{2}[\exp(-j\beta_- l) - \exp(-j\beta_+ l)] \\ &= \frac{E_0}{2} \exp[-j(\beta_- + \beta_+)l/2][u_x\{\exp[-j(\beta_- - \beta_+)l/2] + \exp[j(\beta_- - \beta_+)l/2]\} \\ &\quad + ju_y\{\exp[-j(\beta_- - \beta_+)l/2] - \exp[j(\beta_- - \beta_+)l/2]\}] \\ &= E_0 \exp[-j(\beta_- + \beta_+)l/2][u_x \cos(\beta_+ - \beta_-)l/2 - u_y \sin(\beta_+ - \beta_-)l/2]. \end{aligned} \tag{8.67}$$

Equation (8.67) represents a linearly-polarized wave that has undergone a phase delay of $(\beta_- + \beta_+)l/2$. The new plane of polarization makes an angle

$$\theta = \tan^{-1}(E_y/E_x) = \tan^{-1}[-\tan(\beta_+ - \beta_-)l/2] = -(\beta_+ - \beta_-)l/2 \tag{8.68}$$

with respect to the x-axis. When $\omega < \omega_0$, i.e., below the ferrite resonance frequency, β_+ is greater than β_-, and the plane of polarization rotates counter clockwise, looking in the direction of propagation. The rate of rotation is $(\beta_+ - \beta_-)/2$ radians per metre. Rotations of 100° or more per centimetre is obtained in ferrites at a frequency of 10,000 MHz.

If the direction of propagation is reversed, it can be shown, by a similar analysis, that the plane of polarization continues to rotate in the same direction. Let the wave propagate from $z = l$ to $z = 0$ in the reverse direction. Then, the original polarization direction is not restored, but the wave will arrive at $z = 0$ polarized at an angle 2θ relative to the x-axis. The wave becomes, at $z = 0$,

$$E = (u_x + ju_y)\frac{E_0}{2} \exp(-2j\beta_- l) + (u_x - ju_y)\frac{E_0}{2} \exp(-2j\beta_+ l) \tag{8.69}$$

so that the new direction of polarization at $z = 0$ makes an angle

$$2\theta = -(\beta_+ - \beta_-)l.$$

Hence, the Faraday rotation is a nonreciprocal effect.

The foregoing analysis was made on the assumption that the ferrite is lossless. But a practical ferrite medium has finite losses, and this has a significant influence on propagation. The propagation constants $\gamma_+ = j\beta_+ + \alpha_+$ and $\gamma_- = j\beta_- + \alpha_-$ for circularly-polarized waves have unequal attenuation constants as well as unequal phase constants. When there are losses, the propagation constants are given by Eq. (8.62) if β^2 is replaced by $-\gamma^2$. Hence

$$\gamma_+ = j\omega\sqrt{\mu_0\epsilon}(1 + X' - jX'' + K' - jK'')^{1/2}, \tag{8.70a}$$

$$\gamma_- = j\omega\sqrt{\mu_0\epsilon}(1 + X' - jX'' - K' + jK'')^{1/2}, \tag{8.70b}$$

where X', X'', K', K'' are given by Eqs. (8.55a)–(8.55d). $\beta_\pm$ and $\alpha_\pm$ are now given by

$$\beta_\pm = \frac{\omega\sqrt{\mu_0\epsilon}}{\sqrt{2}}[1 + X' \pm K' + \sqrt{(1 + X' \pm K')^2 + (X'' \pm K'')^2}]^{1/2}, \tag{8.71a}$$

$$\alpha_{\pm} = \omega^2 \mu_0 \epsilon \frac{\chi'' \pm K''}{2\beta_{\pm}} = \frac{\omega^2 \epsilon \mu''_{\pm}}{2\beta_{\pm}}. \tag{8.71b}$$

The permeabilities for circularly-polarized waves are

$$\mu_{\pm} = \mu'_{\pm} - j\mu''_{\pm} = \mu_0(1 + \chi' - j\chi'' \pm K' \mp jK''). \tag{8.72}$$

The values of $\mu'_{\pm}$, $\mu''_{\pm}$, as given by Eqs. (8.55a)–(8.55d) and (8.72), and the propagation factors $\beta_{\pm}$ and $\alpha_{\pm}$ are plotted in Figs. 8.25 and 8.26, for a typical ferrite with parameters $\omega_m = 2\pi \times 5.6 \times 10^9$, $\alpha = 0.05$, as a function of ω_0 at a frequency of 10,000 MHz. It may be noticed that ω_0 equals $2\pi \times 2.8$ MHz/oersted of applied field H_0 and that $4\pi \times 10^{-3}$ oersted is a field strength of 1 amp/m. The value of ω_m chosen corresponds to a saturation magnetization of 2000 gauss, or $\mu_0 M_s$ equals 0.2 weber/m^2.

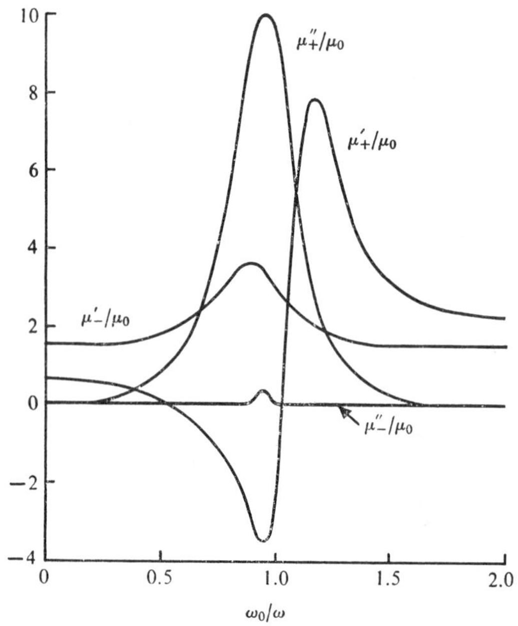

Fig. 8.25 Real and imaginary components of permeability for circularly-polarized waves in a ferrite as function of ω/ω_0 for $\omega/(2\pi) = 10$ kMHz. [$\omega_m/(2\pi) = 5.6$ kMHz, $\alpha = 0.05$.]

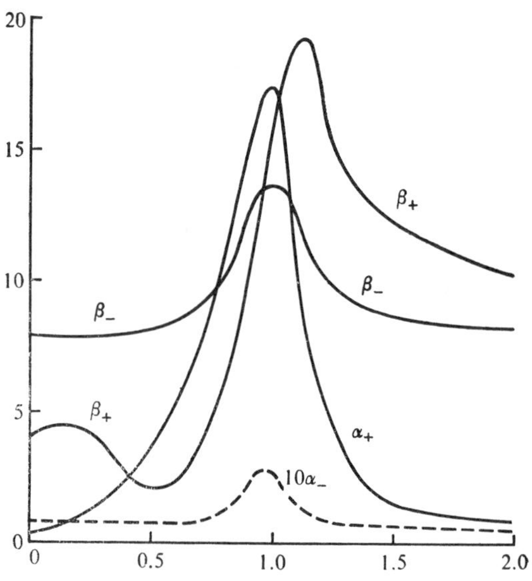

Fig. 8.26 Propagation and attenuation constants for circularly-polarized waves in a ferrite, with parameters given in Fig. 8.25 ($\epsilon = 10\epsilon_0$). (α_- is very small, and hence $10\alpha_-$ is plotted.)

The curves of Fig. 8.26 show that, in the vicinity of the resonant frequency $\omega_0 \approx \omega$, α_- is always very small but α_+ is large. For $\omega_0 \gg \omega$, the attenuation α_+ becomes small, but in this region, β_+ and β_- do not differ greatly so that the rate of Faraday rotation is small. At low values of ω_0, i.e., for small H_0, the rate of rotation is much greater, particularly in the region where β_+ goes through a minimum, which occurs in the range where μ'_+ is negative. Also, $\beta_+ > \beta_-$ when $\omega_0 > \omega$, but $\beta_+ < \beta_-$ for $\omega_0 < \omega$. Hence, the direction of Faraday rotation is different in the two regions above and below the resonant frequency.

8.14 Applications of Microwave Ferrite Devices

Ferrite devices have found a large number of applications at microwave frequencies. We shall describe some of these devices in this section.

Gyrator

A gyrator is a two-port device which has a relative difference in the phase shift of 180° for transmission from port 1 to port 2 as compared with the phase shift for transmission from port 2 to port 1. The principle of Faraday rotation is made use of in the design of a gyrator. Figure 8.27 shows a typical microwave gyrator consisting of a rectangular waveguide with a 90° twist

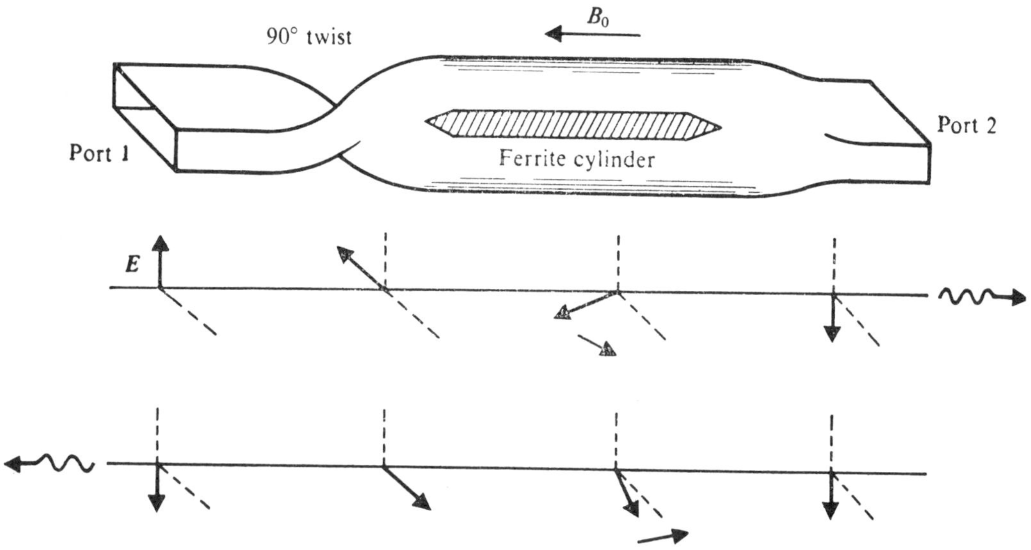

Fig. 8.27 Gyrator with twist.

connected to a circular waveguide, which in turn is connected to a rectangular waveguide at the other end, having the same orientation at the output as the input of the input rectangular waveguide. There is a thin cylindrical ferrite rod with tapered ends inside the cylindrical waveguide. A 90° Faraday rotation of the dominant TE_{11} mode in the cylindrical waveguide is produced by a static magnetic field of suitable magnitude in the axial direction. If a wave is propagating from port 1 to port 2, its plane of polarization is rotated by 90° in a counterclockwise direction in passing through the twist. If the ferrite rod in the cylindrical waveguide produces an additional 90° rotation, the total angle of rotation will be 180°. On the other hand, for a wave propagating in the opposite direction from port 2 to port 1, the Faraday rotation is still 90° in the same sense, but in passing through the twist, the 90° rotation is in the opposite sense and cancels the Faraday rotation. Hence, for transmission from port 1 to port 2, there is a 180° rotation, whereas there is no net rotation for transmission from port 2 to port 1. The 180° rotation in the direction from port 1 to port 2 is equivalent to a phase shift of 180° since it reverses the direction of the electric field. Hence, the device functions as a gyrator.

A gyrator without a 90° twist can also be built, as shown in Fig. 8.28. In this structure, the ferrite rod produces a 90° rotation and the output guide is rotated by 90° in relation to the input

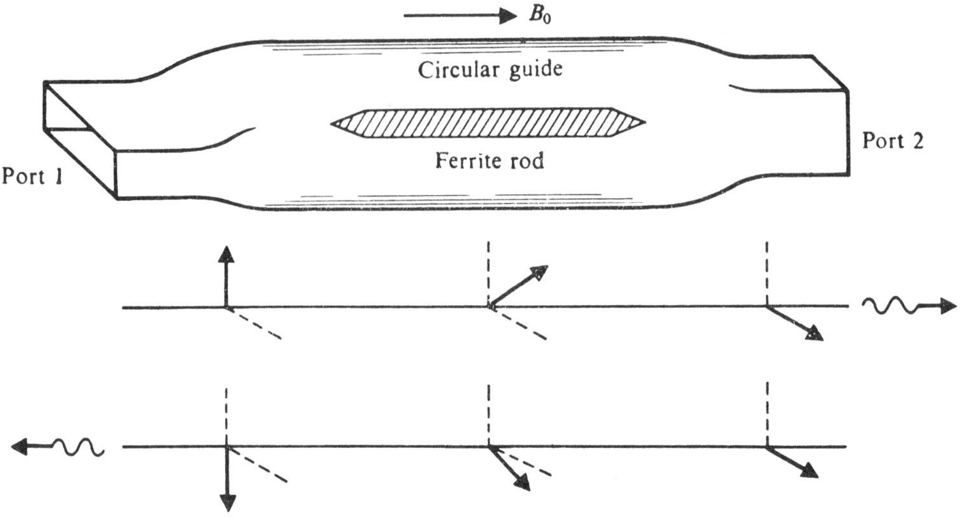

Fig. 8.28 Gyrator without twist.

guide, and the emerging wave has the right polarization to propagate in the output guide. If the propagation is in the opposite direction from port 2 to port 1, then the wave arriving in port 1 changes its polarization by 180°. Therefore, there is a differential phase shift of 180°.

Isolator

An isolator, also called a uniline, is a microwave component which permits transmission from port 1 to port 2, without attenuation or with very little attenuation, but which has very high attenuation for transmission in the opposite direction from port 2 to port 1. If the isolator is used to couple a microwave signal generator to a load, then all the available power is delivered to the load, and no reflections from the load can react with the generator. The generator sees a matched load, and power output variation and frequency pulling with variations in the load impedance can be avoided.

The construction of the isolator is similar to that of the gyrator, with a 45° twist section and a 45° Faraday rotation. Thin resistive cards are placed in the input and output guides to absorb the polarized field, with the electric vector parallel to the wide side of the guide (see Fig. 8.29). A wave travelling from port 1 to port 2 has its polarization rotated by 45° counter clockwise by the twist and 45° clockwise by the ferrite rod, due to Faraday rotation. This wave arrives at port 2 with the right polarization to propagate in the output waveguide. On the other hand, a wave travelling from port 2 to port 1 has its plane of polarization rotated by 90°, and hence, when it arrives at port 1, its electric field becomes parallel to the resistance card, and thus it will be absorbed. In the absence of the resistance cards, the wave will be reflected from port 1, because its polarization is not the right one to enable it to propagate in the waveguide of port 1. This leads to multiple reflections from both ports, and transmission takes place in both directions. Hence, the use of resistance cards is absolutely necessary for satisfactory performance of the isolator. An isolator, with a forward transmission loss of less than 1 db and reverse attenuation of 20 to 30 db and bandwidth of operation about 10 per cent, is typical.

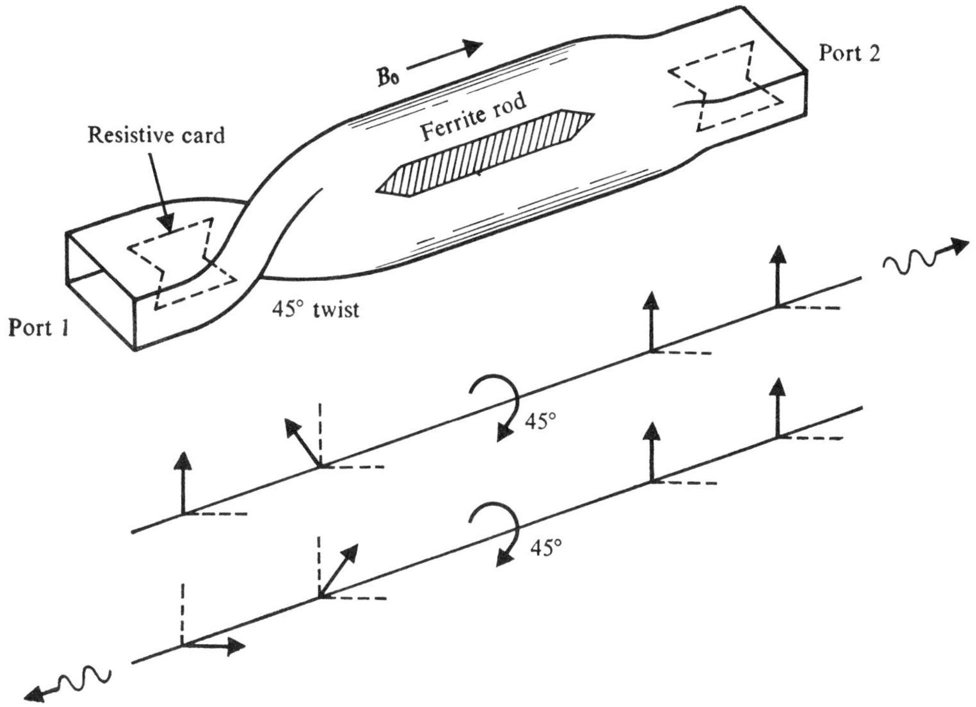

Fig. 8.29 Isolator.

Resonance Isolator

From Fig. 8.26, we see that, in the vicinity of the resonance point $\omega_0 \approx \omega$, the attenuation constant for negative circular polarization is always very small, whereas the attenuation constant for positive circular polarization is very large. A resonance isolator may be constructed by using this property. A negative circularly-polarized wave is used for transmission in the low-loss direction and a positive circularly-polarized wave is used for transmission in the reverse high-attenuation direction. The field components of the dominant TE_{10} mode in a rectangular waveguide are

$$
\begin{aligned}
E_y &= \sin\left(\frac{\pi x}{a}\right) \exp(\pm j\beta z), \\
H_x &= \frac{\pm j\beta}{j\omega\mu_0} \sin\left(\frac{\pi x}{a}\right) \exp(\pm j\beta z), \qquad (8.73) \\
H_z &= -\frac{1}{j\omega\mu_0}\frac{\pi}{a} \cos\left(\frac{\pi x}{a}\right) \exp(\pm j\beta z).
\end{aligned}
$$

Here, $e^{j\omega t}$ is omitted. H_x and H_z differ in phase by 90°, and hence circular polarization occurs when $|H_x| = |H_z|$, or at $x = x_1$,

$$\tan\left(\frac{\pi x_1}{a}\right) = \pm\frac{\pi}{\beta a} = \pm\frac{\lambda_g}{2a}. \qquad (8.74)$$

In the range $0 < x_1 < a/2$,

$$\frac{H_x}{H_z} = \mp j \tag{8.75a}$$

and in the range $a/2 < x_1 < a$,

$$\frac{H_x}{H_z} = \pm j. \tag{8.75b}$$

The solution given by Eq. (8.75a) provides a negative circularly-polarized field with respect to the y-axis for propagation in the $+z$-direction and a positive circularly-polarized wave for propagation in the $-z$-direction; the directions are reversed for the solution given by Eq. (8.75b).

This property of the TE_{10} mode is used in the resonance isolator shown in Fig. 8.30. Here, a thin ferrite slab is placed in a rectangular waveguide at a position where the r-f magnetic field is circularly polarized (see Fig. 8.30a). Sometimes, two ferrite slabs are placed, as can be seen

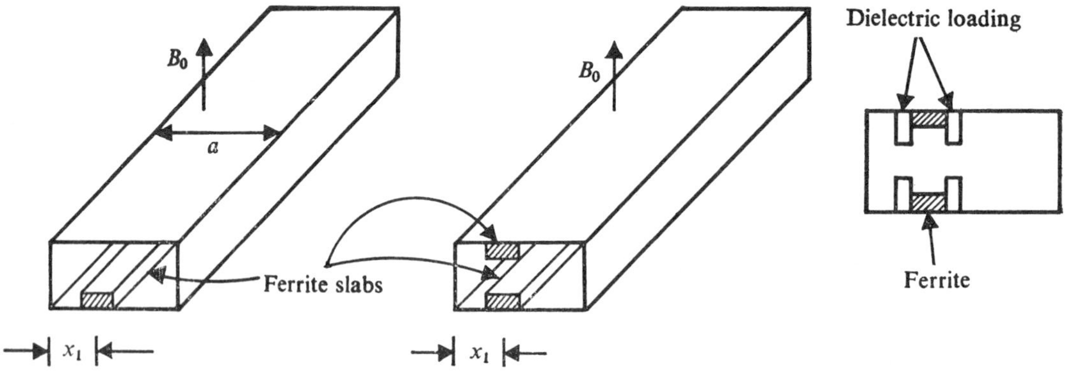

(a) Single slab (b) Double slab (c) Dielectric loaded

Fig. 8.30 Resonance isolators.

from Fig. 8.30b. For propagation in one direction, the magnetic field is negatively polarized, and hence there is very little attenuation, whereas for propagation in the reverse direction, it is positively polarized, and hence there is very high attenuation. The forward loss can be made smaller than 0.5 db/in at $\lambda_0 = 3$ cm, and the reverse loss can be as high as 6 to 10 db/in. Dielectric loading (see Fig. 8.30c) gives an improved reverse-to-forward attenuation ratio.

8.15 Ferrite Four-Port and Three-Port Circulators

Figure 8.31 shows a schematic diagram of a four-port circulator. A wave incident in port 1 is coupled to port 2 only, a wave incident in port 2 is coupled to port 3 only, and so on. An ideal circulator is a matched device with all its ports except one terminated in matched loads. The input impedance of the unmatched port is equal to the characteristic impedance of its input line, and hence the unmatched port also represents a matched load.

Using two magic-tees and a gyrator, a four-port circulator can be constructed as shown in Fig. 8.32. The electrical path lengths of the paths from b to a, from c to d, and from d to c are all equal, and the gyrator produces an additional phase shift of 180° for propagation in the direction from a to b. If a wave is incident in port 1, it gets divided into two equal amplitude waves which are in phase and propagate in the side arms of the hybrid junction. These two waves arrive at points a and c in phase and will add and emerge from port 2. If a wave is incident in port 2, it divides into two waves, one arriving at d with a phase ϕ and the other

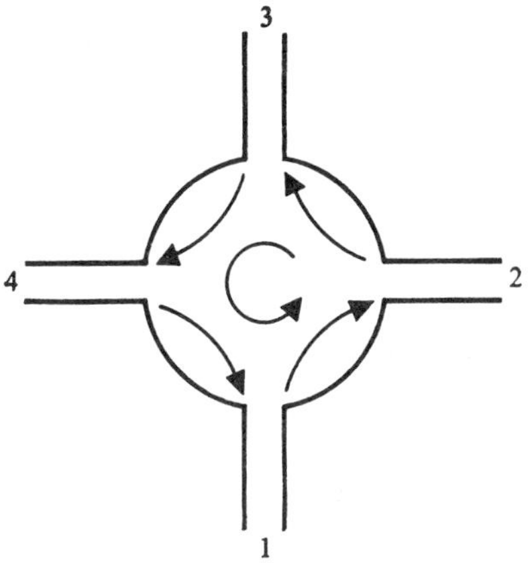

Fig. 8.31 Four-port circulator (schematic diagram).

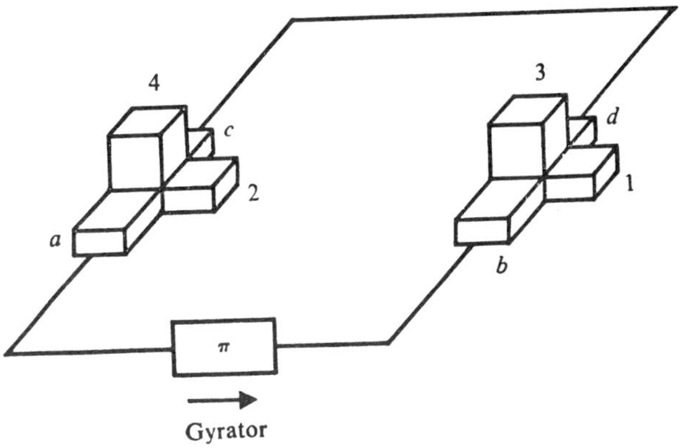

Fig. 8.32 Four-port circulator.

arriving at b with a phase $\phi + \pi$ due to its travel through the gyrator. These two waves have the correct phase relationship to add and emerge from port 3. A wave incident in port 3 divides into two waves which have equal amplitude but differ in phase by 180°, and these two waves arrive at the other hybrid junction with the correct phase difference to add and emerge from port 4. A wave incident in port 4 splits into two equal waves differing by 180° in phase, but the gyrator restores equality of phase, and the two waves combine to add and emerge from port 1. This is how the device works as a circulator. It can be similarly argued that there can be no transmission in the reverse circular direction.

By using 3-db side-hole directional couplers and rectangular waveguide nonreciprocal phase shifters, it is possible to construct a more compact form of a four-port circulator, as illustrated in Fig. 8.34. The nonreciprocal phase shifter consists of a thin ferrite slab placed in a rectangular

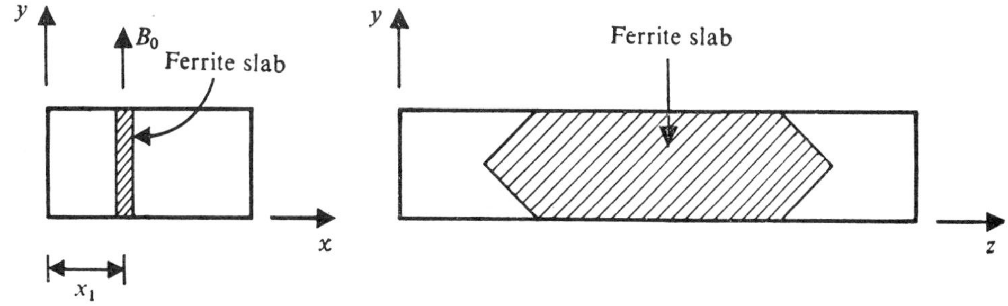

(a) Transverse cross-section (b) Longitudinal cross-section

Fig. 8.33 Nonreciprocal phase shifter.

waveguide at a point where the a-c magnetic field of the TE_{10} mode is circularly polarized (see Fig. 8.33). The biassing field B_0 is applied in the y-direction. The effect of the ferrite slab is different for the two different directions of propagation, because the a-c magnetic field is right circularly polarized at x_1 for one direction, and is left circularly polarized for the opposite direction. Therefore, the propagation constant β_+ for forward propagation is different from the propagation constant β_- for backward propagation. The length of the ferrite slab is chosen to get $(\beta_+ - \beta_-)l = \pi/2$ and a differential phase shift of 90° for the two directions of propagation.

Figure 8.34 shows a four-port circulator using two 90° nonreciprocal phase shifters. By using permanent magnets, as shown in Fig. 8.34b, the phase shifters are biassed by static magnetic

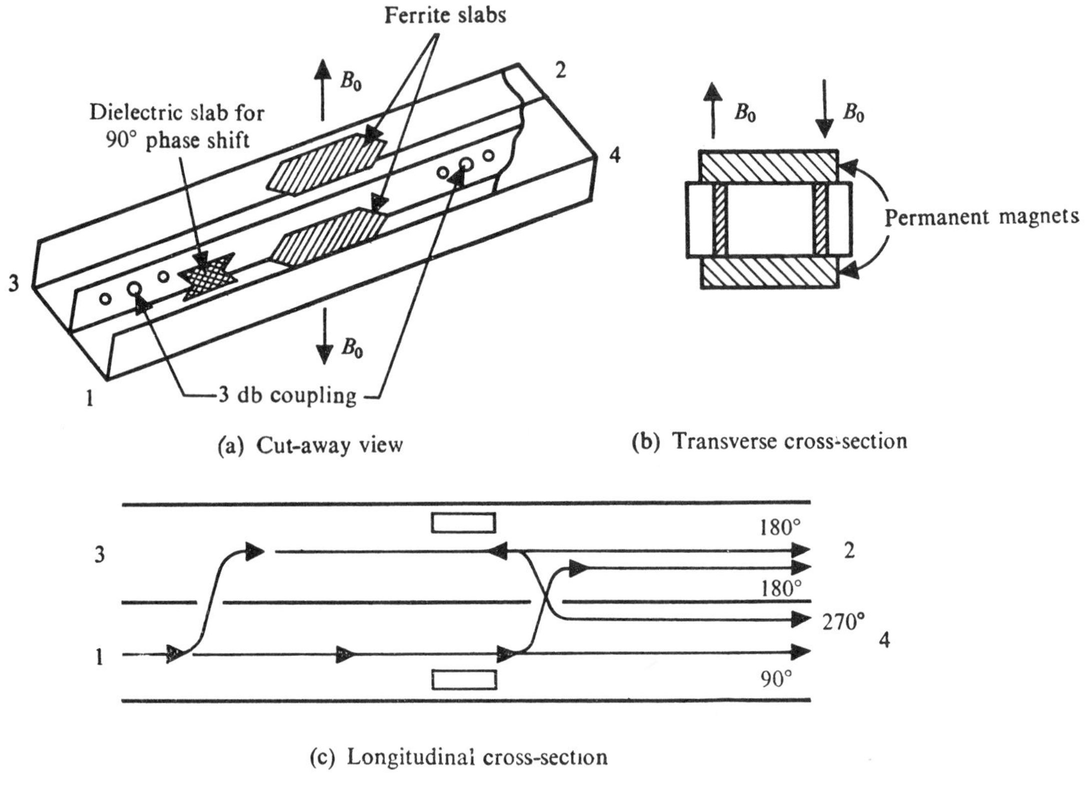

Fig. 8.34 Four-port circulator using two 90° nonreciprocal phase shifters.

fields in opposite directions. One of the rectangular waveguides contains also a dielectric slab to provide an additional 90° reciprocal phase shift. The coupling holes provide a 3-db coupling and a 90° change in phase. A wave incident in port 1 gets divided into two waves by the first 3-db coupler. The wave in the upper guide suffers a 90° phase change because of the transmission properties of the aperture. This wave in the upper guide arrives at the second coupler with a relative phase of 180°, whereas the wave in the lower guide arrives with a relative phase of 90°. The second coupler again splits these two waves in a manner similar to the first coupler. This results in two waves arriving in port 2 in phase and two waves arriving in port 4 out of phase.

Hence, there is a transmission in the direction from port 1 to port 2, but not from port 1 to port 4. It can similarly be reasoned out that a wave arriving in port 2 will emerge from port 3, and so on, resulting in the sequence $1 \to 2 \to 3 \to 4$.

It may be noted that a lossless, matched, nonreciprocal three-port microwave junction is a perfect three-port circulator.

A typical three-port circulator can be constructed as a symmetrical Y-junction of three identical waveguides or *stripline* type of transmission lines, in which an axially magnetized ferrite rod or disc is placed at the centre, as shown in Fig. 8.35. Suitable tuning elements are

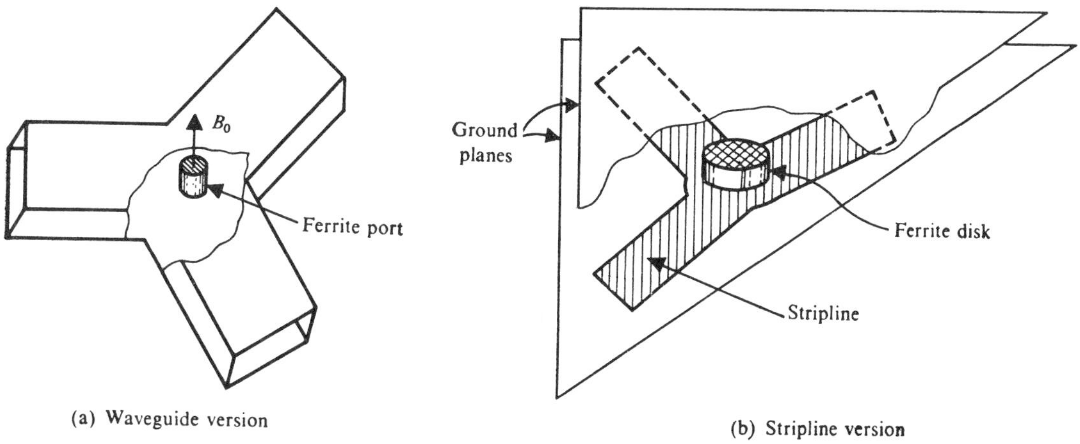

Fig. 8.35 Three-port ferrite circulator.

placed in all the arms (which are identical because of the three-fold symmetry) so that the junction is matched. In a practical device, losses can be minimized, but cannot be made zero, and this limits the performance of the circulator. In such a device, insertion loss of less than one db can be obtained in the transmission direction, whereas isolation of about 40 db can be obtained in the reverse direction; input reflection coefficients less than 0.2 can be attained.

PROBLEMS

1 A small circular aperture is made in the common side wall of two rectangular waveguides, as shown in Fig. P. 8.1. If the incident TE_{10} mode has an amplitude A_0, show that the amplitude

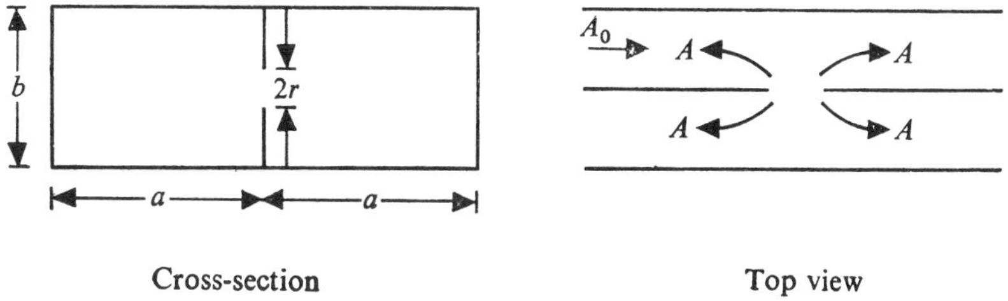

Fig. P. 8.1

of the TE_{10} mode radiated in both directions in the second guide is given by

$$A = -\frac{j\omega\mu_0}{abk_0Z_0\beta}\left(\frac{\pi}{a}\right)^2 \tfrac{4}{3}r^3A_0.$$

Show that the reflection coefficient in the input guide is A/A_0.

2 Design a Bethe-hole directional coupler with a centred circular aperture. The waveguide size is $a = 2.5$ cm and $b = 1.25$ cm. The centre frequency is 10 GHz. The required coupling is 22 db. Find the aperture radius and the frequency band over which the directivity remains greater than 44 db.

3 In the rotary phase changer, show that if the output quarter-wave plate, transition, and rectangular guide are rotated by an angle θ, an additional phase change of θ_1 is produced in the transmitted wave.

9 Periodic Structures, Delay Lines

9.1 General Remarks

Periodic structures are waveguides and transmission lines loaded at periodic intervals with identical obstacles like a reactive element such as a diaphragm. Such structures are also known as slow-wave structures or delay lines. These structures are of interest because of two basic properties, namely, (i) pass-band stop-band characteristics and (ii) support of waves with phase velocities much less than the velocity of light. The pass-band stop-band characteristics are due to the existence of frequency bands throughout which a wave propagates unattenuated (such a band is called a pass-band), except for conductor losses, along the structure separated by frequency bands throughout which the wave is cut off and does not propagate (such a band is called a stop-band). These characteristics are very helpful in the use of a periodic structure as filter.

The property of many periodic structures, namely, that of supporting a wave having a phase velocity much less than that of light, is of basic importance for travelling-wave tube circuits. The interaction between the electron beam and the travelling wave in a travelling-wave tube is efficient only if the phase velocity is very nearly equal to the beam velocity. Since it is difficult to obtain a beam velocity more than 10 to 20 per cent of the velocity of light, the electromagnetic wave has to be slowed down considerably. There are periodic structures which are suitable for use in travelling-wave tubes and they are usually called slow-wave structures or delay lines. This chapter deals with these structures. The principle of operation of travelling-wave tubes is discussed in Chapter 10.

9.2 General Properties of Delay Lines or Slow-Wave Structures Used in Travelling-Wave Tubes

In travelling-wave tubes (see Chapter 10) using slow-wave structures or delay lines, the electron velocity $\bar{v}$ and the phase velocity v_p of the travelling wave guided by the line should be approximately equal both in magnitude and direction. This leads to the relations

$$\left.\begin{aligned} v_p &\approx \bar{v} = 5.93 \times 10^5 \bar{V}^{1/2}, \\ \frac{c}{v_p} &= 505\bar{V}^{1/2} \qquad \text{(in } O\text{-type tubes)}, \end{aligned}\right\} \tag{9.1}$$

$$\left.\begin{aligned} v_p &\approx \bar{v} = \frac{\bar{E}}{\bar{B}} \\ \frac{c}{v_p} &= 3 \times 10^8 \frac{\bar{B}}{\bar{E}} \end{aligned}\right\} \qquad \text{(in } M\text{-type tubes)}, \tag{9.2}$$

where $\bar{B}$ is the d-c crossed magnetic field, $\bar{E}$ is the d-c transverse electric field, and $\bar{V}$ is the accelerating voltage of the electron beam. Here, c/v_p is known as the retardation ratio. The wavelength in the line is given by

$$\lambda_g = \frac{\lambda_0 v_p}{c} \tag{9.3}$$

and the phase constant of the wave by

$$\beta = \frac{\omega}{v_p} = \frac{2\pi}{\lambda_g} = \frac{2\pi c}{\lambda_0 v_p}. \tag{9.4}$$

The phase of the wave changes with distance according to the relation

$$\phi = \beta z = \frac{\omega z}{v_p}, \tag{9.5}$$

where z is the direction of propagation.

In this chapter, we consider only lossless lines in the absence of electron beams. The direction of the group velocity is always taken to be in the positive z-direction, so that the group velocity is always positive and is given by

$$v_g = (\frac{d\beta}{d\omega})^{-1} = v_p(1 - \frac{\omega}{v_p}\frac{dv_p}{d\omega})^{-1} = [\frac{1}{v_p} - \lambda_0 \frac{d(1/v_0)}{d\lambda_0}]^{-1}. \tag{9.6}$$

If v_p depends on ω, then $v_g \neq v_p$, and the line is dispersive. The pass-band of a line lies between two frequencies at which v_g becomes zero (cut-off frequencies). Negative frequencies are also included; in other words, for a low-pass filter, v_g does not necessarily have to be zero at $\omega = 0$. In such a case, the lower cut-off frequency defined by $v_g = 0$ is negative.

Figure 9.1 shows some β versus ω curves, where only the solid curves are physically significant since, for these curves, $d\beta/d\omega > 0$, and hence $v_g > 0$. The dashed curves have no

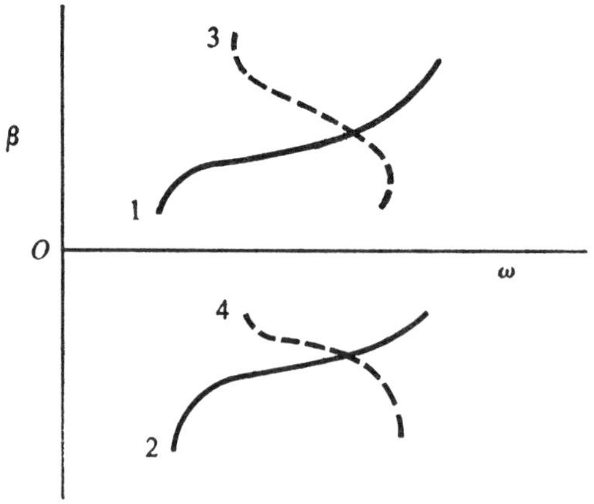

Fig. 9.1 β versus ω curves for delay lines. (Since $v_g > 0$, curves 1 and 2 are allowed but not curves 3 and 4. Curve 1, forward wave; curve 2, backward wave.)

physical meaning. A measurement of $\phi(\omega)$ as well as $\beta(\omega)$ leaves the sign of ϕ and β still in doubt. However, since $d\beta/d\omega > 0$, we can decide from the shape of the $\beta(\omega)$-curve whether β and ϕ (and thus also v_p) are positive or negative. A wave with v_g and v_p having the same sign is called a *forward wave* ($v_g > 0$, $v_p > 0$), and a wave with v_g and v_p having opposite signs is called a *backward wave* ($v_g > 0$, $v_p < 0$).

The phase transit time of a signal is given by

$$\tau_p = \frac{\phi}{\omega} = \frac{\beta z}{\omega} = \frac{z}{v_p} \tag{9.7}$$

and the group transit time by

$$\tau_g = \frac{d\phi}{d\omega} = \frac{z\, d\beta}{d\omega} = \frac{z}{v_g}. \tag{9.8}$$

For distortionless transmission, v_g must be independent of ω. From Eq. (9.6), this condition is fulfilled when β is a linear function of ω.

The coupling impedance K is defined as

$$K = \hat{E}_z\hat{E}_z^*/(2P\beta^2), \tag{9.9}$$

where $\hat{E}_z$ is the amplitude of the electric field component in the direction of wave propagation at the given point and P is the power carried by the wave. P is given by

$$P = \tfrac{1}{2}\,\mathrm{Re} \int_A (\hat{E} \times \hat{H}^*)\, dS, \tag{9.10}$$

where A is the total cross-section of the field.

Experimental methods for measuring K are described by Cutler[1] and Jaynes[2]. If W' is the energy stored per unit length of the line, then

$$P = W' v_g \tag{9.11}$$

provided v_g, the group velocity, is identical with the velocity of energy transfer.

In order to suppress oscillations in travelling-wave amplifiers, it is usually necessary to introduce attenuation either along part or along all of the line. The value of the attenuation should be such that it suppresses the oscillations but does not unduly decrease the amplification and efficiency. It is not very easy to calculate the attenuation of the delay lines used in practice, and a suitable value of attenuation is usually determined experimentally.

The matching of the delay line to the generator and the load, as also the matching of the attenuator to the delay line, is usually a difficult problem. This gives rise to distortions in travelling-wave amplifiers.

9.3 Classification of Delay Lines

There are two different methods of slowing down waves guided by transmission lines. These are

(i) By introducing a material which has $\epsilon_r\mu_r > 1$ in part of the space through which the wave travels.

[1] C. C. Cutler, Experimental determination of helical-wave properties, *Proc. I.R.E.*, **36**, 230, 1948.

[2] E. T. Jaynes, The concept and measurement of impedance in periodically loaded waveguides, *J. Appl. Phys.*, **23**, 1077, 1952.

(ii) By means of periodic structures in which the walls are suitably shaped.

If $\boldsymbol{F}$ is a vector quantity, which characterizes any field quantity representing the field of a line, for example, $\boldsymbol{E}$, $\boldsymbol{H}$, $\boldsymbol{A}$, $\boldsymbol{B}$, $\boldsymbol{D}_1 \ldots$, then it satisfies the vector wave equation

$$\nabla^2 \boldsymbol{F} - (\epsilon_r \epsilon_0)(\mu_r \mu_0) \frac{\partial^2 \boldsymbol{F}}{\partial t^2} = 0. \tag{9.12}$$

In Eq. (9.12), $\epsilon_r \mu_r > 1$ for method (i) and $\epsilon_r \mu_r = 1$ for method (ii). The equation therefore becomes

$$\nabla^2 \boldsymbol{F} - \epsilon_0 \mu_0 \frac{\partial^2 \boldsymbol{F}}{\partial t^2} = 0, \tag{9.13}$$

the delay effect being obtained by a suitable choice of the boundary conditions. Figure 9.2 is an

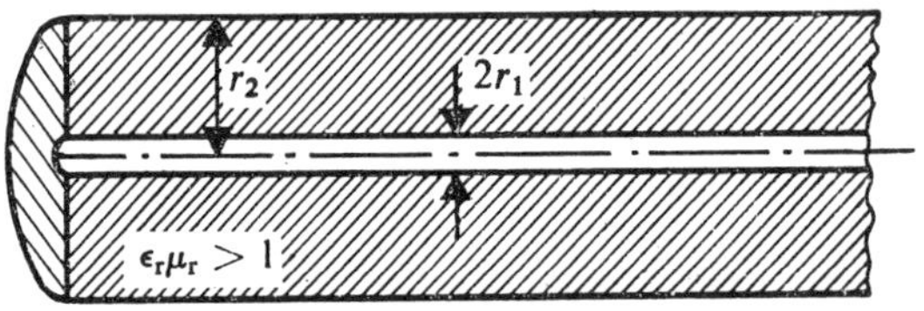

Fig. 9.2 Cylindrical waveguide filled with material having $\epsilon_r \mu_r > 1$, and hence $v_p/c < 1$.

example of a delay line which uses method (i), namely, a cylindrical waveguide containing a material for which $\epsilon_r \mu_r > 1$ and having an axial hole for the electron beam. If $r_1 \ll r_2$, the lower cut-off frequency of the TM_{01} mode is given by

$$f_0 = \frac{3 \times 10^8}{2.6 r_2 (\epsilon_r \mu_r)^{1/2}} = \frac{1.16 \times 10^8}{r_2 (\epsilon_r \mu_r)^{1/2}}, \tag{9.14}$$

where f_0 is in cycles per second and r_2 in metres.

For frequencies very much greater than the cut-off frequency, the retardation ratio is equal to

$$\frac{c}{v_p} \approx (\epsilon_r \mu_r)^{1/2}. \tag{9.15}$$

Lines for which $\epsilon_r \mu_r > 1$ usually have low coupling impedance and high dispersion, and hence are normally not used in microwave tubes. On the other hand, lines in which the retardation is produced by a suitable choice of boundary conditions are very useful in microwave tubes. Such lines are periodic in the axial direction, and are equivalent to a number of elementary four-terminal networks of length L, connected in cascade along the axis. These lines are called periodic or inhomogeneous delay lines. If $L = 0$, then they become homogeneous.

9.4 Comparison between Delay Lines and Ordinary Waveguides

In metal waveguides of rectangular or circular cross-section or in parallel-plane waveguides, the phase velocity v_p for all modes is greater than the velocity of light c. On the other hand, in delay lines, it is necessary that v_p be less than c.

In waveguides, each of the doubly infinite set of TE_{mn} and TM_{mn} modes can propagate independently. In delay lines, this is usually not true. Even if TE or TM modes exist independently, the field consists of the summation of an infinite number of TE or TM modes. If the walls of the delay line contain skew elements (i.e., elements having edges forming an angle other than 0° or 90° with the direction of propagation), then no pure TE or TM mode can exist, and the field will have both E_z and H_z components, where z is the direction of propagation. Also, the relationship between all the six components of the electric and magnetic fields depends only on the geometry of the line and remains unchanged even when the wave is amplified by the action of the electron beam on the wave.

In lines where the field depends only on one or two coordinates, it is possible to derive other differences between metal waveguides and metal delay lines. In the wave equation

$$\nabla^2 \boldsymbol{E} = \epsilon_0 \mu_0 \frac{\partial^2 \boldsymbol{E}}{\partial t^2} = 0, \tag{9.16}$$

if $\partial/\partial x = 0$ and $\boldsymbol{E} = \boldsymbol{E}_0 \exp[j(\omega t - \beta z)]$, then

$$\frac{\partial^2 \boldsymbol{E}}{\partial y^2} + (\omega^2 \epsilon_0 \mu_0 - \beta^2)\boldsymbol{E} = \frac{\partial^2 \boldsymbol{E}}{\partial y^2} + (k^2 - \beta^2)\boldsymbol{E} = 0, \tag{9.17}$$

where $k^2 = \omega^2 \mu_0 \epsilon_0$. In ordinary metal waveguides,

$$\eta^2 = k^2 - \beta^2 = \frac{\omega^2}{c^2} - \frac{\omega^2}{v_\mathrm{p}^2} > 0. \tag{9.18}$$

In delay lines,

$$\eta^2 = \frac{\omega^2}{c^2} - \frac{\omega^2}{v_\mathrm{p}^2} < 0, \tag{9.19}$$

and hence

$$-\eta^2 = \zeta^2 > 0. \tag{9.20}$$

For $\eta^2 > 0$, Eq. (9.17) has the general solution

$$\boldsymbol{E} = [\boldsymbol{A} \sin(\eta y) + \boldsymbol{B} \cos(\eta y)] \exp[j(\omega t - \beta z)]; \tag{9.21}$$

for $\eta^2 < 0$,

$$\boldsymbol{E} = [\boldsymbol{A} \sin(\zeta y) + \boldsymbol{B} \cos(\zeta y)] \exp[j(\omega t - \beta z)]. \tag{9.22}$$

While the trigonometric functions have zeros, the hyperbolic functions have no zeros except possibly at $y = 0$. Hence, in delay lines, the boundary conditions alone (and not the excitation) uniquely determine the field components of the wave. $\boldsymbol{E}$ can vary only monotonically (exponentially) in the y-direction perpendicular to the z-direction of propagation. On the other hand, in metal waveguides, an infinite number of modes can exist. The situation is similar in lines of circular cross-section in which $\partial/\partial\phi = 0$. The wave equation, say, for E_z, is of the form

$$\frac{1}{\rho}\frac{\partial}{\partial \rho}\left(\rho \frac{\partial E_z}{\partial \rho}\right) + \eta^2 E_z = 0. \tag{9.23}$$

For waveguides, $\eta^2 > 0$, and the solution of Eq. (9.23) is

$$E_z = [AJ_0(\eta\rho) + BY_0(\eta\rho)] \exp [j(\omega t - \beta z)] \tag{9.24}$$

and for delay lines, $\eta^2 < 0$, and hence the solution is

$$E_z = [AI_0(\zeta\rho) + BK_0(\zeta\rho)] \exp [j(\omega t - \beta z)]. \tag{9.25}$$

The Bessel functions J_0, J_1, Y_0, Y_1 are shown in Fig. 9.3a and the modified Bessel functions I_0, I_1, K_0, K_1 are shown in Fig. 9.3b. It can be seen from these figures that the modified Bessel

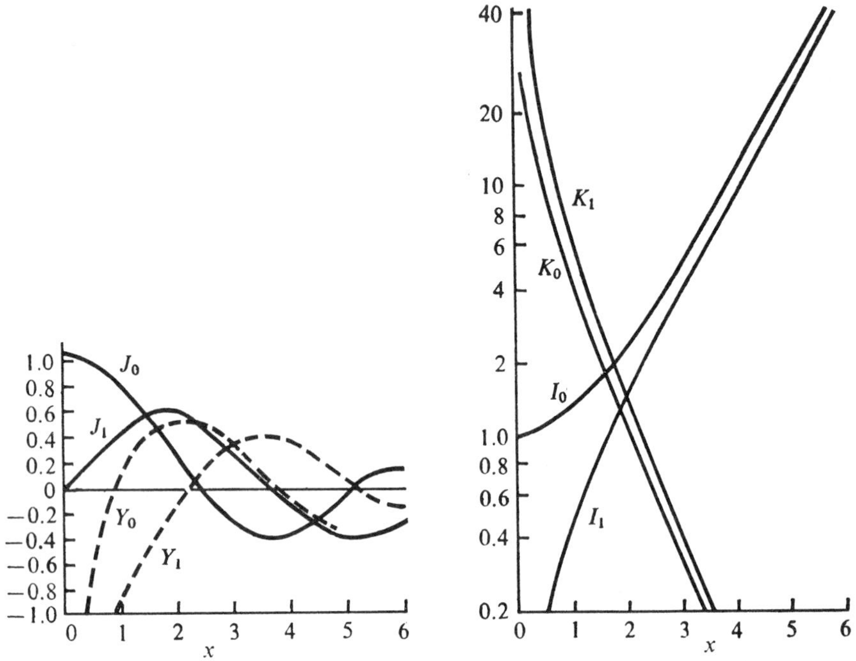

(a) J_0, J_1, Y_0, Y_1

(b) Modified Bessel functions I_0, I_1, K_0, K_1

Fig. 9.3 Bessel functions.

functions I_0 and K_0 have no zeros, and hence for delay lines, there are no nodes except at the boundary. On the other hand, since the Bessel functions J_0 and Y_0 have an infinite number of zeros, there are an infinite number of modes differing in the number of zeros of E_z in the ρ-direction for constant z. Thus, while TM_{0n} modes with $n = 1, 2, 3, \ldots$ can exist in waveguides, only TM_{01} mode can exist in a delay line.

If $\partial/\partial x \neq 0$ and $\partial/\partial\phi \neq 0$, then delay lines too can have different modes, and the shape of the boundary walls at $x = a$, $x = b$ (or $\rho = a$, $\rho = b$) determines whether waves having nodes in the region $a < x < b$ (or $a < \rho < b$) can exist. Such lines are not of much practical interest, and the analysis is complicated.

The lines considered just now have the wave propagating in the z-direction. For lines with the wave propagating in the azymuthal direction (as in the cylindrical magnetron) and for $\partial/\partial z = 0$, the wave equation for H_z becomes

$$\frac{1}{\rho}\frac{\partial}{\partial\rho}\left(\rho\frac{\partial H_z}{\partial\rho}\right) + \frac{1}{\rho^2}\frac{\partial^2 H_z}{\partial\phi^2} - \epsilon_0\mu_0\frac{\partial^2 H_z}{\partial t^2} = 0 \qquad (9.26)$$

which has the solution

$$H_z = [AJ_\gamma(k\rho) + BY_\gamma(k\rho)] \exp [j(\omega t - \gamma\phi)] \qquad (9.27)$$

for $\partial/\partial t = j\omega$, $k = \omega/c$. The angular velocity of the wave is

$$\frac{d\phi}{dt} = \omega_{\text{ph}} = \frac{\omega}{\gamma}. \qquad (9.28)$$

For such delay lines, the boundary conditions do not determine the type of function in the solution, and the solution has the same character for waveguides with $v_p > c$ and for delay lines with $v_p < c$.

9.5 Sheath Helix

The helix shown in Fig. 9.4 is most commonly used in travelling-wave tubes. The retardation process can be understood qualitatively by considering an electromagnetic wave propagating

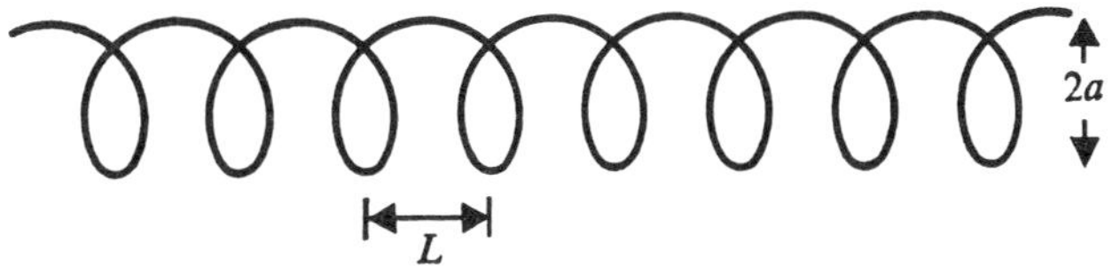

(a) Simple helix

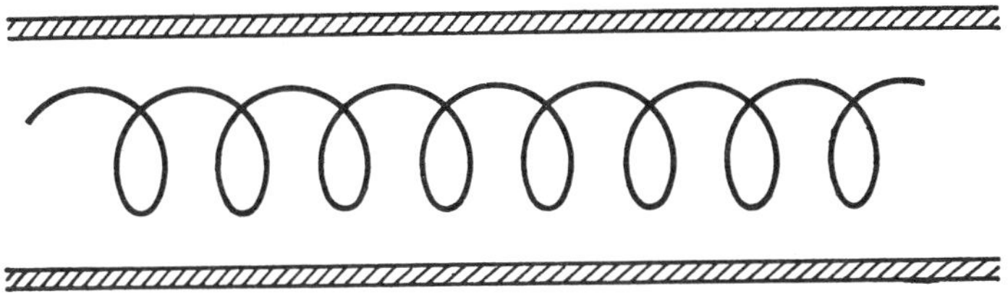
(b) Shielded helix

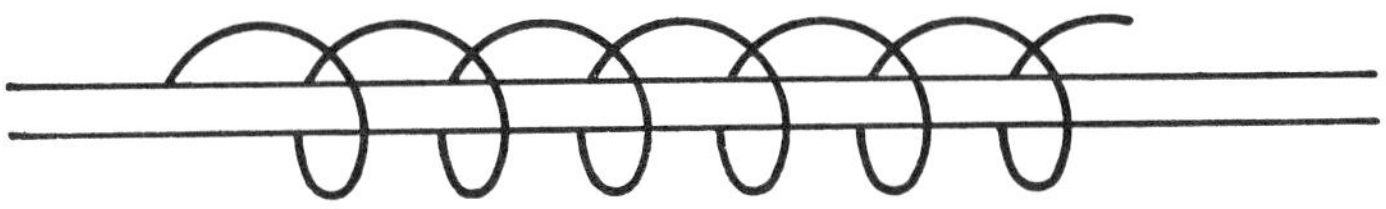
(c) Helix with inner conductor

Fig. 9.4 Types of helix.

along the turns of the helix with a velocity $v_p \approx c$. If a is the radius of the helix and L is the pitch, then the retardation in the axial direction is approximately equal to the ratio of turn length to pitch, i.e.,

$$c/v_p \approx 2\pi a/L. \qquad (9.29)$$

The crude model (Fig. 9.4) and the expression (9.29) often give quite good results.

A better understanding of the behaviour of helix is possible by taking as a model a cylinder which has a radius equal to the mean radius of the helix and which is conducting only in the direction that makes an angle ψ with the plane normal to the axis (see Fig. 9.5). This

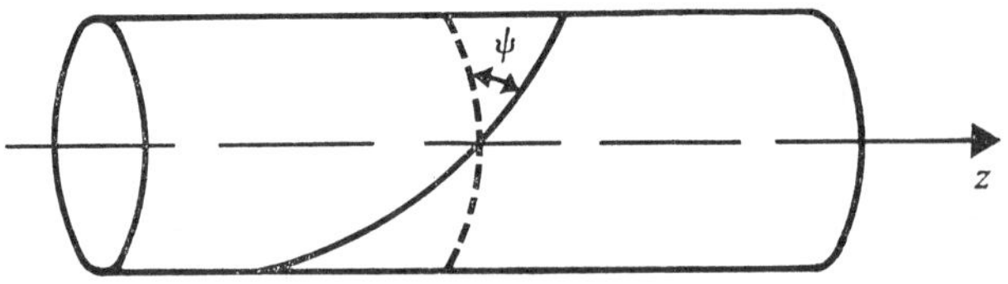

Fig. 9.5 Sheath helix.

model neglects the periodic structure of the practical helix and corresponds to a homogeneous delay line. The diameter of the helix wire is also neglected. Though a practical helix is supported by insulators (quartz, glass, and rods or tubes) which affect the field of the wave and its phase velocity v_p, this is not taken into account in this model.

The replacement of a periodic delay line by a model similar to the one given in Fig. 9.5 serves to simplify calculations. Useful results are obtained only if the phase change $|\phi_L| = |\beta|L$ over each element of length L of the periodic structure is considerably less than π. This corresponds to the condition

$$\lambda_g \gg 2L.$$

The results cease to be valid if c/v_p approaches the value $\lambda_0/(2L)$.

Since the sheath helix conducts in a skew direction, it is to be expected that TM and TE modes always appear together. The solutions of the wave equation

$$\frac{1}{\rho}\frac{\partial}{\partial\rho}(\rho\frac{\partial E_z}{\partial\rho}) + \eta^2 E_z = 0 \tag{9.30}$$

differ inside and outside the helix because, for $\rho \to 0$, we have $K_0(\zeta\rho) \to \infty$, and for $\rho \to \infty$, we have $I_0(\zeta\rho) \to \infty$. Hence, the axial field components inside the helix are given by

$$\begin{aligned} E_z^i &= AI_0(\zeta\rho) \exp[j(\omega t - \beta z)], \\ H_z^i &= CI_0(\zeta\rho) \exp[j(\omega t - \beta z)], \end{aligned} \tag{9.31a}$$

where $\zeta^2 = -\eta^2$, and outside the helix by

$$\begin{aligned} E_z^e &= BK_0(\zeta\rho) \exp[j(\omega t - \beta z)], \\ H_z^e &= DK_0(\zeta\rho) \exp[j(\omega t - \beta z)]. \end{aligned} \tag{9.31b}$$

Using Maxwell's equations, the remaining field components can be obtained. The components inside the helix are

$$\begin{aligned} &E_z^i = AI_0, \quad E_\rho^i = Aj(\beta/\zeta)I_1, \quad H_\phi^i = Aj[k/(\zeta Z_w)]I_1, \\ &H_z^i = CI_0, \quad H_\rho^i = Cj(\beta/\zeta)I_1, \quad E_\phi^i = -Cj(kZ_w/\zeta)I_1 \end{aligned} \tag{9.32a}$$

and those outside the helix by

$$E_z^e = BK_0, \qquad E_\rho^e = -Bj(\beta/\zeta)K_0, \qquad H_\phi^e = -Bj[k/(\zeta Z_w)]K_1,$$
$$H_z^e = DK_0, \qquad H_\rho^e = -Dj(\beta/\zeta)K_1, \qquad E_\phi^e = Dj(kZ_w/\zeta)K_1, \tag{9.32b}$$

where $Z_w = (\mu_0/\epsilon_0)^{1/2} = 377$ ohms.

The boundary conditions to be satisfied at $\rho = a$ are as follows.

(i) E_ϕ and E_z must be continuous across the boundary. This gives

$$AI_0 - BK_0 = 0, \tag{9.33}$$

$$CI_1 + DK_1 = 0. \tag{9.34}$$

(ii) The electric field should disappear in the direction of current flow (infinite conductivity), i.e.,

$$E_z^i \sin\psi + E_\phi^i \cos\psi = 0,$$

and hence

$$AI_0 \sin\psi - Cj(kZ_w/\zeta)I_1 \cos\psi = 0. \tag{9.35}$$

(iii) The component of magnetic field in the direction of conduction must be continuous across the boundary since no current flows in the direction perpendicular to it, so that

$$H_z^i \sin\psi + H_\phi^i \cos\psi = H_z^e \sin\psi + H_\phi^e \cos\psi,$$

and hence

$$(CI_0 - DK_0)\sin\psi + j[k/(\zeta Z_w)](BK_1 + AI_1)\cos\psi = 0 \tag{9.36}$$

and therefore

$$(CI_0 - DK_0)\sin\psi + j[k/(\zeta Z_w)](BK_1 + AI_1)\cos\psi = 0. \tag{9.37}$$

From Eqs. (9.33)–(9.37), it is possible to express the constants B, C, D in terms of the constant A, and so on. That is,

$$B = AI_0/K_0,$$

$$C = -Aj\frac{\zeta}{kZ_w}\frac{I_0}{I_1}\tan\psi, \tag{9.38}$$

$$D = Aj\frac{\zeta}{kZ_w}\frac{I_0}{K_1}\tan\psi.$$

If Eqs. (9.38) are introduced in Eq. (9.37), a relation which no longer contains any arbitrary constants is obtained. This relation is

$$(\zeta a)^2\frac{I_0(\zeta a)K_0(\zeta a)}{I_1(\zeta a)K_1(\zeta a)} = (ka\cot\psi)^2, \tag{9.39}$$

where $\cot\psi = 2\pi a/L$. Equation (9.39) is called the dispersion equation. It determines ζ, and hence the phase constant $\beta = (\zeta^2 + k^2)^{1/2}$.

Figure 9.6 gives the results of the solution of Eq. (9.39). The solid curves show c/v_p as a function of $(\omega/c)a \cot \psi$. The dashed curves refer to a helix with an external shield as obtained by Pierce[1]. The closer the shield to the helix, the less the v_p depends on frequency. For large values

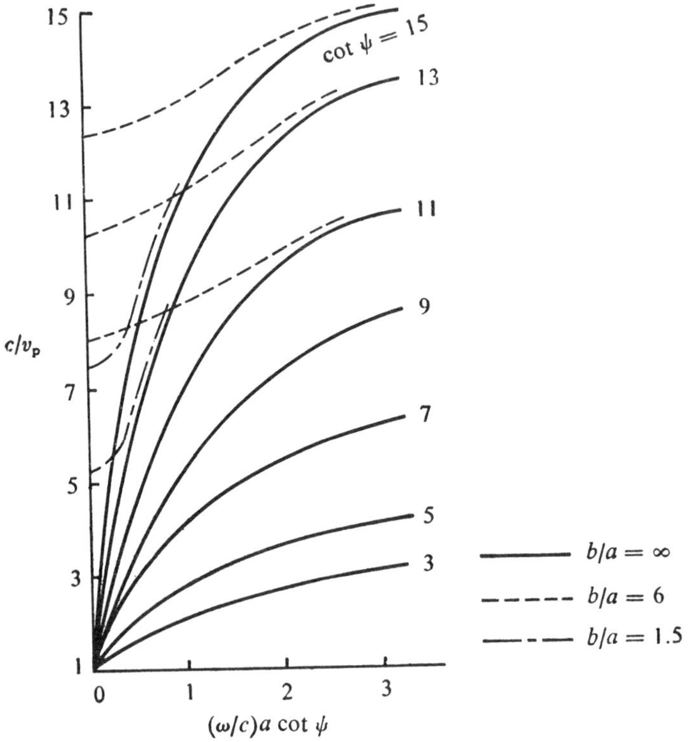

Fig. 9.6 Retardation c/v_p of a shielded helix (a, helix radius; b, shield radius).

of ζa, $I_0K_0/I_1K_1 \to 1$, so that $\zeta^2 = \beta^2 - k^2 \to \cot^2 \psi$, and from Eq. (9.39),

$$\beta \approx \frac{k}{\sin \psi} = k\frac{2\pi a}{L}. \tag{9.40}$$

Hence, for high frequencies, the wave effectively propagates with the velocity ω/β given by Eq. (9.40) along the wire. For $\zeta a \to 0$, we obtain $\beta = k$, $v_p = c$. Thus, for low frequencies, the field of the wave extends far into the outer space; the wave propagation is then hardly affected by the presence of the helix.

To calculate the coupling impedance

$$K = \frac{\hat{E}_z \hat{E}_z^*}{2P\beta^2}, \tag{9.41}$$

[1]J. R. Pierce, Theory of the beam-type travelling-wave tube, *Proc. I.R.E.*, **35**, 111, 1947.

we first calculate P, i.e.,

$$P = \pi \operatorname{Re} \int_{\rho_1}^{\rho_2} (\hat{E}_\rho \hat{H}_\phi^* - \hat{E}_\phi \hat{H}_\rho^*)\rho \, d\rho, \tag{9.42}$$

where ρ_1 and ρ_2 are the limits of the radius between which the power is calculated. For the helix, the integration has to be done separately over the inner and outer regions. Hence

$$P = \pi \operatorname{Re} \left[\int_0^a (\hat{E}_\rho^i \hat{H}_\phi^{i*} - \hat{E}_\phi^i \hat{H}_\rho^{i*})\rho \, d\rho + \int_a^\infty (\hat{E}_\rho^e \hat{H}_\phi^{e*} - \hat{E}_\phi^e \hat{H}_\rho^{e*})\rho \, d\rho \right], \tag{9.43}$$

where the field components are given by Eqs. (9.32). The integrations can be carried out by using the relations

$$\int_0^a I_1^2(\zeta\rho)\rho \, d\rho = \tfrac{1}{2}a^2[I_1^2(\zeta a) - I_0(\zeta a)I_2(\zeta a)], \tag{9.44}$$

$$\int_\infty^a K_1^2(\zeta\rho)\rho \, d\rho = \tfrac{1}{2}a^2[K_1^2(\zeta a) - K_0(\zeta a)K_2(\zeta a)], \tag{9.45}$$

$$\int_0^a I_1(\zeta\rho)K_1(\zeta\rho)\rho \, d\rho = \tfrac{1}{4}a^2[2I_1(\zeta a)K_1(\zeta a) - I_0(\zeta a)K_2(\zeta a) - I_2(\zeta a)K_0(\zeta a)]. \tag{9.46}$$

The expression finally obtained for P contains AA^* as a constant of proportionality, in addition to the Bessel functions of argument ζa. In Eq. (9.41), $\hat{E}_z \propto A$ and, in particular, for $\rho = 0$ (i.e., on the axis of the helix), $\hat{E}_z(0) = A$ since $I_0(0) = 1$. If we assume that $v_p \ll c$ and $\zeta \approx \beta$ (which are usually true in practice), then we obtain, for $\rho = 0$,

$$K = \frac{\hat{E}_z(0)\hat{E}_z^*(0)}{2P\beta^2} = \frac{\beta}{2k}\,\frac{240}{\beta a}\,\frac{K_0}{I_0}\left(\frac{I_1}{I_0} - \frac{I_0}{I_1} + \frac{K_0}{K_1} - \frac{K_1}{K_0} + \frac{4}{\beta a}\right)^{-1} \text{ohms}, \tag{9.47}$$

where the argument of the Bessel function is ζa. This expression is approximately equal to

$$K = 184(\beta/k) \exp(-2\beta a) \text{ ohms}. \tag{9.48}$$

In Fig. 9.7, K is shown as a function of βa with c/v_p as a parameter.

The measurement of the phase velocity v_p as a function of frequency is fairly easy, and the group velocity v_g can be found by using relation (9.6). On the other hand, experimental determination of K is quite difficult so that, by knowing v_p as well as β, K can be found by using Eq. (9.48).

For $v_p \ll c$, the fields of the wave can be derived from a scalar potential V which must satisfy the Laplace equation

$$\nabla^2 V = 0. \tag{9.49}$$

In cylindrical systems with only one conductor $\rho = a$, where $V = V_a$, the solutions of Eq. (9.49) are

$$V = V_a \frac{I_0(\beta\rho)}{I_0(\beta a)} \exp[j(\omega t - \beta z)] \quad \text{for } 0 \leq \rho \leq a, \tag{9.50}$$

$$V = V_a \frac{K_0(\beta\rho)}{K_0(\beta a)} \exp[j(\omega t - \beta z)] \quad \text{for } a \leq \rho < \infty. \tag{9.51}$$

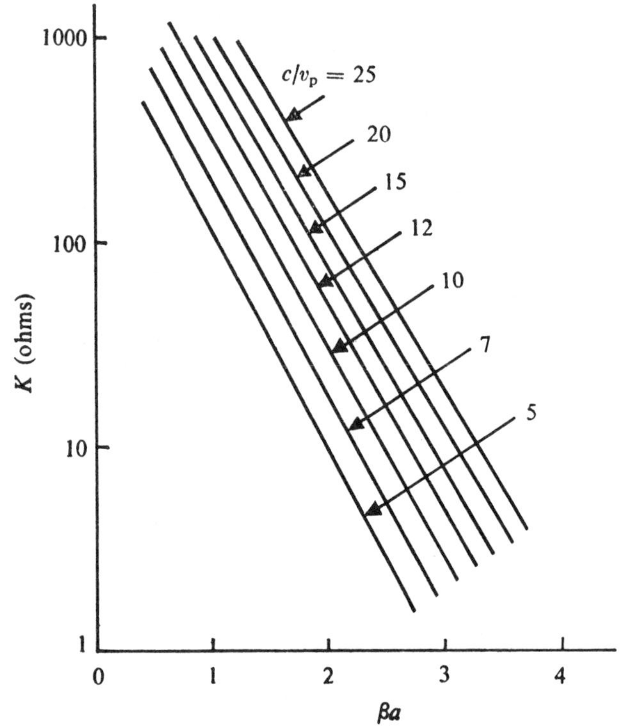

Fig. 9.7 Coupling impedance K on axis of sheath helix.

Hence, for $0 \leqslant \rho \leqslant a$,

$$E_\rho = -\frac{\partial V}{\partial \rho} = -V_a \frac{\beta I_1(\beta\rho)}{I_0(\beta a)} \exp\left[j(\omega t - \beta z)\right], \tag{9.52}$$

$$E_z = -\frac{\partial V}{\partial z} = jV_a \frac{\beta I_0(\beta\rho)}{I_0(\beta a)} \exp\left[j(\omega t - \beta z)\right] \tag{9.53}$$

and for $a \leqslant \rho < \infty$,

$$E_\rho = -\frac{\partial V}{\partial \rho} = V_a \frac{\beta K_1(\beta\rho)}{K_0(\beta a)} \exp\left[j(\omega t - \beta z)\right], \tag{9.54}$$

$$E_z = -\frac{\partial V}{\partial z} = jV_a \frac{\beta K_0(\beta\rho)}{K_0(\beta a)} \exp\left[j(\omega t - \beta z)\right]. \tag{9.55}$$

The energy stored per unit length of the line, W', is equal to twice the stored electric energy since the stored magnetic and electric energies are equal. Hence

$$W' = \pi\epsilon_0 \int_0^\infty (\hat{E}_\rho \hat{E}_\rho^* + \hat{E}_z \hat{E}_z^*)\rho \, d\rho. \tag{9.56}$$

After substituting for E_ρ and E_z from Eqs. (9.52)–(9.55) and using the recurrence formulae (9.44) and (9.45), we obtain

$$W' = \frac{\pi\epsilon_0 a}{\beta}\left(\frac{I_1}{I_0} + \frac{K_1}{K_0}\right)\hat{E}_z(0)\hat{E}_z^*(0). \tag{9.57}$$

However,

$$P = W'v_g, \qquad \epsilon_0 = \frac{1}{cZ_w} = \frac{1}{120\pi c},$$

so that we obtain, on the axis,

$$K = 60\left(\frac{c}{v_g}\right)I_0(\beta a)K_0(\beta a) \text{ ohms}. \tag{9.58}$$

Expressions (9.47) and (9.58) give very nearly the same values for K, and the difference between them is generally smaller than that between the measured and calculated values.

In general, the fundamental equations (9.50)–(9.56) hold for delay lines of circular cross-section of only one conductor, at $\rho = a$. Hence, this simple method of determining coupling impedance (for known v_p and v_g) can be used quite generally. For different lines, the integration limits in Eq. (9.56) may have to be altered. $K \propto 1/v_g$ is always obtained as the final relation. For comparable lines and a given c/v_p, the coupling impedance K increases as v_g decreases with respect to the phase velocity v_p. Thus, an increase in dispersion leads to tighter coupling and hence to greater amplification in travelling-wave devices. High dispersion means small bandwidth and phase distortion.

9.6 Parallel-Plate Delay Line

A simple type of delay line is a parallel-plate delay line consisting of two parallel-plane conductors of large extension in the x-direction as compared to their separation d in the y-direction. Let one conductor be solid and the other suitably periodically constructed (for example, in the form of a wire with hair-pin bends) to slow down the wave (see Fig. 9.8).

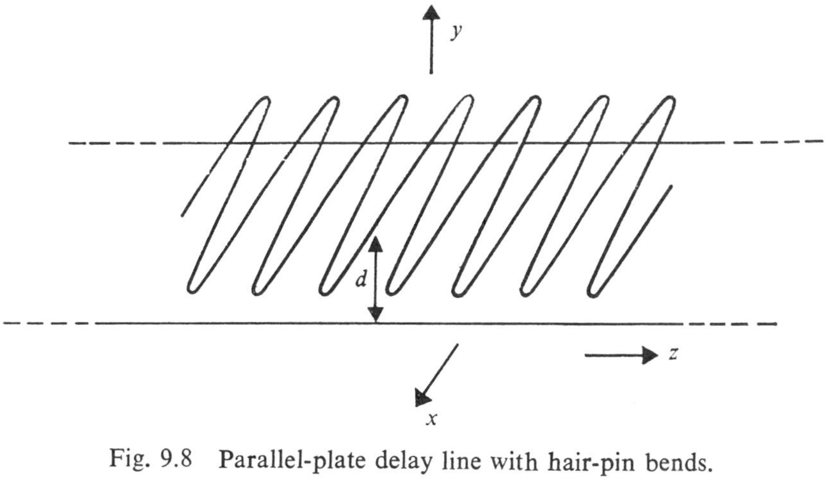

Fig. 9.8 Parallel-plate delay line with hair-pin bends.

Using the same type of approximations as used in Section 9.5, K can be calculated from the measured values of v_p and v_g. A forced sinusoidal wave for which $v_p \ll c$ can be derived from a potential function

$$V = f(y) \exp [j(\omega t - \beta z)]. \tag{9.59}$$

The solutions of Laplace's equation for a two-dimensional plane system, with the boundary conditions $V = V_d \exp [j(\omega t - \beta z)]$, when $y = d$ for $0 \leqslant y \leqslant d$, are

$$V = V_d \frac{\sinh (\beta y)}{\sinh (\beta d)} \exp [j(\omega t - \beta z)], \tag{9.60}$$

$$E_z = j\beta V_d \frac{\sinh (\beta y)}{\sinh (\beta d)} \exp [j(\omega t - \beta z)], \tag{9.61}$$

$$E_y = -\beta V_d \frac{\cosh (\beta y)}{\sinh (\beta d)} \exp [j(\omega t - \beta z)] \tag{9.62}$$

and for $d \leqslant y < \infty$,

$$V = V_d \exp [-\beta(y - d)] \exp [j(\omega t - \beta z)], \tag{9.63}$$

$$E_z = j\beta V_d \exp [-\beta(y - d)] \exp [j(\omega t - \beta z)], \tag{9.64}$$

$$E_y = \beta V_d \exp [-\beta(y - d)] \exp [j(\omega t - \beta z)]. \tag{9.65}$$

For a line of width Δx, the energy stored per unit length is

$$W' = \tfrac{1}{2}\epsilon_0 \Delta x \int_0^\infty (\hat{E}_y \hat{E}_y^* + \hat{E}_z \hat{E}_z^*)\, dy \tag{9.66}$$

in which the integration is carried out separately for $y < d$, $y > d$. Consequently,

$$W' = \tfrac{1}{2}\epsilon_0 \Delta x \beta |V_d|^2 (\coth \beta d + 1) \tag{9.67}$$

and since $P = W' v_g$,

$$K = \frac{|V_d|^2}{2P\beta^2} = \frac{1}{\beta \epsilon_0 v_g \Delta x} \frac{\sinh^2 (\beta y_0)}{\sinh^2 (\beta d)(\coth \beta d + 1)}$$

$$= \frac{120\pi}{\beta \Delta x} \frac{c}{v_g} \frac{\sinh^2 (\beta y_0)}{\sinh (\beta d) \exp (\beta d)} \text{ ohms}, \tag{9.68}$$

where y_0 is the distance between the solid electrode and the point at which K is calculated. (In a travelling-wave tube, this point is the position of the electron beam.)

9.7 Inhomogeneous Delay Lines

Figure 9.9 shows a few different forms of delay lines with periodic structures. The delay line of Fig. 9.9a is often used in linear accelerators. The interdigital line of Fig. 9.9e is used in carcinotrons and magnetrons. The magnetron delay line of Fig. 9.9f has an angular period and is closed on itself to form a ring.

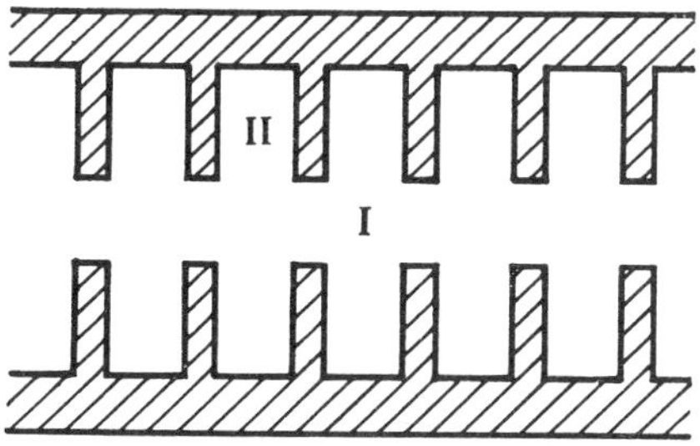

(a) Corrugated cylindrical waveguide

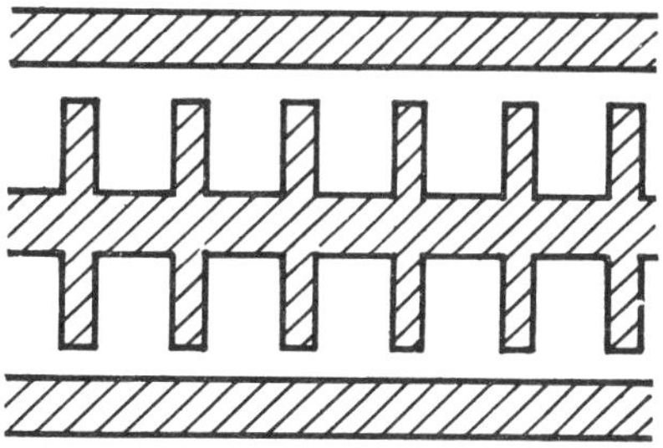

(b) Coaxial line with corrugated conductor

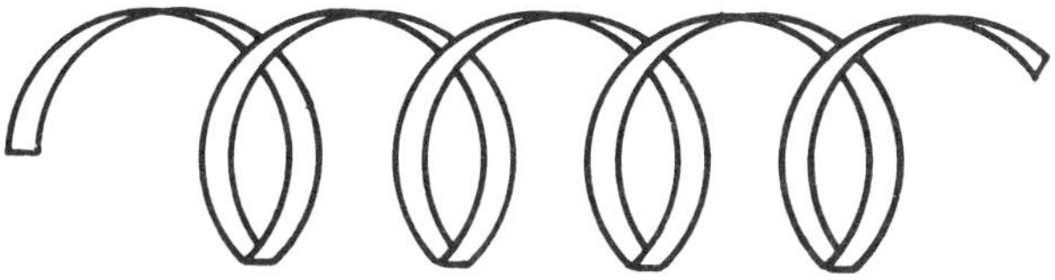

(c) Tape helix

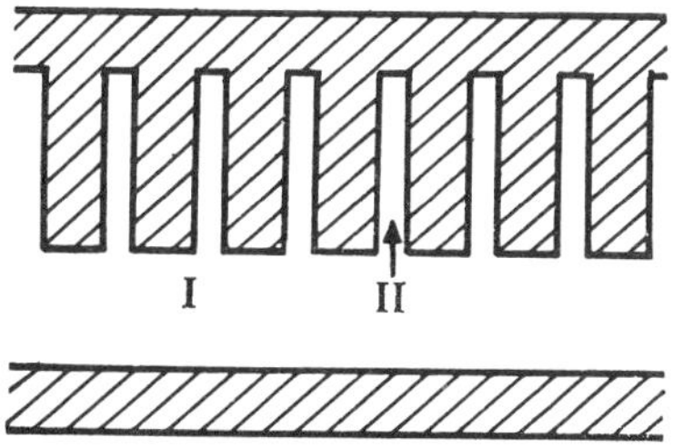

(d) Parallel-plane line with one corrugated conductor

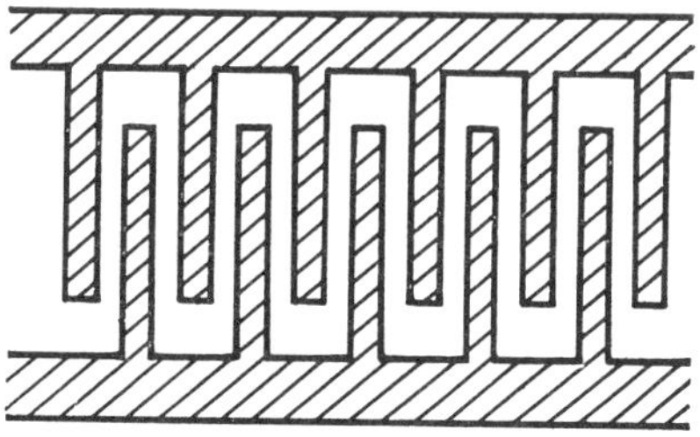

(e) Interdigitated line

(f) Travelling-wave magnetron with slotted anode

Fig. 9.9 Periodic inhomogeneous delay lines.

9.8 Equivalent Circuits of Inhomogeneous Delay Lines or Periodic Structures

Equivalent circuits containing lumped susceptances for the elementary four-terminal networks give a good qualitative picture of inhomogeneous delay lines. The properties of these equivalent circuits are known from filter theory. Figure 9.10 shows such an equivalent circuit.

The pass-band of a four-terminal network is given by

$$-1 \leqslant \cos \phi_L = 1 + \frac{B_2}{2B_1} \leqslant 1. \tag{9.69}$$

The term $\cos \phi_L = 1$ or $\phi_L = 0$ corresponds to the resonant frequency of the shunt susceptance

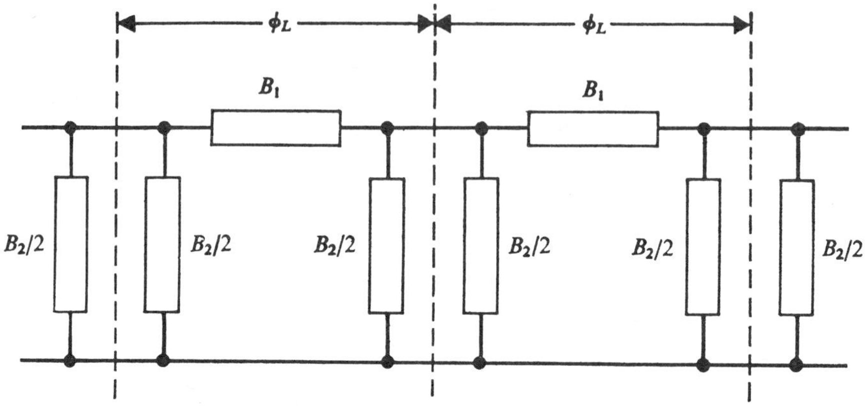

Fig. 9.10 Equivalent circuit of delay line in elementary four-terminal network form.

$B_t = B_2$ and to the cut-off frequency f_0; $\cos \phi_L = -1$ or $|\phi_L| = \pi$ corresponds to the resonant frequency of the susceptance $B_l = B_1 + B_2/4$ and to the cut-off frequency f_π. From Eq. (9.69), it can be seen that

$$\cos \phi_L = \frac{4B_l + B_t}{4B_l - B_t}. \tag{9.70}$$

For real values of ϕ_L, Eq. (9.70) can be satisfied only if B_l and B_t have opposite signs. Frequency bands in which B_l and B_t have the same sign are the stop-bands of the filter.

Since $\cos \phi_L = \cos(-\phi_L)$, the sign of ϕ_L cannot be determined from Eq. (9.69) or (9.70). The sign is obtained from the convention that the group velocity v_g is always assumed to be positive. Hence, $d\phi/d\omega$ is always positive.

According to this convention, it is possible to show that there are two types of filters: (i) those with $\phi > 0$ (forward-wave filters) and (ii) those with $\phi < 0$ (backward-wave filters). Figure 9.11 shows two basic four-terminal filter networks: one a forward-wave structure and the other a backward-wave structure. In Fig. 9.11a, the current $\hat{I}_1$ entering the series admittance lags behind the voltage $\hat{V}_1$, and the voltage $\hat{V}_2$ in turn lags behind the current $\hat{I}_1$, so that $\hat{V}_2$ lags behind $\hat{V}_1$. Since the energy at point 2 also lags behind the energy at point 1, the phase velocity v_p and the group velocity v_g have the same sign. In the filter of Fig. 9.11b, however, $\hat{I}_1$ leads $\hat{V}_1$ and $\hat{V}_2$ leads $\hat{I}_1$ so that $\hat{V}_2$ leads $\hat{V}_1$. Hence, the phase velocity has the opposite sign to the group velocity and, since by definition, $v_g > 0$, v_p must be less than zero. Therefore,

Fig. 9.11a is a forward-wave structure and 9.11b is a backward-wave structure. Also, the former is a low-pass filter and the latter a high-pass filter. Similar considerations also hold good for band-pass filters where tuned circuits exist in place of L and C.

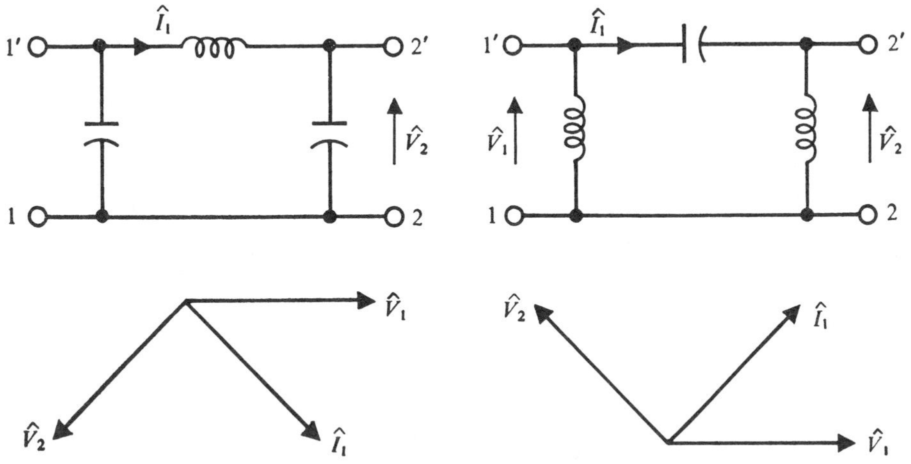

(a) Forward-wave structure (b) Backward-wave structure

Fig. 9.11 Basic four-terminal filter networks and vector diagrams.

According to Eq. (9.70), the susceptances in the series and shunt arms must have opposite signs. In general, we may conclude that the filter is a forward-wave filter if B_l is inductive and B_t capacitive, and a backward-wave filter if B_l is capacitive and B_t is inductive.

The foregoing considerations can be applied qualitatively to the two structures shown in Fig. 9.12. The adjacent wires of both lines are coupled inductively and capacitively. The capacitive

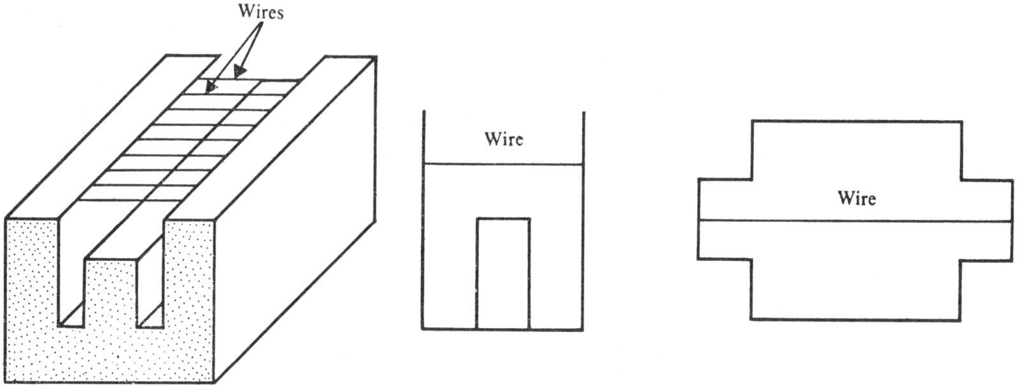

(a) Karp circuit and cross-section of simplified model of the line (b) Section through delay line with transverse wires (backward-wave structure)

Fig. 9.12 Slow-wave structures.

coupling is increased if additional capacitance is introduced at a point of high field strength. This is provided in Fig. 9.12a by the central ridge which increases the transverse field E_y at this point. This corresponds to a capacitive shunt susceptance B_t in the equivalent circuit. Hence, this line is a forward-wave (low-pass) filter. In the line of Fig. 9.12b, however, the ridges are at points where the field strength is low. For this reason, the ridges produce practically no increase in the capacity of the shunt branch of the equivalent circuit, but they reduce the inductance of the series branch to the extent of making B_l capacitive. Figure 9.12b corresponds to a backward-wave (high-pass) filter.

9.9 Wave Propagation in Lines of Periodic Structure: Space Harmonics

Since inhomogeneous delay lines are periodic structures, the field quantities of the wave at the points z and $z + L$ differ merely by the phase angle $\phi_L = \beta L$; so that

$$F(z + L) = F(z) \exp(-j\phi_L), \tag{9.71}$$

where L is the period or the length of an elementary four-pole network. For a given frequency, this relation is not only satisfied by the phase angle $\phi_{L,0}$, but by any phase angle $\phi_{L,n}$ given by

$$\phi_{L,n} = \phi_{L,0} + 2\pi n, \tag{9.72}$$

where $n = 0, \pm 1, \pm 2, \ldots$. The corresponding phase constants and phase velocities are

$$\beta_n = \frac{\phi_{L,n}}{L} = \frac{\phi_{L,0} + 2\pi n}{L}, \tag{9.73}$$

$$v_{p,n} = \omega/\beta_n = \omega L/(\phi_{L,0} + 2\pi n). \tag{9.74}$$

Relations (9.73) and (9.74) show that, at a given angular frequency ω, the propagation of an infinite number of waves is fundamentally possible in periodic delay lines. Depending on the sign and value of n, these waves can have positive or negative phase velocities, $v_{p,n}$. These waves are called *space harmonics* or *Hartree harmonics*, and together they form a wave group of single group velocity independent of n, i.e.,

$$\frac{1}{v_g} = \frac{d\beta_n}{d\omega} = \frac{d\beta_0}{d\omega}. \tag{9.75}$$

When the a-c power of a given frequency is introduced at the input of a periodic delay line, the boundary conditions will determine whether all the space harmonics or only some or one of them will be excited. If the conducting walls of the delay line have corners, ridges, or slots, it is impossible to satisfy the corresponding boundary conditions by the field of a wave with a single phase velocity, and a wave group of phase velocities $v_{p,0}, v_{p,1}, v_{p,2}, v_{p,-1} \ldots$ is then required; also, an infinite number of space harmonics are required mathematically. This is similar to the Fourier analysis of a periodic curve with sharp corners or discontinuities, where an infinite number of harmonics are required. 'Space harmonics' are not harmonics in the sense of multiple frequencies, but are waves of the same frequency and group velocity and have different phase constants and phase velocities. The wave having $n = 0$ is called the fundamental space harmonic; the waves for which $n \neq 0$ are called higher space harmonics.

The dispersion curves of a group of space harmonics are shown in Fig. 9.13, where $c/v_p = f(\lambda_0/L)$. The pass-band is the same for all space harmonics.

This figure applies to a band-pass delay line in which the fundamental space harmonic ($n = 0$) is a forward-wave structure ($\phi_0 > 0$). Such a structure is said to have positive dispersion in contrast to a line of negative dispersion in which the fundamental space harmonic is a backward-wave structure ($\phi_0 < 0$). Delay lines may have positive dispersion in certain frequency ranges and negative dispersion in some other frequency ranges.

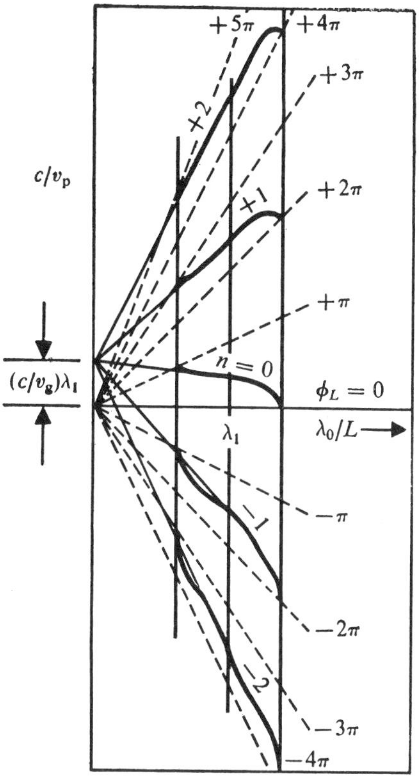

Fig. 9.13 Retardation c/v_p of space harmonics in a periodic delay line with band-pass characteristics. [The fundamental space harmonic ($n = 0$) is a forward wave (line with positive dispersion).]

From the dispersion curves shown in Fig. 9.13, a few general properties of these curves may be observed. From Eq. (9.74), we can derive the relation

$$\frac{c}{v_{p,n}} = \left(\frac{\phi_{L,0}}{2\pi} + n\right)\frac{\lambda_0}{L}. \tag{9.76}$$

For shielded lines in which the field cannot extend to infinity, we have, at the limits of the pass-band, $\phi_{L,0} = 0$ or $|\phi_{L,0}| = \pi$. Hence, each dispersion curve of the separate space harmonics lies within a region bounded by the two straight lines through the origin, i.e.,

$$\frac{c}{v_p} = \frac{n\lambda_0}{L}, \qquad \frac{c}{v_p} = (n + \tfrac{1}{2})\frac{\lambda_0}{L}.$$

These lines are the dashed lines in Fig. 9.13 for different values of $\phi_{L,0} = n\pi$. At the limits of the pass-band, $v_g = 0$ (except at the lower limit of a low-pass filter, where $\lambda_0 = \infty$). From Eq. (9.6), we have

$$\frac{d(c/v_p)}{d(\lambda_0/L)} = \pm\infty \tag{9.77}$$

which means that the dispersion curves have vertical tangents at the ends of the band.

If $c/v_p = f(\lambda_0/L)$ is known for one space harmonic, namely, $n = m$, then c/v_p for all other space harmonics, viz., $n = m + l$, can be found from it. From Eq. (9.76),

$$\frac{c}{v_{p,\,m+l}} = \frac{c}{v_{p,\,m}} + \frac{l\lambda_0}{L}. \tag{9.78}$$

For a given λ_0/L, the tangents to all the dispersion curves cut the ordinate axis at the point c/v_g, where v_g is the group velocity corresponding to the given value of λ_0/L.

In Fig. 9.14, we have $AP = c/v_p$, $OA = \lambda_0/L$, and therefore

$$OC = AP - BP = \frac{c}{v_p} - \frac{\lambda_0 d(c/v_p)}{Ld(\lambda_0/L)}. \tag{9.79}$$

Using Eq. (9.6), we have

$$OC = c[\frac{1}{v_p} - \frac{d(1/v_p)}{d(\lambda_0)}] = \frac{c}{v_g}. \tag{9.80}$$

Let us now examine the position of the pass-band for all space harmonics and the values

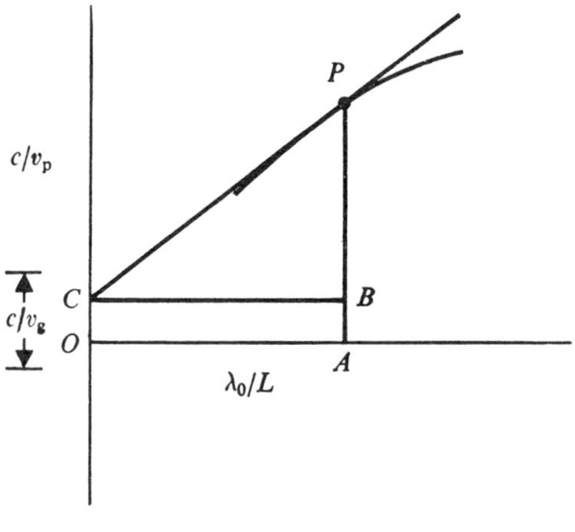

Fig. 9.14 Group velocity of arbitrary space harmonic.

of c/v_p at the band limits. $\phi_{L,0} = 0$ corresponds to $\lambda_g = \infty$. Since

$$\phi_{L,0} = \beta_0 L = \frac{2\pi c L}{\lambda_0 v_p} = \frac{2\pi L}{\lambda_g},$$

at $\phi_{L,0} = 0$ for the fundamental, $c/v_p = 0$. An exception is a low-pass filter where, for $\phi_{L,0} = 0$ and $\lambda_0 = \infty$, we have $c/v_p \neq 0$. $|\phi_{L,0}| = \pi$ always corresponds to $|c/v_p| = \lambda_0/(2L)$. Therefore, c/v_p is always nonzero at this end of the pass-band. For lines which are not shielded, one cut-off frequency is determined by the fact that none of the space harmonics can travel faster than light; otherwise, energy would be radiated from the line into the outer space.

An example of an interdigital line serves to illustrate the foregoing results. This line is shown in Fig. 9.15 and consists of two interpenetrating conductors, each having fingers of length h and possibly a conducting plate parallel to the fingers. The structure can be regarded

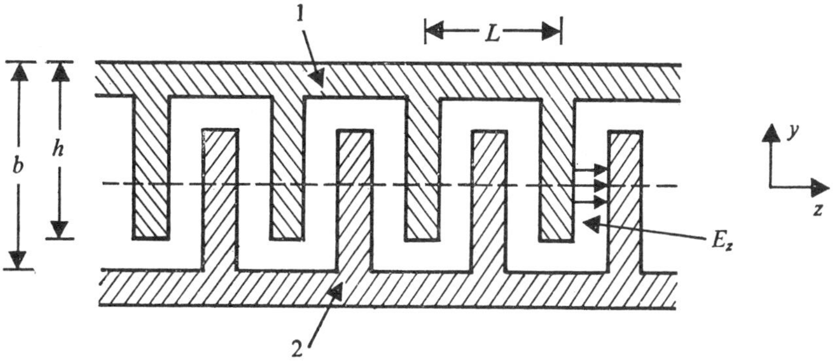

Fig. 9.15 Interdigital line.

as a folded parallel-plate transmission line in which the wave travels along the fingers in the y-direction, with a velocity $v_p = c$. Total reflection takes place at the walls. This results in a phase change $\pm\pi$, and the appropriate sign is selected later on. Such a phase change occurs twice in each period of length L. Then, the phase difference over the distance $z = L$ is

$$\phi_{L,n} = \beta_n L = \frac{\omega L}{v_{p,n}} = (2h + L)\frac{\omega}{c} + 2\pi n \pm 2\pi \tag{9.81}$$

or

$$\frac{c}{v_{p,n}} = 1 + \frac{2h}{L} + (n \pm 1)\frac{\lambda_0}{L}. \tag{9.82}$$

The group velocity of all space harmonics follows directly from Eq. (9.81), i.e.,

$$\frac{c}{v_g} = c\left(\frac{d\beta}{d\omega}\right) = 1 + \frac{2h}{L}, \tag{9.83}$$

and is independent of n and ω. By definition, the fundamental space harmonic ($n = 0$) has the property that the absolute value of its phase velocity is greater than that of any other space harmonic. For large values of λ_0, this is true if the negative sign is chosen in Eq. (9.81) so that

$$\frac{c}{v_{p,0}} = 1 + \frac{2h}{L} - \frac{\lambda_0}{L}. \tag{9.84}$$

Thus, $c/v_{p,0} < 0$ for $\lambda_0 > 2h + L$. The fundamental space harmonic is then a backward wave, and the line has negative dispersion. The pass-band is for $-\pi < \phi_{L,0} < 0$, and therefore from Eq. (9.81), we obtain $\lambda_0 = 2h + L$ and $\lambda_0 = 4h + 2L$ for the band limits. In Fig. 9.16, the dispersion curves are shown for $h = 3L$. The fundamental space harmonic ($n = 0$) has $c/v_p < 0$; the space harmonic $n = 1$ has $c/v_p = \text{constant} > 0$, and hence there is no dispersion. For all higher space harmonics, c/v_p is strongly dependent on n. The tangents to the curves $c/v_p = f(\lambda_0/L)$ at the ends of the band are not vertical. This is so because the calculations which are approximate are not valid at the limits of the band. Hence, deviations in the shape of the curves are to be expected. In curves derived by more exact methods, the tangents at the limits are vertical.

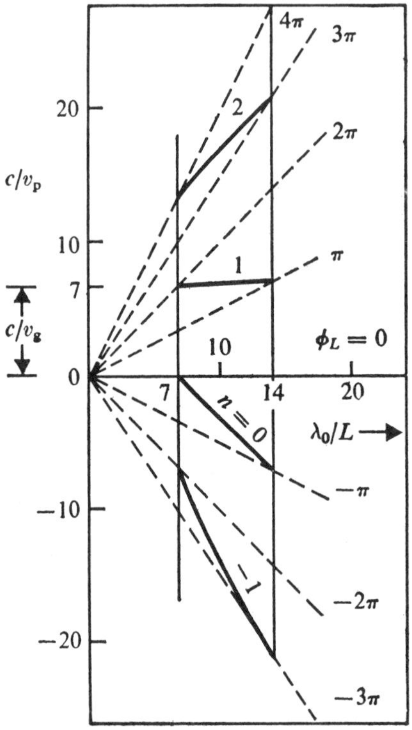

Fig. 9.16 Dispersion curves $c/v_p = f(\lambda_0/L)$ of space harmonics of the interdigital line.

Using the foregoing considerations, the relative amplitudes of the space harmonics of the line for different values of y can be compared qualitatively. In the middle of the fingers at $y = 0$, the line is periodic in the z-direction with period $L/2$, though for some other values of y the line is periodic in L, where L is the distance between two adjacent fingers attached to one of the walls. Therefore, at $y = 0$, by symmetry, the phase changes by $\beta L/2$ in a distance $z = L/2$. Therefore, only the space harmonics with n even are present at $y = 0$. At the edges of the lines $y = \pm d$, this is not possible, and space harmonics with both n even and n odd are present. For each space harmonic, the field pattern in the y-direction varies in a different manner, and the relative amplitudes of the separate space harmonics depend on the value of y.

Delay lines are used in travelling-wave tubes, travelling-wave magnetrons, carcinotrons and so on, which belong to the general class of growing-wave tubes. Amplification and oscillation occur in these tubes when the difference between the electron velocity and the phase velocity of the wave is small. When inhomogeneous delay lines are used for amplification, the electron velocity can be matched to the phase velocity of a single space harmonic so that, for a given frequency, amplification can be obtained for a whole line spectrum of different electron velocities.

In this process, the entire wave group is amplified, and not just the n-th space harmonic having $v_{p,n}$ approximately equal to the electron velocity. This is so because the relative amplitudes of the separate space harmonics depend only on the geometry of the line and are not affected by the amplification. Also, v_g is always greater than zero, even though v_p is negative. Since the net power flow has the direction of the group velocity, a wave can be amplified so that the direction of power increase is opposite to that of the electron beam if the electron velocity is made equal to a negative phase velocity of one of the space harmonics. This is the principle of the mechanism of the backward-wave tube and the M-carcinotron.

Usually, amplification effects in growing-wave type of microwave tubes result from the interaction of the electron beam with the fundamental space harmonic which is the wave having the maximum phase velocity. The amplitudes of the higher space harmonics are smaller than those of the fundamental space harmonic. This means that the beam is more loosely coupled to the higher space harmonics than to the fundamental space harmonic. The ratio of the amplitudes of the higher space harmonics to that of the fundamental space harmonic decreases with $|\phi_{L,0}|$. For $|\phi_{L,0}| \to 0$ and $n = 0$, we have

$$\lambda_g = \frac{2\pi}{\beta_0} = \frac{2\pi L}{\phi_{L,0}} \gg L.$$

Hence, a periodic line can be treated as a homogeneous delay line ($L/\lambda_g \to 0$). The higher space harmonics can then be neglected. As $|\phi_{L,0}| \to \pi$, it is increasingly necessary to take the higher space harmonics into account. The energy produced by deceleration of electrons in the field of one space harmonic is always delivered to the whole wave group, and not to this space harmonic alone. This corresponds to a reduction in the coupling impedance of this space harmonic as compared to the case where no other space harmonics are present or where they are neglected. Therefore, the coupling impedance of each space harmonic depends on the amplitude of all the space harmonics, and the presence of the higher harmonics changes the coupling impedance of the fundamental space harmonic, particularly as $|\phi_{L,0}| \to \pi$.

9.10 General Method of Analyzing Periodic Delay Lines

A rigorous analysis of a delay line consists of the solution of Maxwell's equations to satisfy the proper boundary conditions. In the lines shown in Figs. 9.9a, 9.9b, and 9.9d, the space surrounding the conductors can be divided into two parts: space I in which a travelling wave propagates and space II inside transverse slots or grooves in the conductor. In space II, standing waves exist as a result of reflections at the ends of the slots.

The field equations are usually set up for the two spaces independently, and the arbitrary constants are obtained from the condition that the fields are continuous across the boundary of the two regions. Though this condition is self-evident physically, in mathematical terms, in most cases, it can be satisfied only approximately. The freedom of choice in field matching leads to somewhat different results obtained by different workers. Investigations which take into account only the fundamental give accurate results only for a small phase angle $\phi_{L,0}$.

Field matching at the boundary can often be carried out by using the principle of equality of the admittances of the waves on the two sides of the boundary. Equating the wave admittances on the two sides of the boundary is equivalent to the equality of the field components.

Considerable uncertainty prevails in the analysis of periodic delay lines, and all calculations only represent varying degrees of approximation, depending on different physical assumptions.

To illustrate the method of analysis, we now consider the example of a plane-periodic delay line.

9.11 Analysis of Plane-Periodic Delay Line

Figure 9.17 shows a plane-periodic structure which may be used in a plane magnetron. As an approximation, the edge effects caused by the finite extension in the x-direction may be

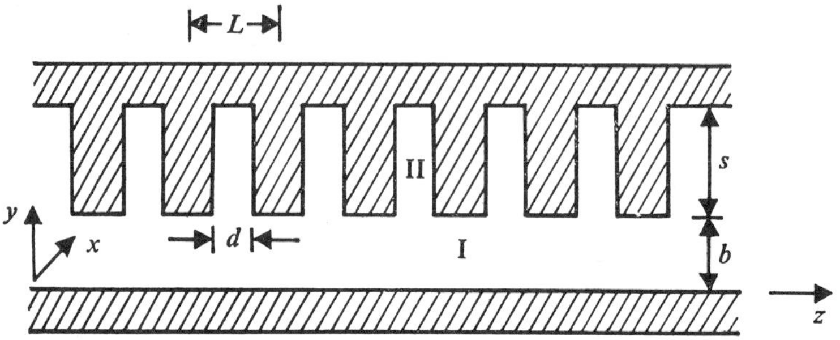

Fig. 9.17 Plane-periodic delay line.

neglected. The wave propagating in space I may be considered to be a summation of an infinite number of space harmonics. For TM waves, the components E_z^{I}, E_y^{I}, and H_x^{I} in space I may be written as

$$E_z^{\mathrm{I}} = \sum_{n=-\infty}^{\infty} A_n \sinh(\zeta_n y) \exp[j(\omega t - \beta_n z)], \tag{9.85}$$

$$E_y^{\mathrm{I}} = j \sum_{n=-\infty}^{\infty} A_n \frac{\beta_n}{\zeta_n} \cosh(\zeta_n y) \exp[j(\omega t - \beta_n z)], \tag{9.86}$$

$$H_x^{\mathrm{I}} = \frac{-j}{Z_{\mathrm{W}}} \sum_{n=-\infty}^{\infty} A_n \frac{k}{\zeta_n} \cosh(\zeta_n y) \exp[j(\omega t - \beta_n z)], \tag{9.87}$$

where

$$\beta_n = \frac{\omega}{v_{\mathrm{p},n}} = (\zeta_n^2 + k^2)^{1/2} \tag{9.88}$$

is the phase constant of the n-th space harmonic associated with a phase change $\phi_{L,n} = \beta_n L$ over the period L, and

$$\phi_{L,n} = \phi_{L,0} + 2\pi n = \beta_n L. \tag{9.89}$$

The phase velocity of the n-th space harmonic is

$$v_{\mathrm{p},n} = \omega/\beta_n = L/(\phi_{L,0} + 2\pi n). \tag{9.90}$$

In space II, the stray fields can be neglected, and assuming that $E_z^{\mathrm{II}} = 0$ at $y = s + b$, the fields in the p-th slot are given by

$$E_z^{\mathrm{II}} = E_z^{\mathrm{II}}(b)\frac{\sin[k(s + b - y)]}{\sin(ks)} \exp[j(\omega t - p\phi)], \tag{9.91}$$

$$E_y^{\mathrm{II}} = 0, \tag{9.92}$$

$$H_x^{\mathrm{II}} = \frac{-j}{Z_{\mathrm{w}}} E_z^{\mathrm{II}}(b) \frac{\cos\,[k(s + b - y)]}{\sin\,(ks)} \exp\,[j(\omega t - p\phi)]. \tag{9.93}$$

To determine the constants A_n, we assume that, at $y = b$, $E_z^{\mathrm{I}}(b)$ is zero between the slots and independent of z (no stray fields) and is equal to $E_z^{\mathrm{II}}(b)$ across the slots. To find A_n, both sides of Eq. (9.85) are multiplied by $\exp\,(j\beta_m z)$ and integrated between $-L/2$ and $+L/2$. Thus, we obtain

$$\int_{-L/2}^{L/2} E_z^{\mathrm{I}} \exp\,(j\beta_m z)\, dz = \int_{-L/2}^{L/2} \sum_{n=-\infty}^{\infty} A_n \sinh\,(\zeta_n b) \exp\,[-j(\beta_n - \beta_m)z]\, dz. \tag{9.94}$$

Due to orthogonality, the integral on the right-hand side of Eq. (9.94) is equal to zero for $m \neq n$; for $m = n$, it is equal to $A_m L \sinh\,(\zeta_n b)$. Introducing the expression for $E_z^{\mathrm{I}}(b)$ from Eq. (9.85) in Eq. (9.94), and integrating only between $z = -d/2$ and $z = +d/2$, because $E_z^{\mathrm{I}}(b) = 0$ in the ranges $-L/2 < z < -d/2$ and $+d/2 < z < +L/2$, we obtain

$$\int_{-d/2}^{d/2} E_z^{\mathrm{I}}(b) \exp\,(j\beta_m z)\, dz = E_z^{\mathrm{II}}(b) \frac{\sin\,(\frac{1}{2}\beta_m d)}{\frac{1}{2}\beta_m d}. \tag{9.95}$$

Therefore

$$A_n = E_z^{\mathrm{II}}(b) \frac{d}{L} \frac{\sin\,(\frac{1}{2}\beta_n d)}{\frac{1}{2}\beta_n d \sinh\,(\zeta_n b)}. \tag{9.96}$$

We have neglected the stray field of the slot [see Eq. (9.92)]. H_x must be matched so that its mean value $\overline{H_x^{\mathrm{I}}(b)}$ at $y = b$ is equal to $H_x^{\mathrm{II}}(b)$ in the slots. From Eqs. (9.87) and (9.96), it follows that

$$\begin{aligned} \overline{H_x^{\mathrm{I}}(b)} &= \frac{-j}{Z_{\mathrm{w}} d} \int_{-d/2}^{d/2} \sum_{-\infty}^{\infty} A_n \frac{k}{\zeta_n} \cosh\,(\zeta_n b) \exp\,(-j\beta_n z)\, dz \\ &= -\frac{j}{Z_{\mathrm{w}}} \frac{d}{L} \sum_{-\infty}^{\infty} E_z^{\mathrm{II}}(b) \frac{k}{\zeta_n} \coth\,(\zeta_n b) [\frac{\sin\,(\frac{1}{2}\beta_n d)}{\frac{1}{2}\beta_n d}]^2. \end{aligned} \tag{9.97}$$

In Eq. (9.97), the value of $E_z^{\mathrm{II}}(b)$ is still not known.

Using Eqs. (9.91)–(9.93) and Eq. (9.97), we can match the wave admittances at the entrance to the slot to find an expression for the dispersion of the line. Thus, we obtain

$$\frac{\cot\,(ks)}{k} = \frac{d}{L} \sum_{n=-\infty}^{\infty} \frac{1}{\zeta_n} [\frac{\sin\,(\frac{1}{2}\beta_n d)}{\frac{1}{2}\beta_n d}] \coth\,(\zeta_n b). \tag{9.98}$$

For a strongly retarded wave, $v_{p,n} \ll c$ and $\zeta_n^2 = \beta_n^2$, and using Eq. (9.89), the relation obtained between $\phi_{L,0}$ and the dimensions of the line is

$$\begin{aligned} \frac{1}{kL \tan\,(ks)} &= \frac{a}{L} \sum_{n=-\infty}^{\infty} [\frac{\sin\,\{(\phi_{L,0} + 2\pi n)d/(2L)\}}{(\phi_{L,0} + 2\pi n)d/(2L)}]^2 \\ &\times \frac{1}{[(\phi_{L,0} + 2\pi n) \tanh\,\{(\phi_{L,0} + 2\pi n)b/L\}]}. \end{aligned} \tag{9.99}$$

For given values of d/L and b/L, the solution of Eq. (9.99) gives $(\phi_{L,0} + 2\pi n)$ as a function of ks. For small values of $\phi_{L,0}$, i.e., for $\lambda_0/L \gg 1$ (when the higher space harmonics have negligible amplitude), Eq. (9.99) reduces to

$$kd \tan (ks) = \beta_0 L \tanh (\beta_0 b). \tag{9.100}$$

Equation (9.100) is a simplified equation for the phase constant β_0 of the fundamental space harmonic. When $\phi_{L,0} \to 0$ and $k \to 0$, we obtain

$$(ks)^2 \approx \frac{s}{d}\frac{b}{L}\phi_{L,0}^2. \tag{9.101}$$

Therefore, for very long waves, the retardation ratio is

$$\left(\frac{c}{v_p}\right)_{(\lambda_0 \to \infty)} = \left(\frac{sd}{bL}\right)^{1/2}. \tag{9.102}$$

The coupling impedance for a given space harmonic at $y = y_0$ is

$$K = \hat{E}^{I}_{z,n}(y_0)\hat{E}^{I*}_{z,n}(y_0)/(2P\beta_n^2), \tag{9.103}$$

where the total power P is given by

$$P = \frac{\hat{E}^{I}_{z}(b)\hat{E}^{I*}_{z}(b)}{Z_w}\frac{kb\Delta x}{L^2}\Sigma\frac{1-\cos(\beta_n a)}{\zeta_n^2\beta_n}\frac{1+\sinh(2\zeta_n b)/(2\zeta_n b)}{\cosh(2\zeta_n b)-1}, \tag{9.104}$$

where Δx is the width of the delay line. Therefore

$$\frac{\hat{E}^{I}_{z,n}(y_0)\hat{E}^{I*}_{z,n}(y_0)}{2} = \frac{\hat{E}^{I}_{z}(b)\hat{E}^{I*}_{z}(b)}{L^2}\frac{1-\cos(2\beta_n a)[\cosh(2\zeta_n y_0)-1]}{\beta_n^2[\cosh(2\zeta_n b)-1]}. \tag{9.105}$$

Substituting Eqs. (9.104) and (9.105) in Eq. (9.103), K can be calculated. Figure 9.18 shows the dispersion curves and the coupling impedance at $y_0 = b$ as a function of λ_0/L for the fundamental space harmonic $n = 0$ and the space harmonics $n = +1$ and $n = -1$. These curves extend to $\lambda_0/L = \infty$, and hence the structure behaves like a low-pass filter. For long waves, $v_{p,0} =$ constant, and the space harmonic with $n = +1$ has a minimum of c/v_p for small values of λ_0. The coupling impedance is a function of y and decreases with distance from the slots, especially for the higher values of $|n|$. It only increases to large values as $|\phi_{L,0}| \to \pi$ [i.e., $c/v_p \to \lambda_0/(2L)$] and becomes infinite at $|\phi_{L,0}| = \pi$. This is to be expected since $|\phi_{L,0}| = \pi$ indicates standing waves. For negligible damping, the field strength would increase to very large values even for very small applied powers. This implies that, at the cut-off frequency, a very strong coupling exists between an electromagnetic wave carried by the line and an electron stream travelling synchronously with the wave. Hence, the π-mode of operation in the travelling-wave magnetron is the most important mode.

The ratio of the coupling impedances for two different space harmonics, namely, $n = p$ and $n = q$, is independent of the total power P and depends only on the dimensions of the line. For $y_0 = b$, from the ratio K_p for $n = p$ to K_0 for $n = 0$,

$$\frac{K_p}{K_0} = \left[\frac{\beta_0^2 \sin(\frac{1}{2}\beta_p d)}{\beta_p^2 \sin(\frac{1}{2}\beta_0 d)}\right]^2. \tag{9.106}$$

Since $\beta_p > \beta_0$, the coupling impedance for the higher space harmonics decreases with increasing p.

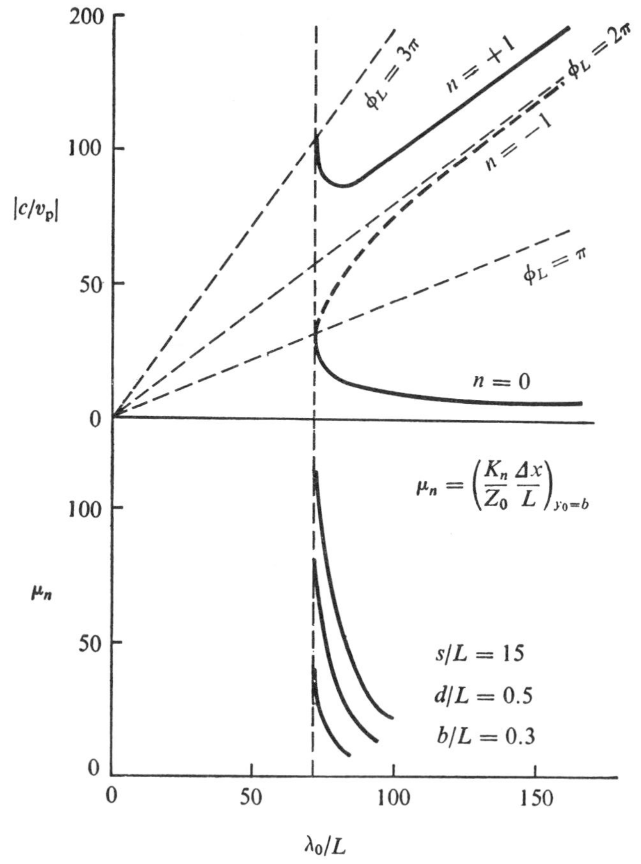

Fig. 9.18 Dispersion curves $|c/v_p|$ and coupling impedance K for space harmonics $n = 0, \pm 1$ versus λ_0/L of the plane corrugated waveguide of Fig. 9.17.

9.12 Tape Helix as Periodic Delay Line

In Section 9.5, the sheath helix has been treated as a homogeneous delay line, and the mathematical analysis has neglected the existence of the space harmonics. The space harmonics can be studied by considering the periodic character of the line. Figure 9.19 shows a tape helix, which is a conductor, infinitely thin in the ρ-direction and wound around a cylinder of radius a with

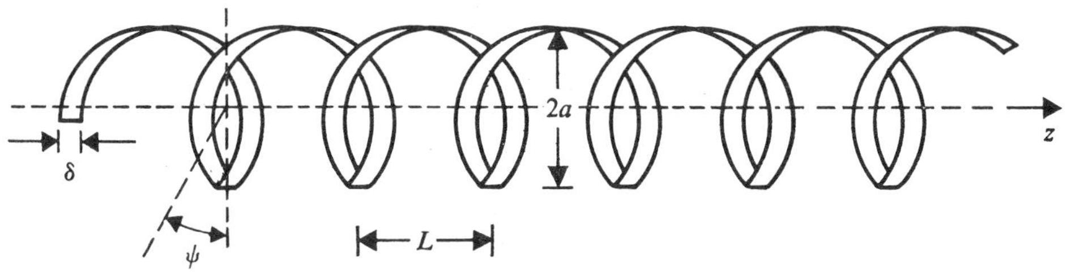

Fig. 9.19 Tape helix.

a constant pitch $L = 2\pi a/\cot \psi$. The model remains unchanged not only when it is displaced a distance L in the z-direction, but also when it is displaced an arbitrary distance Δz and at the same time rotated through an angle $2\pi\Delta z/L$ in the ϕ-direction (the positive sign refers to a left-hand screw and the negative sign to a right-hand screw). Therefore, in addition to the simple periodicity relation given by

$$F(z + L) = F(z) \exp(-j\phi_L), \tag{9.107}$$

there is also the relation

$$F(\rho, \phi \pm 2\pi\Delta z/L, z + \Delta z) = \exp(-j\beta\Delta z)F(\rho, \phi, z). \tag{9.108}$$

Relation (9.108) is satisfied if $\beta = \beta_0$, and

$$F(\rho, \phi, z) = \exp(-j\beta_0 z)G[\rho, z \mp L\phi/(2\pi)], \tag{9.109}$$

where the function G depends only on ρ and $z \mp L\phi/(2\pi)$. $F/\exp(-j\beta_0 z)$ is periodic in z, and hence F is of the form

$$F(\rho, \phi, z) = \exp(-j\beta_0 z) \sum_{n=-\infty}^{\infty} Z_n(\rho) \exp[-jn\{(2\pi z/L) \mp \phi\}]$$

$$= \sum_{n=-\infty}^{\infty} Z_n(\rho) \exp(-j\beta_n z \pm jn\phi), \tag{9.110}$$

where Z_n is a general expression containing the modified Bessel functions. Inside the helix ($\rho \leqslant a$),

$$Z_n = A_n I_n(\zeta_n \rho), \tag{9.111}$$

and outside the helix ($\rho \geqslant a$) when no shield is present,

$$Z_n = B_n K_n(\zeta_n \rho). \tag{9.112}$$

The argument of the Bessel functions must be real because, for $k^2 > \beta_n^2$, the n-th space harmonic from Eq. (9.111) has the character of a radiating cylindrical wave. Radiation involves energy loss, and the band in which it takes place is therefore useless in travelling-wave tubes. Hence, this is the forbidden band. The condition $|\beta_n| > k$ means that all the space harmonics have a phase velocity less than c. Thus, in the $(c/v_p, \lambda_0/L)$ plane, as shown in Fig. 9.13, the dispersion curves of all space harmonics remain above the straight line $c/v_p = 1$. They also remain outside the band $\lambda_0/L < 2$. This is obvious because the individual turns can be regarded as elements of a periodic array of antennas. If the elements are separated by a distance greater than half a wavelength, then the energy flow is not limited to the direction of the periodicity.

The boundary conditions that must be satisfied on the cylindrical surface $\rho = a$ are

(i) Continuity of the tangential components E_z and E_ϕ

(ii) Continuity of the other field components between the turns

(iii) Discontinuity of the magnetic field strength $\boldsymbol{H}$ at the tape helix where the surface current density is $\boldsymbol{S}$.

If S_ϕ, S_z are the components of $\boldsymbol{S}$, the relations

$$H_\phi^e(a) - H_\phi^i(a) = S_z, \qquad H_z^e(a) - H_z^i(a) = -S_\phi \tag{9.113}$$

must therefore be satisfied. All field components can be derived from E_z and H_z. A series is assumed for each of these components both inside and outside the helix, as given by Eqs. (9.110)–

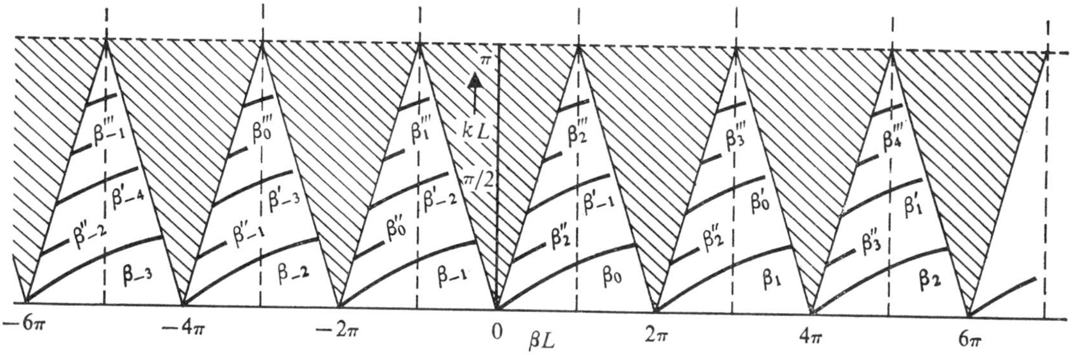

(a) $kL = \omega L/c = f(\beta L)$ (after Sensiper)

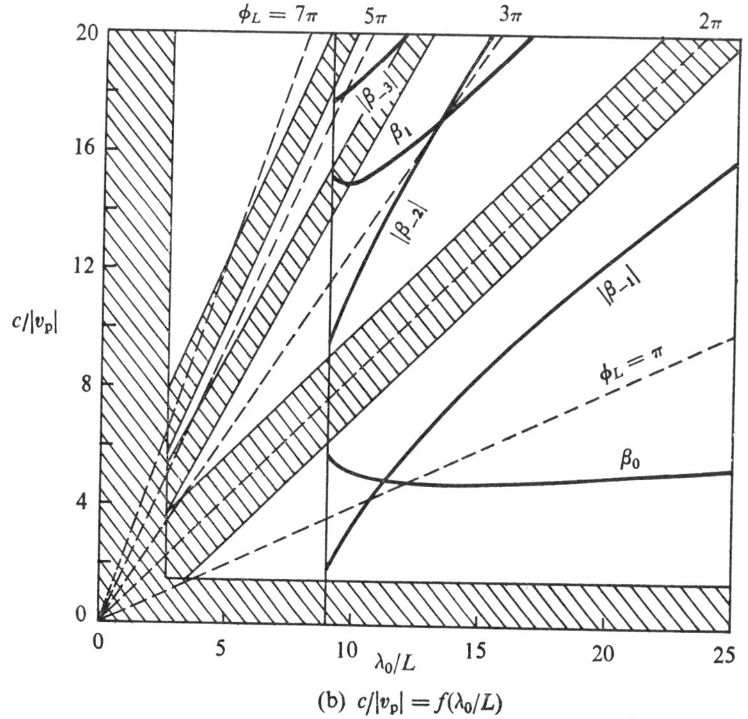

(b) $c/|v_p| = f(\lambda_0/L)$

Fig. 9.20 Dispersion curves of the tape helix of Fig. 9.19, with $\psi = 10°$, $\delta/L = 1/(10\pi)$. (Waves with negative subscripts are backward waves.)

(9.112). From the boundary conditions, the coefficients in the series can be expressed in terms of those in the expansions of

$$S_z = \sum_{n=-\infty}^{\infty} C_n \exp(-j\beta_n z \pm jn\phi), \tag{9.114}$$

$$S_\phi = \sum_{n=-\infty}^{\infty} D_n \exp(-j\beta_n z \pm jn\phi). \tag{9.115}$$

Sensiper[1,2] assumes several possible current distributions across the tape, for example, $S_z = S_0 \sin \psi \exp(-j\beta_0 z)$, $S_\phi = S_0 \cos \psi \exp(-j\beta_0 z)$, that is, S is assumed parallel to the direction of the turns and independent of ϕ. If the width δ of the tape is small compared to the pitch L, then the choice of S does not have any great influence on the results. For a given current distribution, all the boundary conditions are not strictly satisfied. For example, if S is assumed parallel to the turn direction, then only the component of E parallel to the turn direction and along the centre line of the tape is important. By this method, Sensiper calculated the quantity β for a tape helix with $\psi = 10°$ and $\delta = L/(10\pi)$. Figure 9.20a shows the result obtained by him in the form $kL = f(\beta L)$, where the forbidden regions are shown shaded. Since $k = \omega/c$, we have $v_g = (d\beta/d\omega)^{-1} = cd(kL)/d(\beta L)$, and the group velocity is given by the slope of each curve in a plot of this kind. Since, by definition, the group velocity is always taken to be positive, only curves of positive slope are admissible. The phase velocity is $v_p = ckL/\beta L$, so that it is given by the slope of the straight line from the origin to the point on the curve. For waves with positive phase velocity, the curves are on the right side of the origin, and for waves with negative phase velocity, the curves are on the left side of the origin. The space harmonics are indicated by subscripts, the primes referring to different branches of the space harmonics. Many of these space harmonics were found experimentally. For the space harmonic with subscript 0, $E_z \propto I_0(\zeta_0 \rho)$, with no variation in the ϕ-direction. This space harmonic consists of three branches: two forward waves, namely, β_0 and β_0' ($v_p > 0$), and one backward wave, β_0'' ($v_p < 0$). The branch β_0, which has low dispersion, i.e., $v_g \approx v_p$, is the one commonly used in helix-type travelling-wave amplifiers. Its phase velocity decreases slowly with increasing frequency. For a wide band of frequencies, $v_p \approx c \sin \psi$.

In Fig. 9.20b, $c/|v_p|$ for the solid branches of the different space harmonics is shown as a function of λ_0/L. Here too, the forbidden regions are shown shaded. For $\lambda_0/L > 11.3$, the space harmonic β_0 (with axially symmetrical field E_z) has the greatest phase velocity. For values of λ_0/L between 7.4 and 11.3, the backward β_{-1} wave is faster than the forward β_0 wave. Sensiper[1,2] has also shown that, in this frequency region, the amplitude of the backward β_{-1} wave is greater than that of the forward β_0 wave. For all space harmonics, namely, β_m, β_m', β_m'', with $m \neq 0$, the field component E_z vanishes on the axis $z = 0$, because $I_m(x)$ associated with the longitudinal field have an m-th order zero for $x = 0$. For maximum interaction with such space harmonics, the beam must be close to the line.

9.13 Closed-Ring Periodic Delay Lines: General Properties

The delay lines used in travelling-wave magnetrons are usually closed-ring periodic lines. Figure 9.21 gives examples of such delay lines. The travelling wave propagates in the azimuthal direction in such delay lines, the radial lines being the lines of constant phase. As the delay line is closed on itself, there are no clearly defined input and output points, and the closed circuit provides a feedback. Thus, when such a delay line is used in a microwave tube, the tube is always an oscillator, and not an amplifier. One important property of this delay line is that only certain

[1]S. Sensiper, Electromagnetic Wave Propagation on Helical Conductors, MIT Electronics Research Laboratory Technical Report, No. 194, Cambridge, Mass., 1951.

[2]S. Sensiper, Electromagnetic wave propagation on helical structures: A review and survey of recent progress, *Proc. I.R.E.*, **43**, 149, 1955.

number of discrete frequencies are excited, and these frequencies depend on the dimensions of the line. There is a mode corresponding to each of these frequencies, but whether any of these modes will be excited will depend on the properties of the electron stream. Any mode is excited if its angular phase velocity ω_{ph} is approximately equal to the angular velocity of the electrons. This is known as the condition of synchronism.

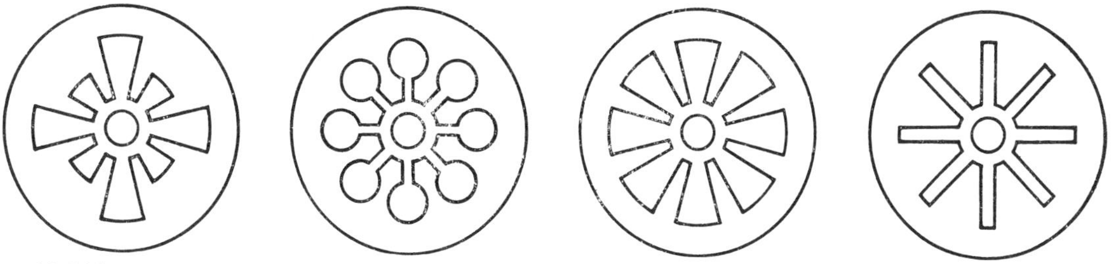

Fig. 9.21 Examples of closed-ring delay lines used in travelling-wave magnetrons.

The modes of ring-shaped delay lines are determined from the condition that no phase jump occurs anywhere along the line. Therefore, the number of wavelengths along the circumference must be an integer. If M is the number of slots or their equivalent four-terminal networks and $\phi_{L,0}$ the corresponding phase change of the fundamental space harmonic per section, then

$$M\phi_{L,0} = 2\pi q, \tag{9.116}$$

where q is an integer. If Eq. (9.116) is satisfied, then there can be no phase jump for the higher harmonics either since

$$\phi_{L,n} = \phi_{L,0} + 2\pi n.$$

We thus obtain

$$M\phi_{L,n} = M\phi_{L,0} + M \cdot 2\pi n = 2\pi(q + nM), \tag{9.117}$$

where q, n, M are integers. Hence, $M\phi_{L,n}$ is an integral multiple of 2π. In the pass-band of a line, $|\phi_{L,0}|$ is always less than π. Therefore, from Eq. (9.116), q can only take the values 0, 1, 2, . . . , $M/2$. Thus, it follows that only $(M/2 + 1)$ modes can be excited in closed-ring delay lines. The mode for which $|\phi_{L,0}| = \pi$ is called the π-mode, where $q = M/2$. For each of these modes ω_q, an infinite number of space harmonics $(-\infty < n < \infty)$ is obtained, all having different angular phase velocities ω_{ph}. Therefore, the angular phase velocity of a wave in a closed-ring delay line is characterized by $\omega_{ph,n,q}$, where n gives the order of the space harmonic and q the order of the mode, the latter being determined from Eq. (9.116). For most cases in travelling-wave magnetrons, $|\phi| = |\phi_{L,0}| = \pi$ (i.e., the π-mode) so that $n = 0$, $q = M/2$.

The field function of a closed-ring delay line may be written in the form

$$F = F_0 \exp [j(\omega t - \Phi)], \tag{9.118}$$

where $\Phi = \gamma_n \phi$ is the phase change corresponding to the angle ϕ swept out by the wave. (ϕ and Φ are measured from the same origin.) γ_n is the phase change of the wave per radian and is therefore equal to the total phase change round the anode block divided by 2π or equal to the phase change per section, $\phi_{L,n}$, divided by the angle ϕ_L subtended per section. Therefore, from Eq. (9.117), we obtain

$$\gamma_n = \Phi/\phi = \phi_{L,n}/\phi_L = \phi_{L,n}M/(2\pi) = q + nM. \qquad (9.119)$$

Since q and n are integers, γ_n is also an integer, each space harmonic having a series of different values γ_n corresponding to the individual modes q.

The integers q, n, M are not enough to fully characterize the field quantities. They would be sufficient if the slots could be represented by lumped-circuit elements whose admittance and frequency are uniquely related. These slots are transmission lines, and the admittances at their open ends vary periodically with frequency. In an infinite number of discrete frequency bands, they are equivalent to the series inductances necessary for retardation. Hence, it is necessary to add a further parameter to indicate how many current antinodes are present along the slot. This is not very important in practice, because travelling-wave magnetrons are always used under conditions such that there exists only one current antinode (at the end of the slot) of the standing wave and the modes lie below the first resonant frequency (parallel resonance) of the slot.

9.14 Analysis of Closed-Ring Delay Line

Figure 9.22 shows a typical closed-ring delay line. Let the end effects due to the finite length in

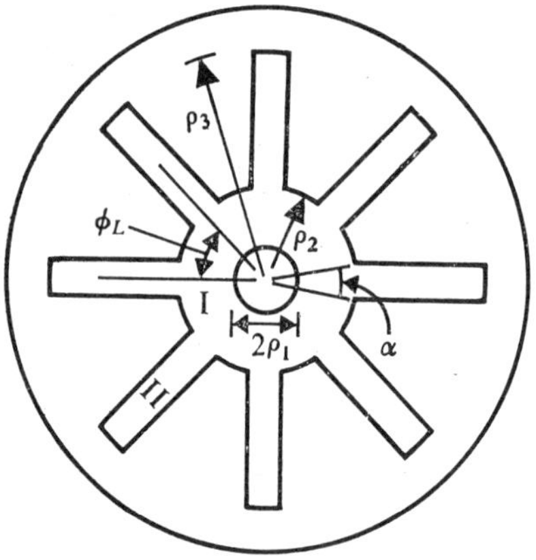

Fig. 9.22 Closed-ring delay line used in travelling-wave magnetron.

the z-direction be neglected. Using Eq. (9.119) and the boundary condition $\hat{E}^{\mathrm{I}}_\phi = 0$ at $\rho = \rho_1$, $\hat{H}^{\mathrm{I}}_z$ can be written in the form

$$\hat{H}^{\mathrm{I}}_z = \sum_{n=-\infty}^{\infty} A_n[Y'_{\gamma_n}(k\rho_1)J_{\gamma_n}(k\rho) - J'_{\gamma_n}(k\rho_1)Y_{\gamma_n}(k\rho)] \exp(-j\gamma_n\phi), \qquad (9.120)$$

where

$$Y' = \frac{dY(k\rho)}{d(k\rho)}, \qquad J' = \frac{dJ(k\rho)}{d(k\rho)}.$$

From Maxwell's equations, the expression for $\hat{E}^{\mathrm{I}}_\phi$ is

$$\hat{E}^{\mathrm{I}}_\phi = jZ_{\mathrm{w}} \sum_{-\infty}^{\infty} A_n[Y'_{\gamma_n}(k\rho_1)J'_{\gamma_n}(k\rho) - J'_{\gamma_n}(k\rho_1)Y'_{\gamma_n}(k\rho)] \exp(-j\gamma_n\phi). \qquad (9.121)$$

Note that the superscript I refers to the region outside the slots and the superscript II refers to the region inside the slots.

In region II inside a given slot with parallel walls when there are standing waves, the field components are

$$\hat{H}_z^{\mathrm{II}} = \frac{j}{Z_{\mathrm{w}}}\hat{E}_\phi^{\mathrm{II}}(\rho_2)\frac{\cos\,[k(\rho_3 - \rho)]}{\sin\,[k(\rho_3 - \rho_2)]}, \tag{9.122}$$

$$\hat{E}_\rho^{\mathrm{II}} = 0, \tag{9.123}$$

$$\hat{E}_\phi^{\mathrm{II}} = \hat{E}_\phi^{\mathrm{II}}(\rho_2)\frac{\sin\,[k(\rho_3 - \rho)]}{\sin\,[k(\rho_3 - \rho_2)]}, \tag{9.124}$$

where $\hat{E}_\phi^{\mathrm{II}}(\rho_2)$ contains the phase term $\exp\,(-js\Phi_{L,n})$ for the s-th slot, $\Phi_{L,n}$ representing the phase change per segment of the n-th harmonic. The field must be continuous across the boundary of regions I and II so that $\hat{E}_\phi^{\mathrm{I}}(\rho_2) = \hat{E}_\phi^{\mathrm{II}}(\rho_2)$ over the opening of the slot.

In order to obtain A_n, multiply both sides of Eq. (9.121) by $\exp\,(j\gamma_m\phi)$ and integrate from $-\phi_L/2$ to $\phi_L/2$. Then, since the integral on the right-hand side is different from zero for $n = m$ (by orthogonality) only, we obtain

$$\int_{-\phi_L/2}^{\phi_L/2} \hat{E}_\phi^{\mathrm{I}}(\rho_2)\exp\,(j\gamma_m\phi)\,d\phi = jZ_{\mathrm{w}}\phi_L A_m[Y'_{\gamma_m}(k\rho_1)J'_{\gamma_m}(k\rho_2) - J'_{\gamma_m}(k\rho_1)Y'_{\gamma_m}(k\rho_2)]. \tag{9.125}$$

We may assume that $\hat{E}_\phi^{\mathrm{I}}(\rho_2) = \hat{E}_\phi^{\mathrm{II}}(\rho_2) =$ constant over the slot and zero elsewhere. We can then write

$$\int_{-\phi_L/2}^{\phi_L/2} \hat{E}_\phi^{\mathrm{I}}(\rho_2)\exp\,(j\gamma_m\phi)\,d\phi = \hat{E}_\phi^{\mathrm{II}}(\rho_2)\int_{-\alpha/2}^{\alpha/2}\exp\,(j\gamma_m\phi)\,d\phi = \hat{E}_\phi^{\mathrm{II}}(\rho_2)\alpha\frac{\sin\,(\frac{1}{2}\gamma_m\alpha)}{\frac{1}{2}\gamma_m\alpha}, \tag{9.126}$$

where α is the angle subtended by a slot. From Eqs. (9.125) and (9.126), and writing $\phi_L = 2\pi/M$, we obtain

$$A_n = \frac{1}{jZ_{\mathrm{w}}}\hat{E}_\phi^{\mathrm{II}}(\rho_2)\frac{\alpha M}{2\pi}\,\frac{\sin\,(\frac{1}{2}\gamma_n\alpha)}{\frac{1}{2}\gamma_n\alpha}[Y'_{\gamma_n}(k\rho_1)J'_{\gamma_n}(k\rho_2) - J'_{\gamma_n}(k\rho_1)Y'_{\gamma_n}(k\rho_2)]^{-1}. \tag{9.127}$$

The magnetic fields are matched by equating $\hat{H}_z^{\mathrm{II}}(\rho_2)$ to the mean value $\overline{\hat{H}_z^{\mathrm{I}}(\rho_2)}$ of $\hat{H}_z^{\mathrm{I}}(\rho_2)$ across the slot. Substituting the value of A_n from Eq. (9.127) in Eq. (9.120) and integrating between the limits $-\alpha/2$ to $\alpha/2$, we obtain

$$\overline{\hat{H}_z^{\mathrm{I}}(\rho_2)} = \frac{-j}{Z_{\mathrm{w}}}\hat{E}_\phi^{\mathrm{II}}(\rho_2)\sum_{n=-\infty}^{\infty}[\frac{\sin\,(\frac{1}{2}\gamma_n\alpha)}{\frac{1}{2}\gamma_n\alpha}]^2[\frac{Y'_{\gamma_n}(k\rho_1)J_{\gamma_n}(k\rho_2) - J'_{\gamma_n}(k\rho_1)Y_{\gamma_n}(k\rho_2)}{Y'_{\gamma_n}(k\rho_1)J'_{\gamma_n}(k\rho_2) - J'_{\gamma_n}(k\rho_1)Y'_{\gamma_n}(k\rho_2)}]e^{j\omega t}. \tag{9.128}$$

Because $\overline{\hat{H}_z^{\mathrm{I}}(\rho_2)} = \hat{H}_z^{\mathrm{II}}(\rho_2)$, from Eq. (9.122), we obtain

$$\cot\,[k(\rho_3 - \rho_2)] = -\frac{\alpha^2 M}{2\pi}\sum_{n=-\infty}^{\infty}[\frac{\sin\,(\frac{1}{2}\gamma_n\alpha)}{\frac{1}{2}\gamma_n\alpha}]^2[\frac{Y'_{\gamma_n}(k\rho_1)J_{\gamma_n}(k\rho_2) - J'_{\gamma_n}(k\rho_1)Y_{\gamma_n}(k\rho_2)}{Y'_{\gamma_n}(k\rho_1)J'_{\gamma_n}(k\rho_2) - J'_{\gamma_n}(k\rho_1)Y'_{\gamma_n}(k\rho_2)}]. \tag{9.129}$$

Equation (9.129) determines k and the resonant or eigen frequencies ω_q as well as the corresponding angular phase velocities $\omega_{\mathrm{ph},n,q} = \omega_q/(q + nM)$ of the space harmonics. The solution of this equation, which can be obtained graphically or by using the computer, shows that the line behaves like a low-pass filter having positive dispersion. (The angular group and phase velocities have the same sign.)

Though the theoretical solution of Eq. (9.129) shows that it behaves like a low-pass filter, the actual experimental dispersion curves show that such a structure behaves like a band-pass filter. The end effects of the finite length in the z-direction probably account for this difference between theory and experiment. In Fig. 9.22, the magnetic lines of force passing through each slot perpendicular to the plane of the paper in the z-direction return through other slots. Each of the slots is therefore inductively coupled to two or more neighbouring slots, and this effect has been neglected in the foregoing theoretical analysis. A more rigorous theoretical treatment (which is a laborious exercise) would be a three-dimensional analysis instead of a two-dimensional analysis.

9.15 Dispersion Curves and Modes of Travelling-Wave Magnetron

We shall now consider the relationship between the shape of the dispersion curves and the modes of closed-ring delay lines used in travelling-wave magnetron oscillators. The assumptions which are usually justified in practice are:

(i) The electron stream interacts with the fundamental wave $n = 0$.

(ii) The travelling-wave magnetron oscillates in the first pass-band of the delay line, i.e., at a frequency below the resonant frequency of the slots.

A dimensionless quantity

$$y = \frac{c}{\rho\omega_{\text{ph}}}$$

is defined as the retardation, where c/ρ is the angular velocity of a wave travelling with the velocity of light round the circle of radius ρ. As

$$L = \frac{2\pi\rho}{M},$$

we have

$$y = \frac{c/\rho}{\omega_{\text{ph}}} = \frac{|\phi_{L,0}|}{2\pi}\frac{\lambda_0 M}{2\pi\rho}. \tag{9.130}$$

The dispersion curves have the same form as the $n = 0$ curve in Fig. 9.13. From Eq. (9.116), we have, for the q-mode $(0 \leqslant q \leqslant M/2)$,

$$2\pi q = M|\phi_{L,0}| = (2\pi)^2\left(\frac{\rho y}{\lambda_0}\right)$$

or

$$y = \frac{c/\rho}{\omega_{\text{ph}}} = \frac{q\lambda_0}{2\pi\rho} = \frac{q\lambda_0}{ML}. \tag{9.131}$$

The curves

$$y = f\left(\frac{\lambda_0}{L}\right)$$

for q/m = constant are straight lines through the origin (see Fig. 9.23). The frequencies of the different modes are determined by the intersections of these straight lines with the dispersion curve

$$\frac{c}{\rho\omega_{\text{ph}}} = f\left(\frac{\lambda_0}{L}\right).$$

The mode with $|\phi_{L,0}| = \pi$, i.e., with $q/M = \frac{1}{2}$, has the advantage of large coupling impedance and is therefore most commonly used.

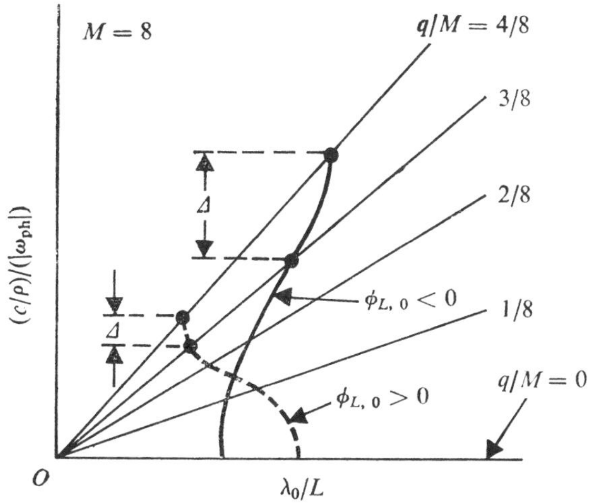

Fig. 9.23 Modes of eight-segment travelling-wave magnetron. (Solid lines indicate negative dispersion and broken lines positive dispersion.)

When the oscillations in a magnetron jump from one mode to a neighbouring mode, the process is called *mode jumping*, which is a disadvantage in travelling-wave magnetrons. The greater the danger, the smaller the difference (Δ) between the corresponding retardations of the two modes. Figure 9.23 shows that, as a result of the shape of the dispersion curve, the separation of the π-mode from the neighbouring modes is better (i.e., larger Δ) for lines with $\phi_{L,0} < 0$ (solid curve) than for those with $\phi_{L,0} > 0$ (dashed curve). This is the reason for using lines with negative dispersion in travelling-wave magnetrons.

Oscillations are sustained when the wave and electron stream are approximately in synchronism. The angular velocity of the electrons is proportional to the ratio $\bar{E}/\bar{B}$ (see Chapter 10). From the dispersion curves, the operating conditions $\bar{E}/\bar{B}$ can be found. Only discrete frequencies corresponding to the modes with $q = 0, 1, 2, \ldots, M/2$ can be excited. An increase in $\bar{E}/\bar{B}$ corresponds to a reduction of $c/\rho|\omega_{\text{ph}}|$. It follows from Fig. 9.23 that the sign of $\phi_{L,0}$ is always the same as that of $\Delta(\bar{E}/\bar{B})/(\Delta\lambda_0)$. Therefore, it is possible to determine from measurements whether the line has positive ($\phi_{L,0} > 0$) or negative ($\phi_{L,0} < 0$) dispersion.

Certain mechanical modifications called straps are introduced to avoid mode jumping in travelling-wave magnetrons. These straps are in the form of rings or bridges that join together the appropriate segments, e.g., alternate segments, which are in phase for the π-mode (see Fig. 9.24). The effect of increasing the number of straps is qualitatively shown in this figure. The difference of the retardation between the π-mode and the adjacent mode increases with the increase in the number of straps. A similar effect can be obtained by periodic variation of the depth of successive slots—the so-called 'rising-sun magnetron'—where the resonant frequencies of neighbouring slots are different.

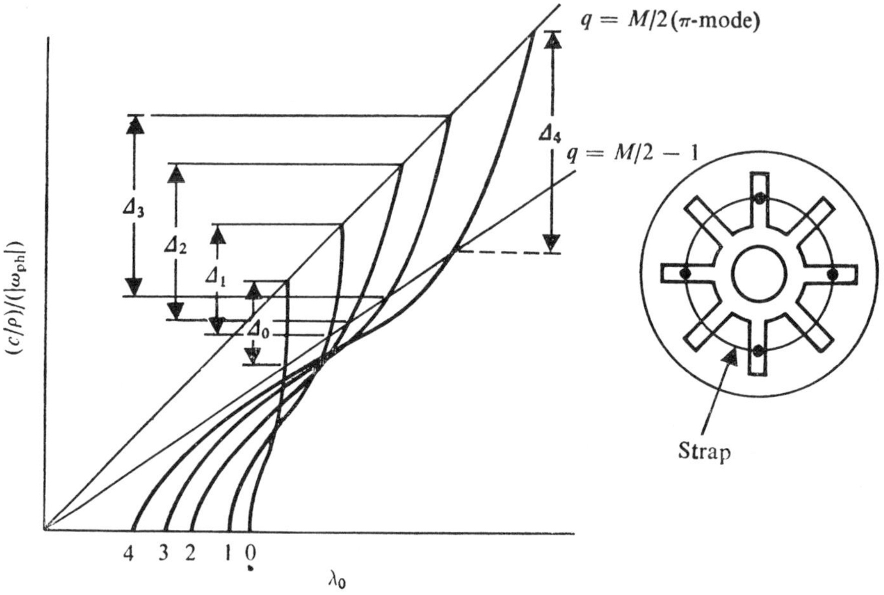

Fig. 9.24 Effect of strapping on the dispersion curves of travelling-wave magnetron with eight slots ($M = 8$). (Curve 0, without straps; curves 1–4, effect of increasing the number of straps.)

PROBLEMS

1 Derive expressions for c/v_p and c/v_g as functions of λ_0/L for the periodic structure shown in

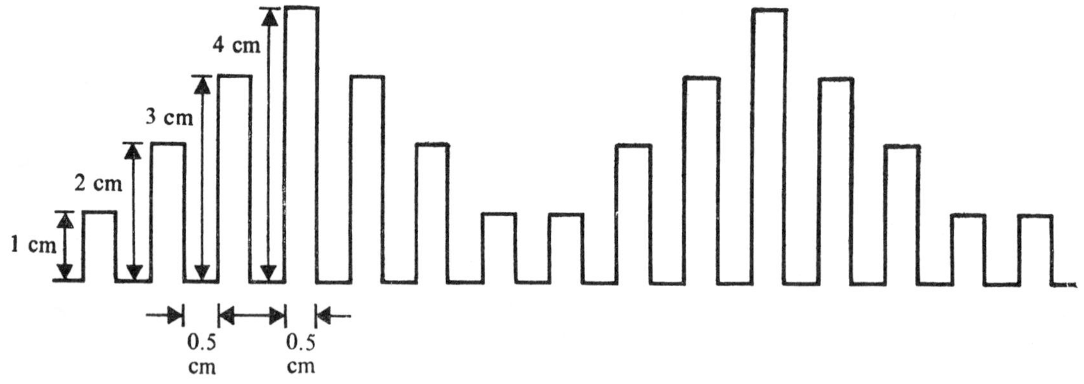

Fig. P. 9.1

Fig. P. 9.1. Calculate c/v_p and c/v_g for the first five space harmonics for the dimensions given at $\lambda_0 = 3$ cm.

2 Solve the characteristic equation

$$kd \tan (ks) = \beta_0 L \tanh (\beta_0 b)$$

for the plane-periodic delay line for $d = 1$ cm, $s = 4$ cm, $L = 5$ cm, $b = 2$ cm, at a frequency of 3000 MHz. From the value of β_0 obtained, calculate c/v_{p_n} for $n = 0, 1, 2, 3, 4$.

10 Microwave Electron Tubes

10.1 High Frequency Limitations of Conventional Tubes

The functions of generation, amplification, detection, modulation, and so on in the microwave frequencies are accomplished by two types of electron devices, namely, the microwave electron tubes and the microwave solid-state electron devices. This chapter deals with microwave electron tubes.

In the early years, attempts to generate coherent electrical power at microwave frequencies were extensions of the space-charge limited vacuum tubes such as triodes, tetrodes, and pentodes. But conventional tubes of this type have two important limitations: (i) parasitic energy-storage circuit elements such as interelectrode capacitances and lead inductances, and (ii) effects due to transit time of electrons between electrodes. These effects were minimized by incorporating the grid-cathode and grid-plate capacitances into resonant cavities and by reducing interelectrode spacings. The lighthouse tube shown in Fig. 10.1 is a result of this. The Western Electric 416 series, with grid cathode spacing of less than 0.0025 cm and operating limit of about 4 GHz, represents the limits of design in extending the conventional type of vacuum tube to microwave frequencies.

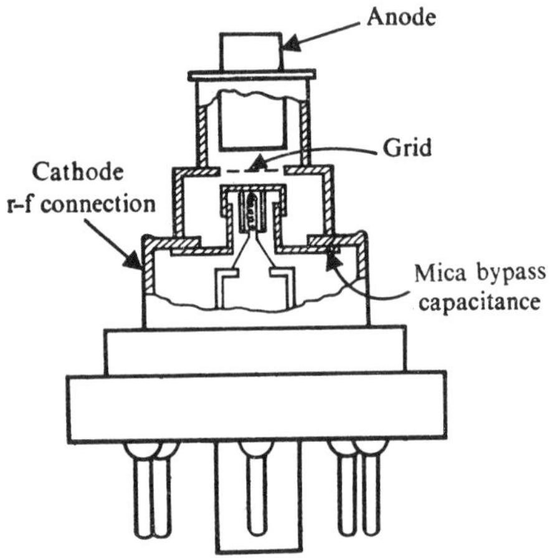

(a) Cut-away view of type 2C 40

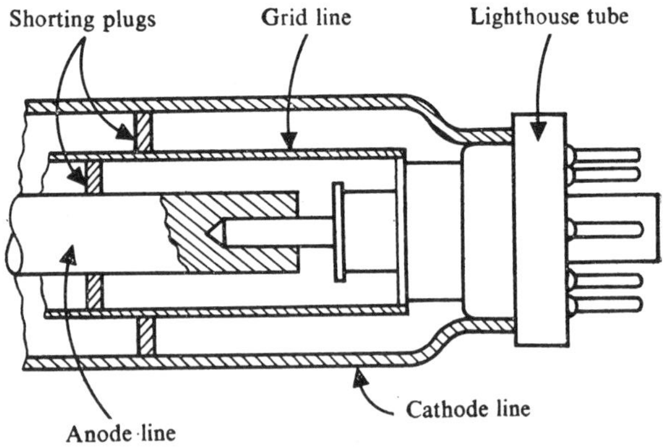

(b) Double concentric line oscillator using lighthouse tube

Fig. 10.1 Lighthouse tube oscillator.

Tubes which are more useful in the microwave frequency range are those that make good use of transit-time effects for their operation. Some examples of such tubes in chronological order are (i) the two-cavity klystron, (ii) reflex klystron, (iii) magnetron, and (iv) the large family of travelling-wave tubes (TWTs), which includes the forward-wave amplifier (FWA), backward-wave amplifier (BWA), backward-wave oscillator (BWO), carcinotron, and the linear magnetron amplifier.

Microwave tubes can be divided into two broad classes, namely, the space-charge controlled tubes and the transit-time tubes.

10.2 Space-Charge Controlled Tubes

Space-charge controlled tubes include diodes, triodes, and tetrodes. The pentode, which has the additional suppressor grid, is not of much value at microwave frequencies. At low frequencies, the suppressor grid makes the output impedance and the voltage swing of a pentode large, but at microwave frequencies, there is no such effect. This is so because, at microwave frequencies, the output impedance of a grid-controlled tube is determined by the losses in the dielectric materials and in the leads of the tubes, and not by the impedance of the electron current. Because of the output losses, the output voltages are rather low when grid-controlled UHF transmitter tubes are adjusted for maximum power loss. Hence, the addition of a suppressor grid is not very useful for increasing the voltage swing. Therefore, the pentode has no application in the microwave region though it is very useful at low frequencies. By using grounded-grid circuits, good decoupling can be obtained between the input and output circuits in both triodes and tetrodes. Also, the addition of a suppressor grid increases the electron transit time.

It is possible to make use of transit-time effects and bunching in intergrid spaces in space-charge controlled tubes at microwave frequencies. The negative dynamic admittances of the electron stream obtained with finite transit times can be used for regeneration and oscillation. But the efficiencies obtained by these methods are very low, and hence these methods have very little practical value. Transit-time effects in grid-controlled tubes are, in general, undesirable as they reduce gain and efficiency.

10.3 Transit-Time Tubes: Classification

Transit-time effects are specifically made use of for amplification or oscillation in transit-time tubes. In space-charge controlled tubes, alternating current flows at the cathode surface. In transit-time tubes, on the other hand, alternating current *does not* flow at the cathode surface. Even if a small amount of alternating current flows at the cathode, it hardly plays any part in the operation of the tube. Table 10.1 gives a classification of different types of transit-time tubes.

In *tubes with r-f field bunching*, there is one tuned circuit and one alternating field, which are oscillators. Due to the velocity modulation and bunching that take place in the alternating field, the electrons give up energy to the field.

The *monotron* is a diode relying only on initial velocities in which there is no steady field but only an alternating field within the tuned circuit. For transit angles θ between 2π and 2.8π, the conductance G_{el} of the electron stream is negative so that the parallel resonant circuit connected to the diode can be excited.

In the *retarding field oscillator*, a static retarding field is present in the tuned circuit between a grid at a positive potential and an anode at a negative potential, with respect to the cathode. Therefore, the electrons turn back within the r-f field of the tuned circuit. Due to this, a negative conductance G_{el} is produced at the terminals of the tuned circuit. The electrons which are in an unfavourable phase can be removed at the anode since they take up energy from the r-f field.

Both the monotron and the retarding field oscillator have only historical interest.

Drift-space tubes, especially the reflex klystron and multicavity klystrons, are of very great practical interest. They are characterized by the fact that, after velocity modulation, bunching takes place in a region where no external r-f fields interfere with the beam. The r-f fields produced by the electron space charge are the only ones present in this region.

Table 10.1 Classification of Transit-Time Tubes

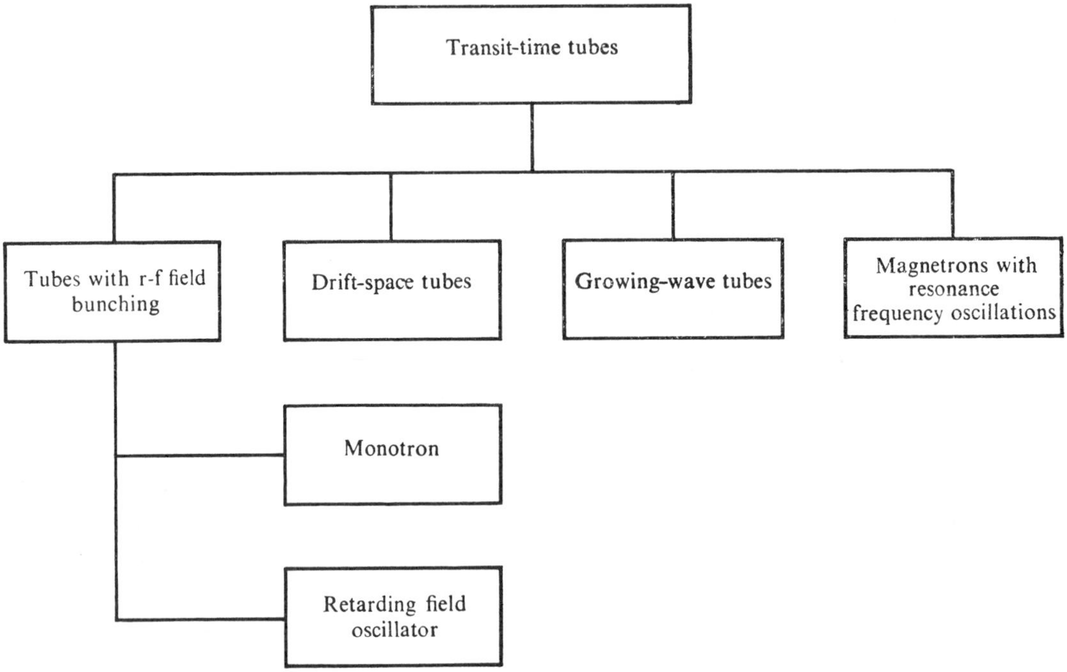

In *growing-wave tubes*, there is interaction between the electric field of a travelling wave guided by a wall of finite impedance and an electron beam. As a result, the electron beam contains at least one travelling wave of convection current having an amplitude growing with distance. The convection current is of the form $I_c = I_c(0) \exp(\alpha z)$, where $\alpha > 0$.

In *magnetrons with resonance frequency oscillations*, there exist mutually perpendicular electrostatic and time-independent magnetic fields in the interaction space. The values of the anode voltage and the magnetic field are such that, when there are no oscillations, the majority of the electrons do not reach the anode, or, in other words, the cut-off condition exists. The anode can be either a complete cylinder or a segmented cylinder. If it is a complete cylinder, the periodic electric field produced by the tuned circuit lies between the cathode and anode. If it is a segmented cylinder, then it lies between adjacent segments. The oscillations are excited at frequencies at which the electron transit time is approximately an integral multiple of the reciprocal of the resonant frequency of the tuned circuit. In this way, the electrons in the unfavourable phase are removed to the anode. This relationship between the electron transit time and the resonant frequency, and the existence of phase selection at the anode are characteristic of magnetrons with resonance frequency oscillations, and help differentiate them from travelling-wave magnetrons. The efficiency of magnetrons with resonance frequency oscillations is very low

($<10\%$) compared to that of travelling-wave magnetrons. Hence, they are not of much practical importance.

10.4 Drift-Space Tubes

There are two types of drift-space tubes, namely, the single-cavity tubes and the multicavity tubes, as shown in Table 10.2. Single-cavity tubes are always oscillators, whereas multicavity tubes may be oscillators or amplifiers. Figure 10.2 illustrates some single-cavity drift-space tubes.

Table 10.2 Different Types of Drift-Space Tubes

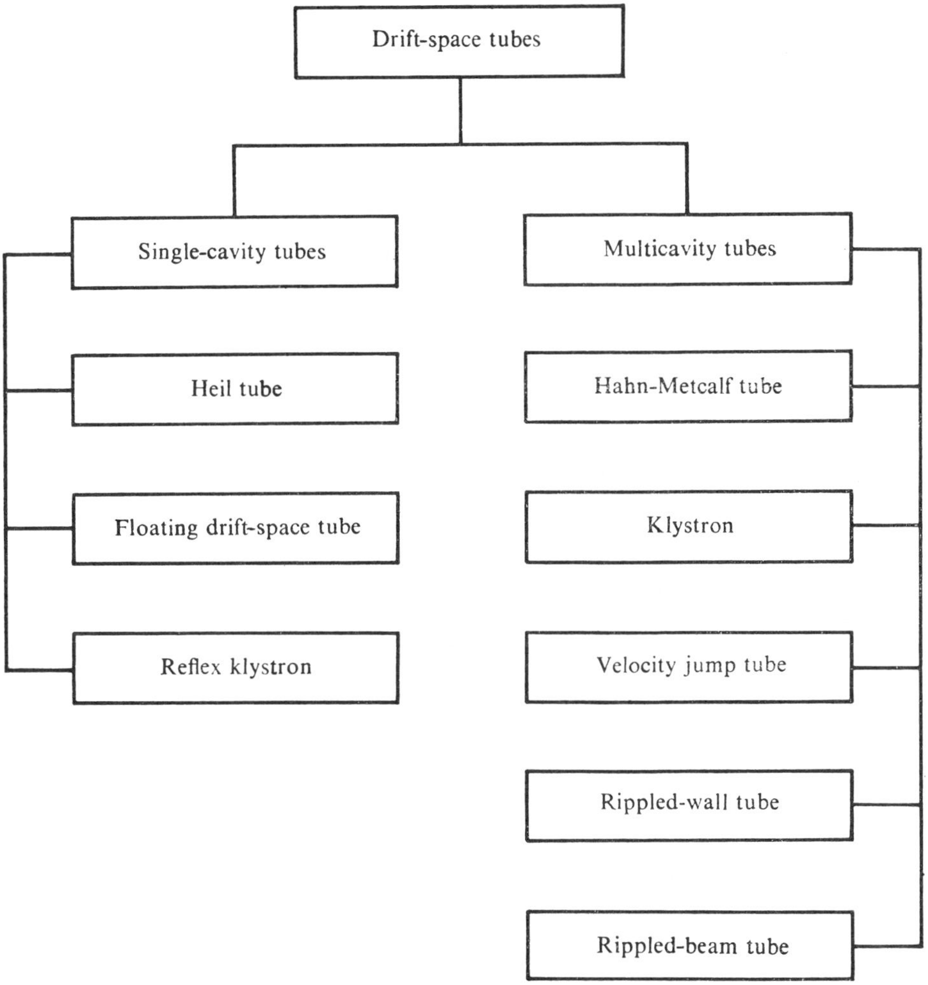

In Fig. 10.2a is shown a Heil tube which consists of a coaxial line in which the drift space is a hole in the inner conductor. Velocity modulation takes place in the field $\tilde{E}_1$, bunching occurs in the drift space, and energy extraction in the field $\tilde{E}_2$. $\tilde{E}_1$ and $\tilde{E}_2$ are out of phase at all times. For maximum energy to be delivered to the circuit, the alternating convection current must pass through the field $\tilde{E}_2$ where it will be retarded most. This condition is satisfied for negligible transit angles in the fields $\tilde{E}_1$ and $\tilde{E}_2$ if the transit angle in the drift space is $\theta = 2\pi(n + 3/4)$, where n is an integer.

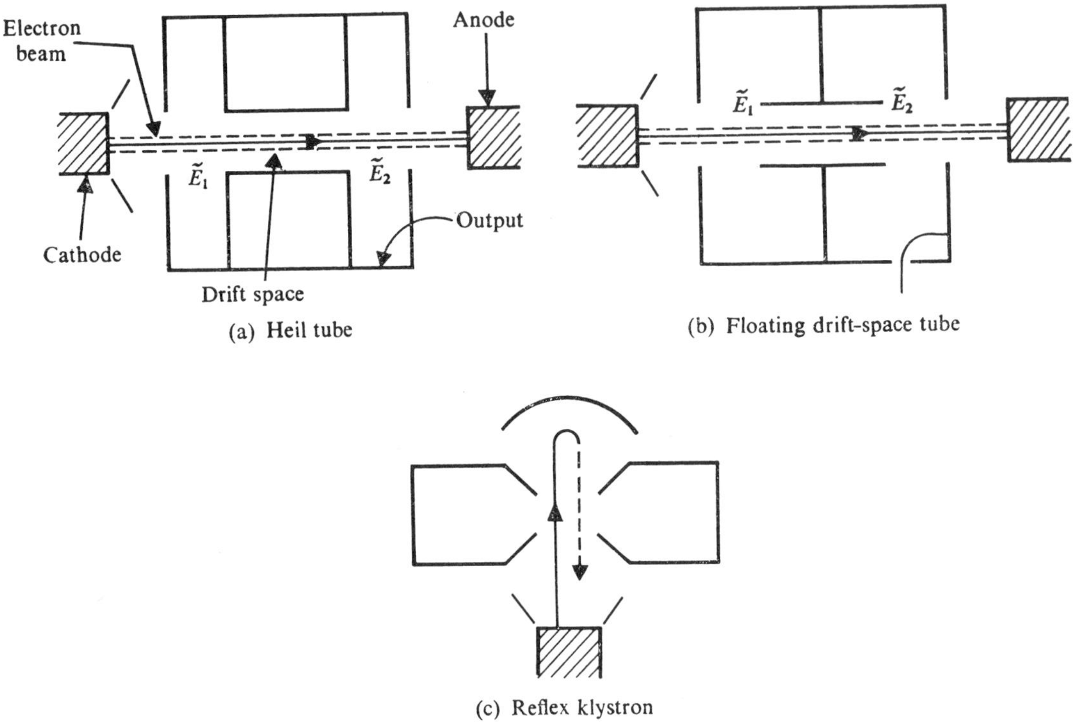

Fig. 10.2 Single-cavity drift-space tubes.

In the floating drift-space tube (Fig. 10.2b), the field-free drift space is a coaxial cylinder inside the resonator, and the phase of the modulating field is the same as that of the output field at all times. Optimum regeneration is obtained if the transit angle θ in the drift space is $2\pi(n + 1/4)$.

The reflex klystron, which is one of the most widely used microwave tubes, is also a single-cavity tube, and is shown schematically in Fig. 10.2c. The electrons are velocity modulated when they pass through the cavity for the first time. After they pass through the cavity, they enter a static retarding field in which they are tuned back. Bunching takes place all the way along both the forward and return paths. When they pass through the cavity in their return path, they are retarded in the periodic field and give r-f power to it. Thus, the same field velocity modulates the electrons and extracts energy from them.

Multicavity drift-space tubes are velocity-modulated amplifiers which can become oscillators if provided with feedback. Figure 10.3 shows various types of such tubes in schematic form. All of them contain input and output cavities. When they are used as amplifiers, the cavities are coupled by the electron beam only, whereas if they are used as oscillators, there is feedback between the input and output cavities.

In the Hahn-Metcalf tube (see Fig. 10.3a), each cavity is in the form of an H tube, and there is a third drift space between the two cavities. This type of tube is only of historical interest.

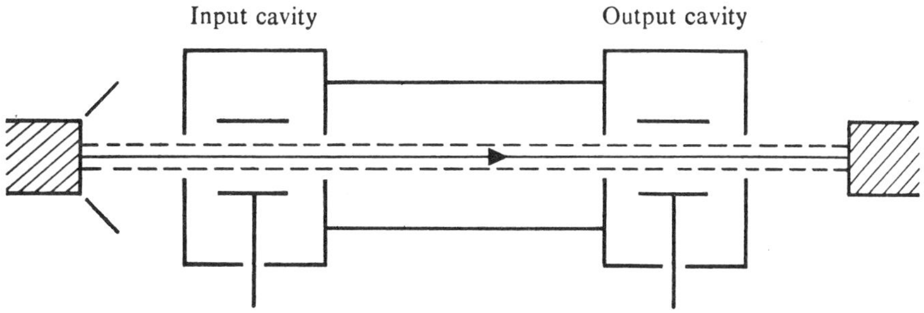

(a) Hahn-Metcalf tube

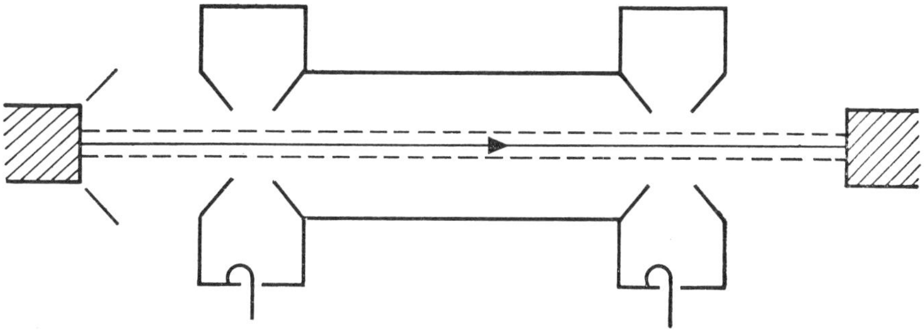

(b) Two-cavity klystron

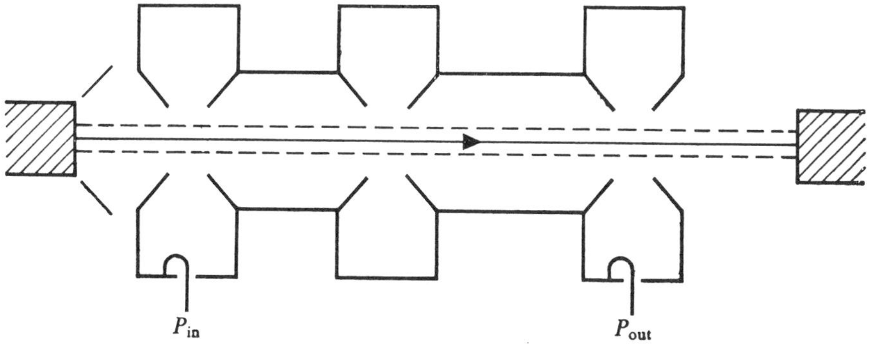

(c) Three-cavity klystron

Fig. 10.3 Multicavity drift-space tubes.

In the multicavity klystrons shown in Figs. 10.3b and 10.3c, which are the most important of this class of tubes, the drift space is at constant potential. The resonant cavities which are introduced along the drift space are floating, that is, they are excited only by the r-f alternating convection current passing through them, and not by r-f voltages. They are tuned at or near the frequency to be amplified. Bunching is increased due to the additional velocity modulation produced by them. This results in a higher gain in the multicavity klystron. By slightly mistuning the circuits, a larger bandwidth can be obtained.

In the velocity-jump tube, rippled-wall tube, and rippled-beam tube, the drift space also acts as a transformer which increases the amplitude of the convection current at the output.

10.5 Growing-Wave Tubes

There are three different types of growing-wave tubes, namely, those without a static transverse field or *O*-type, those with a transverse electrostatic field or *E*-type, and those with a transverse electrostatic field and a crossed magnetostatic field or *M*-type. Figure 10.4 shows schematically some of the *O*-type and *M*-type growing-wave tubes.

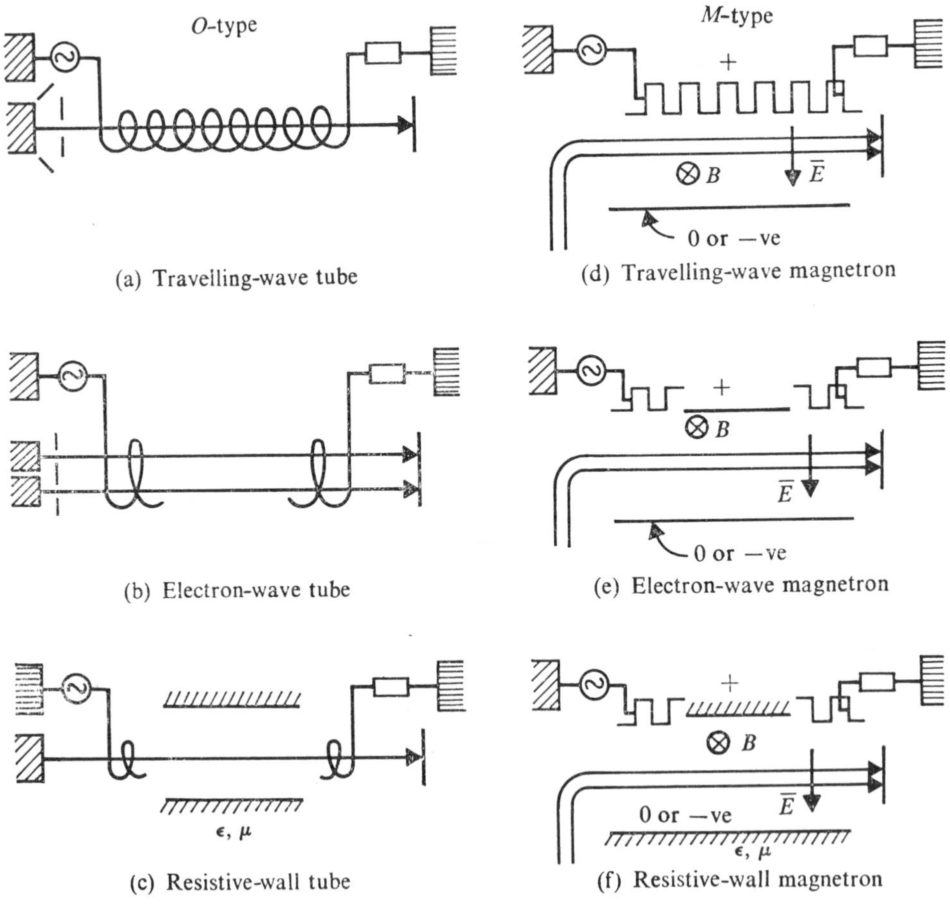

Fig. 10.4 Growing-wave tubes.

In the travelling-wave tube illustrated in Fig. 10.4a, the electrons flow within a delay line or slow-wave structure. Amplification occurs if the electron velocity $\bar{v} = (2\eta\bar{V})^{1/2}$ is equal (or nearly equal) to the phase velocity v_p of the slow wave, where $\bar{V}$ is the accelerating potential. The simplest type of slow-wave structure used is a helix. For a travelling-wave tube, the phase and group velocities of the slow wave must have the same sign or, in other words, it must be a forward-wave structure. Here, both the convection current and the retarded wave grow in the direction of electron flow. The signal is applied to the input of the line, and the amplified power is taken out at the output end of the line. In the backward-wave oscillator, the phase and group velocities of the retarded wave have opposite signs. The slowing down of the electrons in the r-f field of the travelling wave produces r-f energy which flows in the direction opposite to that of the electron beam, resulting in positive feedback and oscillations.

In the electron-wave tube (see Fig. 10.4b), the interaction takes place among several electron beams. In order to obtain amplification, the velocity spectrum of the complete beam should contain two peaks. If there are two beams, it is similar to a travelling-wave tube assuming that the slower of the two beams is equivalent to a delay line.

In the resistive-wall amplifier (Fig. 10.4c), the delay line consists of a conductor of non-zero resistivity ρ and finite dielectric constant ϵ_r enclosing the beam. A resistive layer of thickness comparable to or smaller than the skin depth can be used. Charges are induced on the resistive walls as a result of the alternating components of current and velocity within the electron beam. The resulting periodic electric fields react on the beam and produce a growing wave of convection current.

The travelling-wave tube and backward-wave oscillator are the only *O*-type tubes which are of practical interest.

Figure 10.5 shows a schematic diagram of an *E*-type tube which has a transverse electrostatic field. Both the conductors 1 and 2 are at positive potentials with respect to the cathode, and

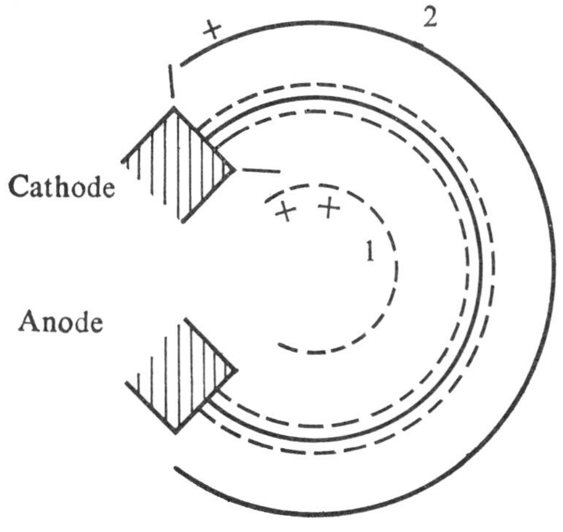

Fig. 10.5 *E*-type growing-wave tube.

a transverse electrostatic field lies between these two conductors. The electrons then move in circular trajectories along which the centrifugal force and the force of the transverse electrostatic field are equal and opposite. Conductor 1 is designed to support a retarded wave with phase velocity approximately equal to the velocity of the electrons. The electric field of this wave interacts with the electron beam, and the electron kinetic energy is converted into electromagnetic field energy. Such a tube is only of academic interest.

In the M-type growing-wave tubes, forms similar to those in the O-type exist. In these tubes, there is a transverse electrostatic field and a crossed static magnetic field which is perpendicular to both the electrostatic field and the direction of electron flow. The drift velocity $\bar{v}$ of the electron is perpendicular in direction to both the electrostatic field $\bar{E}$ and the magnetostatic field $\bar{B}$, and is equal to $\bar{E}/\bar{B}$.

The travelling wave magnetron (Fig. 10.4d) is the most important M-type tube. Here, the delay line consists of two conductors with the transverse electrostatic field between them. The electromagnetic waves are retarded by the conductor which is at the positive potential. Amplification results when the phase velocity of the retarded wave is equal or nearly equal in magnitude to the drift velocity and is in the same direction. Also, the phase and group velocities of the travelling waves must have the same direction. If the line is closed on itself in a ring, feedback is obtained, and the amplifier becomes an oscillator. Such a travelling-wave magnetron is used extensively as a generator of very large amount of power and high efficiency.

The crossed-field backward-wave oscillator or the M-carcinotron results if the phase and group velocities of the electromagnetic wave have opposite signs.

The electron-wave magnetron shown in Fig. 10.4e corresponds to an O-type electron-wave tube in which the walls enclosing the beam have zero impedance. However, amplification results in the electron-wave magnetron if there is a monotonic variation of velocity in the beam cross-section, which always exists because of the space-charge depression of potential in the beam.

In the resistive-wall magnetron shown in Fig. 10.4f, one (or both) of the walls surrounding the electron beam consists of a lossy dielectric material, similar to the resistive-wall amplifier of the O-type.

10.6 Double-Resonator Klystron

Since the klystron is one of the most widely used microwave tubes of the velocity-modulated type, this section deals in detail with the working of the double-resonator klystron. Section 10.7 discusses in detail about the reflex klystron.

In the negative-grid triode used as an oscillator or amplifier, the electrons travel through superimposed d-c and a-c electric fields and are therefore simultaneously accelerated by the d-c field and retarded by the a-c field. If the retarding a-c field is comparable to the d-c field, the electrons never attain a very high velocity and, therefore, the electron transit time may be very large at microwave frequencies. On the other hand, in a klystron, the electrons are initially accelerated to a high velocity by a d-c field before they enter the retarding a-c field so that the electrons have a relatively high velocity as they travel through the a-c field. This considerably reduces the electron transit time.

A schematic diagram of a double-resonator klystron oscillator is shown in Fig. 10.6. It contains an indirectly heated oxide-coated cathode, a control grid, two metal resonators, and a collector anode. The resonator nearest to the anode is called the input or the buncher resonator; the other resonator is called the output or the catcher resonator. The whole assembly is evacuated and the high-frequency electrical connections are made of small coupling loops placed inside the resonators with vacuum-sealed coaxial fittings for external connections. The electrons are accelerated by the d-c field resulting from the potential V_0; consequently, they enter the buncher-grid region with a high initial velocity. If the klystron operates as an amplifier, the buncher resonator is excited at its resonant frequency by the incoming signal which is to be amplified, thus producing an alternating field between the buncher grids. When the electrons pass through this a-c field, they are either accelerated or decelerated, depending on the phase of the buncher voltage during their transit. The electrons which are accelerated emerge with higher velocities than the entering velocity; those that are decelerated emerge with velocities lower than the entering velocity. This variation of the velocities of the electrons in the electron beam is known as *velocity modulation*. In the field-free drift space between the buncher and catcher grids,

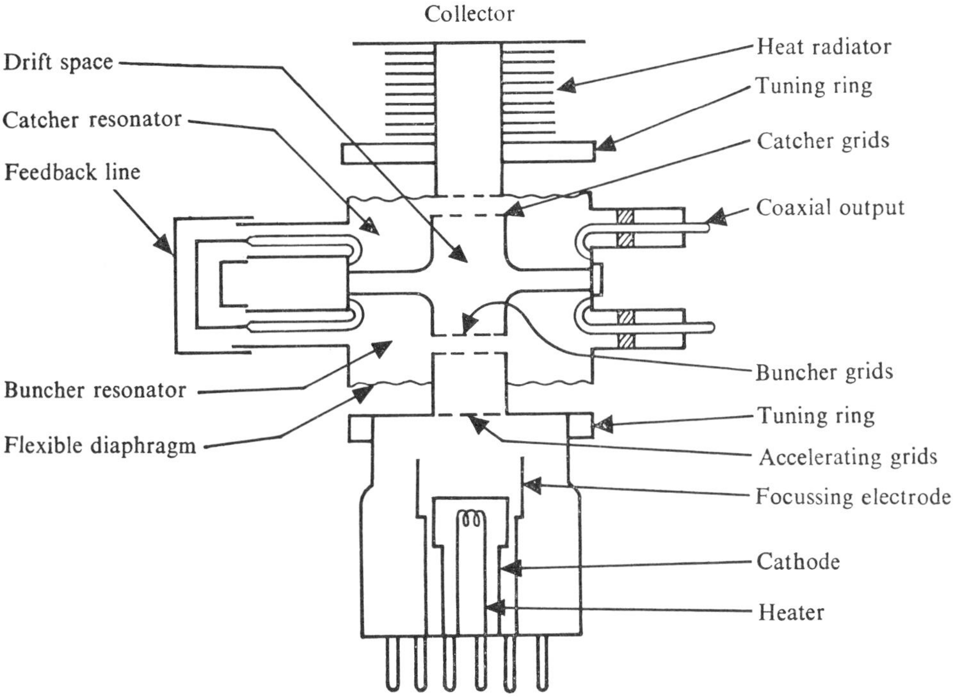

Fig. 10.6 Double-resonator klystron oscillator.

the high-velocity electrons overtake the low-velocity electrons, resulting in a bunching of the electrons as they travel towards the catcher resonator. For optimum performance, maximum bunching should occur approximately mid-way between the catcher grids. The electron bunches pass through the alternating field between the catcher grids during its retarding phase, and thus the electrons transfer energy to the field of the catcher resonator. The electrons finally emerge from the catcher resonator with reduced velocities and are collected by the collector. In a double-resonator klystron, energy is fed back from the catcher to the buncher through a short coaxial line to sustain oscillations.

Figure 10.7 shows the 'Applegate' diagram which illustrates the bunching process. It shows the distance travelled by the electron as a function of time for electrons leaving the buncher at different phases of the buncher voltage. Each line represents the displacement-time relationship for a single electron, and the slope of the line is proportional to the electron velocity. The higher-velocity electrons are represented by lines having steeper slopes. Bunches are formed around the electron which passes through the buncher grids when the buncher voltage is zero and is changing from deceleration to acceleration. This particular electron is called the centre-of-bunch electron.

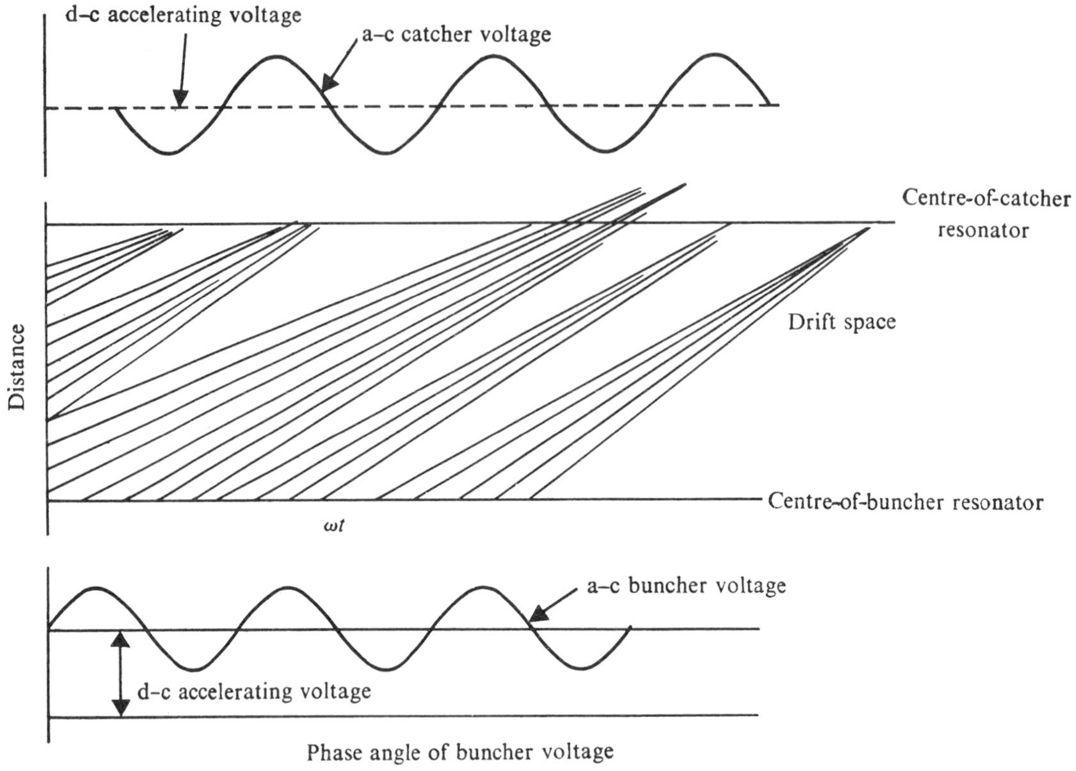

Fig. 10.7 Applegate diagram.

Figure 10.8 gives an equivalent circuit of a double-resonator klystron. The pentode is used in this representation to emphasize the complete isolation of the input and output circuits. The

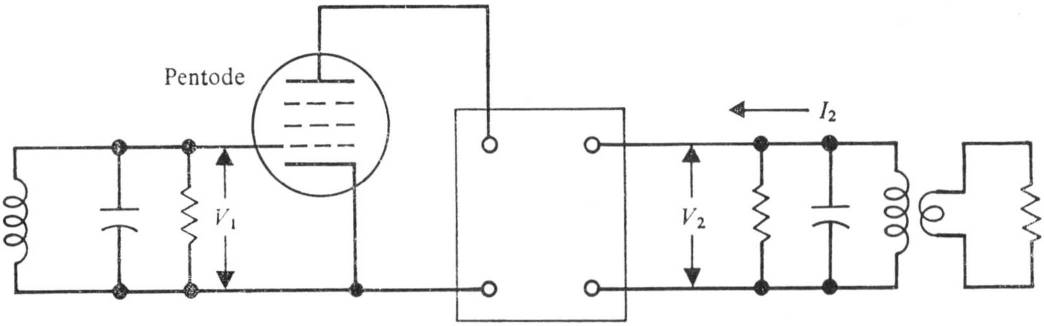

Fig. 10.8 Equivalent circuit of double-resonator klystron.

delay network represents the buncher-to-catcher transit-time delay, and the two tuned circuits represent the buncher and catcher resonators. The current I_2 is the induced current flowing in the catcher resonator.

The buncher or catcher resonator is usually of toroidal shape (Fig. 10.9). The effective capacitance of the resonator is approximately equal to the capacitance of the parallel grids, and the inductance is approximately proportional to the volume of the resonator. The resonant frequency can be varied either by increasing or by decreasing the volume of the resonator or by

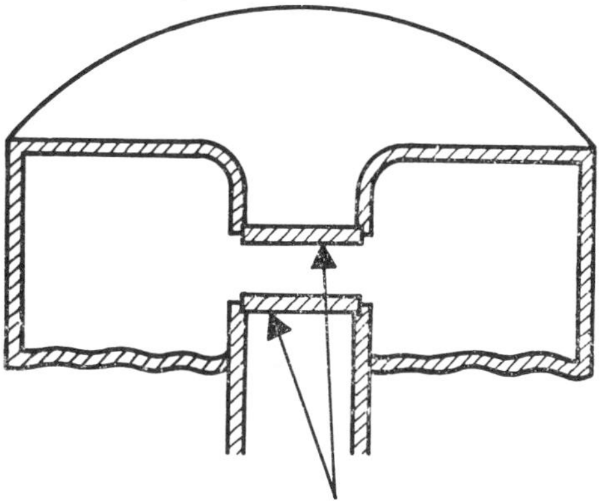

Fig. 10.9 Buncher or catcher resonator.

changing the distance between the two grids. Tuning is sometimes accomplished by making one wall of the resonator a thin corrugated diaphragm, and pressure on the diaphragm changes the effective capacitance of the resonator.

10.7 Analysis of Double-Resonator Klystron

We now give an analysis of the double-resonator klystron. It is based on the following assumptions:

(i) The transit angles of the electrons through the buncher and catcher grids are negligible as compared to a period of the alternating current.

(ii) The alternating voltages between the buncher grids are small as compared to the d-c accelerating voltage.

(iii) Space-charge effects are negligible.

(iv) The electron beam has uniform density in the cross-section of the beam, and all the electrons which leave the cathode pass through the catcher grids.

The foregoing assumptions lead to ideal values of power output and efficiency.

Figure 10.10 gives a schematic diagram of a double-resonator klystron oscillator. The electrons are accelerated to an initial velocity v_0 by the accelerating voltage V_0 before they enter the buncher grids. If it is assumed that the electrons leave the cathode with zero velocity, the kinetic energy and the velocity of the electrons entering the buncher are given by

$$\tfrac{1}{2}m_e v_0^2 = q_e V_0, \tag{10.1}$$

$$v_0 = \left(\frac{2q_e V_0}{m_e}\right)^{1/2}, \tag{10.2}$$

where q_e and m_e are the charge and mass of the electron, respectively. Oscillations in the buncher resonator produce an alternating potential difference $V_1 \sin \omega t$ between the resonator grids so that it can be assumed that the two buncher grids are at the potentials V_0 and $V_0 + V_1 \sin \omega t$, with respect to the cathode. If it is assumed that the voltage of the two buncher grids remains constant during the passage of an electron through the buncher grids, which means that the transit time of the electron through the two buncher grids is negligible, then the kinetic energy and the velocity of the electron as the electron emerges from the buncher grids are

$$\tfrac{1}{2} m_e v_1^2 = q_e(V_0 + V_1 \sin \omega t_1), \tag{10.3}$$

$$v_1 = [\frac{2q_e}{m_e} V_0(1 + \frac{V_1}{V_0} \sin \omega t_1)]^{1/2} = v_0(1 + \frac{V_1}{V_0} \sin \omega t_1)^{1/2}. \tag{10.4}$$

Equation (10.4) is called the equation of velocity modulation.

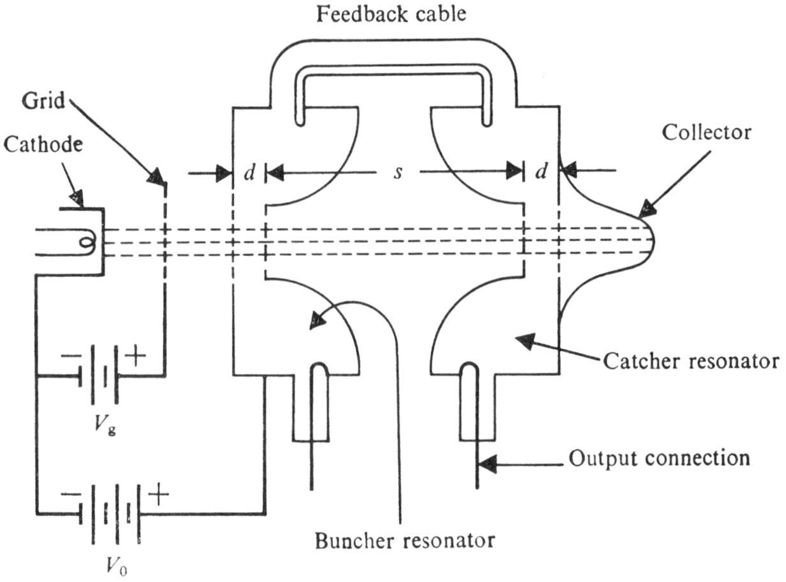

Fig. 10.10 Double-resonator klystron oscillator.

An electron which passes through the buncher grids at $\omega t_1 = n\pi$, where $n = 0, 1, 2, 3, \ldots$, emerges with unchanged velocity. The electrons which pass through the buncher grids during the maximum accelerating and maximum retarding phases of the buncher voltage, i.e., when $\omega t_1 = \pi/2$ and $\omega t_1 = -\pi/2$, have maximum and minimum velocities, respectively, at the exit of the buncher grids. The maximum and minimum velocities are $v_0\sqrt{1 + V_1/V_0}$ and $v_0\sqrt{1 - V_1/V_0}$.

If T is the time required by an electron to travel the distance s in the field-free space between the buncher and catcher grids, then

$$T = \frac{s}{v_1} = \frac{s}{v_0[(1 + V_1/V_0) \sin \omega t_1]^{1/2}}. \tag{10.5}$$

If the buncher alternating voltage is taken as the time reference, then the electron arrives at the catcher grids at time $t_2 = t_1 + T$. The corresponding phase angle ωt_2 is given by

$$\omega t_2 = \omega(t_1 + T) = \omega t_1 + \frac{\omega s}{v_0}(1 + \frac{V_1}{V_0} \sin \omega t_1)^{-1/2}. \tag{10.6}$$

The quantity s/v_0 is the buncher-to-catcher transit time for the electron passing through the buncher when $\omega t_1 = 0$, i.e., the centre-of-bunch electron. The corresponding transit angle is

$$\alpha = \omega s/v_0 . \tag{10.7}$$

Assuming that $V_1/V_0 \ll 1$, Eq. (10.6) can be rewritten approximately as

$$\omega t_2 = \omega t_1 + \alpha(1 - \frac{V_1}{2V_0} \sin \omega t_1). \tag{10.8}$$

Equation (10.8) is a relation between the departure phase angle at the buncher grids and the arrival phase angle ωt_1 at the catcher grids.

To calculate the power output and the efficiency of the klystron, it is assumed that the buncher and catcher voltages are in time phase so that the catcher voltage can be written as $V_2 \sin \omega t$. If the catcher voltage remains constant during an electron's transit time through the catcher grids, the electron gives up an amount of energy $w = -q_e V_2 \sin \omega t_2$, the negative sign signifying energy transfer from the electron to the field. According to this convention, the energy output and power output are positive quantities. Using Eq. (10.8), we obtain the expression for w as

$$w = -q_e V_2 \sin \omega t_2 = -q_e V_2 \sin [\omega t_1 + \alpha(1 - \frac{V_1}{2V_0} \sin \omega t_1)]. \tag{10.9}$$

The average energy per electron transferred to the catcher resonator is given by

$$w_{av} = \frac{1}{2\pi}\int_{\omega t_1=0}^{\omega t_1=2\pi} w d(\omega t_1) = -\frac{q_e V_2}{2\pi}\int_0^{2\pi} \sin [\omega t_1 + \alpha(1 - \frac{V_1}{2V_0} \sin \omega t_1)]d(\omega t_1)$$
$$= -q_e V_2 J_1(x) \sin \alpha, \tag{10.10}$$

where

$$x = \alpha V_1/(2V_0). \tag{10.11}$$

The quantity x is called the bunching parameter.

Now, assume that N electrons are emitted from the cathode per second. Then, the direct current emitted from the cathode is $I_0 = Nq_e$. Assuming that all the electrons emitted from the cathode flow through the catcher resonator, there are N electrons flowing through the catcher per second. The power output P_{ac} at the catcher resonator is

$$P_{ac} = w_{av} \cdot N = -q_e N V_2 J_1(x) \sin \alpha = -I_0 V_2 J_1(x) \sin \alpha. \tag{10.12}$$

The power supplied by the d-c potential source is $P_0 = I_0 V_0$, and hence the conversion efficiency is

$$\eta = \frac{P_{ac}}{P_0} = -\frac{V_2}{V_0} J_1(x) \sin \alpha. \tag{10.13}$$

In deriving Eq. (10.13), it is assumed that the power required for bunching the electrons is negligible. This is true if the transit angle of the electron through the bunching grids is negligible, because then the time-average power required for the bunching operation would be zero since the alternating field of the buncher accelerates some electrons and retards others. In practice, the transit angle is not negligible, and some power is required for the bunching operation. $J_1(x)$ has a maximum value of 0.58 when $x = 1.81$. From Eq. (10.13), it can be seen that maximum efficiency is obtained when

(i) V_2/V_0 has a maximum value,

(ii) $x = 1.84$, $J_1(x) = 0.58$, and

(iii) $\sin \alpha = -1$, or $\alpha = 2\pi n - \pi/2$ ($n = 1, 2, 3, \ldots$).

V_2/V_0 is always $\leqslant 1$ because, if $V_2 > V_0$, then some of the electrons would reverse their direction of travel at the catcher resonator and absorb power from the alternating field. Therefore, the maximum value of V_2/V_0 can be 1. Thus, the maximum theoretical efficiency is 58 per cent.

For maximum power output, the d-c beam current I_0 and the catcher resonator potential V_2 should have maximum values. It will be shown later in this chapter that $\sin \alpha = -1$ is always satisfied if sustained oscillations exist. Then, the expressions for the power output and efficiency become

$$P_{ac} = I_0 V_2 J_1(x), \tag{10.14}$$

$$\eta = \frac{V_2}{V_1} J_1(x). \tag{10.15}$$

The criterion $\sin \alpha = -1$ or $\alpha = 2\pi n - \pi/2$ determines the value of the d-c accelerating voltage required for maximum power output. α is given by

$$\alpha = \frac{\omega s}{v_0} = \frac{\omega s}{(2q_e V_0/m_e)^{1/2}} = 2\pi n - \frac{\pi}{2}, \tag{10.16a}$$

and hence

$$V_0 = \left(\frac{\omega s}{2\pi n - \pi/2}\right)^2 \frac{m_e}{2q_e} = 0.284 \times 10^{-11} \left(\frac{\omega s}{2\pi n - \pi/2}\right)^2. \tag{10.16b}$$

From Eqs. (10.16), it is seen that there are an infinite number of discrete values of the d-c accelerating voltage V_0, which give maximum power output, one for each integral value of n. The highest value of V_0 corresponds to the lowest value of n, i.e., $n = 1$.

The conditions $x = 1.84$ and $\alpha = 2\pi n - \pi/2$, which make the power output as well as the efficiency maximum, determine the optimum ratio of buncher voltage to d-c accelerating voltage V_1/V_0. Substituting the value of α in the equation

$$x = \frac{\alpha V_1}{2V_0} = 1.84,$$

we obtain

$$\left(2\pi n - \frac{\pi}{2}\right)\frac{V_1}{2V_0} = 1.84$$

or

$$\frac{V_1}{V_0} = \frac{3.68}{2n\pi - \pi/2}. \tag{10.17}$$

For each integral value of n, there is an optimum value of V_1/V_0 which gives the maximum power output. When $n = 1$, we obtain $V_1/V_0 = 0.78$. As n increases, both the d-c accelerating voltage V_0 and the ratio V_1/V_0 decrease. The reason for this is as follows: The electron velocity varies as the square root of V_0, and hence, if V_0 is small, the electrons travel more slowly, have more time in which to bunch, and require less bunching voltage.

In the normal operation of the klystron oscillator, electron bunching occurs only at the catcher resonator. It is, however, possible to have the electrons alternately bunch and debunch several times in the drift space between the buncher and catcher resonators. This is given by the larger values of x at which $J_1(x)$ has the second, third, . . . maxima. The value of $x = 1.84$ corresponds to single bunching and gives the highest efficiency.

To obtain the maximum power output, the catcher voltage V_2 should be maximum. The catcher voltage increases with the d-c accelerating voltage. Therefore, the highest permissible accelerating voltage gives the maximum power output when all other conditions are satisfied.

The voltage V_2 also depends on the effective impedance of the catcher resonator. At its resonant frequency, the impedance of the resonator is a pure resistance which is proportional to the Q of the resonator and to its effective L/C ratio. In order to obtain the maximum value of the voltage V_2, the resonator impedance should be as large as possible. Also, the electron transit time through the resonator grids must be minimized, because large transit time results in inadequate bunching, and power is consumed in the bunching process. If the transit time through the catcher resonator is too much, the power output and efficiency are reduced. Figure 10.11

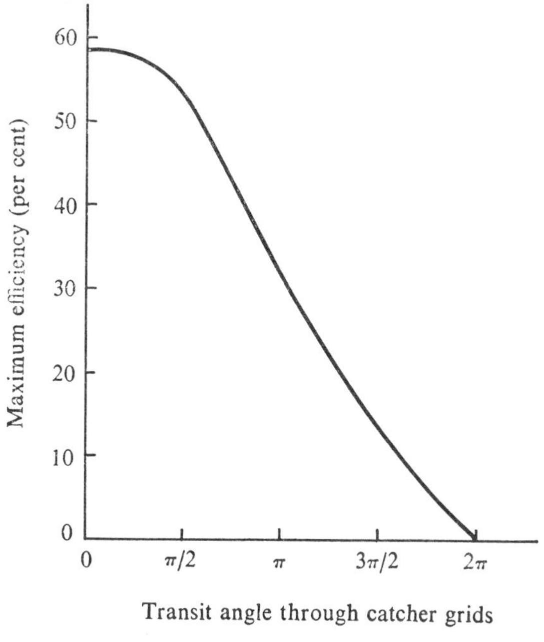

Fig. 10.11 Maximum efficiency of klystron versus catcher transit angle.

shows the maximum efficiency of the klystron as a function of the catcher transit angle. The electron transit time can be minimized by a close spacing of the grids. This, however, increases the capacitance, and hence the L/C ratio increases, resulting in the lowering of the effective impedance of the resonator. So, there should be a compromise between making the spacing between grids smaller and the impedance larger.

We have assumed so far that the buncher and catcher resonators oscillate in time phase. If this is a necessary condition, then it results in extremely critical conditions of operation, because the slightest deviation of the voltages from the values given by Eqs. (10.16) and (10.17) would stop the oscillations. However, in practice, there is a certain amount of flexibility in the phase relationships between the buncher and catcher voltages as a result of phase shift existing in the resonators and the feedback line.

The criterion of oscillation for any oscillator is $\beta K = 1\angle 2\pi n$, where K is the voltage gain and β is the ratio of the feedback voltage to the output voltage (also termed the feedback ratio) and n is any integer, including zero. This requires that the sum of the phase angles of K and β, or the total phase shift around the closed circuit, be equal to $2\pi n$ radians. Applying this criterion to the klystron oscillator, we can determine the phase relationships between the buncher and catcher resonators. Figure 10.12 shows the buncher and catcher voltages out of phase. The centre-of-bunch electron leaves the buncher when its alternating voltage is zero and arrives at the catcher

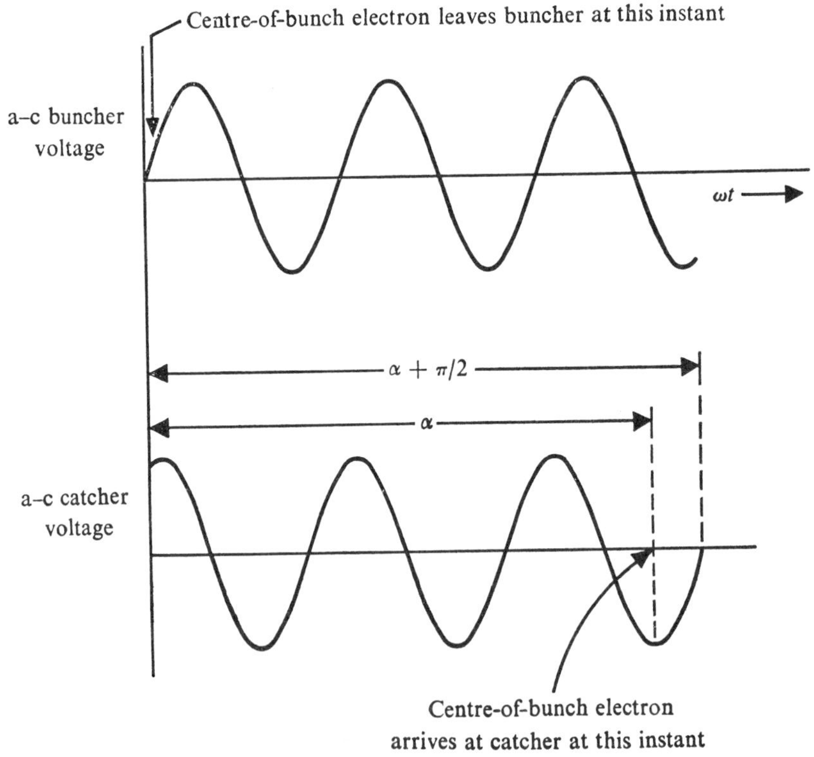

Fig. 10.12 Phase relationships between buncher and catcher voltages in klystron.

resonator when its alternating voltage has the maximum negative value. If α is the transit angle of the electron between the buncher and catcher, then $(\alpha + \pi/2)$ is the phase angle between the zeros of the buncher and c atcher voltages. To obtain the phase shift around the closed circuit of the klystron oscillator, we must include the phase shift in the resonators and in the feedback line. If the resonator is at its resonant frequency, then the phase shift in the resonator is zero. If the frequency of oscillation of a klystron deviates slightly from the resonant frequency of the resonator, then the voltage and current in the resonator are not in phase, and there is a phase shift in the resonator. If θ is the total phase shift in the resonators and the feedback line, then the criterion of oscillation requires that

$$\theta + \alpha + \frac{\pi}{2} = 2\pi n \text{ radians.} \tag{10.18}$$

If the two resonators oscillate in time phase, then $\theta = 0$, and Eq. (10.18) becomes

$$\alpha + \frac{\pi}{2} = 2\pi n. \tag{10.19}$$

This equation is identical to Eq. (10.16a) which is the condition for maximum power output and maximum efficiency. Therefore, when the two resonators are in phase, the condition $\alpha = 2\pi n - \pi/2$ is not only a condition for maximum power output but also a criterion that must exist for sustained oscillations. In general, however, the resonators may not oscillate in time phase. In the operation of a klystron oscillator, if there is a small change in the d-c accelerating voltage, then there is a change in frequency. This is so because a change in the d-c voltage changes the transit angle α, and the frequency of oscillation shifts in such a way that a new value of θ, which satisfies Eq. (10.18), results.

If the two resonators, namely, the catcher and buncher resonators, are overcoupled, a double-peaked resonance curve occurs, as shown in Fig. 10.13. Such an arrangement makes it

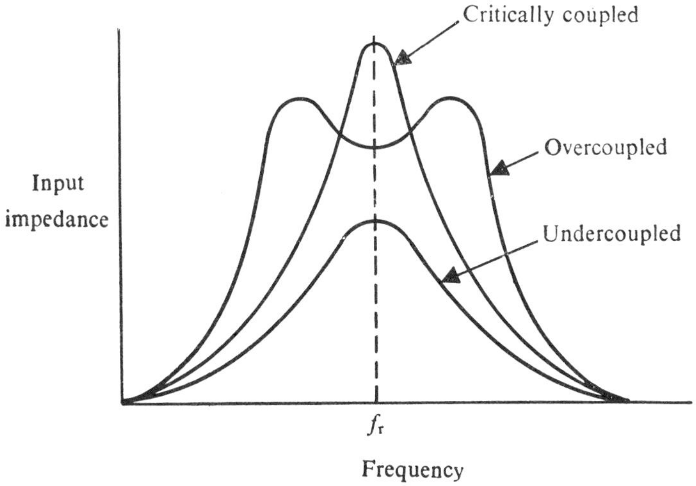

Fig. 10.13 Input impedance of two coupled resonators having same resonant frequency.

possible to obtain oscillations over somewhat wider ranges of d-c voltages. If the resonators are critically coupled, then the klystron has a near straight-line variation in frequency with accelerating voltage. This arrangement is very useful for frequency modulation. By tuning the two resonators to slightly different frequencies, the power output may be held more nearly constant.

The bunching process can be clearly understood by plotting the curves of electron departure time at the buncher against arrival time at the catcher. In Eq. (10.8), substituting for α and using the relation $x = \alpha V_1/(2V_0)$, we obtain

$$\omega\left(t_2 - \frac{s}{v_0}\right) = \omega t_1 - x \sin \omega t_1. \tag{10.20}$$

In Eq. (10.20), the phase angle at which an electron departs from the buncher is given by ωt_1, and the phase angle at which the electron arrives at the catcher is given by ωt_2, both measured with respect to the buncher voltage.

Figure 10.14 shows the curves of ωt_1 as a function of $\omega(t_2 - s/v_0)$ for various values of the bunching parameter x. Here, $\omega s/v_0$ is the buncher-to-catcher transit angle of the centre-of-bunch electron. The quantity $\omega(t_2 - s/v_0)$ is the difference between the arrival phase angle at the catcher and the buncher-to-catcher transit angle of the centre-of-bunch electron. From the figure, it can

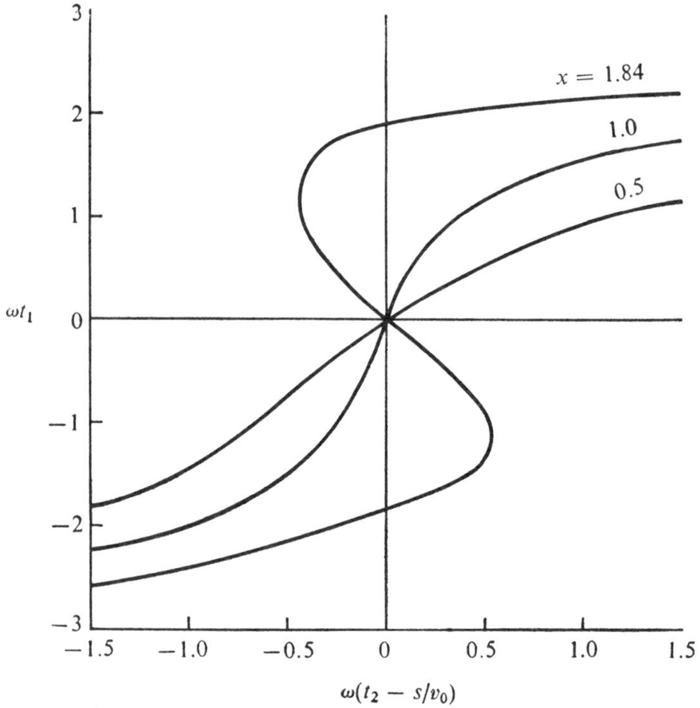

Fig. 10.14 Electron departure phase angle ωt_1 versus $(\omega t_2 - s/v_0)$.

be seen that the electrons arriving at the catcher at any instant of time may have left the buncher at different instants of time. For $x = 1.84$, electrons leaving the buncher at three different times arrive at the catcher simultaneously.

If the space-charge densities at the buncher and catcher resonators are q_1 and q_2 and the corresponding velocities are v_1 and v_2, respectively, then an amount of charge $q_1 v_1\, dt_1$ flows through the unit area at the buncher exit during the time interval dt_1, and an equal amount of charge flows through the unit area at the catcher during a different time interval dt_2, which may be expressed as $q_2v_2\ dt_2$. Therefore

$$q_2v_2\ dt_2 = q_1v_1\ dt_1. \tag{10.21}$$

However, if there is a condition under which some electrons overtake other electrons in the drift space, then the charges arriving at the catcher during the time interval dt_2 would have left the buncher at several different intervals of dt_1. To take care of such a situation, we may write

$$q_2v_2\ dt_2 = \Sigma\ q_1v_1\ dt_1. \tag{10.22}$$

Since the space-charge density q_1 at the buncher is very nearly equal to the space-charge density q_0 just before the buncher, we may put $q_1 = q_0$. Also, since the velocity variation is small, we can put $v_2 \approx v_1 \approx v_0$. Then, Eq. (10.22) becomes

$$q_2v_0\ dt_2 = q_0v_0\ \Sigma\ dt_1$$

or

$$\frac{q_2}{q_0} = \Sigma\frac{dt_1}{dt_2} = \Sigma\frac{d(\omega t_1)}{d(\omega t_2)}. \tag{10.23}$$

Here, $d(\omega t_1)/d(\omega t_2)$ is the slope of the curves shown in Fig. 10.14. A negative slope indicates that electrons arriving at the catcher in one sequence have left the buncher in the reverse sequence. In adding the slopes, we consider only absolute values and discard the sign. Figure 10.15 shows the curves of q_2/q_0 or J_2/J_0 against the arrival phase angle $\omega(t_2 - s/v_0)$.

In examining the curves of Fig. 10.15, we notice that the value of $x = 1.84$, which, as

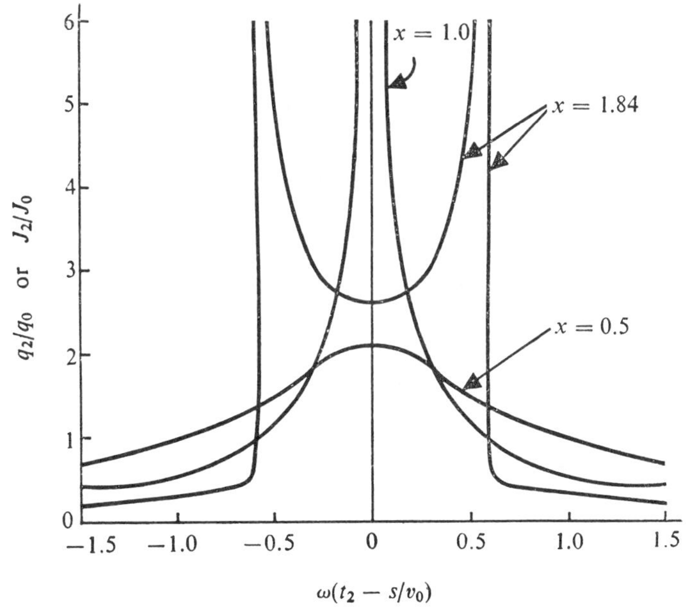

Fig. 10.15 Curves of q_2/q_0 or J_2/J_0 against $\omega(t_2 - s/v_0)$.

already noted, yields the maximum power output, is not the critically bunched condition. The double peak in this curve shows overbunching with two distinct groups of electrons flowing through the catcher resonator a short time interval apart. The critically bunched condition, which is the condition for maximum bunching, corresponds to $x = 1.0$. Since the power output and efficiency vary directly as $J_1(x)$, the maximum theoretical power output and efficiency for the critically bunched condition of $x = 1.0$ are 24 per cent lower than those for the overbunched value of $x = 1.84$.

The convection current densities at the buncher and catcher resonators respectively are

$$J_1 = q_1 v_1 \approx q_1 v_0, \qquad J_2 = q_2 v_2 \approx q_2 v_0.$$

J_1 is approximately equal to J_0 which is the current density before the buncher. Hence

$$\frac{J_2}{J_1} \approx \frac{J_2}{J_0} = \frac{q_2}{q_0}. \tag{10.24}$$

Since the beam current has a high harmonic content, the current induced in the catcher resonator has a similar waveform. By the Fourier analysis, the d-c components and the fundamental and higher-order harmonic a-c components can be obtained. The catcher resonator may be tuned to the fundamental harmonic or to any of the higher harmonics. The a-c components of current of the harmonics have amplitudes proportional to $J_m(mx)$ which is the Bessel function of the m-th order, where m is the order of the harmonic. Hence, the power output for the m-th harmonic is $P_{ac} = I_0 V_m J_m(mx)$, where V_m is the voltage developed across the catcher resonator for the m-th harmonic. If V_m and I_m are the peak values of the m-th harmonic voltage and induced current at the catcher resonator, then the time-average power output is $V_m I_m/2$ which is also equal to $I_0 V_m J_m(mx)$. Hence

$$I_m = 2 I_0 J_m(mx). \tag{10.25}$$

The effective impedance of the resonator at the m-th harmonic is

$$Z_m = V_m / I_m \tag{10.26}$$

which is a resistance if the catcher resonator is operating at its resonant frequency. If the buncher and catcher resonators are tuned to the same frequency, then $m = 1$, and the klystron is an oscillator. If the catcher is tuned to a harmonic of the buncher voltage, then it acts as a frequency multiplier.

The resonant impedance of a resonator depends on its geometry, the loading, and the electronic effects due to the beam current. The impedance is usually increased by increasing the L/C ratio, which requires increasing the volume and decreasing the capacitance between the grids.

10.8 Operation of Double-Resonator and Multiresonator Klystrons

When a double-resonator klystron is to be tuned, several variables have to be simultaneously adjusted. These variables are the grid voltage, the d-c accelerating voltage, and the tuning of the two resonators. The tuning procedure is sometimes simplified by connecting a 50-cycle a-c voltage of about 50–100 volts in series with the d-c accelerating voltage. The accelerating voltage is then adjusted to the approximate value till all the other adjustments are made. The presence of oscillations is seen by means of a crystal detector and microammeter connected to the klystron output. The 50-cycle a-c voltage is removed during the final adjustment.

Power outputs of the order of a fraction of a watt to several kilowatts can be obtained with klystrons at frequencies of the order of 3000 MHz or higher. Actual efficiencies of klystrons are very much less than the theoretical maximum efficiency of 58 per cent. The reasons for this are debunching of the electrons due to space-charge effects, collisions of the electrons with the grids, secondary emission at the grid, power consumed in bunching the electrons, transit-angle delay of the electrons in their passage through the buncher and catcher grids, and losses in the resonators.

The frequency stability of the klystron oscillator depends on the temperature of the resonator and also on the stability of the power-supply voltages. The resonator walls are of metal, and their thermal coefficient of expansion is quite large, and hence the frequency drift with temperature may also be quite large. By using suitable materials, it may be possible to compensate for the thermal expansion of the resonator walls by the thermal expansion of the tuning mechanism. Very well regulated power supplies are also essential for frequency stability.

Figure 10.16 shows the typical experimental curves of power output and frequency deviation as functions of the d-c accelerating voltage for a double-resonator klystron.

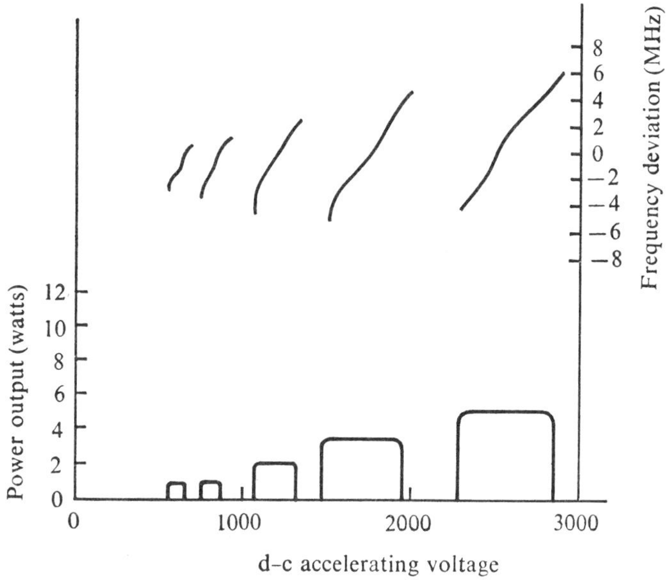

Fig. 10.16 Experimental curves of power output and frequency deviation versus d-c accelerating voltage for double-resonator klystron.

If the input signal is fed into the buncher resonator and the output is taken out of the catcher resonator, then the double-resonator klystron behaves as an amplifier. Regenerative amplification can be obtained by loose coupling between the buncher and catcher resonators. The adjustments of the klystron amplifier are less critical than those of an oscillator, because it is not necessary for the electrons to be perfectly bunched when the klystron is used as an amplifier.

Higher amplification can be obtained by the use of more than two resonators in cascade (see Fig. 10.17). The input signal is fed into the buncher resonator. Partial bunching occurs at the second resonator, and the a-c field produced by the electron bunches passing through this resonator velocity modulates the electron stream in such a manner as to increase the electron bunching. Even if the input signal is very weak, the cumulative bunching of several resonators

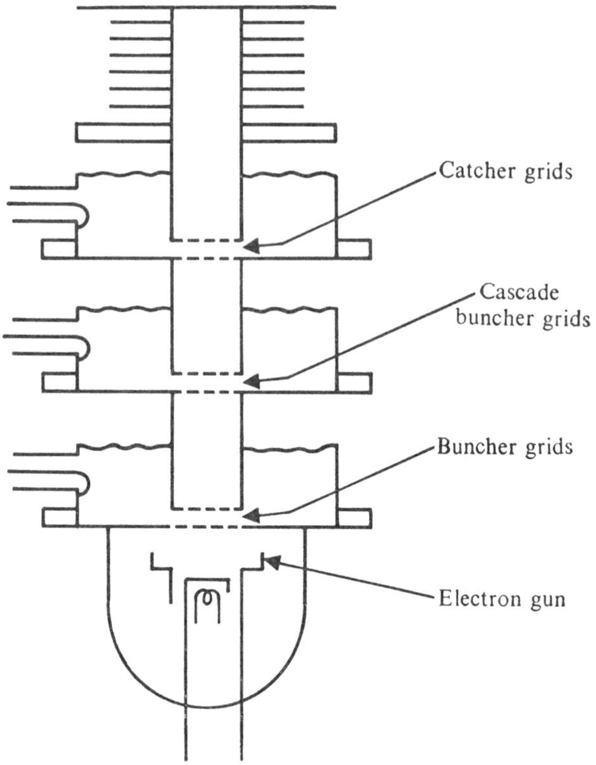

Fig. 10.17 Multiresonator klystron.

makes it possible to approach critical bunching at the final resonator, and power gains of 2 to 20 per stage are possible. Sometimes, a cascade klystron may consist of two stages as oscillators and the third stage as an amplitude-modulated amplifier.

10.9 Reflex Klystron

The reflex klystron is a single-resonator velocity-modulated tube in which the single resonator does the functions of both the buncher and catcher resonators. Figure 10.18 gives a schematic diagram of a reflex klystron. An electrode, called the reflector electrode, is placed a short distance beyond the resonator grids, and is at a negative potential with respect to the cathode. This reflector electrode sets up a retarding d-c field which causes the electrons to reverse their direction of travel and go through the resonator grids. If the tube is oscillating, the electrons are velocity modulated on their first excursion through the resonator grids. When they enter the d-c retarding field in the reflecting space, they are turned back and then pass through the resonator grids in the opposite direction. Electron bunching occurs during transit through the reflecting space. On their return journey, the electron bunches pass through the alternating field between the resonator grids during its retarding phase, and hence energy is transferred from the electrons to the alternating field of the resonator.

The bunching action in the reflex klystron is different from that in the double-resonator klystron in which all the electrons travel the same distance between the buncher and catcher. The centre-of-bunch electron is the one that passes through the buncher grids when its field changes from deceleration to acceleration. On the other hand, in the reflex klystron, the higher-velocity

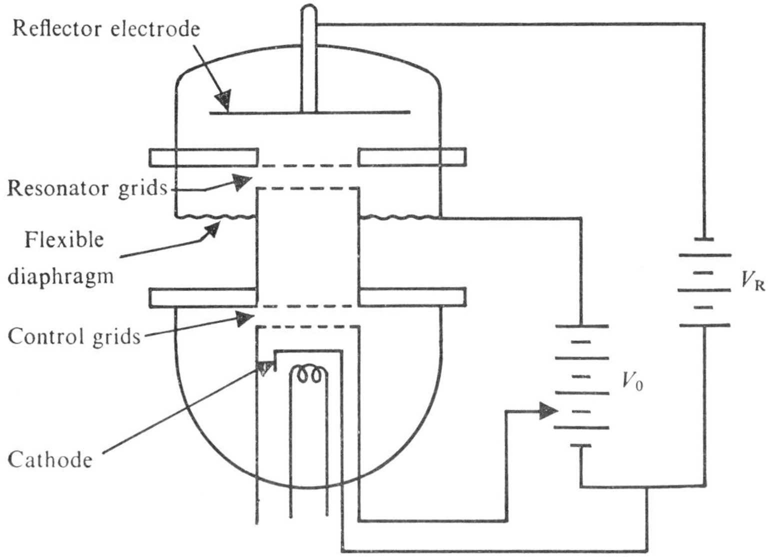

Fig. 10.18 Reflex klystron.

electrons travel further in the reflecting space and take longer time to return. Hence, the bunches centre around the electron which passes through the grids on its first excursion when the resonator alternating potential is zero, changing from acceleration to deceleration. Also, there is a 180° phase difference between the resonator voltage of the reflex klystron and the catcher voltage of the double-resonator klystron at the instant when the bunched electrons pass through the resonator grids, because the returning electrons flow through the resonator grids in a direction opposite to that taken by the double-resonator klystron which travels through the catcher grids.

The bunching that occurs in the double-resonator klystron is called *drift-space bunching*, whereas the bunching in the reflex klystron is called *reflector bunching*. If both types of bunching occur in the same tube, then the two cancel each other and debunching occurs. Because of the single resonator, the tuning operation of a reflex klystron is very much simpler than that of the double-resonator klystron.

Oscillations are obtained by adjusting the accelerating and reflector voltages, and the frequency is varied by changing the spacing between the grids of the resonator by means of a suitable tuning mechanism. By varying either the accelerating voltage V_0 or the reflector voltage V_R, a small variation in frequency can be obtained. The frequency variation obtainable by varying the reflector voltage without retuning the resonator is of the order of 10–20 MHz in a 3000-MHz oscillator.

The reflex klystron has a maximum theoretical efficiency which is less than that of the double-resonator klystron because, in a reflex klystron, a single resonator does the work of bunching the electrons as well as extracting the energy from the returning bunches of electrons. The optimum value of the a-c resonator voltage is therefore a compromise between the value required for optimum bunching and the value required for maximum power transfer from the bunched electrons to the resonator field. However, the reflex klystron has the advantages of simple tuning, simple adjustment and operation, and compactness.

10.10 Power Output and Efficiency of Reflex Klystron

While analyzing the reflex klystron, we shall derive a relation between the time of departure of the electrons from the resonator grids and the return time. This relation in turn will establish a relationship between the various electrode potentials. Such a relationship must be satisfied if oscillating conditions are to be obtained.

Figure 10.19 shows the various electrodes of a reflex klystron and their potentials. An a-c

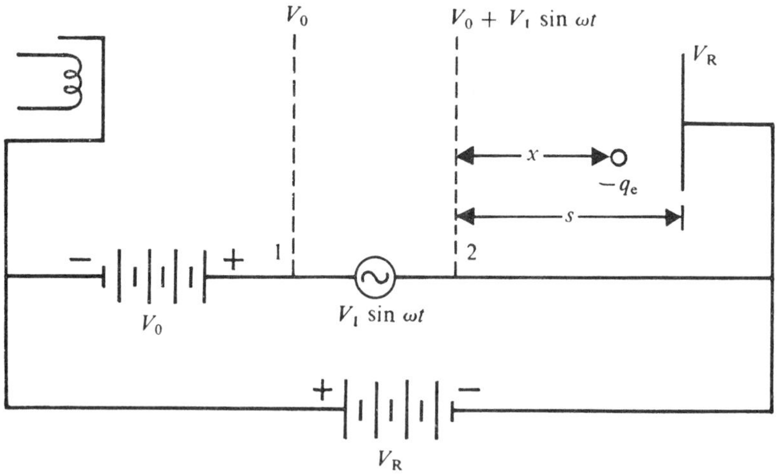

Fig. 10.19 Schematic diagram of reflex klystron.

potential difference $V_1 \sin \omega t$ is assumed to exist between grids 1 and 2 of the resonator. Let V_0 be the accelerating potential and V_R the negative potential of the reflector electrode. Then, the electrons enter grid 1 with a velocity

$$v_0 = (\frac{2V_0 q_e}{m_e})^{1/2} \tag{10.27}$$

and leave grid 2 with a velocity

$$v_1 = v_0(1 + \frac{V_1}{V_0} \sin \omega t)^{1/2}. \tag{10.28}$$

The potential difference between grid 1 and the reflector electrode is $V_R - (V_0 + V_1 \sin \omega t)$. It is assumed that $V_1 \sin \omega t$ is small as compared to $V_R - V_0$, the small signal conditions. Assuming that there is a uniform field in the region between grid 2 and the reflector electrode, the electric field intensity is $E = -(V_R - V_0)/s$, where s is the distance between grid 2 and the reflector electrode.

The force on the electron in the reflecting region is $-q_e E = q_e[(V_R - V_0)/s]$. This force is equal to mass times acceleration, and hence

$$m_e \frac{d^2x}{dt^2} = \frac{q_e(V_R - V_0)}{s} \tag{10.29}$$

Integrating Eq. (10.29) twice, we obtain

$$x = \frac{q_e(V_R - V_0)}{2m_e s}(t - t_1)^2 + v_1(t - t_1), \tag{10.30}$$

where $t = t_1$ is the time at which the electron leaves grid 2 with a velocity v_1. The electron gets decelerated as it approaches the negative reflector electrode, reverses its direction of travel, and travels towards the resonator grids. If t_2 is the time of arrival on its return at the grids, then put $x = 0$ at $t = t_2$ in Eq. (10.30). The total transit time of the electron in the reflector space is $(t_2 - t_1)$ given by

$$(t_2 - t_1) = -\frac{2m_e s v_1}{q_e(V_R - V_0)}. \tag{10.31}$$

Substituting Eq. (10.28) in Eq. (10.31), we obtain the arrival phase angle ωt_2 as

$$\omega t_2 = \omega t_1 - \frac{2\omega m_e s v_0}{q_e(V_R - V_0)}(1 + \frac{V_1}{V_0} \sin \omega t_1)^{1/2}. \tag{10.32}$$

Put

$$\alpha' = -\frac{2\omega m_e s v_0}{q_e(V_R - V_0)},$$

where α' is the round-trip transit angle of the centre-of-bunch electron. Then, Eq. (10.32) can be simplified to

$$\omega t_2 = \omega t_1 + \alpha'(1 + \frac{V_1}{2V_0} \sin \omega t_1). \tag{10.33}$$

The bunching parameter is given by

$$x = \alpha' V_1/(2V_0). \tag{10.34}$$

The value of α' required for sustained oscillation is found by applying the criterion of oscillation. Figure 10.20 gives the Applegate diagram for the reflex klystron. The space-time curve for each electron is a parabola. In this figure, the angle α' is the phase angle between the zero of the resonator alternating potential (when the potential changes from acceleration to deceleration) and the positive peak of the resonator voltage. The phase difference between two zeros of resonator voltage is $\alpha' + \pi/2$; therefore, we have $\alpha' + \pi/2 = 2\pi n$. The relation between accelerating voltage and reflector voltage required for oscillation is found by putting $\alpha' = 2\pi n - \pi/2$, and if Eq. (10.27) is used, then

$$\frac{V_0}{(V_R - V_0)^2} = \frac{[2\pi n - (\pi/2)]^2 q_e}{8\omega^2 s^2 m_e}, \tag{10.35}$$

where $n = 0, 1, 2, \ldots$. There are a number of discrete values of $V_0/(V_R - V_0)^2$, which produce

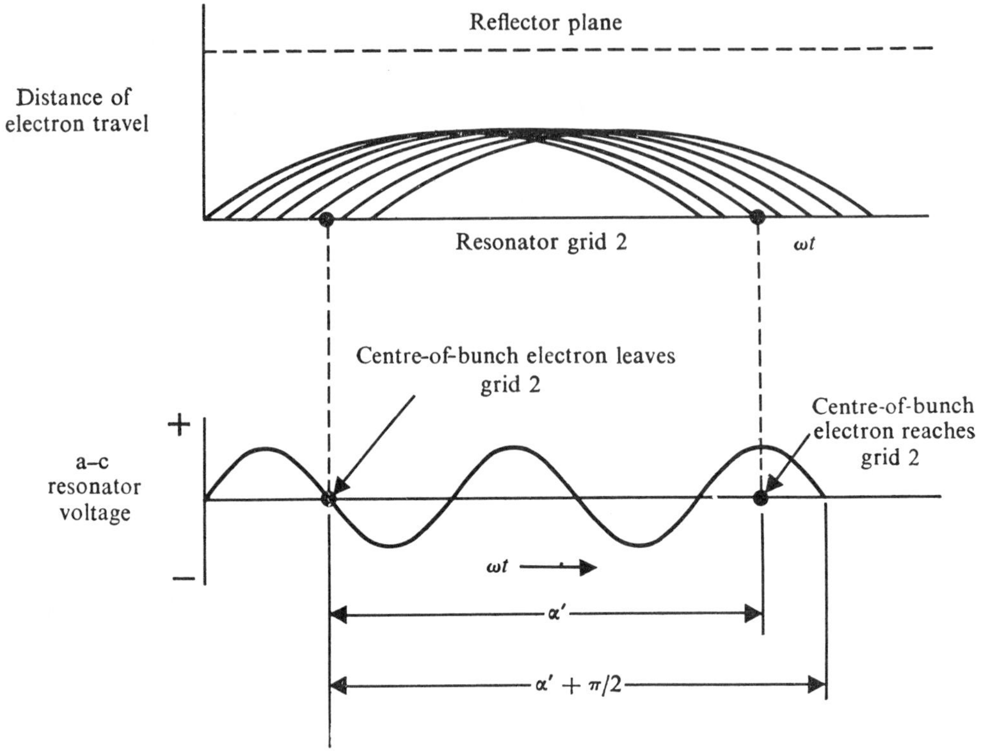

Fig. 10.20 Applegate diagram of reflex klystron.

oscillation. If V_R is the variable, the smaller values of n correspond to higher values of V_R. The ratio V_1/V_0 is given by

$$\frac{V_1}{V_0} = \frac{2x'}{\alpha'} = \frac{2x'}{2\pi n - \pi/2}. \tag{10.36}$$

The power output and efficiency of the reflex klystron are derived on the same lines as those of the double-resonator klystron, and are given by

$$P_{ac} = I_0 V_1 x' J_1(x') \sin \alpha', \tag{10.37}$$

$$\eta = \frac{V_1}{V_0} x' J_1(x'). \tag{10.38}$$

V_1/V_0 increases as n decreases, and hence the power output and efficiency also increase as n decreases. The efficiency is a maximum when the product $x'J_1(x')$ has its maximum value. The maximum theoretical efficiencies of reflex klystrons vary between 20 per cent and 30 per cent.

Some reflex klystrons are so built that the tube and resonator are separate units. The resonator is sometimes a coaxial-line resonator. Tuning is done by simultaneous adjustment of the reflector voltage and the length of the coaxial line. Power output is obtained by coupling either a probe or a loop to the resonant line.

There are various modes of oscillation of the reflex klystron corresponding to the different values of n. In general, the power output increases with decreasing values of n, but the stability of the oscillator under various load conditions becomes poorer. Values of n from 1–5 offer the most stable performance, with $n = 3$ being preferred. The mode of operation is partly determined by the reflector voltage V_R. If continuous tuning by varying of V_R and the length of the coaxial line is attempted, mode jumping may occur, resulting in an abrupt change in frequency and power output. For example, if the resonant line operates as a 1/4 wavelength resonant line, the new frequency may be such that the line operates as a 3/4 wavelength line. This tendency of mode jumping makes it difficult to obtain continuous tuning over a large range of frequencies.

10.11 Multicavity Magnetron Oscillator

The multicavity magnetron was developed during the Second World War as a result of a desperate quest for a suitable microwave generator, capable of delivering large power output at high efficiencies under pulsed conditions, for use in high-definition radar. The British made an early model in 1940, and since then an intensive research programme was launched in many countries to improve the model.

Figure 10.21 shows the cut-away view of a 10-cm multicavity magnetron. It contains an indirectly heated oxide-coated cathode and a laminated copper anode. The anode block is bored to provide for eight identical cylindrical resonators. A pickup loop in one of the resonators is connected to the external circuit by means of a vacuum-sealed coaxial line to provide power output connection. The multicavity magnetron also contains disc-shaped fins for forced air cooling.

Figure 10.22 illustrates a 3-cm magnetron, complete with the associated magnet. The output of the magnetron is coupled to a waveguide through a vacuum-sealed dielectric window. A high d-c potential is applied between the cathode and anode, setting up a radial electric field, and an axial magnetic field is provided by either a permanent magnet or an electromagnet. The forces on the electrons emitted from the cathode consist of a force directed radially outward due to the d-c electric field and a force perpendicular to their instantaneous direction of motion due to the magnetic field. Because of these two forces, the electrons take a spiral path, whose radius of curvature decreases with increasing magnetic field strength. For a given value of the d-c anode potential, there is a critical value of the magnetic field, called the cut-off magnetic field, which causes the electrons to just graze the anode. If the magnetic field strength is greater than the cut-off value, the curvature of the electron trajectory is such that the electrons miss the anode and spiral back to the cathode. Usually, the magnetic field required for oscillation is of the order of one to two times the cut-off value.

When the magnetron is oscillating, the electrical oscillation of the resonators sets up an a-c electric field across the resonator gaps. In the multicavity magnetron, the a-c field in the interaction space is largely tangential. Also, there is usually a phase difference between the successive cavity resonators. This produces a rotating a-c field. The d-c anode potential and the magnetic field strength are adjusted so that the whirling cloud of electrons rotates in synchronism with either the fundamental or a submultiple component of the rotating a-c field.

Some of the electrons rotate in such a phase that they are accelerated by the a-c field, and the others rotate in such a phase that they are retarded by the a-c field. The electrons which are accelerated by the a-c field experience an increase in torque from their motion in the magnetic field, and their paths are so altered that they spiral back to the cathode and are therefore withdrawn from the interaction space. The electrons retarded by the a-c field remain in the interaction space and give energy to the a-c field and thus contribute to the power output of the magnetron.

As a result of this selection process, the electrons form groups resembling the spokes of a wheel, and this spoke-like formation rotates at synchronous speed with respect to a component of the a-c field.

The description just given is applied to the travelling-wave magnetron. The magnetron may also operate as a negative-resistance oscillator or as a cyclotron-frequency oscillator. However, the travelling-wave magnetron is the most widely used one and will therefore be dealt with in more detail.

10.12 Pulsed Operation of Magnetrons

Radar systems and pulse-time modulation systems require pulsed operation of the generators, and the magnetron is very well suited for this purpose. Under pulsed conditions, a high d-c anode potential is applied to the magnetron for a very short interval of time, with a relatively long interval between pulses during which the tube is inoperative. It is thus possible to obtain very high values of peak power output and still remain within the safe limits of average power output and anode dissipation.

For a water-cooled magnetron which can deliver 2.5 megawatts peak power output at a wavelength of 10 cm under pulsed conditions, a typical pulse cycle is a pulse duration of one microsecond, with 1000 cycles per second. The average power output is 2.5 kilowatts, the cathode current during the pulse interval is 140 amperes, and the peak anode voltage is 50 kilovolts. When a magnetron is working under pulsed conditions, there are very severe requirements on the cathode. Current density, as high as 50 amp/cm^2, is required. It has been shown experimentally that the cathode current under pulsed conditions can become very high. This may be due to the fact that some of the electrons which are rotating in an unfavourable phase with respect to the a-c field return to the cathode with relatively high kinetic energy and, consequently, produce secondary electrons. Since some of these secondary electrons emitted from the cathode have an unfavourable phase with respect to the a-c field, these electrons return to the cathode and bring out other secondary electrons. This results in a cumulative bombardment of the cathode, which builds up the emission to very large values. To heat the cathode, magnetrons sometimes depend entirely on this bombardment of the cathode by returning electrons, but no additional power is required to heat the cathode. Sometimes, the back bombardment becomes excessive, and this results in overheating and damage of the cathode.

The very large emission of electrons from the cathode may also be due to the fact that the work function of the cathode is lowered due to (i) very large electric fields at the cathode surface, (ii) ionic conduction, or (iii) electrolytic conduction. It can also be due to the space-charge cloud which accumulates around the cathode during the quiescent period. Cathodes are made of nickel cylinders, nickel wire meshes coated by oxides, or matrix cathodes. Special oxide coatings are used to overcome the limitation of the power output from the small size cathode.

10.13 Electron Motion in Parallel-Plane Magnetron

The force F on an electron, known as the Lorentz force, moving with a velocity $\boldsymbol{v}$ in a magnetic field of flux density $\boldsymbol{B}$ webers per square metre, is given by

$$F = q_e v B \sin \theta, \tag{10.39}$$

where θ is the angle between $\boldsymbol{B}$ and $\boldsymbol{v}$. The force $\boldsymbol{F}$ is in a direction mutually perpendicular to $\boldsymbol{v}$

Fig. 10.21 Cut-away view of 10-cm multicavity magnetron.

Fig. 10.22 Three-centimetre magnetron with magnet and waveguide output.

and $\boldsymbol{B}$ so that

$$\boldsymbol{F} = q_e \boldsymbol{v} \times \boldsymbol{B}. \tag{10.40}$$

Figure 10.23 shows the motion of an electron in a uniform magnetic field. The magnetic field is directed into the page. If the electron moves in a plane perpendicular to the magnetic field so that $\theta = 90°$, the path is a circle.

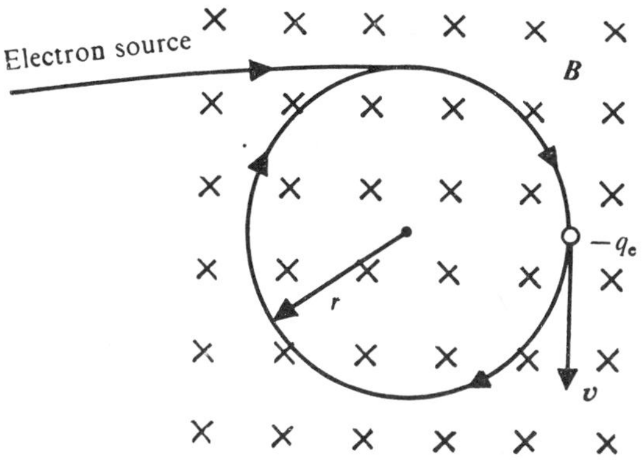

Fig. 10.23 Electron motion in uniform magnetic field.

The Lorentz force, directed in a radially inward direction, is equal and opposite to the centrifugal force $m_e v^2/r$, where r is the radius of the path and m_e is the mass of the electron. Hence

$$q_e v B = m_e v^2/r \tag{10.41}$$

so that

$$r = m_e v/(q_e B), \tag{10.42}$$

$$\omega_e = \frac{v}{r} = \frac{q_e B}{m_e}, \tag{10.43}$$

$$T = \frac{2\pi}{\omega_e} = \frac{2\pi m_e}{q_e B}, \tag{10.44}$$

where ω_e and T are respectively the angular velocity and the period of rotation. The radius of the path is directly proportional to the magnitude of the electron velocity and inversely proportional to the magnetic field strength. The magnitude of the velocity and the kinetic energy of the electron moving in a magnetic field remain constant. Therefore, a stationary magnetic field can neither give energy to nor take energy from an electron moving in the field.

Now, let us consider the motion of an electron in an idealized parallel-plane magnetron, as shown in Fig. 10.24. The electric intensity is in the negative z-direction and the magnetic intensity is in the positive y-direction. The force on an electron moving in such a field is

$$\boldsymbol{F} = -q_e(\boldsymbol{E} + \boldsymbol{v} \times \boldsymbol{B}). \tag{10.45}$$

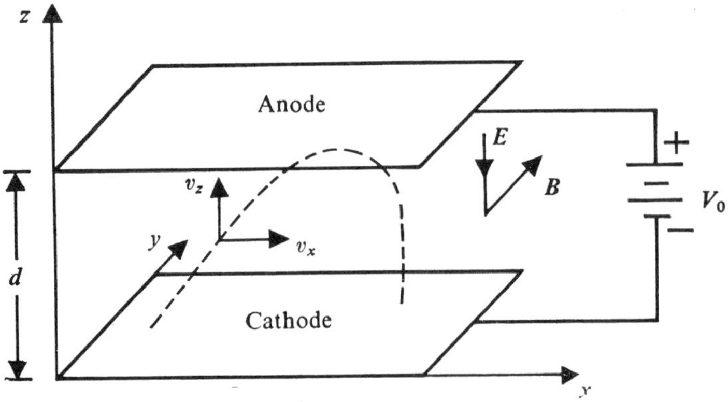

Fig. 10.24 Motion of electron in parallel-plane magnetron.

Splitting Eq. (10.45) into the x- and z-component, we have

$$q_e v_z B = m_e a_x = f_x,$$

$$-q_e(E + v_x B) = m a_z = f_y,$$

where v_x, v_z are the x- and z-component of the velocity v; E is the electric field; and a_x, a_z are the x- and z-component of acceleration. Putting

$$a_x = \frac{dv_x}{dt}, \qquad a_z = \frac{dv_z}{dt}, \qquad E = -\frac{V_0}{d},$$

we obtain

$$dv_x/dt = \omega_e v_x, \tag{10.46}$$

$$\frac{dv_z}{dt} = \frac{q_e V_0}{m_e d} - \omega_e v_x, \tag{10.47}$$

where $\omega_e = q_e B/m_e$. Differentiating Eq. (10.47) with respect to time and using Eq. (10.46), we obtain

$$d^2v_z/dt^2 = -\omega_e^2 v_z \tag{10.48}$$

whose solution is of the form

$$v_z = C_1 \sin \omega_e t + C_2 \cos \omega_e t. \tag{10.49}$$

Assuming that the electron leaves the cathode surface $z = 0$ at time $t = 0$ with zero velocity, we have $C_2 = 0$. Therefore

$$v_z = C_1 \sin \omega_e t, \tag{10.50}$$

and hence

$$dv_z/dt = C_1 \omega_e \cos \omega_e t. \tag{10.51}$$

From Eqs. (10.47) and (10.51) and from the assumption that when $t = 0$, $v_x = 0$, we obtain

$$C_1 = q_e V/(\omega_e m_e d).$$

Therefore

$$\frac{dz}{dt} = v_z = \frac{q_e V_0}{\omega_e m_e d} \sin \omega_e t. \tag{10.52}$$

Using Eq. (10.52) in Eq. (10.47), we obtain

$$\frac{dx}{dt} = v_x = \frac{q_e V_0}{\omega_e m_e d}(1 - \cos \omega_e t) \tag{10.53a}$$

$$= \frac{V_0}{Bd} t - \frac{q_e V_0}{\omega_e m_e d} \cos \omega_e t. \tag{10.53b}$$

Integrating Eqs. (10.52) and (10.53) and assuming that the electron leaves the origin ($x = 0$, $z = 0$) at time $t = 0$, we get

$$z = \frac{q_e V_0}{\omega_e^2 m_e d}(1 - \cos \omega_e t), \tag{10.54}$$

$$x = \frac{q_e V_0}{\omega_e^2 m_e d}(\omega_e t - \sin \omega_e t). \tag{10.55}$$

Equations (10.54) and (10.55) describe the path of the electron. This path is a cycloid which is the locus of a point on the circumference of a circle which is rolling along a straight line. From Eq. (10.54), we can see that the maximum displacement of the electron in the z-direction occurs when $\omega_e t = \pi$, and is given by

$$z_{\max} = \frac{2 q_e V_0}{\omega_e^2 m_e d} = \frac{2 m_e V_0}{q_e B^2 d}. \tag{10.56}$$

From Eqs. (10.52) and (10.53), we can see that the motion of the electron is a combination of uniform translation in the x-direction with velocity $\bar{v}_x = V_0/dB$ and rotation with angular velocity ω_e. When $z_{\max} = d$, where d is the distance between the cathode and anode, the electron just grazes the anode, and cut-off is said to occur. The magnetic flux density B_c necessary for the cut-off condition is given by

$$B_c = \frac{1}{d}\left(\frac{2 m_e V_0}{q_e}\right)^{1/2}. \tag{10.57}$$

From Eq. (10.54), it is seen that the electron leaves the cathode when $\omega_e t = 0$ and returns when $\omega_e t = 2\pi$. Therefore, the transit time for one complete cycle is

$$T = \frac{2\pi}{\omega_e} = \frac{2\pi m_e}{q_e B}. \tag{10.58}$$

Equations (10.57) and (10.58) have been derived for electrons moving in a plane perpendicular to the magnetic field. The equations are more complicated if the electron has a component of velocity in the direction of the magnetic field. The uniform translation velocity $\bar{v}_x = V_0/dB$ is called the *drift velocity*.

The parallel-plane magnetron just described has very little practical value, because successive cycles of the electron motion will carry it out of bounds in the x-direction.

10.14 Electron Motion in Cylindrical Magnetron

The cylindrical magnetron is a more practical structure. We shall now study the behaviour of electrons in a cylindrical magnetron under d-c operating conditions. Figure 10.25 shows a cylindrical magnetron, in which the electric field $\boldsymbol{E}$ is radially inward in the negative ρ-direction and the magnetic flux density $\boldsymbol{B}$ is in the axial z-direction. The radii of the cathode and anode are a and b, respectively.

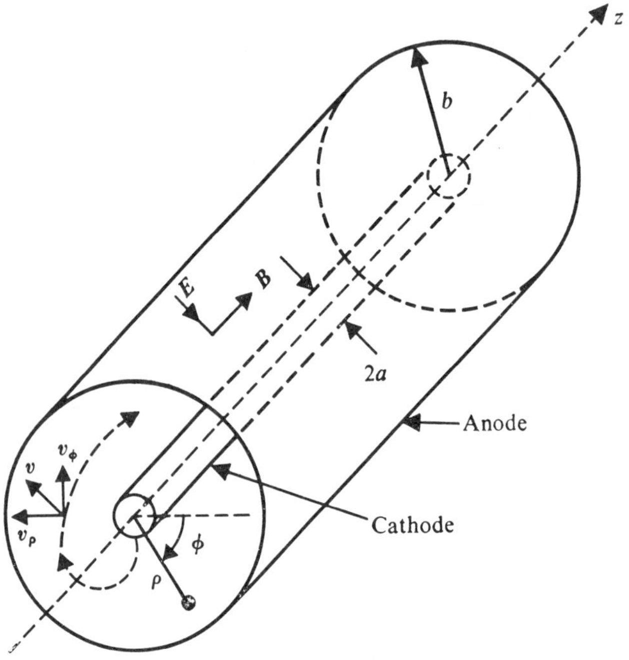

Fig. 10.25 Cylindrical magnetron.

The Lorentz force on an electron, due to the motion of the electron in a magnetic field of flux density $\boldsymbol{B}$, is $\boldsymbol{F} = -q_e(\boldsymbol{v} \times \boldsymbol{B})$. The ϕ-component of this force is $F_\phi = q_e v_\rho B$, and the corresponding torque is $\rho F_\phi = \rho q_e v_\rho B$. The moment of inertia of the electron is $m_e\rho^2$, and the angular momentum is $m_e\rho^2(d\phi/dt)$, where $d\phi/dt$ is the angular velocity.

Equating the torque to the time rate of the angular momentum, we obtain

$$q_e B\rho\frac{d\rho}{dt} = \frac{d}{dt}\left(m_e\rho^2\frac{d\phi}{dt}\right). \tag{10.59}$$

Integrating Eq. (10.59) with respect to t, we obtain

$$\frac{q_e B\rho^2}{2} = m_e\rho^2\frac{d\phi}{dt} + C_1. \tag{10.60}$$

Assuming that the angular velocity $d\phi/dt$ of the electron at the cathode $\rho = a$ is zero, we get $C_1 = q_e Ba^2/2$ so that, from Eq. (10.60),

$$\frac{d\phi}{dt} = \frac{q_e B}{2m_e}(1 - \frac{a^2}{\rho^2}) = \omega_e(1 - \frac{a^2}{\rho^2}), \tag{10.61}$$

where

$$\omega_e = q_e B/(2m_e). \tag{10.62}$$

Equation (10.61) shows that the angular velocity $d\phi/dt$ increases with ρ, starting with zero at $\rho = a$ and approaching an asymptotic value of $d\phi/dt = \omega_e$ at points $\rho \gg a$. The value of ω_e for the cylindrical magnetron is one-half of the value for the parallel-plane magnetron.

An electron in motion in an electric field has a kinetic energy of $\frac{1}{2}m_e v^2$ and potential energy of $-q_e V$. The kinetic and potential energies of the electron are not affected by the magnetic field B. If the fields are stationary, the sum of the kinetic and potential energies remains constant as the electron moves through the field. If the cathode is assumed to be at zero potential, and if the electron is emitted with zero velocity, the sum of the kinetic and potential energies at the cathode is zero. Therefore, at any other point in space, we have $\frac{1}{2}m_e v^2 - q_e V = 0$. The resultant velocity v is equal to $(v_\rho^2 + v_\phi^2)^{1/2}$, where v_ρ and v_ϕ are the ρ- and ϕ-component. Hence

$$q_e V = \tfrac{1}{2}m_e v^2 = \tfrac{1}{2}m_e(v_\rho^2 + v_\phi^2) = \tfrac{1}{2}m_e[(\frac{d\rho}{dt})^2 + \rho^2(\frac{d\phi}{dt})^2]. \tag{10.63}$$

Substituting Eq. (10.61) in Eq. (10.63), we obtain

$$q_e V = \tfrac{1}{2}m_e[(\frac{d\rho}{dt})^2 + \omega_e^2\rho^2(1 - \frac{a^2}{\rho^2})^2]. \tag{10.64}$$

At the anode, $\rho = b$, $V = V_0$, and therefore Eq. (10.64) becomes

$$q_e V_0 = \tfrac{1}{2}m_e[(\frac{d\rho}{dt})^2 + \omega_e^2 b^2(1 - \frac{a^2}{b^2})^2]. \tag{10.65}$$

Cut-off occurs when the radial velocity $d\rho/dt$ is zero at the anode. Therefore, the value of the magnetic flux density B_c for cut-off is given by

$$V_0 = \frac{m_e\omega_c^2 b^2}{2q_e}(1 - \frac{a^2}{b^2})^2 = \frac{q_e B_c^2 b^2}{8m_e}(1 - \frac{a^2}{b^2})^2 \tag{10.66}$$

or

$$B_c = \frac{1}{b(1 - a^2/b^2)}(\frac{8m_e V_0}{q_e})^{1/2}. \tag{10.67}$$

In many practical magnetrons, $b \gg a$, and hence Eq. (10.67) becomes approximately

$$B_c = \frac{1}{b}(\frac{8m_e V_0}{q_e})^{1/2} = \frac{6.75 \times 10^{-6}}{b}\sqrt{V_0}. \tag{10.68}$$

The relations [(10.66)–(10.68)] derived for the cylindrical magnetron are valid for any degree of space-charge density. An electron moving in a cylindrical magnetron under d-c operating conditions describes approximately an epicycloidal path, which is the locus of a point on the circumference of a circle that rolls along the circumference of another circle (the cathode). Figure 10.26 shows the electron paths for different values of magnetic field strength. Figure 10.26b gives the cut-off condition ($B = B_c$); Fig. 10.26c corresponds to $B > B_c$.

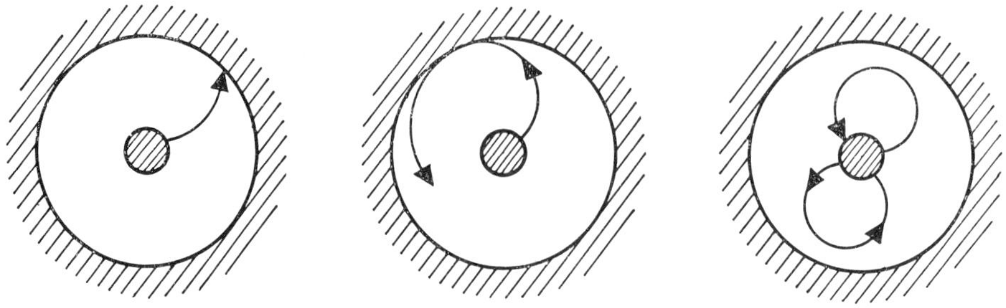

(a) $B < B_c$ (b) $B = B_c$ (c) $B > B_c$

Fig. 10.26 Electron paths in cylindrical magnetron.

Equations (10.67) and (10.68) give the critical values of the magnetic flux density B above which the anode current would fall abruptly to zero. But the actual measured anode currents as a function of B indicate a gradual decrease (see Fig. 10.27). This is possibly due to several reasons such as emission of electrons from the cathode with random electron velocities, electron collisions, and space-charge field effects.

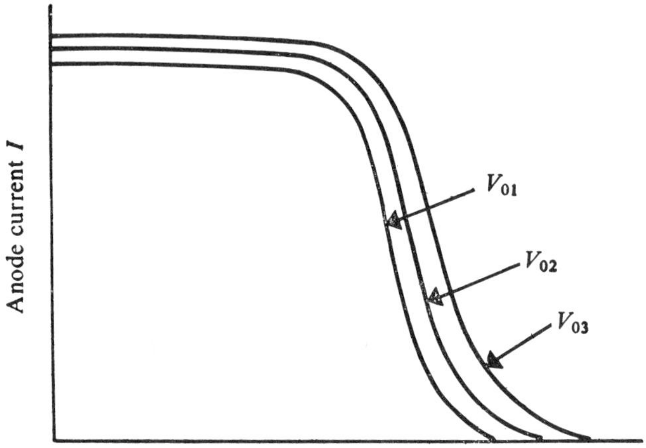

Fig. 10.27 Anode current of cylindrical magnetron as function of magnetic flux density and anode voltage.

10.15 Negative-Resistance Magnetron and Cyclotron-Frequency Magnetron

The *negative-resistance magnetron* has a split-anode construction, with two anode segments. Such a magnetron is used only below 800 MHz, and the resonant circuit is usually mounted external to the tube. The resonant circuit usually consists of a transmission line, with one end connected to the two anode segments and with an adjustable short-circuiting bar at the other end. There is a negative-resistance characteristic between the anode voltage and anode current, due to the fact that, if the two anode segments have different values of d-c potential with respect to the cathode, then the electron orbits are such that a majority of the electrons travel to the least positive anode within a certain range of potentials. This negative-resistance characteristic can be used to make the magnetron oscillate when connected to an external resonant circuit. Power outputs of several hundred watts at efficiencies as high as 50–60 per cent have been obtained by this means. The period of electrical oscillation depends only on the external resonant circuit, and not on the electron transit time which should be small as compared to the period of electrical oscillation. Figure 10.28 shows a typical electron trajectory in a split-anode magnetron having different values of d-c potential applied to the two anode segments. Figure 10.29 shows the negative-resistance characteristic of split-anode magnetron.

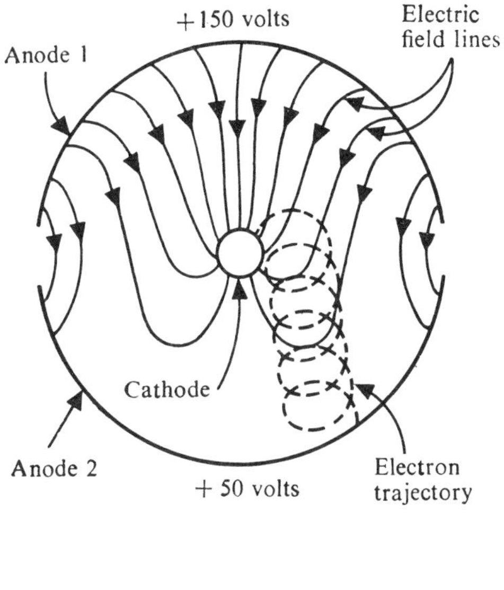

Fig. 10.28 Electron path in split-anode magnetron.

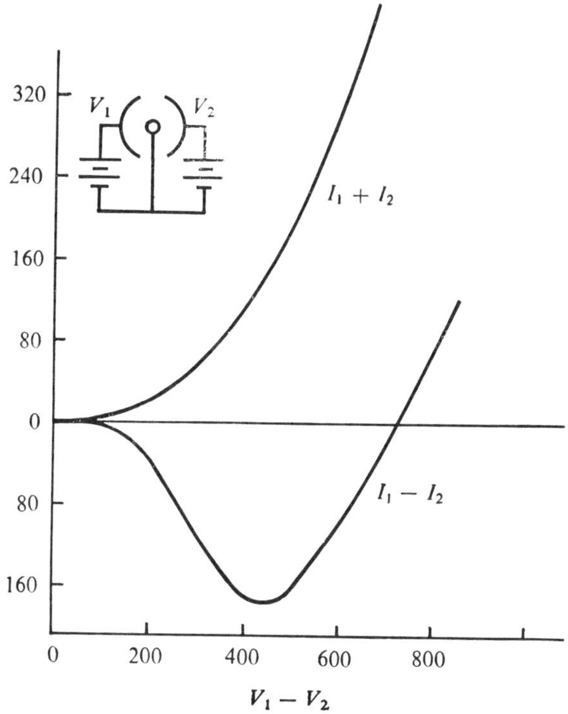

Fig. 10.29 Negative-resistance characteristic of split-anode magnetron.

In a *cyclotron-frequency magnetron*, the period of electrical oscillation is approximately equal to the electron transit time from the cathode to the vicinity of the anode and back to the cathode. This transit time is known as the cyclotron period of the electron, and hence the oscillation is known as the cyclotron-frequency oscillation. For this oscillation, λH is a constant, and is given by

$$\lambda H = 12{,}000 \text{ (MKS units)}$$

$$= 15{,}000 \text{ (emu units)},$$

where λ is the wavelength and H is the magnetic field intensity.

Cyclotron-frequency oscillations can occur either in a single-anode magnetron or in a split-anode magnetron. In a single-anode magnetron, the resonant circuit is connected between the cathode and anode, whereas in a split-anode magnetron, it is connected between the two anodes. In the operation of the cyclotron-frequency oscillator, the unfavourable electrons which take energy from the a-c field have their orbits altered so that they return to the cathode and withdraw from the field. On the other hand, the favourable electrons which give energy to the field have their orbits altered so that they remain longer in the interaction space. This is how the selection process works. Figure 10.30 illustrates the two cases for a parallel-plane magnetron.

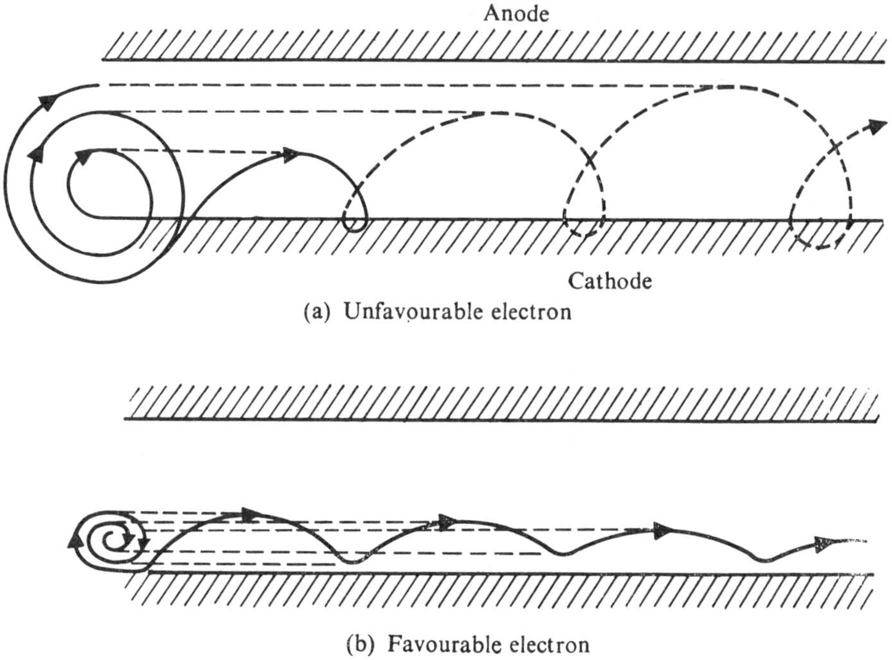

Fig. 10.30 Electron trajectories in parallel-plane magnetron. (Spirals represent projections of electron path.)

For the unfavourable electron, the radius of curvature of the electron path continuously increases, but for a favourable electron, it goes on continuously decreasing. The cylindrical magnetron can be thought of as a parallel-plane magnetron rolled into a cylinder, allowing the cathode radius to decrease. In the cyclotron-frequency type of oscillation, the electrons in the interaction space gradually fall out of phase with the a-c field, and it is necessary to remove them before they begin to take energy from the a-c field. This can be done by tilting the magnetron with respect to the magnetic field, so that the electrons are given an axial component of velocity and hence of motion, enabling them to spiral down the interaction space and finally spiral out of the end of the tube. Another method consists of having two end plates perpendicular to and insulated from the cathode. These end plates are at a positive potential so that they attract the electrons. The difficulty of removing the electrons at the appropriate time in their orbits, before they start taking energy from the a-c field, constitutes the principal disadvantage of the cyclotron-frequency oscillator.

Also, the efficiency of such an oscillator is very much less than that of the travelling-wave magnetron, it being only about 10–15 per cent.

10.16 Travelling-Wave Magnetron Oscillator

Most magnetrons used today are of the travelling-wave type. Such a magnetron is a multicavity magnetron (see Figs. 10.31 and 10.21). In the travelling-wave type of oscillation, the phase

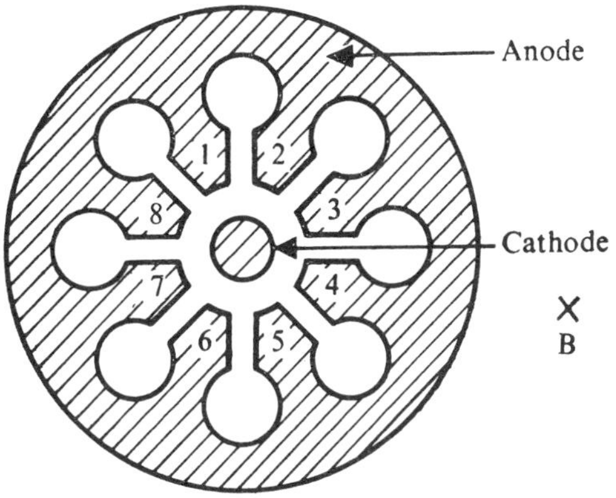

Fig. 10.31 Multicavity magnetron.

difference between the electrical oscillations of successive resonators is such as to produce a rotating a-c field or a travelling wave in the interaction space. The electron space-charge cloud whirls around in the interaction space with a mean angular velocity equal to the angular velocity of a component of the rotating field. Those electrons which rotate in such a phase as to take energy from the a-c field have their paths so altered that they spiral back to the cathode and are therefore withdrawn from the interaction space. The other electrons which give energy to the field are grouped in a spoke-like formation, the spokes rotating synchronously with the a-c field and in such a phase that the electrons are retarded by the tangential component of the a-c field. The electron trajectories are approximately epicycloids, and the electrons having a favourable phase eventually terminate at the anode.

The electrons are retarded by the tangential component of the a-c field, and may gradually fall into an unfavourable phase. But this is prevented in the travelling-wave type of oscillation by the *phase-focussing effect* due to the radial component of the a-c field which tends to keep the electrons rotating synchronously with respect to the rotating a-c field. If an electron *leads* the retarding a-c field, the force due to the radial component of the a-c field is directed inward the cathode. The resulting radial component of velocity in the magnetic field produces a torque in such a direction as to decrease the angular velocity of the electron. On the other hand, an electron that *lags* behind the retarding a-c field experiences a force directed radially outward because of the radial component of the a-c field, and due to a torque caused by the magnetic field, which tends to increase the angular velocity. Thus, the phase-focussing action tends to retard the leading electrons and speed up the lagging electrons, thereby keeping the space-charge cloud rotating in synchronism with the retarding a-c field.

In the cyclotron-type of oscillation, the a-c field is radial, but in the travelling-wave type of oscillation, it is mainly tangential. The electron trajectories of the unfavourable electrons are tilted towards the cathode, whereas those of the favourable electrons are tilted towards the anode, so that the unfavourable electrons terminate at the cathode and the favourable ones terminate at the anode.

The rotating space-charge bunches induce emf's and currents in the resonators which are analogous to the emf's and currents induced in the stator windings of a polyphase a-c generator by the rotor field. The resulting electrical oscillation of the resonators sets up a rotating electric field in the interaction space, which interacts with the electrons and slows them down. This is analogous to the reaction of the stator field back upon the rotor of the generator, thus tending to retard the rotor in the polyphase generator. The electron bunches tend to lead the retarding a-c field, whereas the phase-focussing effect tends to keep the electrons rotating at synchronous speed with respect to the rotating field. The electrons gain energy from the rotating field as they spiral outward. This is similar to the mechanical power supplied to the rotor to overcome the retarding force due to the field of the armature windings on the stator.

Since the electrons rotate at synchronous speed with respect to a component of the a-c field, the analysis of the travelling-wave magnetron involves the evaluation of the angular velocities of the electrons in terms of the magnetic field strength, d-c potential, and so on, and also in terms of the resonant frequencies of the resonant systems and the travelling waves which are set up by the resonator fields.

Figure 10.32 gives a schematic diagram of a developed view of the magnetron shown in Fig. 10.31. Each curve represents a plot of the a-c component of potential difference between the cathode and a point on the anode circle. The various curves represent potential distributions at different instants of time. The potential is constant across the face of the anode and varies approximately linearly across the gap.

The travelling-wave type of oscillation has a number of different modes of oscillation which correspond to the various resonant frequencies of the resonant system as well as to the various angular velocities of the electrons that react favourably with the field to produce oscillation. The mode shown in Fig. 10.32 is known as the π-mode which is characterized by a phase difference of π radians between the electrical oscillations of successive resonators. The electric intensity at any point is proportional to the negative of the slope of the potential curve, and hence the force on the electron is proportional to the slope. Thus, a negative slope of the potential curve shows that an electron in that position is being retarded by the a-c field. An electron that so travels as to be always retarded by the a-c field is called a favourable electron. The dashed lines in Fig. 10.32 show the progress of the favourable electrons. These electrons give energy to the a-c field and thus contribute to the power output of the magnetron. The angular velocity of the electron is inversely proportional to the slope of the dashed line which represents the progress of the electron. An electron may have any one of a number of discrete angular velocities corresponding to the various dashed lines. It may also be retarded by the a-c field of successive resonators.

A relationship can now be derived between the electrical frequency of oscillation and the angular velocity of the electron. The angular velocity of the electron is $d\phi/dt$, and it travels once around the interaction space in a time of $2\pi/(d\phi/dt)$ seconds. If the magnetron has N anodes (or N resonators), the time T_1 taken by the electron to travel from the centre A of one anode gap to the centre B (see Fig. 10.32) of the next adjacent anode gap is

$$T_1 = \frac{2\pi}{(d\phi/dt)N}. \tag{10.69}$$

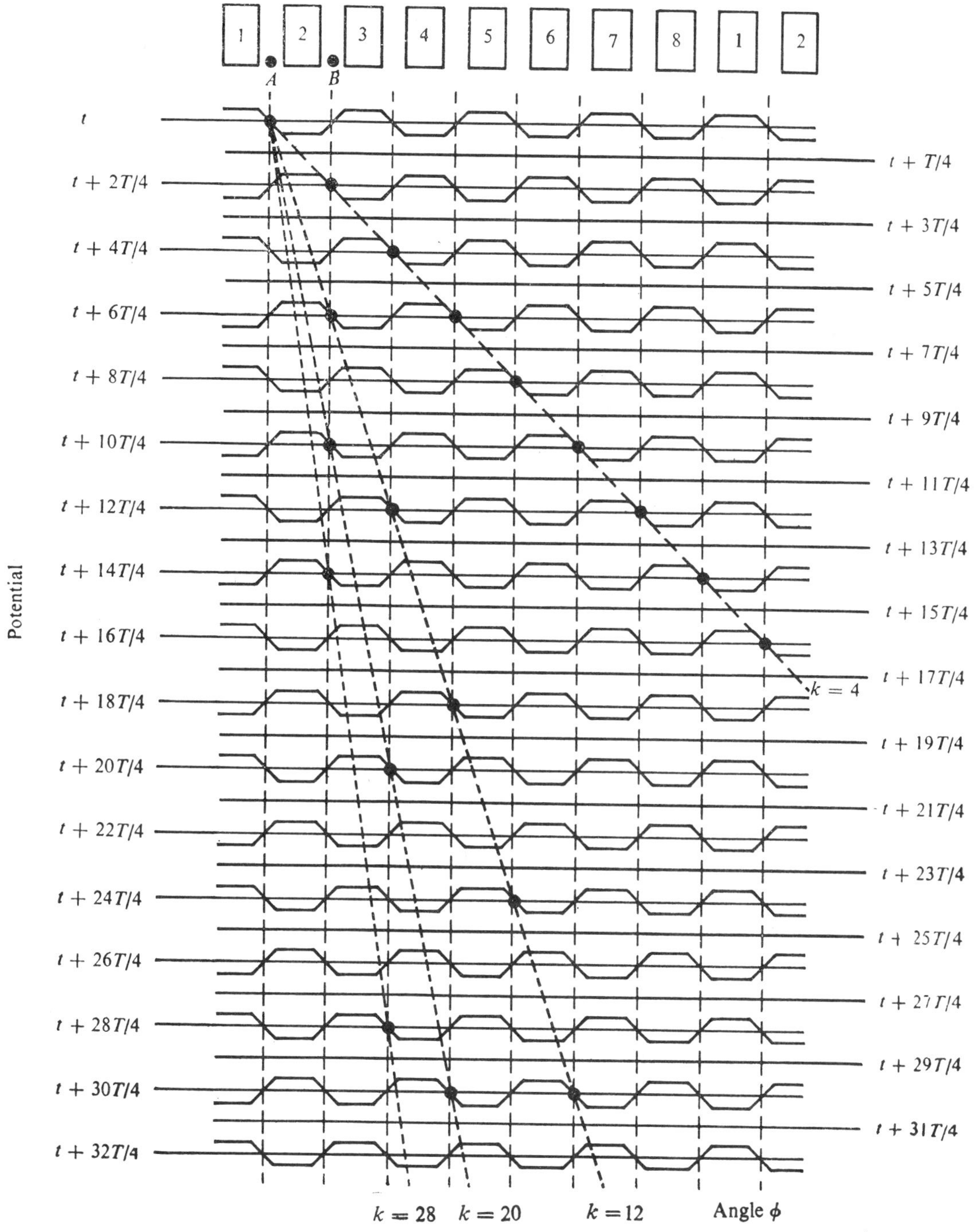

Fig. 10.32 Potential distribution in eight-resonator magnetron oscillating in π-mode at different instants of time. (Dashed lines represent progress of electrons travelling in favourable phase.)

Since the field at any point in the interaction space is single valued, the total phase shift around a closed path is $2\pi n$ radians, where n is an integer representing the number of cycles of phase shift around the closed path. Since the number of resonators is N, the phase difference between the oscillations of two adjacent resonators is

$$\theta = 2\pi n/N, \tag{10.70}$$

where n may take any integral value from zero to $N/2$. Assuming that the electron has the same direction of rotation as the a-c field, the travelling electron sees a phase difference of $(\omega T_1 - \theta)$ radians between corresponding points A and B of two successive resonators. Since the electron has to experience a maximum retarding force in both positions, this phase shift must be an integral multiple of 2π radians, or

$$\omega T_1 - \theta = 2\pi p, \tag{10.71}$$

where $p = 0, \pm 1, \pm 2, \ldots$. Using Eqs. (10.69) and (10.70) in Eq. (10.71), we obtain

$$\omega = |(p + \frac{n}{N})N| \frac{d\phi}{dt} = |k| \frac{d\phi}{dt}, \tag{10.72}$$

where

$$k = (p + \frac{n}{N})N. \tag{10.73}$$

Here, k may be positive or negative. Equation (10.72) gives an expression for the angular frequency ω of oscillation in terms of the angular velocity of the electron. The various values of k represent the different modes of oscillation.

The value of ω, as given by Eq. (10.72), must also coincide with a resonant frequency of the resonant system. The travelling-wave magnetron has a number of cavities, as shown in Fig. 10.31. The cavities are coupled with each other, and hence the magnetron has a number of different resonant frequencies. Hence, ω may have any value equal to one of these resonant frequencies.

From Eq. (10.61), it can be seen that, under d-c operating conditions, the angular velocity of the electrons increases from a value of zero at the cathode to a limiting value of $d\phi/dt = q_e B/(2m_e)$ at distances where $\rho \gg a$. Under dynamic conditions, however, most of the electrons rotate with constant angular velocity, in synchronism with the a-c field. Since the favourable electrons are retarded by the a-c field, the synchronous angular velocity must be less than the limiting value for d-c operating conditions.

To obtain an approximate expression for $d\phi/dt$, it is assumed that at some radius ρ', the radial forces on an electron due to the electric and magnetic fields are equal and opposite. The force on the electron due to the d-c field, neglecting the variation of the field with radius, is $F = -q_e E = q_e V_0/(b - a)$. The radial force due to the motion of the electron in the magnetic field is $F = q_e v_\phi B = q_e B\rho'(d\phi/dt)$. Equating these two forces, we have

$$\frac{d\phi}{dt} = \frac{V_0}{\rho'(b - a)B}. \tag{10.74}$$

Equation (10.72) gives the angular velocity at radius ρ'. Using Eq. (10.74) in Eq. (10.72), we obtain

$$\frac{V_0}{B} = \frac{\omega\rho'(b - a)}{|k|} \tag{10.75}$$

which gives the value of V_0/B for a given frequency of oscillation. However, ρ' is still the undetermined radius at which the radial forces on the electron are equal and opposite. As a first approximation, we may assume that this occurs at the mid-point between the cathode and anode, i.e., $\rho' = (a + b)/2$. Then

$$\frac{V_0}{B} = \frac{\omega(b^2 - a^2)}{2|k|}. \tag{10.76}$$

Equation (10.76) gives the approximate frequency of oscillation if V_0 and B are known.

A more accurate derivation obtained by Hartree, based on the assumption that the electrons just reach the anode for an infinitesimal amplitude of the a-c voltage, gives

$$V_0 = \frac{\omega b^2 B}{2|k|}[1 - (\frac{a}{b})^2] - \frac{m_e}{2q_e}(\frac{\omega b}{|k|})^2 \tag{10.77}$$

which is a quadratic equation in the frequency ω. If the term

$$\frac{m_e}{2q_e}(\frac{\omega b}{|k|})^2$$

is small as compared to the first term in Eq. (10.77), it reduces to Eq. (10.76).

Equations (10.76) and (10.77) show that, from an operational viewpoint, it is necessary to adjust the magnetic flux density B and the d-c anode potential V_0 to obtain the proper value of $d\phi/dt$ so that Eq. (10.72) for the desired mode is satisfied. A large number of modes are possible in a magnetron, as seen from Eqs. (10.76) and (10.77), although only a few of these modes have practical significance, because many of the modes yield low power output and low efficiency. Magnetron oscillations have a tendency to jump from one mode to another under slight changes in operating conditions. This results in an abrupt change in frequency of oscillation and power output. Hence, the unwanted modes have to be avoided by some means or the other. This will be described later in this chapter.

10.17 Different Modes of Oscillation of Travelling-Wave Magnetron

The various modes of oscillation of the travelling-wave magnetron are given by the different possible values of k. The most common mode is the π-mode, for which the phase shift θ between successive resonators is π radians. From Eq. (10.70), $\theta = 2\pi n/N$, and hence for the π-mode, $n = N/2$. Using this value of n in Eq. (10.73), we obtain

$$k = (p + \tfrac{1}{2})N \tag{10.78}$$

which gives the allowed values of k for different values of p.

Usually the π-mode is so operated that the electrons rotate synchronously with respect to the fundamental component of potential. This requires that $p = 0$, and hence $k = N/2 = n$. Using this value of k in Eq. (10.72), we get the angular velocity of the electrons as $d\phi/dt = 2\omega/N$.

Considering the a-c field of the π-mode, as shown in Fig. 10.32, we see that the potential distribution as a function of angle is approximately trapezoidal. If the centre of one of the anode pole faces is chosen as the reference point, then the Fourier series representing the a-c potential as a function of ϕ contains only the cosine terms, and hence

$$V_{ac} = \sum_{k=1}^{k=\infty} V_k \cos(\omega t) \cos(k\phi), \tag{10.79}$$

where V_k is the amplitude of the k-th space harmonic. Note that the anode structure is a periodic slow-wave structure and its field can be expressed as the sum of an infinite number of space harmonics, as discussed in Chapter 9. In the series given in Eq. (10.79), k must have integral values from zero to infinity. However, only those terms in the series which react favourably with the electrons are of interest. Hence, only those values of k which satisfy Eq. (10.78) are considered.

Equation (10.79) can now be rewritten as

$$V_{ac} = \sum_{k=1}^{\infty} \frac{V_k}{2}[\cos(\omega t - k\phi) + \cos(\omega t + k\phi)]. \tag{10.80}$$

The first term in Eq. (10.80) is a potential wave travelling in the positive ϕ-direction with an angular velocity of ω/k radians per second, whereas the second term represents a wave travelling in the negative ϕ-direction with the same velocity. The potential distribution given by Eq. (10.79) or (10.80) and shown in Fig. 10.32 represents a standing wave which consists of two travelling waves of the same amplitude and velocity travelling in opposite directions. For the magnetron operation, the electrons may rotate synchronously with respect to any one of these two travelling waves in either direction of travel.

For the electrons to react favourably with the fundamental component of the potential wave, $p = 0$, and hence $k = N/2 = n$. Then, the angular velocity of the electrons is

$$\frac{d\phi}{dt} = \frac{\omega}{k} = \frac{2\omega}{N}.$$

The electrons may rotate in either direction and thus travel approximately synchronously with respect to either one of the travelling waves corresponding to the fundamental space harmonic component. When this happens, the electrons react favourably with the higher-order space harmonics of the potential wave. The space harmonics of the potential wave with which the electrons can react favourably are known as the *Hartree harmonics*.

On an average, the electron transfers power only to that particular component of the field which has approximately the same angular velocity as the electron. The power transfer to all other field components rapidly alternates between positive and negative values, averaging zero.

The π-mode is the most important mode in the travelling-wave magnetron, because the coupling impedance of the slow-wave structure is maximum, as was discussed in Chapter 9, and maximum power is transferred from the electrons to the field.

The mode corresponding to $n = 0$ is the cyclotron-frequency mode, which has been already discussed. For this mode, $\theta = 0$, and the potential of all the anodes rises and falls in time phase, and the a-c field is essentially radial. The value of k is $k = pN$, and the value of ω is $\omega = pN(d\phi/dt)$.

In the more general case, for which θ is neither π nor zero, the potential distribution can be represented also by a Fourier series consisting of travelling waves in both directions, with unequal amplitude and phase, as

$$V_{ac} = \sum_{k=1}^{k=\infty} [A_k \cos(\omega t - k\phi + \delta) + B_k \cos(\omega t + k\phi + \gamma)], \tag{10.81}$$

where δ and γ are arbitrary phase constants. In the π-mode, the electrons have no preferred direction. Either direction gives the same power output. However, in the more general case, the preferred direction is the direction of rotation of the stronger potential wave. When the electrons travel in the direction of rotation of the weaker potential wave, they are referred to as driving a *reverse mode*.

Figure 10.33 represents a potential distribution as a function of angle ϕ and time for a magnetron having 8 resonators, and an arbitrary value θ different from π or zero, and n equal to 2. The dashed lines represent favourable electron paths corresponding to various modes. The electrons corresponding to $-k$ modes are driving a reverse mode. For the π-mode, the positive

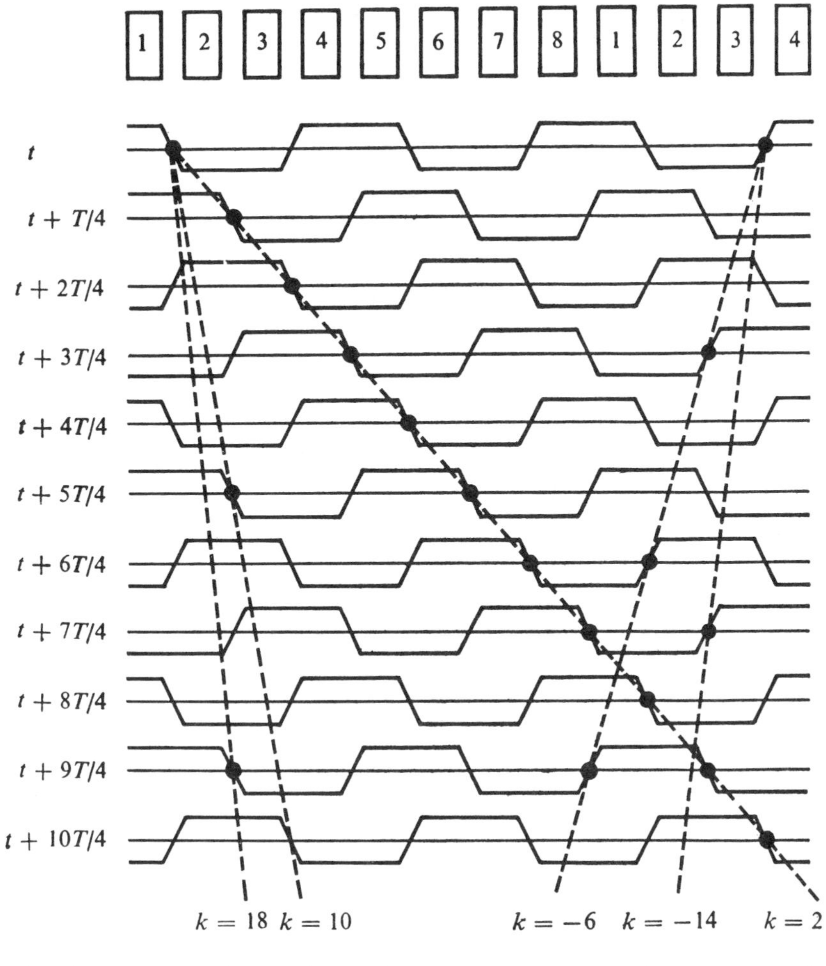

Fig. 10.33 Potential distribution as function of angle and time corresponding to $n = 2$. (Dashed lines represent paths of favourable electrons. Negative values of k correspond to electrons driving reverse mode.)

and negative values of p in Eq. (10.73) give the same sequence of values of k. On the other hand, for other modes, the sequence of values of k is different for positive and negative values of p. For example, if $N = 8$, $n = 2$, the positive values of p give $k = 2, 10, 18, \ldots$, and the negative values of p give $k = -6, -14$, and -22.

The electric and magnetic field distributions for various modes in a magnetron having eight resonators are shown in Fig. 10.34. The output coupling loops are represented by the arrows pointing outward the circular drawings and by the bar at the centre of each developed view. The sine

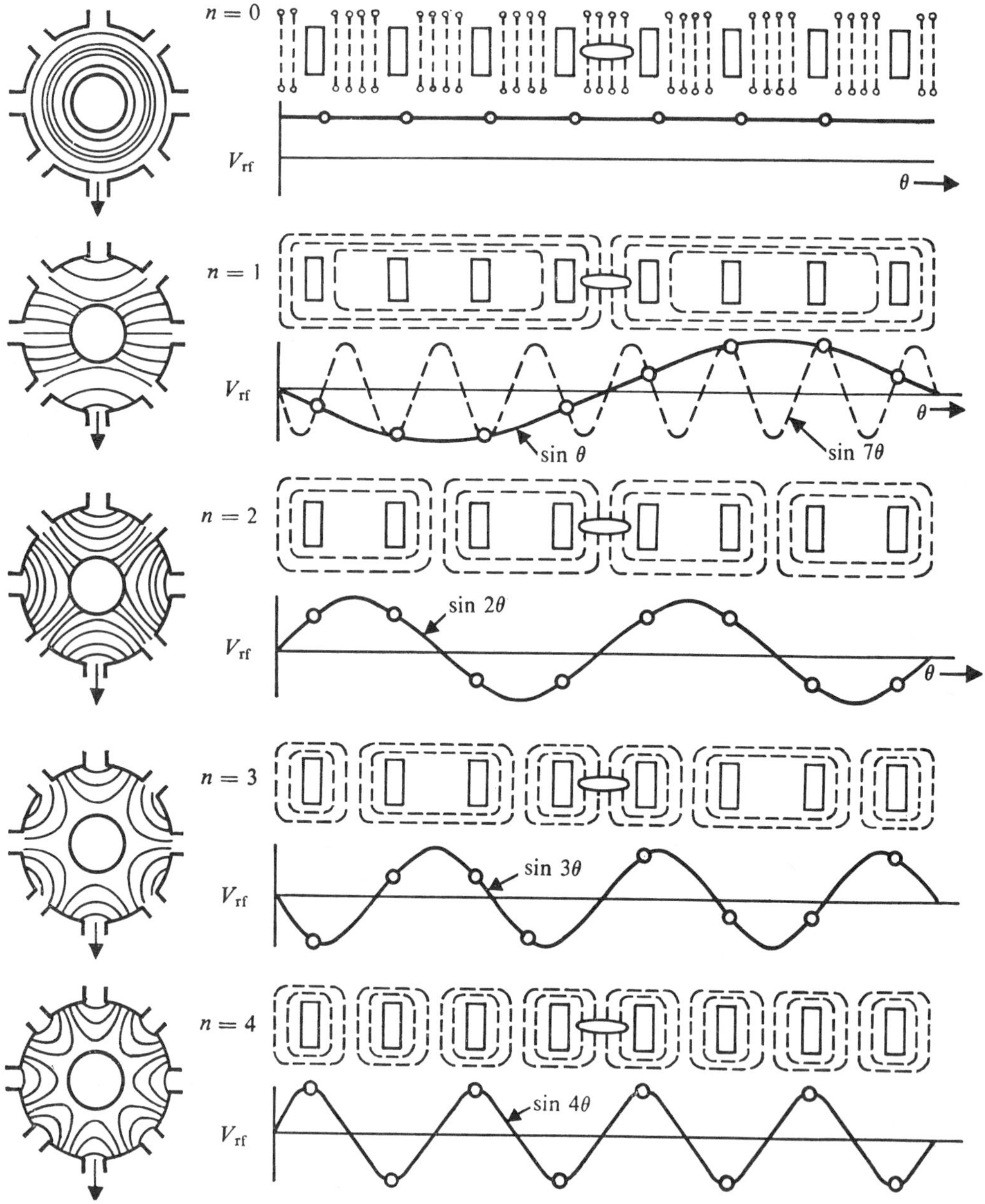

Fig. 10.34 Electric and magnetic field distributions in eight-resonator magnetron for various modes. ($n = 4$ is π-mode. Solid lines, electric field lines; dashed lines, magnetic field lines.)

wave curves represent the fundamental component of potential as a function of θ. For the mode $n = 1$, the potential distribution for the second Hartree harmonic ($p = -1$, $k = -7$) is also shown.

10.18 Resonant Frequencies of Multicavity Travelling-Wave Magnetron

A single resonator of the multicavity magnetron can be represented by a parallel *LC* circuit. Then, the equivalent circuit of the system of *N* resonators would consist of a closed chain of *N* parallel *LC* circuits, with mutual inductance between adjacent circuits.

If two identical parallel *LC* circuits are coupled together and shock excited, the system will oscillate simultaneously at two slightly different frequencies, which produce a beat effect, with part of the energy surging back and forth between the two coupled circuits at the beat frequency. The two frequencies of oscillation are due to the fact that the mutual inductance can either add to or subtract from the self-inductances. The two circuits corresponding to two slightly different resonant frequencies may oscillate either in phase or π radians out of phase.

Similarly, the closed chain of *N* identical resonators in a ring-like configuration in the multicavity magnetron has *N* resonant frequencies, spaced slightly apart on the frequency scale. The phase difference between successive resonators is not restricted to 0 or π radians, as in the case of the two-resonator system, but may have any value, provided the sum of the phase differences around the closed system is equal to an integral multiple of 2π radians. Instead of having the mutual inductance either add to or subtract from the self-inductance, in the more general case, there is a phase difference between the effects of self-inductance and mutual inductance, which makes new resonant frequencies possible. As the coupling between resonators decreases, the various resonant frequencies draw closer together, finally converging on the resonant frequency of a single resonator.

10.19 Mode Separation in Multicavity Travelling-Wave Magnetron

It has been already mentioned that, in a magnetron, there is a tendency for mode jumping. To avoid such a tendency, it is necessary to provide some means of compelling the magnetron to oscillate in a single mode. This can be accomplished, to a certain extent, by using light coupling between the resonators so that the resonant frequencies are separated as far apart as possible on the frequency scale.

A very common method of preventing mode jumping is to connect the anode segments together by means of conducting straps so that a fixed phase relationship between the oscillations of the various resonators is maintained. Figure 10.35 shows a magnetron with two ring straps. Each strap is connected electrically to alternate anodes so as to compel the potentials of the alternate anodes to oscillate in time phase. This restricts the possible modes of oscillation to the π-mode or the mode $n = 0$ (for which the potentials of all anode segments oscillate in time phase). Each of the straps is broken at one point in order to allow for the asymmetry of the field produced by the coupling loop. The straps add a capacitive reactance, thus causing a shift in the mode-frequency distribution. It is necessary to shield the straps from the interaction space in order to minimize the possibility of the $n = 0$ mode, since the strap situated closest to the cathode tends to set up a radial a-c field between the cathode and anode, giving rise to the $n = 0$ mode. Shielding can be done by milling grooves in the anode structure and embedding the straps in the grooves.

Another method employed for obtaining mode separation without the use of straps is by using the "rising-sun" resonator system, as depicted in Fig. 10.36. In this system, there are alter-

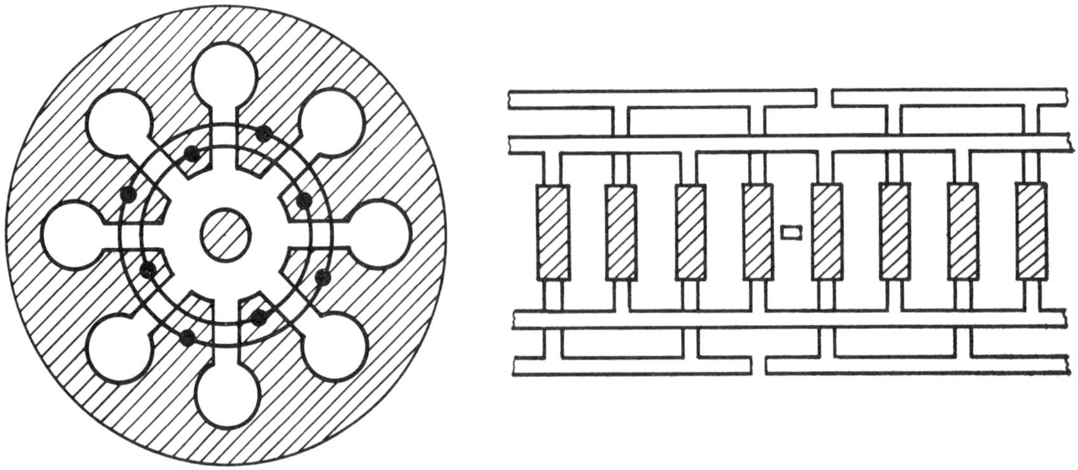

Fig. 10.35 Strapping arrangement of π-mode.

nately large and small size resonators. The two sizes of resonators have widely different resonant frequencies. When the resonators are coupled together, the π-mode resonant frequency lies between the two individual resonant frequencies. There are also a number of other resonant modes present in the coupled system. But, if the two sizes of resonators differ appreciably, there will be sufficient mode separation to ensure relatively stable operation in the π-mode.

Fig. 10.36 Rising-sun resonator system.

Figure 10.37 gives a comparison of the mode separation for different resonator systems. The resonant frequencies of the rising-sun resonator fall into two groups, the π-mode lying approximately half-way between the two groups. Though the mode separation for the rising-sun resonator system is not as large as that of the heavily strapped magnetron, still this system has its compensating advantages. In the strapped magnetron, the mode separation decreases as the axial length of the anode increases. This is rather disadvantageous in short-wavelength magnetrons, because the restriction on the anode length imposes a serious limitation on the power output obtainable from the magnetron. In the rising-sun magnetron, on the other hand, the mode

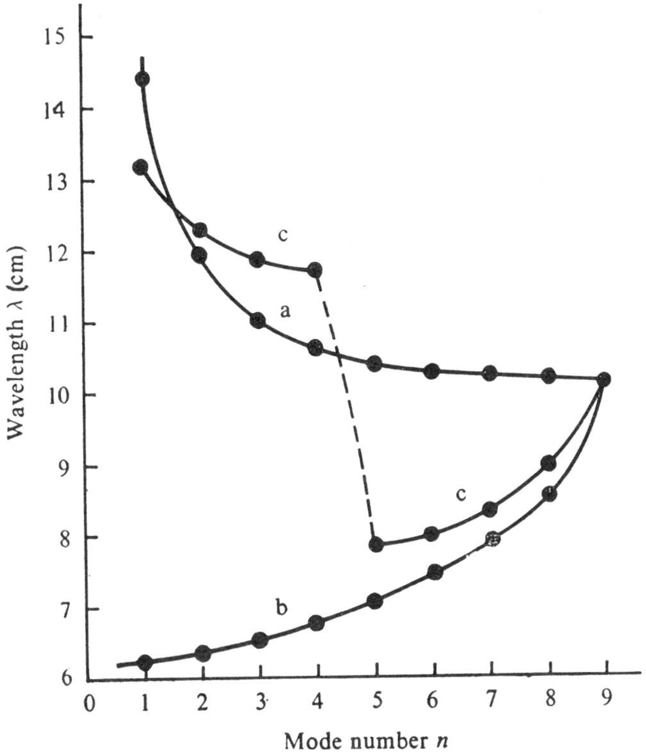

Fig. 10.37 Mode separation in magnetrons having 18 resonators: plot of wavelength versus mode number. a, unstrapped resonator; b, heavily strapped resonator; c, rising-sun resonator. (π-mode corresponds to $n = 9$.)

separation is not seriously impaired by anode lengths up to about 3/4 of a wavelength, which is greater than the allowable length for the strapped magnetron. Also, in the rising-sun magnetron, a large number of resonators may be used and still a reasonable degree of mode separation can be maintained, whereas in the strapped magnetron, the modes tend to fall close together when a large number of resonators are used. Further, the copper losses in the rising-sun magnetron are somewhat less than those in the strapped magnetron, and hence the efficiency of the rising-sun magnetron is higher.

One main disadvantage of the rising-sun magnetron, however, is the tendency to operate in the zero mode ($n = 0$). Because of the asymmetry of the resonators, the a-c field strength across the gap of the large resonators differs from that of the small resonators. Since all the resonators of one size oscillate in time phase, the excess field contributes to a zero-mode field. The magnetron can be designed so as to minimize the possibility of zero-mode oscillation by avoiding the value $\lambda B = 15{,}000$, where λ is in centimetres and B in gauss. This is the value of λB required for zero-mode oscillation.

10.20 Graphical Representation of Performance Characteristics of Magnetrons

There are various types of graphical representation of the performance characteristics of magnetrons. Figure 10.38 is one such representation, which shows the contours of constant power output, constant efficiency, and constant magnetic field strength as functions of the d-c anode voltage and anode current, for an eight-resonator, 10-cm magnetron, pulsed with a pulse of one

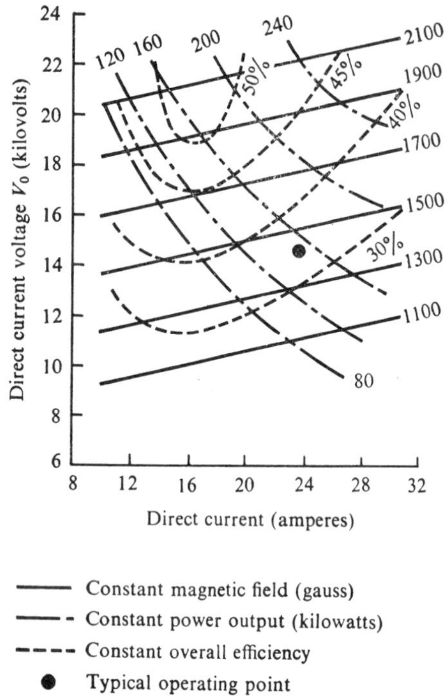

Fig. 10.38 Contours of constant magnetic field strength, constant power output, and constant overall efficiency as functions of anode voltage and anode current for an eight-resonator, 10-cm pulsed magnetron.

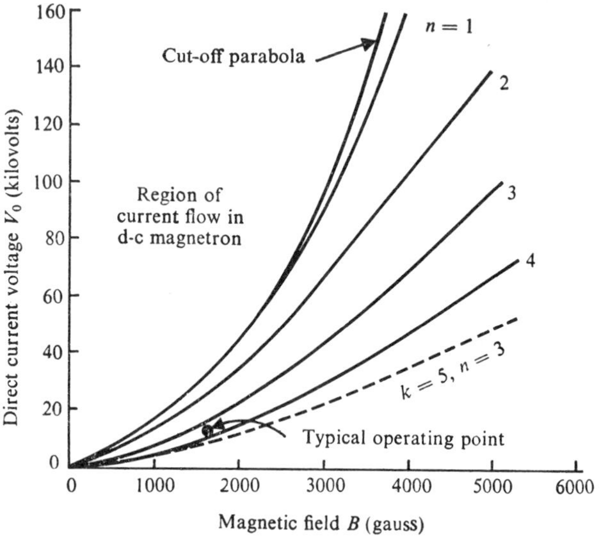

Fig. 10.39 d-c anode potential versus magnetic field strength for various modes of oscillation in eight-cavity magnetron.

microsecond duration and 1000 pulses per second. The typical operating point represented by the solid dot at the centre of the figure corresponds to a peak power output of 135 kilowatts at 42 per cent efficiency, requiring an anode potential of 16 kilovolts.

Figure 10.39 shows a plot of d-c anode potential versus magnetic field for various modes in an eight-cavity magnetron.

The cut-off parabola is a plot of Eq. (10.68), and is given by

$$B_c = \frac{6.75 \times 10^{-6}}{b}\sqrt{V_0} \text{ (MKS units)},$$

whereas the straight lines are a plot of the Hartree equation (10.77), for the same magnetron whose performance characteristics are shown in Fig. 10.38. The magnetic field strengths required for the various modes of operation are somewhat greater than the cut-off value, as indicated by the typical operating point in Fig. 10.39. Also, the range of values of V_0 and B in this figure is far greater than that used in ordinary practice.

Another method of graphical representation is the Riecke diagram which is a method of representing the performance characteristics of a magnetron in terms of load impedance (or admittance) on a Smith chart. Figure 10.40 gives a Riecke diagram consisting of contours of constant power output and constant frequency, plotted on an admittance Smith chart. These data are obtained experimentally by varying the load impedance and observing the power output and frequency. The Riecke diagram enables us to select the optimum load impedance.

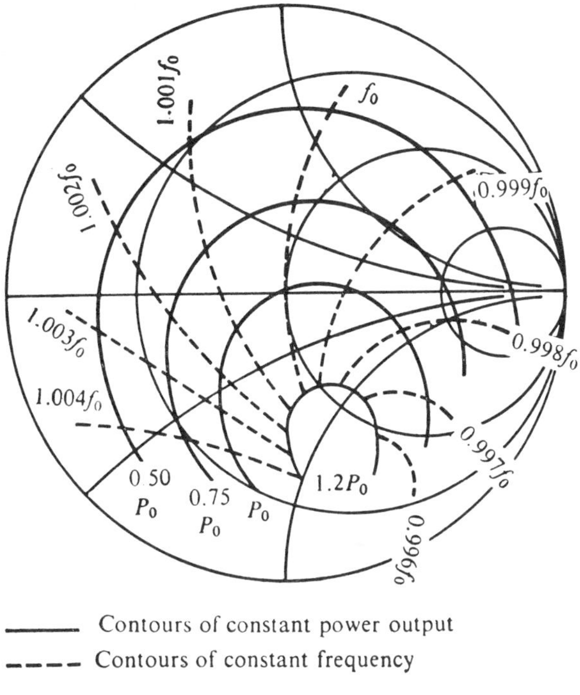

Fig. 10.40 Riecke diagram showing contours of constant power output and constant frequency, plotted on a Smith chart.

10.21 Equivalent Circuit of Magnetron

A criterion of oscillation is the admittance criterion which states that oscillation will occur if the sum of the admittances looking both ways at any pair of terminals is zero. If we apply this criterion to the magnetron, we may take the junction at the extremities of the anode gap. One of the admittances is then the circuit admittance looking into the anode at the junction. The other admittance is the admittance of the electron stream. If these are represented by Y_c and Y_e, respectively, the criterion of oscillation is

$$Y_c + Y_e = 0. \tag{10.82}$$

Let the electron admittance be

$$Y_e = G_e + jB_e. \tag{10.83}$$

The magnetron will be represented by the equivalent circuit shown in Fig. 10.41. In this equi-

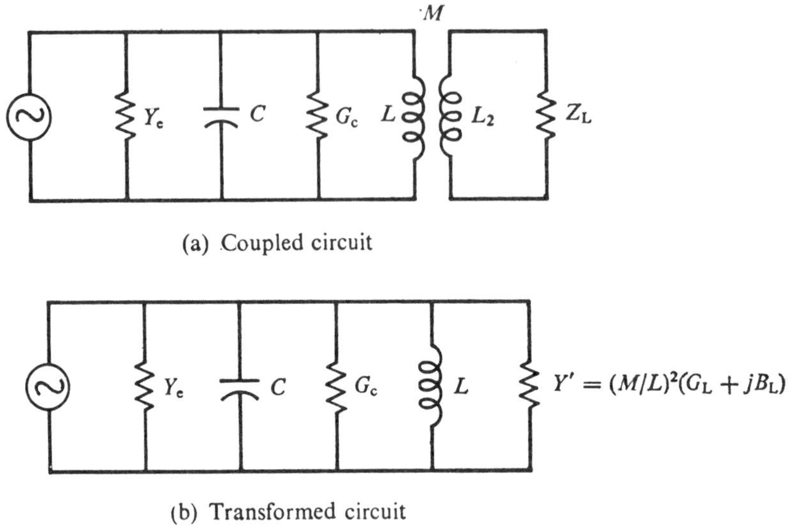

Fig. 10.41 Equivalent circuit of magnetron.

valent circuit, L and C represent the combined inductance and capacitance of N resonators in parallel. L is the inductance of a single resonator multiplied by $1/N$ and C is the capacitance of a single resonator multiplied by N. The shunt conductance G_c represents the loss in the resonator walls. The inductance of the coupling loop is represented by L_2 and the mutual inductance between the resonator and coupling loop is represented by M. The external load impedance is Z_L (see Fig. 10.41a).

If the transformer is assumed to be ideal, the secondary admittances may be transferred to the primary by multiplying the former by the factor $(M/L)^2$, and hence

$$Y'_L = \left(\frac{M}{L}\right)^2 \frac{1}{jX_2 + Z_L} = \left(\frac{M}{L}\right)^2 (G'_L + jB'_L). \tag{10.84}$$

Here, $X_2 = \omega L_2$, and G'_L and B'_L respectively denote the conductance and susceptance of the secondary circuit.

The circuit admittance Y_c may be regarded as the ratio of the induced alternating current at the anode junction to the a-c voltage developed across the anode gap. Y_c may be expressed in terms of the primary circuit parameters (shown in Fig. 10.41b) and the reflected secondary circuit parameters, as

$$Y_c = \frac{1}{j\omega L} + j\omega C + G_c + Y'_L. \tag{10.85}$$

If $\omega_0 = 1/\sqrt{LC}$ and $Y_0 = \sqrt{C/L}$, where Y_0 is the characteristic admittance of the resonator, then

Eq. (10.78) becomes

$$Y_c = j\sqrt{\frac{C}{L}}\left(\frac{\omega}{\omega_0} - \frac{\omega_0}{\omega}\right) + G_c + Y'_L \approx 2jY_0\left(\frac{\omega - \omega_0}{\omega_0}\right) + G_c + Y'_L. \tag{10.86}$$

Using Eq. (10.84) in Eq. (10.86), we get

$$Y_c = 2jY_0\left(\frac{\omega - \omega_0}{\omega_0}\right) + G_c + \left(\frac{M}{L}\right)^2(G'_L + jB'_L). \tag{10.87}$$

Oscillation occurs if the sum of the circuit admittance and the electron admittance is zero. Adding Eqs. (10.83) and (10.87), we obtain

$$G_e = -G_c - \left(\frac{M}{L}\right)^2 G'_L, \tag{10.88}$$

$$B_e = -2Y_0\left(\frac{\omega - \omega_0}{\omega_0}\right) - \left(\frac{M}{L}\right)^2 B'_L \tag{10.89}$$

which are the criteria of oscillation. Equation (10.88) states that the equivalent conductance of the electron stream must be equal in magnitude but opposite in sign to that of the circuit conductance. Since the circuit conductance is always positive, the electronic conductance must be negative so that oscillations can occur. Equation (10.89) implies that the electronic and circuit susceptances must be equal in magnitude but opposite in sign. This equation determines the frequency of oscillation.

The foregoing relations are useful in visualizing adjustments that are likely to take place under conditions of variable load impedance. For example, if the load susceptance B_L is varied, it is possible for the angular frequency of oscillation, ω, to so change as to satisfy Eq. (10.89) without appreciably altering the value of B_e. This accounts for the frequency drift of the magnetron with variable loading. The electron admittance $Y_e = G_e + jB_e$ is not constant: it varies with the operating conditions of the magnetron. The frequency may be stabilized by using a high-Q resonator system or by increasing the characteristic admittance Y_0 of the resonator.

The variation in frequency with variation in load impedance of a magnetron is called *frequency pulling*; the variation in frequency with operating voltages and currents is called *frequency pushing*.

Magnetrons can be tuned by varying the effective inductance or the effective capacitance of the resonators. The inductance can be varied by sliding a conducting pin into or out of the

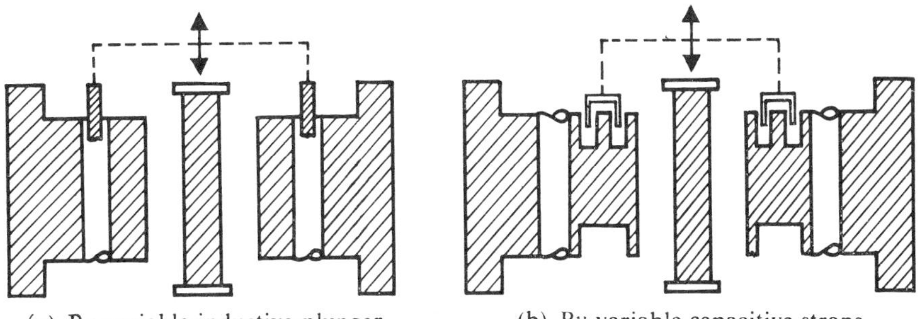

(a) By variable inductive plunger (b) By variable capacitive straps

Fig. 10.42 Tuning of magnetrons.

resonator; the capacitance can be varied by moving an annular ring into or out of grooves in the anode. This is shown in Fig. 10.42.

10.22 Efficiency of Travelling-Wave Magnetron

Figure 10.43 is a schematic diagram of a linear travelling-wave magnetron, showing the electric field lines of the travelling waves for the π-mode. From Eq. (10.53b), we have seen that, in crossed electric

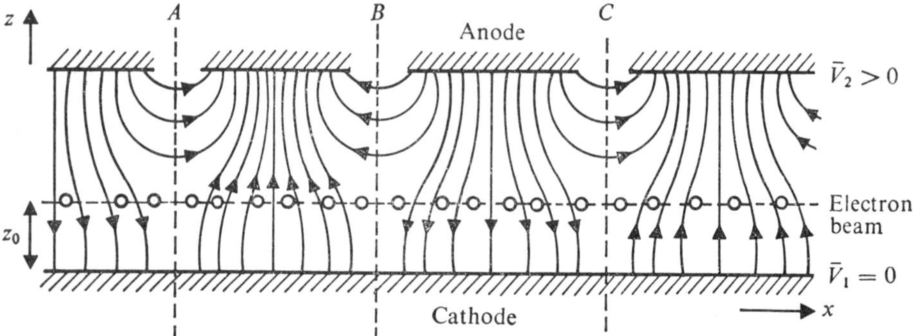

(a) Electric lines of force (arrows indicate negative of electric field)

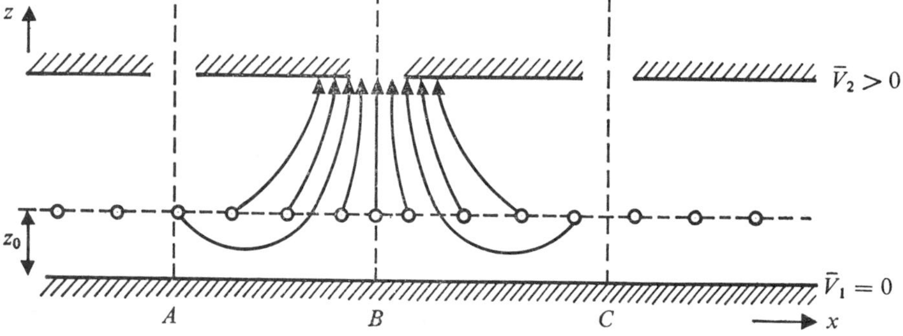

(b) Electron trajectories for reference frame moving synchronously with wave and electron

Fig. 10.43 Schematic diagram of linear travelling-wave magnetron.

and magnetic fields E and B, the electrons have a translational velocity $V_0/dB = -E/B = \bar{v}_x$ in the x-direction. This velocity is called the drift velocity. The electric field component in the x-direction is zero at the cathode and increases as the electron travels towards the anode. The arrows shown in Fig. 10.43a are the negative of the electric field lines and indicate the force on the electrons. The transverse a-c component E_y of the electromagnetic wave and the electrostatic field E have the same direction in the region AB and opposite direction in the region BC. If v is the a-c velocity of the electrons, then $v + \bar{v}_x$ is the total velocity of the electrons. In the region AB, $v + \bar{v}_x > \bar{v}$, and $v + \bar{v}_x < \bar{v}$ in the region BC. Therefore, bunching of the electrons is obtained near the point B. The electrons approach the anode, while their kinetic energy remains practically unchanged, because $v \ll \bar{v}_x$, i.e., the a-c velocity of the electrons is very much less than the drift velocity $\bar{v}_x$. The resulting electron paths are shown schematically in Fig. 10.43b.

As the electrons approach the anode, they reach points of higher potential and, therefore, lose potential energy without a corresponding gain in kinetic energy. The energy $q_e \Delta V$, corresponding to the change of electron potential is transferred to the wave which increases in amplitude and power. These simple energy considerations are sufficient to provide an estimate of the efficiency. If the system is sufficiently long or the r-f field great enough, the electrons reach the anode at a potential $\bar{V}_2$ with a velocity approximately equal to $\bar{v}_x$. Hence, the ratio of the energy gained by the wave to the direct current energy supplied, i.e., $q_e \bar{V}_2$, the electronic efficiency, is given by

$$\eta_{el} \approx \frac{q_e \bar{V}_2 - \frac{1}{2} m_e \bar{v}_x^2}{q_e \bar{V}_2}. \tag{10.90}$$

However, $\bar{v}_x = E/B$. Hence

$$\eta_{el} \approx 1 - 2.84 \frac{(E/B)^2}{\bar{V}_2}, \tag{10.91}$$

where E is in volts and B in gauss. In a practical magnetron, $E = 10^6$ volts/metre, $B = 1000$ gauss, $\bar{V}_2 = 2840$ volts, and hence $\eta_{el} = 0.9$.

The foregoing simple analysis for a linear travelling-wave magnetron shows that the electronic efficiency is very high and may approach 100 per cent for large values of B. However, in practical cylindrical magnetrons, the actual efficiency is lower than the theoretical efficiency, but may be as high as 50 or 60 per cent.

The main reason for the high efficiency in a travelling-wave magnetron is that the potential energy of the electrons is transferred to the potential energy of the electromagnetic wave, while the kinetic energy remains more or less constant. In a klystron, the efficiency is very much lower, because the kinetic energy of the electrons is converted into the energy of the electromagnetic field. This is also true for a travelling-wave tube, as we shall see later.

10.23 Travelling-Wave Amplifier Tube

The travelling-wave amplifier tube (TWT) is probably the most important type of amplifier tube of the class of growing-wave tubes. In a klystron amplifier, resonant circuits are used to generate large r-f voltages which can efficiently interact with electrons to produce electronic gain. Amplifiers of this sort have a basic limitation, i.e., they have an inherent gain-bandwidth product that cannot be exceeded. Since the information capacity of an amplifier is proportional to its bandwidth, the electronic gain must decrease as the information capacity increases. The travelling-wave tube avoids an inherent gain-bandwidth limitation by using a distributed slow-wave circuit instead of resonant circuits. An electromagnetic wave propagates along the distributed slow-wave circuit with a phase velocity in the axial direction, which is small compared with the velocity of light. An axial electron beam, with a velocity approximately equal to the phase velocity of the travelling wave, interacts with the wave. The cumulative velocity modulation, which results from this interaction, produces a current modulation on the beam. This current modulation in turn induces additional currents on the circuit, and the electric fields associated with this induced wave tend to retard the electron bunches which give up their kinetic energy to the wave. This process is cumulative and results in an exponential growth of the wave along the circuit as long as the beam velocity is not decreased below a value necessary to maintain synchronism between the beam and wave. Figure 10.44 gives a schematic illustration of the bunching process in a TWT. Maximum gain of the TWT occurs when the beam velocity is slightly greater than the phase velocity of the travelling wave, thus placing the electron bunches in a decreasing field. Hence, the electron bunches move forward slightly, as shown in Fig. 10.44. The gain of the TWT can be as high as 35 db, but even gains of 60 db have been reported for commercially available tubes.

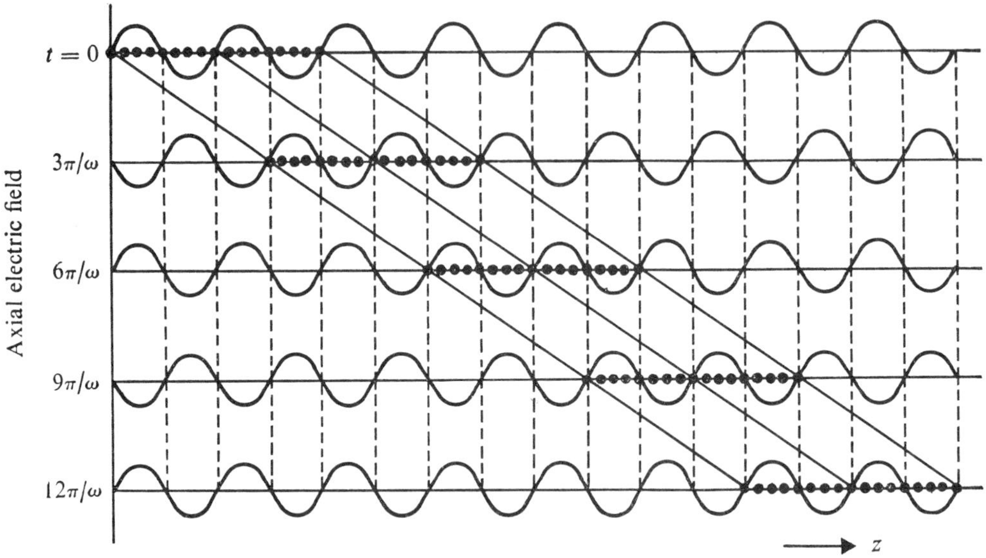

Fig. 10.44 Schematic diagram illustrating bunching of electrons in TWT. (Arrow indicates direction of force on electrons.)

The bandwidth of a TWT is limited by the frequency range over which the slow-wave circuit propagates a constant velocity wave and by the impedance match between the slow-wave circuit and the transmission line (or waveguide) connecting it to other parts in the system. Power reflected from both the output match and the input match are amplified by the TWT, resulting in regeneration. If the loop gain equals unity, oscillations result. To reduce the loop gain, distributed attenuation, which greatly attenuates the backward-travelling wave, is placed near the circuit. If the attenuation is located several guide wavelengths from the tube input after appreciable modulation of the beam has taken place, the forward-travelling wave is not attenuated as much as the backward-travelling wave, resulting in a great improvement in the performance of the TWT.

Travelling-wave tubes have noise figures of about 10 db; noise figures less than 6 db are common in the 2–4 GHz band. The noise figure increases with frequency.

Travelling-wave tubes are of two types, the *O*-type and the *M*-type, as discussed earlier. In the *O*-type, the electrons move parallel to the axis of the slow-wave circuit by some method of focussing like that by a strong magnetic field parallel to the beam and circuit axis, or by periodic magnetic or electric focussing by means of a series of magnetic or electric lenses that reconverge the electron beam which tends to diverge because of the electronic space charge. A typical example of the *O*-type tube is the *travelling-wave amplifier tube* shown in Fig. 10.45. In the *M*-type travelling-wave tube, the beam moves through crossed electric and magnetic fields at a drift

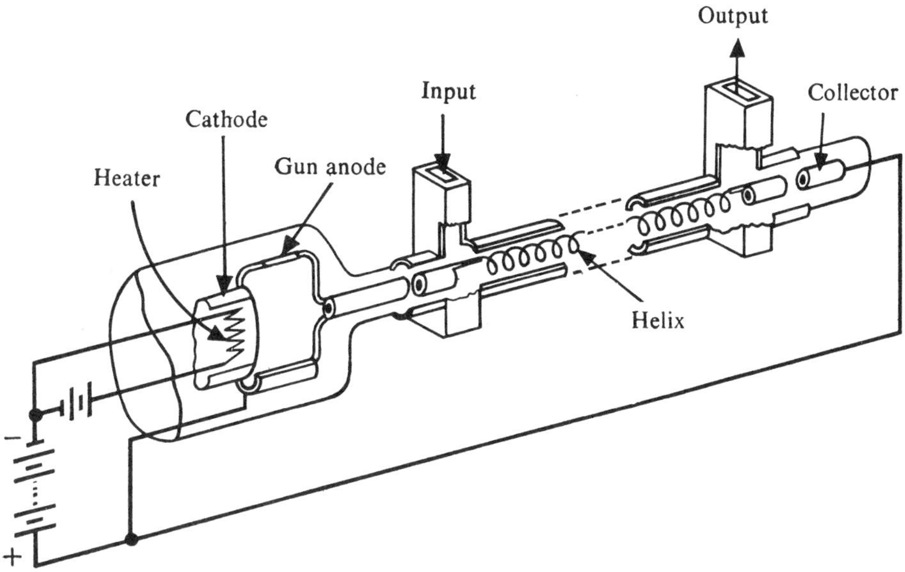

Fig. 10.45 Travelling-wave tube (O-type).

velocity given by the ratio of $\boldsymbol{E}$ to $\boldsymbol{B}$. In this type of tube, the interaction of travelling wave with electrons is much more complicated. A typical example of the M-type tube is called the *crossed-field amplifier* or *linear magnetron* given in Fig. 10.46.

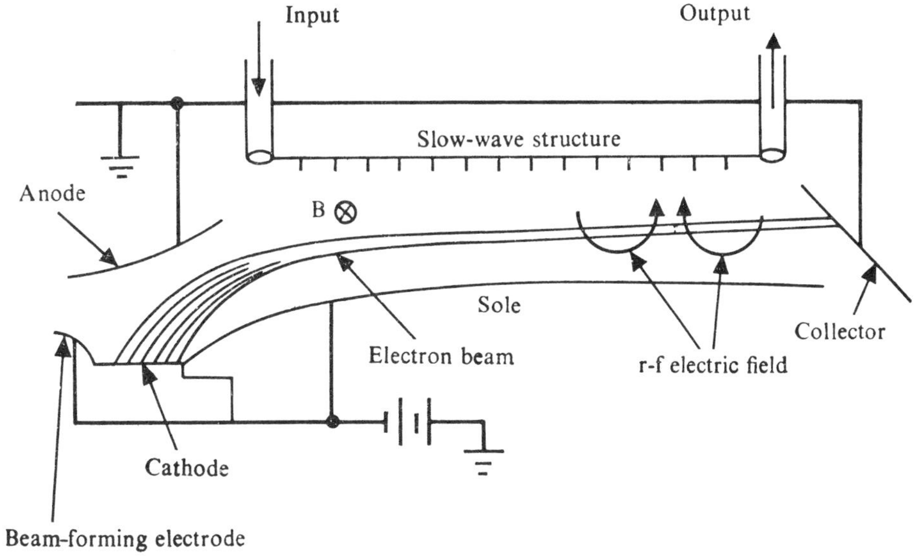

Fig. 10.46 Crossed-field travelling-wave amplifier tube (M-type).

There is one important difference between the fundamental energy relations in the two types of tubes. The O-type travelling-wave tube amplifies the electromagnetic wave as the kinetic energy of the electrons is transferred to the field. If the efficiency of the tube is to be high, the electrons must be slowed down to a fraction of their initial velocity. Hence, unless special precautions are observed, the electron bunches will lose synchronism with the circuit wave and the cumulative interaction of the travelling wave will break down. Though attempts have been made to decrease the circuit velocity near the output end and to decrease the potential of the collector below the circuit potential to minimize the energy dissipated as heat, the travelling-wave tube is still not a very efficient device. Typical efficiencies of high-power TWTs range from 20 to 40 per cent. On the other hand, the crossed-field amplifier tube amplifies the travelling wave supported by the slow-wave structure as the potential energy of the electrons is transferred to the wave. When an electron is slowed down, it moves closer to the anode (see Fig. 10.46), gains potential energy from the electric field which increases its velocity back to the synchronous velocity, and remains in step with the wave. Hence, high-power crossed-field devices are capable of high efficiencies ranging from 40 to 70 per cent, and in this they are similar to the travelling-wave magnetron oscillators. To obtain large power outputs, the electrons are allowed to approach very close to the slow-wave structure, and many of them strike it. As the frequency of operation increases, the dimensions of the circuit decrease and the ability of the circuit to dissipate power also decreases. The size of the electron beam must decrease, and to obtain the same power output, the power density in the beam must increase. Thus, as the slow-wave circuit can dissipate less power per unit area, the demands on it are increased. Hence, the power output of these devices is limited. The same reasoning applies to the O-type travelling-wave tube also, except that the beam does not inherently approach the circuit more closely as its energy is extracted. By using stronger focussing to keep the beam from striking the slow-wave circuit, more power output can be obtained from the tube.

The weight of the solenoid which supplies the focussing magnetic field makes the travelling-wave tubes rather heavy. To overcome this, extensive light-weight packaging has been developed, and output powers of several watts can now be obtained from tubes of less than 1 lb weight. This weight, however, does not include the weight of the power supply.

10.24 Slow-Wave Circuits for Travelling-Wave Tubes

Different types of slow-wave circuits have been studied intensively in Chapter 9. For the O-type travelling-wave tube amplifier, the most common type of slow-wave circuit used is the helix because of its ease of fabrication, high interaction impedance, and large bandwidth. Other circuits, which have higher power handling capacities, or which are easier to scale to higher frequencies, have also become popular in recent years. Here, we shall consider only the helix. The phase velocity v_p of this circuit, normalized to the velocity of light, is shown in Fig. 10.47. At low frequencies, the phase velocity varies rapidly with frequency or, in other words, it is highly dispersive. At higher frequencies, the phase velocity is almost uniform over a wide band; in this frequency range, the wave can be considered as travelling along the wire of the helix with the velocity of light. For small pitch angles, the wave travels along the wire approximately one circumference $2\pi a$ to advance one pitch p of the helix, or $v_p/c \approx p/(2\pi a)$, where v_p is the phase velocity and c the velocity of light. The ω-β (or Brillouin) diagram of the helix is shown in Fig. 10.48. In this figure, the slope $\partial\omega/\partial\beta$ of the curve gives the group velocity v_g of the waves.

The electric field at any point along any periodic structure can be written in terms of the space harmonics as

$$E_1(x, y, z) = \sum_{n=-\infty}^{\infty} E_{1n}(x, y) \exp[-j(2\pi nz/p)] \exp[j(\omega t - \beta_0 z)]$$

$$= \sum_{n=-\infty}^{\infty} E_{1n}(x, y) \exp[j\{\omega t - (\beta_0 + 2\pi n/p)z\}], \tag{10.92}$$

where β_0 is the phase constant of the wave travelling in the z-direction and p the periodicity of

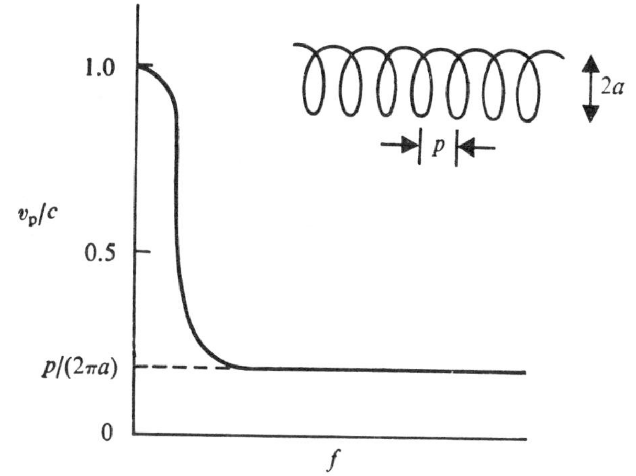

Fig. 10.47 Normalized phase velocity of helix as function of frequency.

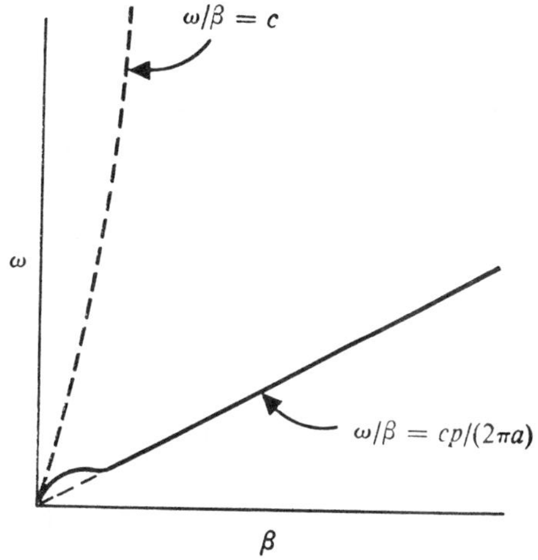

Fig. 10.48 Brillouin or ω-β diagram of helix.

the structure. From Eq. (10.92), we see that the propagation constant of the n-th term is

$$\beta_n = \beta_0 + \frac{2\pi n}{p}. \tag{10.93}$$

This n-th order term is called the n-th space harmonic. If the electric field E is known, the complex vector amplitudes of any space harmonic can be determined by the formula

$$E_{1n}(x, y) = \frac{1}{p}\int_z^{z+p} E \exp[-j(\omega t - \beta_n z)]\, dz. \tag{10.94}$$

Figure 10.49 shows the Brillouin diagram for several space harmonics of the helix. It can be noticed that the space harmonic corresponding to $n = -1$ has a negative phase velocity, but the slope of the ω-β curve is positive and identical to the slopes of all other spatial harmonics at the same frequency, and hence its energy flow is in the positive z-direction. For travelling-wave interaction to occur, the electron beam must move with the phase velocity of the travelling wave. If the energy on the helix moves in the forward direction, and if the beam velocity is nearly equal to the phase velocity of the $n = -1$ space harmonic, i.e., if the beam is moving in the negative direction, then the beam carries energy in one direction and the helix carries the energy in the opposite direction. When this happens, feedback occurs, resulting in backward-wave oscillation. The frequency of this oscillation can be varied by changing the electron velocity and thus the intercept A in Fig. 10.49. Such electronic tuning is possible over an octave range in

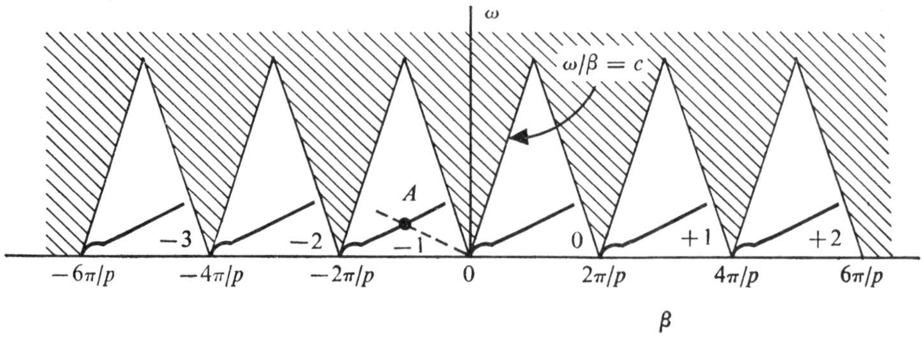

Fig. 10.49 Brillouin diagram of the space harmonics of helix. (Propagation is forbidden in shaded region.)

some tubes. It is much more rapid than mechanical tuning and is extremely useful in some microwave systems. Propagation is forbidden in the shaded regions which occur at intervals of $2\pi/p$ along the β-axis. The reason for this is as follows: The helix is an open structure, and a field analysis of any open periodic structure shows that, if the axial phase velocity of propagation of any space harmonic is greater than the velocity of light, the structure radiates energy. The nondispersive nature of the fundamental helix mode, together with the presence of the forbidden regions, implies that the helix will radiate if the circumference is equal to or greater than a free-space wavelength of the frequency. This observation, based on the properties of space harmonics, has also been experimentally verified.

10.25 Small-Signal Analysis of Travelling-Wave Tube

A small-signal analysis of the travelling-wave tube can show that the signal propagating the slow-wave structure grows exponentially with distance in the presence of the electron beam. Such an analysis also explains the many simple assumptions necessary to derive this result and gives a surprisingly good quantitative accuracy.

Given a periodic structure of finite length propagating a slow electromagnetic wave and an electron beam of finite diameter entering the circuit with velocity u_0 parallel to the axis of the circuit, we have to determine the electric field, beam velocity, and beam current everywhere in the system, for a given power input at the start of the periodic circuit.

The force on a single electron is

$$m_e \frac{d\boldsymbol{v}}{dt} = -q_e(\boldsymbol{E} + \boldsymbol{v} \times \boldsymbol{B}), \qquad (10.95)$$

where $\boldsymbol{v}$ is the velocity of the electron and $\boldsymbol{E}$ and $\boldsymbol{B}$ are the electric field intensity and magnetic flux density. Equation (10.95) can now be rewritten as

$$\frac{d\boldsymbol{v}}{dt} = -\eta(\boldsymbol{E} + \boldsymbol{v} \times \boldsymbol{B}), \qquad (10.96)$$

where $\eta = q_e/m_e$. The convection current density $\boldsymbol{J}$ is given by

$$\boldsymbol{J} = \rho \boldsymbol{v}. \tag{10.97}$$

Equations (10.96) and (10.97) give all the information necessary for solving for the exact fields if the boundary conditions of the circuit and beam are known. However, in practice, it is difficult to solve these equations, and hence the following approximations are made.

(i) The electron beam flows only in the axial direction and, in practice, a large axial magnetic field is used for restricting transverse motion of the electrons. This is a good assumption. Hence, we may put

$$\boldsymbol{v} = \boldsymbol{u}_z v, \qquad \boldsymbol{J} = J\boldsymbol{u}_z. \tag{10.98}$$

Due to this assumption, the $\boldsymbol{v} \times \boldsymbol{B}$ term in Eq. (10.96) cannot accelerate the electrons in the z-direction, and hence the z-component of the acceleration can be neglected.

(ii) All quantities consist of a time-invariant part and a small-amplitude time-varying part which has an exponential variation both in time and distance. Therefore

$$J = -J_0 + J_1 \exp(j\omega t - \Gamma z), \tag{10.99a}$$

$$v = u_0 + v_1 \exp(j\omega t - \Gamma z), \tag{10.99b}$$

$$\rho = \rho_0 + \rho_1 \exp(j\omega t - \Gamma z), \tag{10.99c}$$

$$E_z = E_1 \exp(j\omega t - \Gamma z), \tag{10.99d}$$

where Γ is, in general, a complex number $\alpha + j\beta$. The minus sign is used for J_0 in the expression for J so that J_0 may be a positive number. (It must be remembered that the current flow in a TWT is in the z-direction.) Using Eqs. (10.99) in Eq. (10.97), we obtain

$$J_0 = \rho_0 u_0, \tag{10.100a}$$

$$J_1 = (\rho_0 v_1 + \rho_1 u_0) \exp(j\omega t - \Gamma z). \tag{10.100b}$$

(iii) A time-varying electric field can be expressed in terms of an electric scalar potential and a magnetic vector potential. For slow-wave propagation along a slow-wave structure, by using Maxwell's equations, it can be shown that the electric field contribution from the magnetic vector potential is $(v/c)^2$ times the contribution from the scalar potential. Since $(v/c) \ll 1$, it is enough if the electric field is derived only from the scalar potential, which can be related to an alternating voltage of the slow-wave circuit. The circuit voltage is introduced to bring about the analogy between the travelling-wave tube circuit and a transmission line.

Let an electric field with an axial amplitude E_1 propagate along the slow-wave circuit. Then, the force equation (10.96) can be rewritten as

$$\frac{dv}{dt} = \left(\frac{\partial}{\partial t} + \frac{\partial z}{\partial t}\frac{\partial}{\partial z}\right)v_1 = (j\omega - \Gamma u_0)v_1 = -\eta E_1, \tag{10.101}$$

where the second-order terms have been neglected in accordance with the small-signal approximation. The continuity equation can be written as

$$\nabla \cdot \boldsymbol{J} + \frac{\partial \rho}{\partial t} = -\Gamma J_1 + j\omega \rho_1 = 0. \tag{10.102}$$

Equations (10.100)–(10.102) involve four unknowns. Eliminating the charge density and velocity between these equations, the current density may be expressed in terms of the electric field as

$$J_1 = j\frac{\omega}{u_0}\frac{\eta J_0}{(j\omega - u_0\Gamma)^2}E_1. \tag{10.103}$$

If the electric field E_1 has uniform amplitude over the cross-section of the electron beam, the alternating current i will be proportional to the direct current I_0 with the same proportionality constant that exists between J_1 and J_0. Then, the alternating current in the electron beam is

$$i = j\frac{\beta_e I_0}{2V_0(j\beta_e - P)^2}E_1, \tag{10.104}$$

where $\beta_e = \omega/u_0$ may be thought of as the phase shift per unit length of a disturbance that travels with the electron velocity. Equation (10.104) states that, if an electric field E_1 propagating on the slow-wave circuit modulates the electron beam, the alternating current on the beam can be obtained provided all the factors in the proportionality constant, including Γ, are known. If Γ is known, the variation of all parameters with distance is determined and a given problem with specified boundary conditions can be solved.

To study the effect of the electron beam on the circuit, an equivalent circuit of the form shown in Fig. 10.50 is used for the forward-wave travelling-wave tube. The lossless transmission line with series inductance L per unit length and shunt capacitance C per unit length is coupled

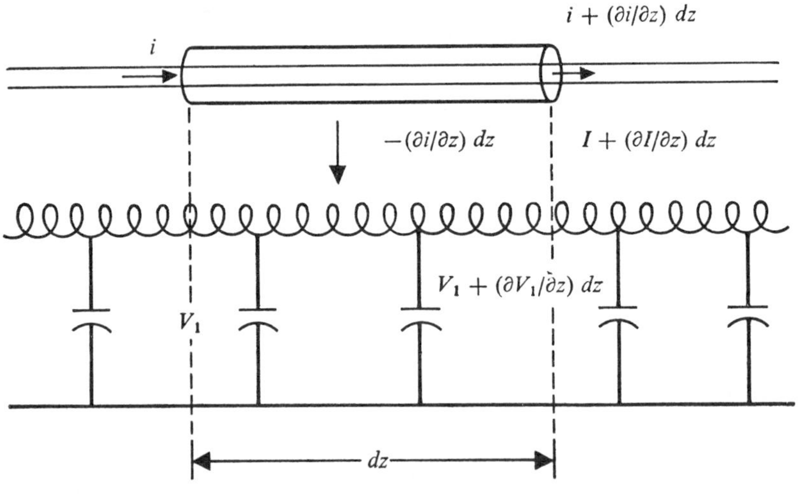

Fig. 10.50 Schematic diagram of electron beam coupled to slow-wave circuit represented by transmission line.

to the electron beam which induces a current in the transmission line. The physical mechanism by which the current is induced can be seen by considering the closed cylindrical surface about the electron beam shown in the figure. The currents flowing into the left-hand end and out of the right-hand end of this cylindrical surface are i and $i + (\partial i/\partial z)\ dz$, respectively, where dz, being the length of the cylinder, is very short as compared to a wavelength of the wave along the transmission line. If $(\partial i/\partial z)\ dz$ is positive, then more convection current comes out of the cylinder than the current that goes in, and the alternating charge density inside the cylinder is negative. Field lines from the transmission line to this negative charge density constitute a displacement current, the value of which can be calculated. Kirchhoff's law can be applied to the total current into the cylinder, which consists of the displacement and convection currents. If the total current into the cylinder is zero, then $-(\partial i/\partial z)\,dz$ flows out of the cylinder's sides and into the transmission line, assuming that no displacement current goes through the ends of the cylinder. The current from length dz of the inductance, due to voltage V across the capacitance $C\,dz$, is $C\,dz(\partial V/\partial t)$, and therefore, Kirchhoff's current law for the transmission-line section of length dz gives, after simplification,

$$\frac{\partial I}{\partial z} = -C\frac{\partial V}{\partial t} - \frac{\partial i}{\partial z}. \tag{10.105}$$

Kirchhoff's voltage law for the closed loop gives V from ground to the inductance at z, $-L\,dz(\partial I/\partial t)$ from z to $z + dz$, and $-[V + (\partial V/\partial z)\,dz]$ from the inductance to ground at $z + dz$. Therefore

$$\frac{\partial V}{\partial z} = -L\frac{\partial I}{\partial t}. \tag{10.106}$$

Eliminating the circuit current I between Eqs. (10.105) and (10.106), we obtain

$$\Gamma^2 V = -\omega^2 LCV - j\omega L\Gamma i. \tag{10.107}$$

If $i = 0$, i.e., if the electron beam is not present, then Eq. (10.107) reduces to the usual wave equation. The propagation constant is then defined as Γ_0 and, therefore, putting $i = 0$ in Eq. (10.107), we get

$$\Gamma_0 = \omega\sqrt{LC}. \tag{10.108}$$

Hence, the characteristic impedance of the line is

$$\frac{V}{I} = \sqrt{L/C} = Z_0. \tag{10.109}$$

When the beam is present, Eq. (10.107) can be rewritten as

$$i = -\frac{(\Gamma^2 - \Gamma_0^2)V}{\Gamma\Gamma_0 Z_0}. \tag{10.110}$$

Comparing Eq. (10.110) with Eq. (10.104) which relates the beam alternating current i to the electric field at the beam, we can see that the circuit voltage V and the electric field E_l are related. If the field is nearly independent of the transverse dimension of the circuit, the axial electric field E_l for slow waves on the helix is given by $-\partial V/\partial z = \Gamma V$. Using this relation and also Eqs. (10.110) and (10.104), we obtain

$$\frac{i}{V} = -\frac{\Gamma^2 - \Gamma_0^2}{\Gamma_0 \Gamma Z_0} = j\frac{\beta_e \Gamma I_0}{2V_0(j\beta_e - \Gamma)^2}, \tag{10.111}$$

where $\beta_e = (\omega/u_0)$ is the equivalent phase shift per unit length of an electron with velocity u_0. The two ratios of i/V have been equated in Eq. (10.111), because the beam current produced by the circuit voltage is the same current that influences the circuit voltage. Equation (10.111) is a quadratic equation in the propagation constant Γ. This equation can be rewritten in the form

$$(\Gamma^2 - \Gamma_0^2)(j\beta_e - \Gamma)^2 = -j\frac{\Gamma^2\Gamma_0\beta_e I_0 Z_0}{2V_0}. \tag{10.112}$$

If $I_0 = 0$, then the roots of Eq. (10.111) are $\pm\Gamma_0$ and $\Gamma = j\beta_e$, which are the normal propagation constants of the circuit and the propagation constant of the electron stream.

When I_0 is finite, the exact solution of Eq. (10.111) or (10.112) can be found by using the numerical methods or by using a computer. However, an approximate solution can be obtained when the electron beam velocity and the circuit phase velocity are equal. Then, $\Gamma_0 = j\beta_e$, and the resulting form of Eq. (10.112) is

$$(\Gamma - j\beta_e)^3(\Gamma + j\beta_e) = 2\Gamma^2\beta_e^2\frac{I_0Z_0}{4V_0} = +2C^3\Gamma^2\beta_e^2, \tag{10.113}$$

where C is the travelling-wave tube gain parameter defined as

$$C = (\frac{I_0Z_0}{4V_0})^{1/2}. \tag{10.114}$$

For small values of C, the solution of Eq. (10.113) is nearly the same as the solutions obtained when C (or Z_0) is zero. If

$$\Gamma = j\beta_e - \beta_e C\delta, \tag{10.115}$$

where $C\delta \ll 1$, and if this expression is substituted in Eq. (10.113), then we obtain the equation

$$-(\beta_e C\delta)^3(j2\beta_e - \beta_e C\delta) = 2\beta_e^2 C^3(-\beta_e^2 - 2j\beta_e^2 C\delta + \beta_e^2 C^2\delta^2). \tag{10.116}$$

Since $C\delta \ll 1$, Eq. (10.116) reduces to

$$\delta = -j^{1/3} = \exp\,[\frac{-j(\pi/2 + 2\pi n)}{3}]. \tag{10.117}$$

Equation (10.117) has three roots, namely, δ_1, δ_2, δ_3, given by

$$\delta_1 = e^{-j\pi/6} = \frac{\sqrt{3}}{2} - \frac{j}{2}, \qquad \delta_2 = e^{-j5\pi/6} = -\frac{\sqrt{3}}{2} - \frac{j}{2}, \qquad \delta_3 = e^{-j3\pi/2} = j. \tag{10.118}$$

By neglecting certain terms in Eq. (10.116), it has been reduced to a third-order equation, and hence one root of the fourth-order polynomial has been lost. By letting $\Gamma = -j\beta_e - \beta_e C\delta_4$, the root which corresponds to the backward-travelling circuit wave is also found, and is given by

$$\delta_4 = -j\frac{C^2}{4}. \tag{10.119}$$

Therefore, the four values of Γ are

$$\Gamma_1 = j(\beta_e + \beta_e\frac{C}{2}) - \beta_e C\frac{\sqrt{3}}{2}, \qquad \Gamma_2 = j(\beta_e + \beta_e\frac{C}{2}) + \beta_e C\frac{\sqrt{3}}{2},$$
$$\Gamma_3 = j\beta_e(1 - C), \qquad \Gamma_4 = -j\beta_e(1 - \frac{C^3}{4}). \tag{10.120}$$

Each of these values of Γ gives the propagation constant of a wave, and the original single travelling wave, in the absence of the electron beam, has now been split into four travelling waves. The waves corresponding to Γ_1 and Γ_2 propagate more slowly than the circuit wave when the beam is absent. The wave corresponding to Γ_1 increases exponentially with distance, whereas the wave corresponding to Γ_2 decays exponentially with distance. The third wave corresponding to Γ_3 is unattenuated, but is faster than the first two waves, while the fourth wave is the backward-circuit wave which is only slightly perturbed.

The boundary conditions on the beam at the input and output of the helix can be specified, and the amplitude of the wave as a function of distance can be calculated. The amplitude of the wave at the output can be calculated in terms of the amplitude of the wave at the input, and the gain of the TWT can be derived. It can be assumed that the output of the TWT is matched to the load so that there is no reflected wave travelling in the $-z$-direction. This means that the wave corresponding to Γ_4 is not excited. Let the input voltage be V. Let each of the three waves associated with Γ_1, Γ_2, Γ_3 have voltages V_1, V_2, V_3 so that

$$V(z) = V_1 \exp(-\Gamma_1 z) + V_2 \exp(-\Gamma_2 z) + V_3 \exp(-\Gamma_3 z)$$

$$= \sum_{i=1}^{3} V_i \exp(-\Gamma_i z). \qquad (10.121)$$

The input velocity and input current provide the other boundary conditions. From Eq. (10.111), the current in each of these three partial waves is given by

$$i_i = \frac{-I_0}{2V_0C^2}\frac{V_i}{\delta_i^2} \exp(-\Gamma_i z) \qquad (10.122)$$

and the total current is

$$i = -\sum_{i=1}^{3} \frac{I_0}{2V_0C^2}\frac{V_i}{\delta_i^2} \exp(-\Gamma_i z). \qquad (10.123)$$

From Eq. (10.101), the velocity can be shown to be

$$v_1 = -\sum_{i=1}^{3} j\frac{u_0}{CV_0}\frac{V_i}{\delta_i} \exp(-\Gamma_i z), \qquad (10.124)$$

where v_1 is the a-c component of velocity of the total wave. Equations (10.122)–(10.124) are the three relations between V_1, V_2, and V_3 in terms of i and v_1 at the circuit input. Usually, v_1 and i can be assumed to be zero at the circuit input. Then at $z = 0$, these three equations become

$$V = V_1 + V_2 + V_3, \qquad (10.125)$$

$$i = -\frac{I}{2V_0C^2}\left(\frac{V_1}{\delta_1^2} + \frac{V_2}{\delta_2^2} + \frac{V_3}{\delta_3^2}\right) = 0, \qquad (10.126)$$

$$v_1 = -j\frac{u_0}{2CV_0}\left(\frac{V_1}{\delta_1} + \frac{V_2}{\delta_2} + \frac{V_3}{\delta_3}\right) = 0. \qquad (10.127)$$

Solving Eqs. (10.125)–(10.127) for V_1, V_2, V_3, we obtain

$$V_1 = V_2 = V_3 = \frac{V}{3}, \qquad (10.128)$$

and hence from Eq. (10.121), we have

$$V(z) = \sum_{i=1}^{3} \frac{V}{3} \exp(-\Gamma_i z). \tag{10.129}$$

If the length L of the circuit is so large that $\beta_e C\delta_1 L$ is large compared to unity, then the first wave predominates, and

$$V(L) = \frac{V}{3} \exp\left(\frac{\beta_e C\sqrt{3}}{2} L\right) \exp\left[-j\left(\beta_e + \frac{\beta_e C}{2}\right)L\right]. \tag{10.130}$$

Usually, $\beta_e L$ is put equal to $2\pi N$, where N is the circuit length in electronic wavelength λ_e which is put equal to u_0/f. The power output is proportional to $|V/L|^2$; similarly, the power input is proportional to $|V(0)|^2$. The power gain G in decibels is

$$G = 10 \log \frac{|V(L)|^2}{|V(0)|^2} = 10 \log \left[\frac{\exp(2\pi\sqrt{3}CN)}{9}\right] = A + BCN, \tag{10.131}$$

where

$$A = 10 \log \tfrac{1}{9} = -9.54 \text{ db},$$

$$B = 10 \times 2\pi\sqrt{3} \log e = 47.3 \text{ db}.$$

Equation (10.131) shows that there is an initial loss of 9.54 db at the input of the circuit. This loss arises because the input voltage splits into three waves of equal magnitude and the growing-wave voltage is only one-third of the total voltage at the input. The term BCN shows that the gain is proportional to the length of the TWT and proportional to the gain parameter C which is a measure of the strength of interaction between the beam and slow-wave circuit. A similar procedure can be used for the calculation of the gain of the backward-wave oscillator.

The noise figure of a TWT amplifier is calculated by knowing the amplitude of the velocity and current modulation on the beam due to statistical variations in the emission velocities of the individual electrons and the number of electrons emitted per unit time. These quantities are affected by the potential variation with distance in the electron gun. Hence, by adjusting the gun potential, the values of v_1 and i at the input can be adjusted to make the TWT noise figure a minimum.

10.26 Limitations of Conventional Types of Tubes at Microwave Frequencies

For low power applications, though solid-state devices are being used far more frequently at the lower microwave frequency, for high power applications, conventional types of tubes are still used in the VHF and UHF range. In this section, we discuss briefly about the limitations of the conventional types of tubes such as triodes and pentodes at higher frequencies, due to parasitic capacitances and inductances, induced currents, and transit time.

Figure 10.51a shows a pentode circuit with a long cathode lead. L_k is the parasitic inductance of the cathode lead. Figure 10.51b gives the equivalent circuit. To calculate the input admittance of the tube, consider the equation

$$e = \text{input voltage} = e_{g_1} + e_k, \tag{10.132}$$

where e_k is the cathode voltage given by

$$e_k = j\omega L_k g_m e_{g_1} \tag{10.133}$$

because it may be assumed that $\omega L_k \ll \dfrac{1}{\omega C_{g_1 k}}$. The input current is given by

$$i = j\omega C_{g_1 g_2} e + j\omega C_{g_1 k} e_g. \tag{10.134}$$

From Eqs. (10.132) and (10.133),

$$e_{g_1} = e(1 + j\omega L_k g_m)^{-1}. \tag{10.135}$$

In a practical circuit, $\omega L_k g_m \ll 1$, and hence

$$i = (j\omega C_{g_1 g_2} + j\omega C_{g_1 k} + \omega^2 L_k C_{g_1 k} g_m) e = Y_e e, \tag{10.136}$$

where Y is the input admittance given by

$$Y = j\omega(C_{g_1 k} + C_{g_1 g_2}) + \omega^2 L_k C_{g_1 k} g_m. \tag{10.137}$$

Equation (10.137) shows that the input conductance $\omega^2 L_k C_{g_1 k} g_m$ increases with the square of the

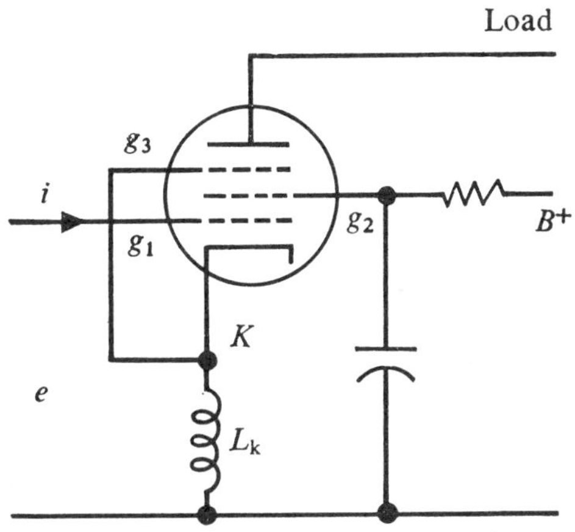

(a) Pentode circuit with long cathode lead

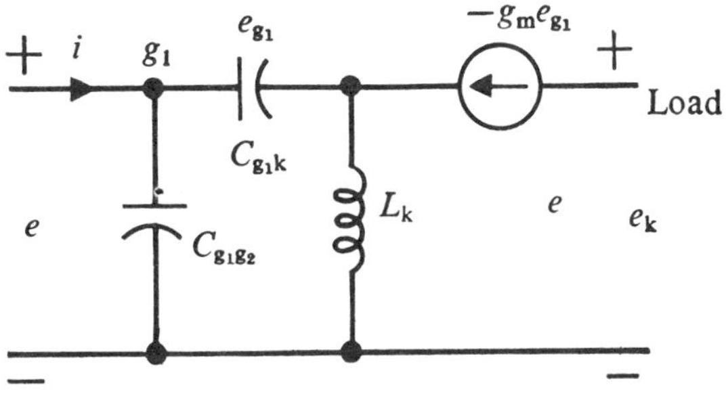

(b) Equivalent circuit

Fig. 10.51 Pentode circuit and its equivalent circuit.

frequency, the input susceptance increases with frequency, and that even small wiring capacitances add to this effect in the 50 to 500 MHz range.

Let us now consider the concept of induced currents. In Fig. 10.52, we consider the induced current when a single electron moves between two parallel planes. From Fig. 10.52a, it can be seen that current flows in the external circuit during all the time that the charge is moving between planes. When the charge arrives at the anode, the current falls to zero at the instant the electron is collected (neglecting secondary electrons). The current flow versus time is somewhat similar to that shown in Fig. 10.52b. The current induced is

$$i = i(t) = \frac{qv(t)}{d}, \tag{10.138}$$

where d is the diode spacing, q the charge, and $v(t)$ the velocity of the charge. If V_0 is the voltage across the planar diode, then the instantaneous power output is

$$P(t) = V_0 i(t). \tag{10.139}$$

This power is equal to the rate of change of the kinetic energy of the charges produced by the acceleration of the charge by the electric field, and hence

$$P(t) = \frac{d}{dt}\left(\frac{m_e v^2}{2}\right) = m_e v \frac{dv}{dt} = v q_e V_0/d. \tag{10.140}$$

Equating relations (10.139) and (10.140), we obtain Eq. (10.138). For a multielectrode system, the current $i(t)$ to the n-th electrode, induced by a charge q moving with vector velocity $\boldsymbol{v}$ in the

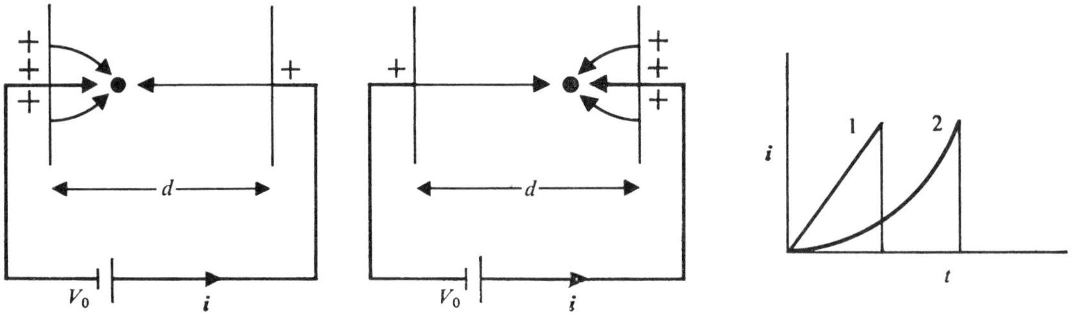

(a) Various positions of charge moving from cathode to anode

(b) Induced current at anode (1, temperature-limited; 2, space-charge-limited)

Fig. 10.52 Induced current between two parallel planes.

neighbourhood, is given by

$$i_n(t) = q\boldsymbol{v} \cdot \boldsymbol{E}_n, \tag{10.141}$$

where $\boldsymbol{E}_n$ is the electric field at the position of the charge when the charge is removed, and the n-th electrode is held 1 volt above ground and all other electrodes are grounded.

From the foregoing concept of induced currents, it can be seen that, if electrons take an appreciable portion of a cycle to travel between electrodes, the induced currents flow during all this time. Hence, the induced currents as well as the acceleration of the charges have effects on the characteristics of space-charge controlled tubes when the transit time between electrodes is an appreciable fraction of the r-f period. When the transit time τ becomes quite large, the induced currents due to charges which leave the cathode at different times do not add with the same phase, and hence both the small-signal a-c conductance and the a-c susceptance are functions of the transit angle $\omega\tau$. This is shown in Fig. 10.53. It can be noticed from the figure that, for transit angles between 2π and 3π, the a-c conductance becomes negative. This negative conductance can be used to make the diode oscillate. The transit angle is also a function of the anode voltage, and hence any variation in the anode voltage can change the a-c capacitance and thus detune the circuit.

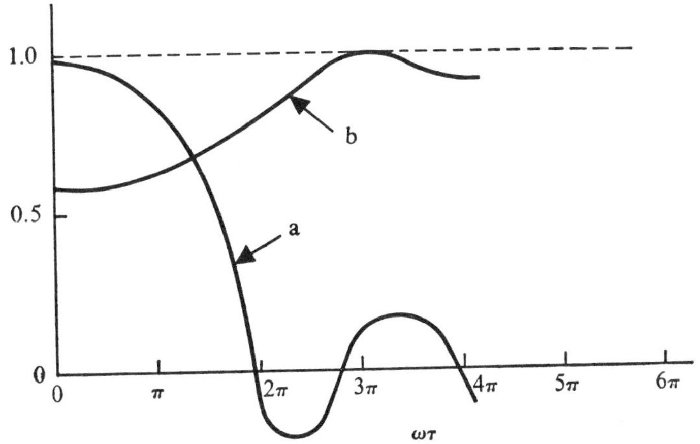

Fig. 10.53 a-c conductance and a-c susceptance of space-charge limited diode as functions of transit angle. (a, a-c capacitance/capacitance with no electrons; b, a-c conductance/low frequency a-c conductance.)

Figure 10.54 shows the induced currents in a negative-grid triode. The net charge delivered to the cathode over a cycle is zero, whereas the net charge delivered to the anode is equal to the

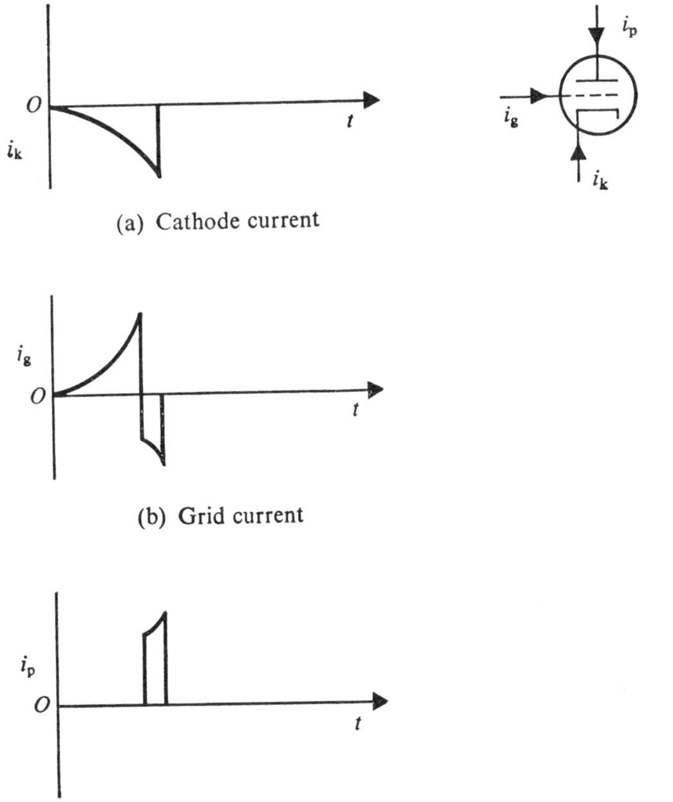

Fig. 10.54 Induced currents in a negative-grid triode.

net charge which leaves the cathode. Hence

$$\int_0^{\tau_g} i_k \, dt = -q_e, \qquad \int_0^{\tau_p} i_g \, dt = 0, \qquad \int_{\tau_g}^{\tau_p} i_p \, dt = q_e, \tag{10.142}$$

where τ_g is the cathode-to-grid transit time and τ_p is the cathode-to-anode transit time. At no instant of time does cathode-to-anode current flow, but cathode-to-grid current flows for $0 < t < \tau_g$ and grid-to-anode current flows for $\tau_g < t < \tau_p$.

Figure 10.55 shows a grounded-grid circuit which is used very often at high frequencies. Here, the grid isolates the cathode and anode circuits. Hence, the input admittance is almost like the diode admittance shown in Fig. 10.53. The input admittance of a grounded-cathode circuit is different, because the induced current in the grid-anode space flows through the input generator, as shown in Fig. 10.56.

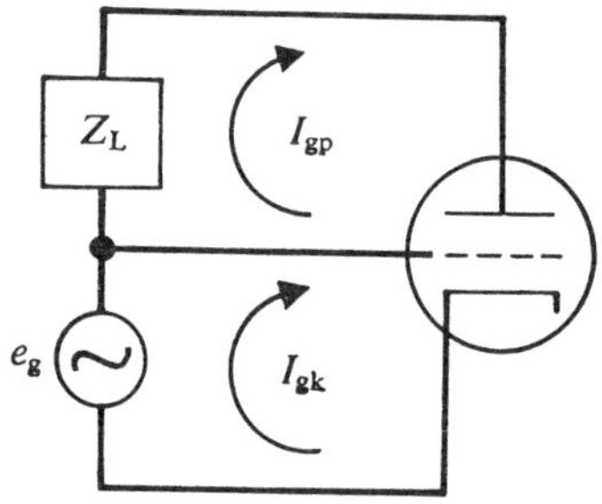

Fig. 10.55 Grounded-grid circuit.

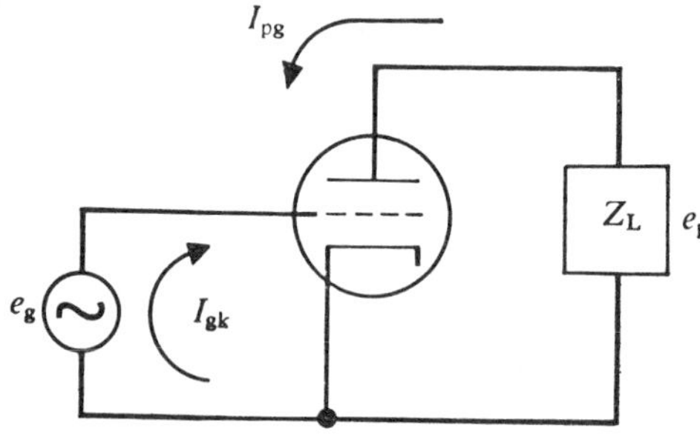

Fig. 10.56 Grounded-cathode circuit.

If q_{ind} is the induced charge on the grid due to the electrons flowing from cathode to anode in the negative-grid triode, then the total charge on the grid due to the impressed voltage e_g is

$$q_g = q_{ind} + C_{gk}e_g + C_{gp}(e_g - e_p) \tag{10.143}$$

assuming only small-signal alternating quantities. If

$$e_g = E_g e^{j\omega t}, \tag{10.144}$$

then the induced charge q_{ind} is given by

$$q_{ind} = K\tau_g g_m E_g \exp\,[j\omega(t - \tau_g)], \tag{10.145}$$

where τ_g is the cathode-to-grid transit time under d-c conditions and K is a proportionality constant nearly equal to unity. The current induced in the grid is then the time derivative of q_{ind}, and hence

$$i_{ind} = jK\omega\tau_g g_m E_g \exp\,[j\omega(t - \tau_g)]. \tag{10.146}$$

Figure 10.57 shows i_{ind} against time. If the transit angle $\omega\tau_g$ is $\ll 1$, then

$$i_{ind} \approx jK\omega\tau_g g_m(1 - j\omega\tau_g + \ldots E_g e^{j\omega t} = Y_{eg} = (\omega^2\tau_g^2 g_m K + j\omega\tau_g g_m K)e_g. \tag{10.147}$$

The term $(\omega^2\tau_g^2 g_m K + j\omega\tau_g g_m K)$ is the input admittance of a grounded-cathode triode amplifier for small transit angles. This admittance consists of a conductance and a capacitance.

Because of the phase lag of the induced grid current with respect to the grid voltage, all the electrons hitting the anode do not have low kinetic energies. Some electrons may also return to the cathode if they do not pass through the grid plane in time, and this may cause power

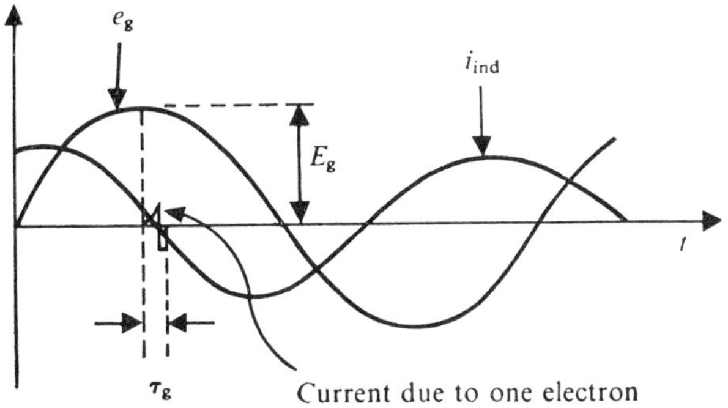

Fig. 10.57 Induced current in grid versus time.

loss as well as cathode damage. Hence, the efficiency of a class C amplifier decreases. Also, there may be adverse effects due to secondary emission.

PROBLEMS

1 A lighthouse tube has parallel-plane electrodes, with cathode-to-grid spacing of 0.0105 cm and grid-to-plate spacing of 0.035 cm. The cathode and plate have areas of 0.8 cm^2 each. The amplification factor and interelectrode capacitances are

$$\mu = 50, \qquad C_{kg} = 2.5 \times 10^{-12} \text{ farad},$$

$$C_{gp} = 1.60 \times 10^{-12} \text{ farad}, \qquad C_{kp} = 0.05 \times 10^{-12} \text{ farad}.$$

If the tube is operated with space-charge limited emission, with a grid voltage of 3 volts and plate voltage of 200 volts, and it is used as an oscillator at a frequency of 2700 MHz with small amplitude operation, calculate the cathode-to-grid transit angle.

2 A double-resonator klystron is tuned to a frequency of 3500 MHz. The gap between the buncher grids and catcher grids is 0.12 cm and the length of the drift space is 2.1 cm. The beam current is 24 mA, and the ratio of V_2/V_0 is 0.35. If the two resonators are assumed to oscillate in time phase, calculate the d-c accelerating voltage, buncher voltage, and catcher voltage required for maximum power output for integral values of n, from 1 to 4. Calculate the power output, power input, and maximum theoretical efficiency of the klystron, for $n = 1, 2, 3$.

3 A reflex klystron is tuned to a frequency of 6000 MHz. The length of the drift space is 0.4 cm. What values of $V_0/(V_R - V_0)^2$ are required for oscillation corresponding to $n = 1$ to $n = 4$? Specify the values of V_0 and V_R that produce oscillations.

4 An electron beam in the z-direction has radius a, velocity v, and space-charge density ρ. Show that it produces a magnetic field H_ϕ given by

$$H_\phi = -\frac{r\rho v}{2} \quad \text{for } 0 \leqslant r \leqslant a,$$

$$H_\phi = -\frac{a^2\rho v}{2r} \quad \text{for } r \geqslant a.$$

Show that the Lorentz force $-q_e v \times B_\phi$ is much less than the radial outward force due to the space-charge electric field. (Solve Poisson's equation.) Hence, the necessity for focussing of electron beams in microwave tubes.

5 A magnetron operating in the π-mode has the following characteristics: $N = 8$, $a = 0.35$ cm, $b = 0.9$ cm, $l = 2.5$ cm, $f = 2500$ MHz, $V_0 = 16$ kV, $B = 0.15$ weber/m^2. Calculate (i) the angular velocity of the electrons, (ii) the radius at which the radial forces due to the electric and magnetic fields are equal and opposite, and (iii) the value of V_0/B required for a frequency of 2500 MHz in the modes (a) $n = 0$, $p = \pm 1$, (b) $n = 1$, $p = 0$, ± 1, and (c) $n = 2$, $p = 0$, ± 1.

6 If a central cylindrical anode and a concentric cathode are used, and if electrons are projected into the interelectrode space so as to miss the anode, write the equations of motion of the electrons. The electrons rotate around the anode in the same way as they rotate around the nucleus in an atom.

11 Microwave Measurements

11.1 Concepts of Microwave Measurements

This chapter deals with the basic methods of electrical measurements used in the microwave region of the electromagnetic spectrum. The topics discussed here provide a background to all microwave measurements. The measurement technique is vastly different from that of more conventional radio practice, largely because, in microwave studies, our attention is necessarily focussed on the electromagnetic fields. The measurement methods are based on the wave character of high-frequency currents rather than on the low-frequency technique of direct determination of current or voltage. For example, the measurement of power flow in a system specifies the product of the electric and magnetic fields, whereas the measurement of impedance determines their ratio. Thus, these two measurements indirectly describe the distribution of the electric and magnetic fields in the system and provide its complete description. In fact, this is the approach to most of the measurements carried out in the microwave region of the electromagnetic spectrum.

11.2 Fundamental Characteristics of Hollow Metal Waveguides

The electrical characteristics of a transmission line for most applications are described by

(i) the phase constant β (radians per unit length) which is equal to $2\pi/\lambda_g$, where λ_g denotes the transmission line guide wavelength,

(ii) the characteristic impedance Z_0, and

(iii) the attenuation constant α (nepers per unit length).

The propagation of waves in conventional transmission lines and waveguides can be treated mathematically in a similar way due to the fact that, in both these cases, a simple harmonic disturbance is propagated longitudinally with a phase factor $\exp(\pm j\beta z)$ which indicates repetition of the wave pattern each time there occurs a change of 2π in the angle βz, the corresponding change in the direction z being called the guide wavelength λ_g. Hence, $\beta = 2\pi/\lambda_g$.

11.3 Guide Wavelength

In waveguide transmission, the guide wavelength λ_g is related to the free-space wavelength λ_0 as

$$\lambda_g = \lambda_0[1 - (\lambda_0/\lambda_c)^2]^{-1/2}, \tag{11.1}$$

where the cut-off wavelength λ_c is the value of λ_0 for which the wave fails to propagate in the guide. The value of λ_c depends on the mode of propagation and guide dimension. For example, for the TE_{10} mode in a rectangular waveguide, $\lambda_c = 2a$, where a denotes the broad dimension of the guide. It is to be noted that, for all waveguide modes, $\lambda_g > \lambda_0$.

11.4 Characteristic Wave Impedance of Waveguide

In a waveguide, the electric field $\boldsymbol{E}$ and the magnetic field $\boldsymbol{H}$ vary over the cross-section of the guide, but ***E***/***H*** remains unaltered over the section. Hence, the ratio ***E***/***H***, where ***E*** is transverse to ***H*** and both ***E*** and ***H*** are transverse to the direction of propagation, is interpreted as the waveguide analogue of characteristic impedance and is called the wave impedance Z_w which varies for different modes. For hollow metal rectangular waveguides supporting TE or TM modes, the wave impedances are given respectively by

$$Z_{h,nm} = Z_w^{TE} = \eta_0[1 - (\lambda_0/\lambda_c)^2]^{-1/2} \text{ ohms,} \tag{11.2}$$

$$Z_{e,nm} = Z_w^{TM} = \eta_0[1 - (\lambda_0/\lambda_c)^2]^{1/2} \text{ ohms,} \tag{11.3}$$

where η_0 represents the intrinsic impedance of the medium, which is equal to 376.7Ω for plane waves in free space. The variation of Z_w with respect to λ_0/λ_c is shown in Fig. 11.1. Because there are no unique currents and voltages in hollow waveguides, Z_w cannot be determined so easily.

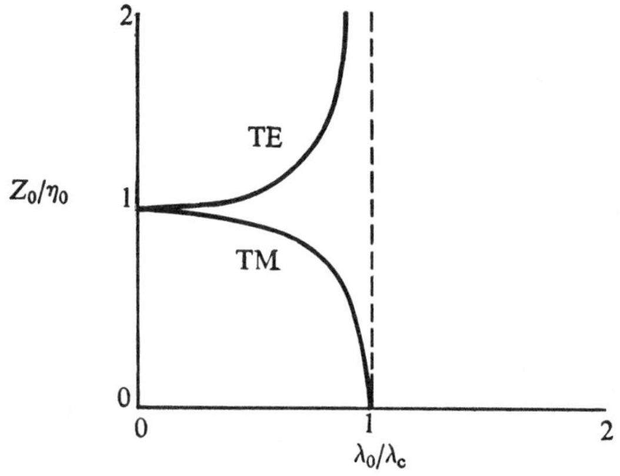

Fig. 11.1 Wave impedance versus λ_0/λ_c.

The fundamental quantities for waveguide work are standing-wave ratio, reflection coefficient, and propagation constant from which the normalized impedance at any point, as also the performance characteristics, such as the scattering parameters and insertion loss of a complete waveguide system, can be determined.

11.5 Standing-Wave Patterns

In general, electromagnetic waves can propagate in waveguides in both directions ($+z$ and $-z$), and hence the electromagnetic field in a waveguide can be resolved into components travelling $E_{forward}$, $H_{forward}$ and $E_{backward}$, $H_{backward}$, i.e.,

$$\begin{aligned} E &= E_{forward} + E_{backward}, \\ H &= H_{forward} + H_{backward}, \end{aligned} \tag{11.4}$$

where forward and backward indicate waves travelling in $+z$- and $-z$-direction, respectively. Assuming the propagation vectors to be represented by exp ($\mp j\beta z$) in the forward and backward directions, respectively, the resultant transverse E and H fields at any point inside the guide can be represented in terms of the values of the forward- and backward-travelling waves at their respective origin, as follows

$$E = E_0 e^{-j\beta z} + E_0 e^{+j\beta z},$$
$$H = H_0 e^{-j\beta z} + H_0 e^{+j\beta z}. \tag{11.5}$$

The backward wave can be regarded as a reflected wave.

The reflection coefficient Γ is defined in terms of electric field components as the complex ratio

$$\Gamma = \frac{E_{\text{backward}}}{E_{\text{forward}}} = \Gamma_0 e^{j2\beta z} \tag{11.6}$$

which can be represented as

$$\Gamma = |\Gamma| e^{j\psi}, \tag{11.7a}$$

where

$$\psi = \psi_0 + 2\beta z \tag{11.7b}$$

with ψ_0 denoting the phase change on reflection; the reflection coefficient at the origin is $\Gamma_0 = |\Gamma_0| \exp(j\psi_0)$.

The expression for the universal standing-wave pattern can be easily derived from Eqs. (11.7) as

$$|\underline{E}| = |E|[1 + |\Gamma|^2 - 2|\Gamma| \cos(2\beta s)]^{1/2}, \tag{11.8}$$

where s denotes the distance of any point along the axis of the guide from the nearest minimum of the standing-wave pattern. The reflection coefficient $|\Gamma|$ in a waveguide system can be determined by measuring the VSWR (voltage standing-wave ratio) ρ from the relation

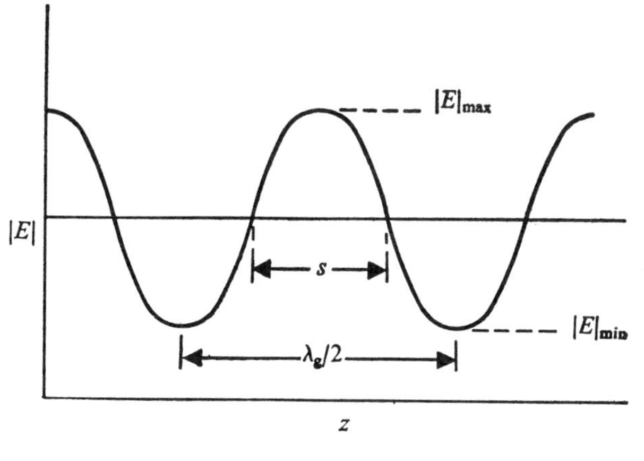

Fig. 11.2 Typical standing-wave envelope.

$$|\Gamma| = \frac{\rho - 1}{\rho + 1}, \tag{11.9}$$

where VSWR is defined as

$$\rho = \frac{|E_{\text{forward}}| + |E_{\text{backward}}|}{|E_{\text{forward}}| - |E_{\text{backward}}|} = \frac{|E|_{\text{max}}}{|E|_{\text{min}}}. \tag{11.10}$$

The different quantities such as β and s are represented in relation to a typical standing-wave pattern (envelope) (see Fig. 11.2).

11.6 Standing-Wave Indicator

A standing-wave indicator consists of a section of uniform, very low-loss transmission line in coaxial or rectangular waveguide configuration, having a longitudinal thin, uniform slot, several half wavelengths long and tapered at each end, and milled parallel to the lines of current flow so that it does not perturb significantly the original mode field configurations. The standing-wave indicator designed to be used with the TE_{01} guide has its slot cut along the middle of one of the broad faces. A sliding carriage carries an electric field pick-up broadband probe, which is inserted into the slot and is connected by a suitable transmission line to a crystal or a bolometer. The probe length inside the guide is adjusted by a screw mechanism to provide loose coupling to the electric field, and the electromotive force so generated is transformed into direct current by the detector if the r-f power is unmodulated. To transfer maximum power from the probe to the crystal, a tuning plunger with an adjustable short circuit is provided to effect matching between the probe and crystal. The sliding carriage is so designed as to maintain a fixed orientation of the probe with respect to the slot all along its motion. The probe position is determined with the aid of a scale attached to the transmission line. For further constructional details, the reader is advised to refer to Sucher and Fox (1963).

11.7 Standing-Wave Measurements

The measurement of standing-wave pattern in a waveguide system, as also the measurement of the voltage standing-wave ratio ρ, forms an essential part of microwave techniques. The measurement of standing-wave pattern is carried out by sampling the field at different positions along the guide axis. The standing-wave pattern is obtained by plotting the crystal current versus the corresponding positions of the sliding carriage along the guide. The emf induced in the probe is proportional to the field strength $|E|$ in the immediate vicinity of the probe. The current at the output of the crystal is proportional to $|E|^2$ if it has a square-law characteristic. The maxima (I_{max}) and minima (I_{min}) of the crystal current, corresponding to the maxima (E_{max}) and minima (E_{min}) positions of the field in the guide, will be indicated. The VSWR is

$$\rho = \left(\frac{I_{\text{max}}}{I_{\text{min}}}\right)^{1/2}$$

which is equal to $E_{\text{max}}/E_{\text{min}}$. In this case, the standing-wave pattern is sinusoidal.

The accuracy with which the VSWR ρ can be measured depends on the accuracy with which I_{min} can be measured, especially when ρ is small. It also depends on the current (I)-voltage (V) characteristic of the crystal.

11.8 Calibration of Crystal Detector (Experiment 1)

The crystal used for detection of microwaves in the X-band is IN 23B, whose current-voltage characteristic follows the law

$$I = KV^p, \tag{11.11}$$

where K is a constant. The object of the experiment is to determine the exponent p. The experimental set-up is shown in Fig. 11.3. The same set-up, with some modifications, can be used for other experiments. The klystron reflector voltage is adjusted so as to operate it at the top of the

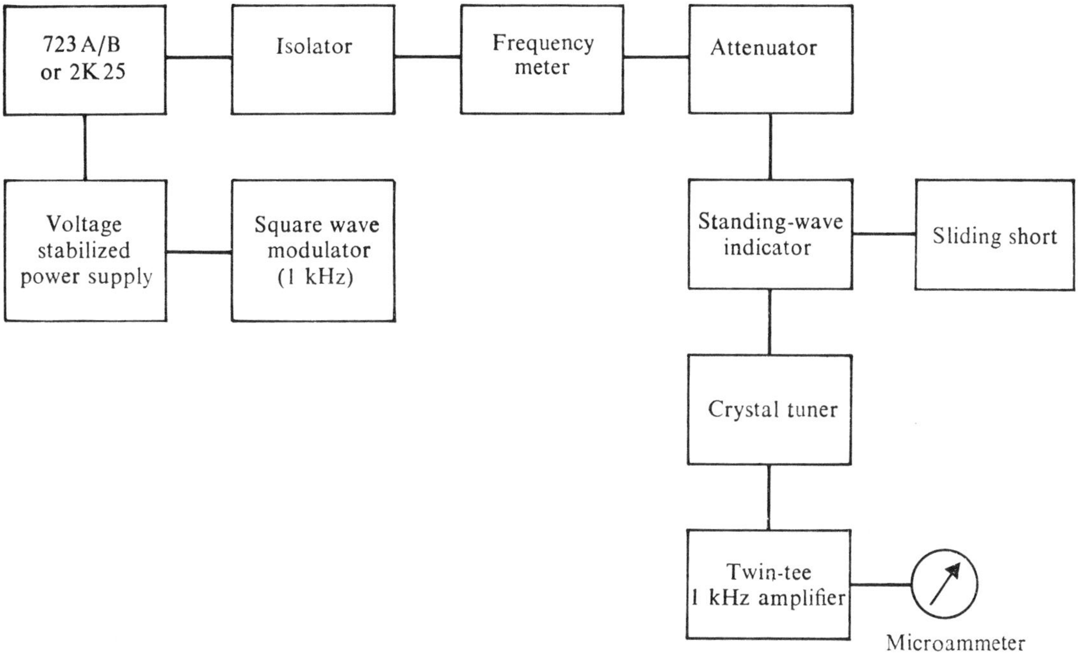

Fig. 11.3 *X*-band experimental set-up.

mode curve. Adjustments of the standing-wave detector and the amplifier are made to get maximum output with the lowest possible gain of the amplifier and about 15 db attenuation in the transmission system. Readings of the microammeter for different positions of the probe are taken, and the standing-wave pattern, from which $\lambda_g/2$ is determined as the distance between any two successive minima, is plotted. Then, if s represents the distance of any point of the probe from any reference minimum, the voltage induced in the probe at that point is proportional to $\sin(2\pi s/\lambda_g)$. The slope of the curve (a straight line) I versus $\sin(2\pi s/\lambda_g)$ on a log-log graph paper gives the value of the exponent p. The crystal will possess a square-law characteristic if $p = 2$.

If the law of response differs from the square law, then a correction is required for the VSWR. If the detector response law is given by

$$p = 2(1 \pm \gamma), \tag{11.12}$$

where $\gamma \ll 1$, the VSWR is

$$\rho = \frac{E_{max}}{E_{min}}(1 + 0.693\gamma). \tag{11.13}$$

The measurement of VSWR requires a special technique if the VSWR is very low ($\ll 2$) or very high ($\gg 10$). In such cases, since the location of exact minima becomes difficult, it should be done by adopting the following procedure.

11.9 Location of Standing-Wave Minima

In order to locate the minimum, z_{min}, the positions z_1 and z_2 corresponding to equal microammeter reading on either side of z_{min} are found and the average is taken (see Fig. 11.4). That is, $z_{min} = (z_1 + z_2)/2$.

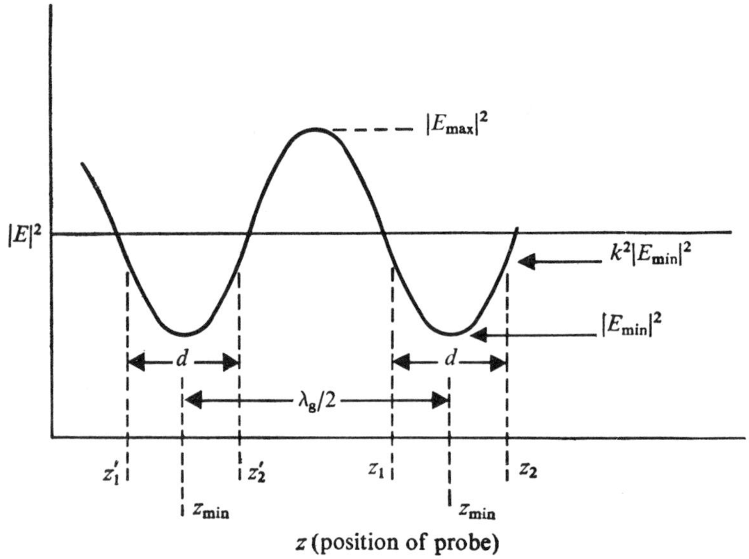

Fig. 11.4 Location of standing-wave minima.

11.10 Measurement of Guide Wavelength (Experiment 2)

In order to measure the guide wavelength λ_g, the experimental set-up (Fig. 11.3) is used. Then, the distance between two successive minima is found, and the guide wavelength is obtained as (see Fig. 11.4)

$$\lambda_g = 2[(z_1 + z_2) - (z_1' + z_2')]. \tag{11.14}$$

The frequency of the source is measured and λ_0 is determined. The frequency meter reading corresponding to the maximum dip of the microammeter indicates the frequency of the source. The cut-off wavelength λ_c is calculated from the relation

$$\lambda_c = \lambda_0[1 - (\frac{\lambda_g}{\lambda_0})^2]^{-1/2} \tag{11.15}$$

and is verified to be $\lambda_c = 2a$.

11.11 Measurement of Low VSWR (Experiment 3)

In order to measure a low VSWR, measurements are carried out in the neighbourhood of minima (see Fig. 11.4) by measuring the distance $d = z_2 - z_1$ between the points z_1 and z_2 which have equal readings $k^2E_{min}^2$ on either side of z_{min}. The VSWR is then given by the relation (see Ginzton 1957)

$$\rho = \frac{(k^2 - \cos^2 \theta)^{1/2}}{\sin \theta}, \tag{11.16}$$

where $\theta = \pi d/\lambda_g$. The accuracy of the measurement depends on the accuracies with which d and λ_g are determined. The minima positions rather than the maxima positions are chosen since minima are sharper and the detector is not overloaded, and hence d and λ_g can be measured more accurately in the minima positions.

11.12 Measurement of High VSWR (Experiment 4)

The direct measurement of maxima for high VSWR causes overloading of the detector. To avoid the difficulty due to overloading, the double-minima method (see Fig. 11.4) is adopted. If the detector obeys square law, and if $k^2 = 2$ and θ is small, then Eq. (11.16) reduces to

$$\rho \approx \frac{\lambda_g}{\pi d} \tag{11.17}$$

since $\cos^2 \theta \approx 1$ and $\sin \theta \approx \theta$.

If the detector obeys a law such that the exponent is given by $p = 2(1 \pm \gamma)$, then (see Ginzton 1957)

$$\rho \approx [\frac{2^{(1\pm\gamma)^{-1}} - \cos^2 \theta}{\sin \theta}]^{1/2}. \tag{11.18}$$

For small θ, this equation reduces to

$$\rho \approx \frac{\lambda_g}{\pi d}[2^{(1\pm\gamma)^{-1}} - 1]^{1/2}. \tag{11.19}$$

11.13 Errors in Standing-Wave Measurements

The slotted-section method of measuring standing wave is susceptible to the following errors[1-3] (see also Montgomery 1947):

(i) Frequency instability of the source

(ii) Errors due to standing-wave indicator besides personal errors, which can be minimized by taking the average of several readings of maxima and minima.

11.14 Errors due to Frequency Instability

The frequency instability of the source, as also the instability in the value of λ_0, may arise mainly from insufficient regulation of the reflector voltage of the klystron, and may arise also from

[1]W. Altar, F. B. Marshall, and L. P. Hunter, Probe error in standing-wave detectors, *Proc. I.R.E.*, **34**, 33, 1946.

[2]K. Tomiyasu, Loading and coupling effects of standing wave detectors, *Proc. I.R.E.*, **37**, 1405, 1949.

[3]A. A. Oliner, The calibration of the slotted section for precision microwave measurements, *Rev. Sci. Instrum.*, **25**, 13, 1954.

frequency pulling if the klystron is not sufficiently isolated from the load. For 2K25 or 723 A/B klystron, which is generally used for laboratory measurements, a change of ± 1 volt in the reflector voltage causes a change of approximately 7 MHz in the klystron frequency. The instability due to frequency pulling can be reduced by using an isolator or an attenuator with at least 15 db between the klystron and the slotted section.

11.15 Errors due to Standing-Wave Indicator

The measurements of VSWR and the location of the voltage minima are affected by errors inherent in slotted section, i.e., by detector-response law error, probe error, slot error, and other discontinuity effects inherent in the equipment.

11.16 Probe Error

The pattern of the standing wave measured with a slotted section is affected by reflections from the probe and by variation of the probe length penetrating inside the guide as the carriage carrying the probe is moved along the guide. The variation of the depth of probe penetration occurs because of the irregularity of the upper surface of the top plate.

Reflections from the probe can be minimized by reducing the depth of penetration of the probe inside the guide. But smaller depth of penetration increases probe error due to the variation of probe length. For example, if the irregularity in the smoothness of the upper surface of a slotted section is ± 0.001 in, the variation in probe penetration depth will introduce an error of 1.6 per cent in the voltage induced in the probe if the depth of penetration is 1/16 in, and hence an error of about 2.56 per cent in the output current of the crystal if it possesses square law. But the error in voltage will increase by 3.2 per cent and, therefore, the error in crystal current will be 10.24 per cent if the depth of penetration is 1/32 in. Generally, at X-band the depth of penetration of the probe is set to not more than 5 to 10 per cent of the inside dimension of the guide in the direction of the probe, i.e., for an X-band guide of 1 in $\times 1/2$ in, the probe penetration is usually 1/20 in.

The effects of the probe on the standing-wave pattern alter the value of ρ and the locations of maxima and minima. The changes in ρ and the shift of the pattern can be calculated by considering the probe as a shunt admittance $Y_p = G_p + jB_p$, where G_p and B_p represent equivalent conductance and susceptance. If the load terminating the slotted section is real and has an equivalent conductance G_L, the measured VSWR ρ_m normalized with respect to the correct VSWR ρ is given by the relation

$$\frac{\rho_m}{\rho} = \frac{1 + G_L + G_p G_L}{1 + G_L + G_p}. \tag{11.20}$$

If $G_L \leqslant 1$ and $B_p = 0$, which is achieved by proper tuning of the probe, the shift in minima (Δ_{min}) and maxima (Δ_{max}) is given by

$$\frac{\Delta_{min}}{\lambda_g} = \frac{B_p G_L^2}{(1 + G_L)^2}, \tag{11.21a}$$

$$\frac{\Delta_{max}}{\lambda_g} = \frac{B_p}{(1 + G_L)^2} \tag{11.21b}$$

provided $G_L \leqslant 1$ and G_p and B_p are very small. In general, Δ_{min} is much less.

11.17 Slot Error

If the slot is narrow and cut symmetrically and with precision along the centre of the broad face of the guide, then the error due to radiation is negligibly small[1]. But the presence of the slot introduces a loading effect and slightly changes the characteristic impedance and wavelength from the values of a waveguide without a slot. The end of the slot introduces a discontinuity effect which gives rise to errors in the measurement of VSWR and in locating a voltage minimum. The slotted section can be calibrated to eliminate these errors by adopting the following procedure.

(i) Short circuit the end of the slotted section with a smooth flat metal plate. Locate the voltage minimum nearest the termination. This is the reference minimum $d_{r\,min}$.

(ii) Remove the plate and connect in its place a variable short. Keeping the probe at the reference minimum position, move the variable short till the minimum is obtained. This is the reference position s_r of the short.

(iii) Measure λ_g in the slotted section and also λ_{gs} in the adjustable short.

(iv) Vary the position s of the short and, for each position of s, track the location of the voltage minimum d_{min} in the slotted section.

(v) Plot a graph $\Delta d = -(d + s\lambda_g/\lambda_{gs})$ versus d (see Fig. 11.5), and draw a line through the average of this curve.

The correct value $d_{c\,min}$ of the location of voltage minimum is obtained by finding the correction factor Δd from the calibration curve. If, in an actual system, the measured VSWR and the position of voltage minimum are ρ_m and d_{min}, respectively, the correct values ρ_c and $d_{c\,min}$ for ρ_m and d_{min} are given by

$$\rho_c = \rho_m + \frac{4\pi\rho_m}{\lambda_g}\Delta d', \tag{11.22}$$

$$d_{c\,min} = [d_{min} + \Delta d(\frac{\rho_m^2 + 1}{\rho_m^2 - 1})]\frac{\lambda_{gs}}{\lambda_g}, \tag{11.23}$$

where $\Delta d'$ is the correction factor obtained from Fig. 11.5 at $d_{min} - \lambda_g/8$.

11.18 Measurement of Impedance with Slotted Line (Experiment 5)

The normalized wave resistances at maximum and minimum electric field points are given respectively by the relations

$$\frac{R_{max}}{Z_0} = \rho, \qquad \frac{R_{min}}{Z_0} = \rho^{-1}. \tag{11.24}$$

In order to find the wave impedance of an unknown load, $Z = R + jX$, seen by the slotted line, use the experimental set-up (Fig. 11.3), replacing the adjustable short by the unknown impedance. Find ρ and d_{min} of the first minimum from the input end of the load, which may be regarded as the flange of the slotted line. Then, use the Smith chart (Fig. 11.6) to determine the load impedance by first fixing the point ρ on the R/Z_0 axis and move round the circle that represents the

[1]A. A. Oliner, The calibration of the slotted section for precision microwave measurements, *Rev. Sci. Instrum.*, **25**, 13, 1954.

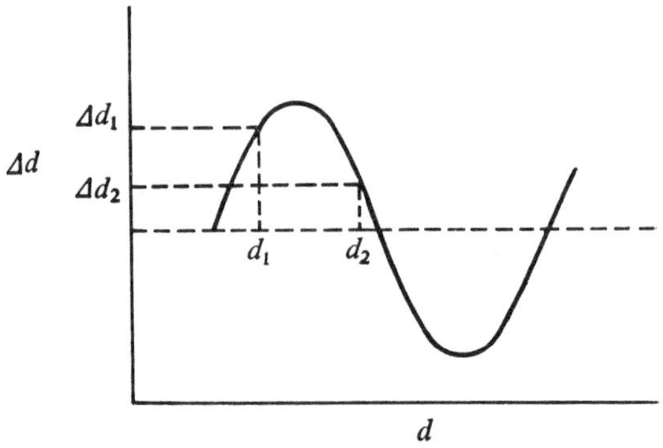

Fig. 11.5 Calibration curve for slotted section.

reflection coefficient Γ corresponding to ρ for an angular distance $4\pi d_{min}/\lambda_g$. The normalized impedance Z/Z_0 is then obtained at the point P corresponding to the appropriate resistance and reactance circles. The use of the Smith chart does not yield correct results if the unknown

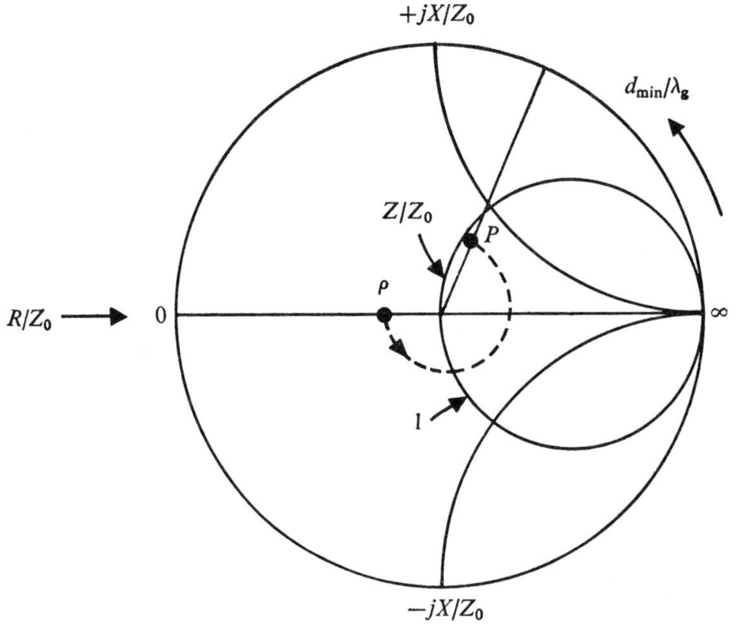

Fig. 11.6 Load impedance.

impedance is small. The unknown impedance can be calculated from the measured values of λ_g, d_{min}, and ρ as follows. The complex reflection coefficient is given by

$$|\Gamma| \exp (j\phi) = \frac{Z - Z_0}{Z + Z_0}.$$

If $Z \ll Z_0$, $|\Gamma| \approx 1$ and ϕ is small, and hence $\exp (j\phi) \approx 1 + j\phi$. Since $|\Gamma| = (\rho - 1)/(\rho + 1)$ and

ρ is high, we have

$$\frac{Z - Z_0}{Z + Z_0} \approx (1 - \frac{2Z}{Z_0}).$$

When $|\Gamma| \approx 1$, it is obvious that

$$\frac{\rho - 1}{\rho + 1} \approx 1 - \frac{2}{\rho}.$$

Therefore

$$(1 - \frac{2Z}{Z_0}) = (1 - \frac{2}{\rho})(1 + j\phi)$$

which yields

$$\frac{Z}{Z_0} \approx \frac{1}{\rho} - j\frac{\phi}{2}, \tag{11.25a}$$

$$\phi = 2\beta d_{\text{min}} - \pi. \tag{11.25b}$$

11.19 Attenuation Constant

The attenuation constant of a waveguide can be measured with the help of a slotted section.

11.20 Standing-Wave Method of Measurement of Attenuation Constant (Experiment 6)

The guide is connected to the output end of the slotted section (see Fig. 11.3) and is terminated by a short circuit. The standing wave is plotted, and the distance δ between the two points on either side of the minimum, which indicates double the value of the minimum output, is measured. The phase constant $\bar{\beta}$ is calculated from the measured value of λ_g. The waveguide is disconnected, the slotted line is shorted by a metal plate, and the distance δ' between the two points on either side of the minimum of the standing wave, where the slotted-section output reading is double that of the minimum reading, is determined. The attenuation constant α of the waveguide of length l is calculated (see Sucher and Fox 1963):

$$\alpha = \frac{1}{2l} \ln \left[\frac{(2 - \cos \bar{\beta}\delta) + [(2 - \cos \bar{\beta}\delta)^2 - 1]^{1/2}}{(2 - \cos \bar{\beta}\delta') + [(2 - \cos \bar{\beta}\delta')^2 - 1]^{1/2}}\right]. \tag{11.26}$$

11.21 Attenuation Constant of Attenuator (Experiment 7)

The attenuation constant of an attenuator can be measured by using the impedance measurement method which employs a slotted line[1]. The total attenuation constant α_T is due to the reflection loss α_R at the input of the attenuator and the dissipative loss α_D in the attenuator. The relations for these are

$$\alpha_R = 10 \log_{10} \frac{(\rho_m + 1)^2}{4\rho_m}, \tag{11.27a}$$

[1]R. W. Beatty, Determination of attenuation from impedance measurements, *Proc. I.R.E.*, **38**, 895, 1950.

$$\alpha_D = 10 \log_{10} \frac{1}{R}, \tag{11.27b}$$

$$\alpha_T = 10 \log_{10} \frac{(\rho_m + 1)^2}{4\rho_m R}, \tag{11.27c}$$

where R is the radius of the reflection coefficient circle and ρ_m the measured VSWR.

The attenuator terminated by a matched load is connected to the slotted line (Fig. 11.3) and ρ_m is measured. The attenuator terminal is then reversed and is terminated by an adjustable short. The VSWR ρ_n and the minima d_{min} for various positions, with an interval of $\lambda/16$ over a distance of half wavelength, are measured. The reflection coefficient Γ_n corresponding to each ρ_n and d_{min} is calculated as

$$|\Gamma_n| \angle \psi_n = \frac{\rho_n - 1}{\rho_n + 1} \angle (2\beta d_{min} + \pi) \tag{11.28}$$

and plotted on a polar graph with $|\Gamma_n|$ as the radius and the phase change as angle. The radius R of the reflection coefficient circle is measured and thus α_D is determined.

11.22 Measurement of Electromagnetic Field

The measurement of microwave fields, such as diffracted, scattered, and radiation fields, plays a significant role in the understanding and application of microwaves, especially in connection with radar cross-section of targets, development of antennas, and propagation phenomena. If an object is illuminated by an electromagnetic field characterized by the electric vector $\boldsymbol{E}_{inc}$, the diffracted field $\boldsymbol{E}_d$ in the vicinity of the object is represented by

$$\boldsymbol{E}_d = \boldsymbol{E}_{inc} + \boldsymbol{E}_s \tag{11.29}$$

provided there is no interaction between the source and the secondary waves $\boldsymbol{E}_s$ generated by displacement currents and conduction currents induced in the object by the incident field $\boldsymbol{E}_{inc}$.

11.23 Measurement of Radiation Field

The measurement of the radiation field is done at distances which are large as compared to the size of the aperture. The field vectors are entirely transverse to the direction of propagation and vary inversely as the distance from the aperture. The spatial distribution of a far field is therefore described as a function of an angular coordinate at fixed radius. The radiation field measurement is important in connection with antenna characteristics. It yields information on the structure of the antenna pattern, viz, beam width of main lobe, side-lobe level with respect to the main lobe, and gain.

The radiation pattern of an antenna is the same whether it is used as a transmitting or receiving antenna provided the system does not include nonlinear components. This nonlinearity condition must be fulfilled to satisfy the reciprocity theorem. The measurement of radiation pattern is done either by measuring the transmitted field intensity of the antenna along an arc of fixed radius or by measuring the angular variation of the response of the antenna to an incoming wave. The measurement can be done either in free space or inside a microwave anechoic chamber.

11.24 Free-Space Measurements of Electromagnetic Field

The antenna test range should satisfy the free-space criteria for separation R between the test antenna and source antenna, aperture A of the antenna, directivity requirements or, equivalently, the uniformity of illumination of the aperture of the test antenna when it is used as a receiver, and the Rayleigh roughness criteria of the intervening ground of the two antennas[1,2]. The free-space criteria are as follows.

Distance requirements. It is necessary to satisfy the plane wave condition, and this requires the phase deviation across the aperture of the test antenna to be less than $\pi/8$, which in turn requires that the separation R between the transmitting and receiving antenna fulfill the condition

$$R \geqslant \frac{2A^2}{\lambda_0}. \tag{11.30}$$

Directivity requirements. For the field distribution across the aperture of the test antenna to be uniform, the source aperture A_s must satisfy the condition

$$A_s \leqslant \frac{R\lambda_0}{4A}. \tag{11.31}$$

Rayleigh criteria. For the intervening terrain between the test and source antenna to produce a pure specular reflection, the height H of the irregularities on the surface of the ground should be limited to

$$H \leqslant \frac{\lambda_0}{16\theta}, \tag{11.32}$$

where θ denotes the grazing angle of incidence.

The effects of ground reflection on the antenna pattern can be reduced to the minimum by suitably adjusting the heights of the antennas. The effects of ground can be tested by measuring the radiation pattern of an antenna at two slightly different frequencies. The effects of ground reflections will be revealed if there is a shift in the pattern maxima and minima.

If the free-space criteria are satisfied, the radiation pattern will be independent of the distance R at which the pattern is measured. Then, the field in any direction (θ, ϕ) is given by

$$E = 376.7H = \frac{F(\theta, \phi)}{R}, \tag{11.33}$$

where 376.7Ω is the characteristic wave impedance of the TEM wave in free space. The radiation pattern function is denoted by $F(\theta, \phi)$, which is often normalized to have maximum value of unity, and the side lobes are expressed in relation to F_{max}.

The radiation pattern is usually determined in two mutually perpendicular planes, namely, (ϕ-plane, $\theta = 0°$) and (θ-plane, $\phi = 0°$), though the complete radiation pattern of any antenna is three dimensional.

To measure the radiation pattern in the θ-plane, the test aerial mounted on a rotatable mast is fed with constant power from a microwave source. A receiving aerial, which is usually a

[1]C. C. Cutler, A. P. King, and W. E. Kock, Microwave antenna measurements, *Proc. I.R.E.*, **35**, 1462, 1947.
[2]W. A. Cumming, Radiation measurements at radio frequencies: A survey of current techniques, *Proc. I.R.E.*, **47**, 705, 1959.

pyramidal horn, receives power proportional to $F^2(\theta)$. If the output of the receiving horn is passed through a square law crystal, then the crystal output I is proportional to the received power, i.e., $I \propto F^2(\theta)$, which gives the power pattern. The field intensity pattern is given by $F(\theta) \propto I^{1/2}$. A typical antenna test range used for the measurement of radiation characteristics of an antenna is shown in Fig. 11.7.

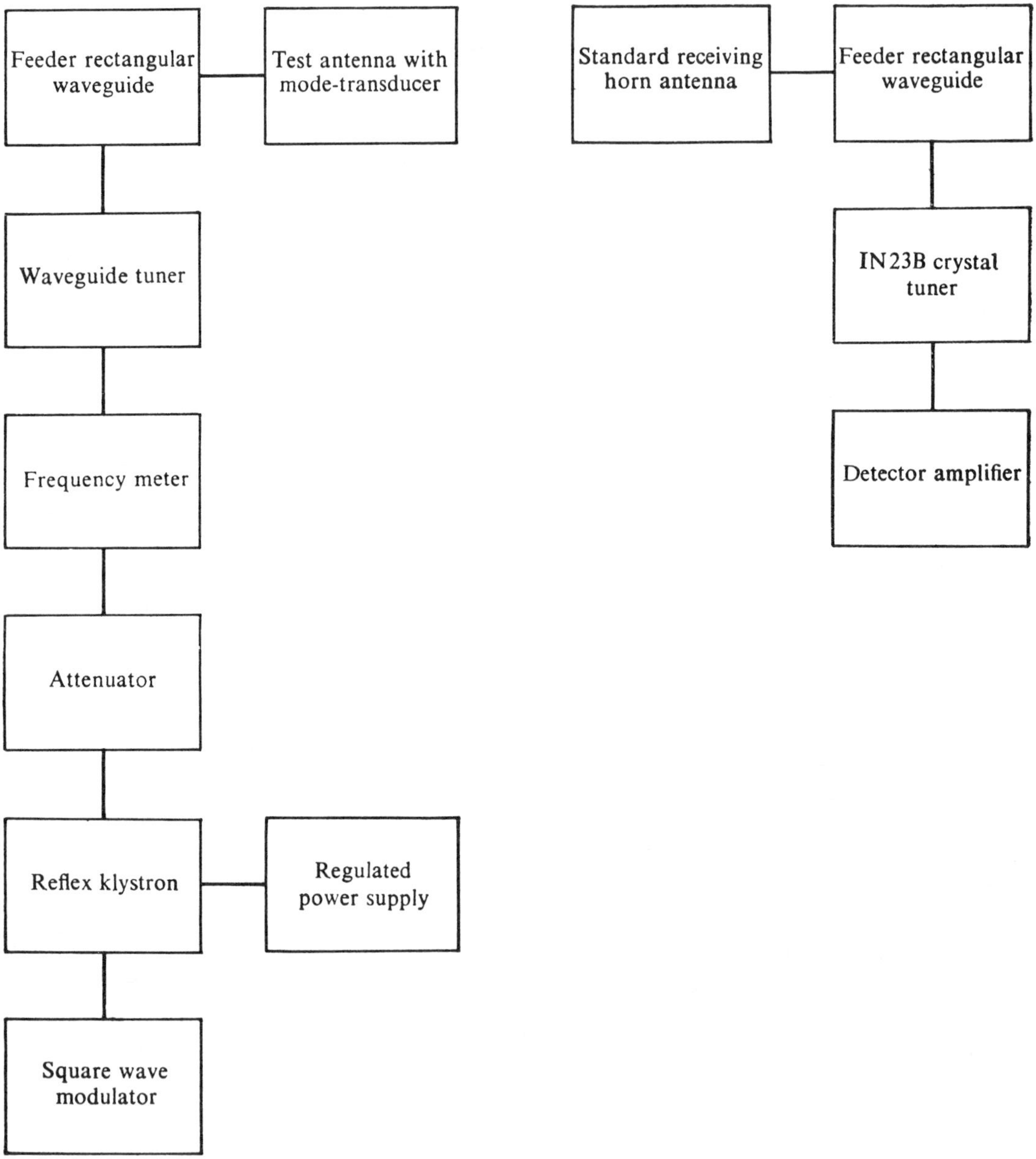

Fig. 11.7 Antenna test range.

11.25 Radiation Patterns of Dielectric Rod Antennas (Experiment 8)

A typical radiation pattern of a dielectric rod antenna of diameter d and length L excited in the

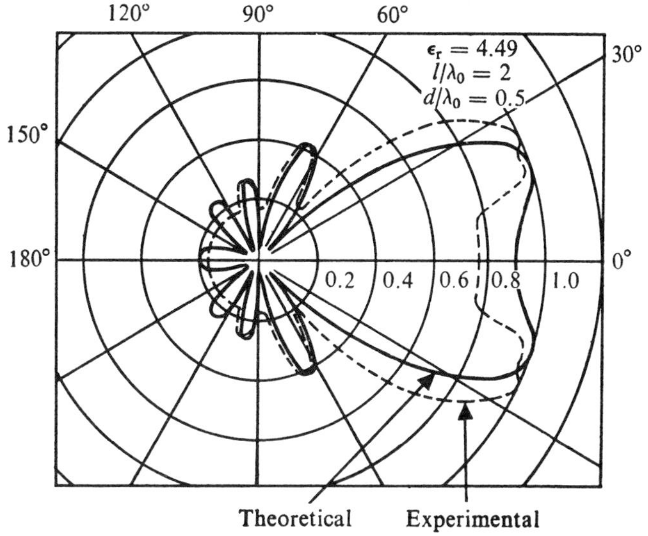

Fig. 11.8 Theoretical and experimental radiation patterns of dielectric rod antenna excited in HE_{11} mode.

HE_{11} mode[1] is shown in Fig. 11.8. The theoretical pattern given for the sake of comparison is calculated from the relations[2]

$$E_P = \boldsymbol{u}_\phi \omega\mu_0 K_1^H [\frac{\sin\{(\beta - k\cos\theta)L\}}{\beta - k\cos\theta}] 2\pi J_1[(kd/2)\sin\theta]$$

$$+ \boldsymbol{u}_\theta k K_1^E [\frac{\cos\{(\beta - k\cos\theta)L\} - 1}{\beta - k\cos\theta}] \frac{2\sqrt{2\pi}}{[(kd/2)\sin\theta]^{1/2}} J_{1/2}[(kd/2)\sin\theta] \tag{11.34}$$

in the $\phi = 0°$ plane, and

$$E_P = \boldsymbol{u}_\phi \omega\mu_0 K_1^H [\frac{\cos\{(\beta - k\cos\theta)L\} - 1}{\beta - k\cos\theta}] \frac{2\sqrt{2\pi}}{[(kd/2)\sin\theta]^{1/2}} J_{1/2}[(kd/2)\sin\theta]$$

$$+ \boldsymbol{u}_\theta k K_1^E [\frac{\sin\{(\beta - k\cos\theta)L\}}{\beta - k\cos\theta}] 2\pi J_1(\frac{kd}{2}\sin\theta) \tag{11.35}$$

in the $\phi = 90°$ plane, where

$$K_1^H = \frac{dBk_1^2 J_1(k_1 d/2)}{8\pi\omega\mu_1}, \qquad K_1^E = \frac{dbk_1^2 J_1(k_1 d/2)}{8\pi\omega\epsilon_1},$$

$$\frac{B}{b} = \frac{j\omega}{\gamma}\frac{\mu_1}{\epsilon_1}\frac{x_1^2 x_2^2}{(x_1^2 - x_2^2)}[\frac{\epsilon_1}{x_1}\frac{J_1'(x_1)}{J_1(x_1)} - \frac{\epsilon_2}{x_2}\frac{H_1^{(1)\prime}(x_2)}{H_1^{(1)}(x_2)}],$$

[1] S. K. Chatterjee and R. Chatterjee, Some investigations at microwave frequencies, *Jour. I.T.E.*, **6**, 168, 1960.
[2] R. Chatterjee and S. K. Chatterjee, Some investigations on dielectric aerials, Pt. I, *J. Ind. Inst. Sci.*, **38B**, 93, 1956.

with

$$x_1 = k_1 d/2, \qquad x_2 = k_2 d/2,$$

$$k_1^2 = \omega^2 \mu_0 \epsilon_1 + \gamma^2, \qquad k_2^2 = \omega^2 \mu_0 \epsilon_0 + \gamma^2.$$

11.26 Measurement of Phase of Electromagnetic Field (Experiment 9)

The measurement of phase of the wavefront of the radiated waves can be done by comparing it with a reference phase (see Fig. 11.9). Here, the signals from the two paths are equalized in magnitude with the help of the attenuator and the phase is adjusted so that the phase difference between the two paths is an odd number of half wavelengths, the output of the mixer being zero.

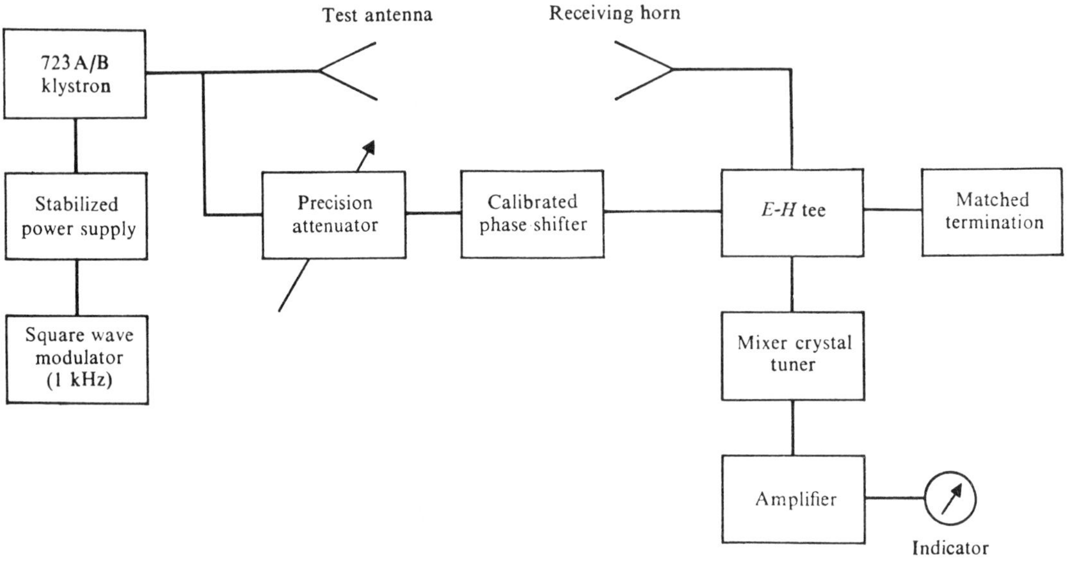

Fig. 11.9 Phase measurement.

11.27 Measurement of Antenna Gain

The absolute antenna gain G (in decibels) is defined in terms of the gain function $g(\theta, \phi)$ as

$$G = 10 \log_{10} |g(\theta, \phi)|_{\max}, \tag{11.36}$$

where the gain function is defined in terms of the pattern functions $F(\theta, \phi)$ in the θ- and ϕ-plane, as

$$g(\theta, \phi) = \frac{4\pi[|F_\theta(\theta, \phi)|^2 + |F_\phi(\theta, \phi)|^2]}{\int_{\theta=0}^{\theta=2\pi} \int_{\phi=0}^{\phi=\pi} [|F_\theta(\theta, \phi)|^2 + |F_\phi(\theta, \phi)|^2] \sin\theta \, d\theta \, d\phi}. \tag{11.37}$$

The directive gain is defined as

$$G_0 = \frac{4\pi P(0, 0) F_\theta^2(\theta, \phi)}{\int_{4\pi} P(0, 0)[|F_\theta(\theta, \phi)|^2 + |F_\phi(\theta, \phi)|^2] \, d\Omega}. \tag{11.38}$$

The directivity D, defined as $G_0|_{max}$, is given by

$$D = \frac{4\pi}{\int_{4\pi} [|F_\theta(\theta, \phi)|^2 + |F_\phi(\theta, \phi)|^2]\, d\Omega}. \tag{11.39}$$

Hence, the directivity can be found from the measured radiation patterns. The gain of an aerial can be found by (i) comparison with a standard microwave aerial, e.g., a horn or (ii) by Purcell's method. We now discuss these methods.

Comparison Method (Experiment 10)

The gain of an antenna can be measured with the same experimental set-up as that used for radiation pattern measurement. The test antenna is used as a receiver and its output is noted. It is replaced by a standard horn antenna whose gain is known, and its output is equalized to the same value as that of the test antenna, with the aid of a calibrated precision attenuator placed in line with the source antenna. The resulting gain is then the gain of the test antenna plus or minus the inserted attenuation, as the case may be.

Purcell's Method (Experiment 11)

In this method of measurement, the antenna is matched in free space to its feeder so that the VSWR ρ measured by a slotted line inserted between the antenna and feeder is almost unity[1]. At a distance R from the antenna, a plane metallic sheet, large enough to completely intercept the main lobe of the antenna, is inserted such that the distance R satisfies the free-space criterion $R \geqslant 2A^2/\lambda_0$. The VSWR in the feeder line is changed from unity to ρ. Considering the image antenna as a substitute for the second antenna, the absolute gain of the antenna is given in terms of the VSWR by the relation

$$G = \frac{8\pi R}{\lambda_0}\left(\frac{\rho - 1}{\rho + 1}\right). \tag{11.40}$$

Writing Eq. (11.40) in terms of the reflection coefficient $\Gamma = (\rho - 1)/(\rho + 1)$, the relation for gain becomes

$$\frac{1}{|\Gamma|} = \frac{8\pi R}{\lambda_0 G}. \tag{11.41}$$

A plot of $1/|\Gamma|$ versus R is expected to give a straight line passing through the origin, the slope of the line giving the absolute gain. It is, however, found that the $|\Gamma|$-R relation obtained experimentally gives a cyclic curve. This, as pointed out by Pippard et al[2], is due to the secondary waves generated from the interaction between the reflector and test antenna. Due to successive reradiation and reflection, the voltage reflection coefficient is modified to the form

$$|\Gamma| = \frac{\alpha}{R}\left|\frac{1}{1 - (\beta/R)\exp(j\phi)}\right| \tag{11.42}$$

[1]E. M. Purcell, A method for measuring the absolute gain of microwave antennas, MIT Radiation Laboratory Series, Report No. 168, Cambridge, Mass., 1943.

[2]A. B. Pippard, O. J. Burrell, and E. E. Cromie, The influence of reradiation on measurements of the power gain of an aerial, *J. I.E.E.*, **93**, Pt. IIIA, 720, 1946.

which is an oscillatory function of R that completes one cycle as R changes by $\lambda_0/2$. $|\Gamma|$ attains maximum values $\alpha/(R-\beta)$ as R assumes values for which $\exp(j\phi) = +1$, and it attains minimum values $\alpha/(R+\beta)$ as R assumes values for which $\exp(j\phi)$ attains a value of -1. Hence, the two extreme values of the oscillatory curve are bounded by two straight lines, i.e.,

$$\frac{1}{|\Gamma|} = \frac{R}{\alpha} \pm \frac{\beta}{\alpha}, \tag{11.43}$$

where

$$\alpha = \frac{G\lambda_0}{8\pi}, \qquad \beta = \left(\frac{\sigma' G'}{16\pi}\right)^{1/2}, \tag{11.44}$$

σ' and G' denoting the cross-section and power gain of the aerial for reradiation. The line drawn half-way between these two lines is expected to pass through the origin of the plot $1/|\Gamma|$ versus R.

It is difficult to apply this method to an aerial which has multilobes close to the main lobe. The method may give an accuracy of 5 per cent in gain if proper adjustment is made.

11.28 Characteristics of Reflex Klystrons

The characteristics of microwave generators may be specified in terms of their frequency and power output to a given load. The most common microwave generator used in the laboratory is the reflex klystron. The important characteristics of reflex klystrons are mode curve, variation of frequency with reflector voltage (electronic frequency tuning) and the Riecke diagram.

11.29 Mode Curve of Reflex Klystrons

The mode curve of a klystron exhibits the variation of power output of the klystron delivered to the load with respect to reflector voltage V_R at a constant beam voltage V_0. The power delivered to the load, P_L, is given in terms of the beam voltage V_0, reflector voltage V_R, beam current I_0, bunching parameter x, load resistance R_L, and the shunt resistance R_S of the cavity resonator, as follows:

$$P_L = \frac{V_0 I_0 \cos\phi}{\pi N} \cdot \frac{R_S}{R_S + R_L} x J_1(x), \tag{11.45}$$

where ϕ, the phase angle of bunching, is given by

$$\cos\phi = \cos\left[2\pi(n - \tfrac{1}{4}) - 8\pi d f \frac{\{V_0[m_e/(2q_e)]\}^{1/2}}{V_0 + V_R}\right], \tag{11.46}$$

n being an integer, f the frequency of oscillation, and d the reflector spacing. Hence, P_L depends on V_R for a constant value of V_0. The curve P_L versus V_R for a constant value of V_0 is called the *mode curve*.

11.30 Frequency Tuning Characteristics

The frequency deviation parameter Δf involving Q_L, the loaded Q, is given by

$$2Q_L\frac{\Delta f}{f} = \tan\left[2\pi(n - \tfrac{1}{4}) - 8\pi f d \frac{\{[m_e/(2q_e)]V_0\}^{1/2}}{V_0 + V_R}\right]. \tag{11.47}$$

Hence, the klystron can be electronically tuned by varying V_R and keeping V_0 constant.

11.31 Experimental Determination of Mode Curve and Electronic Frequency Tuning Curve of Reflex Klystron (Experiments 12, 13)

The experimental set-up for plotting the mode curve and frequency tuning curve of klystron is shown in Fig. 11.10. Keeping V_0 and the load constant, the reflector voltage is varied and fre-

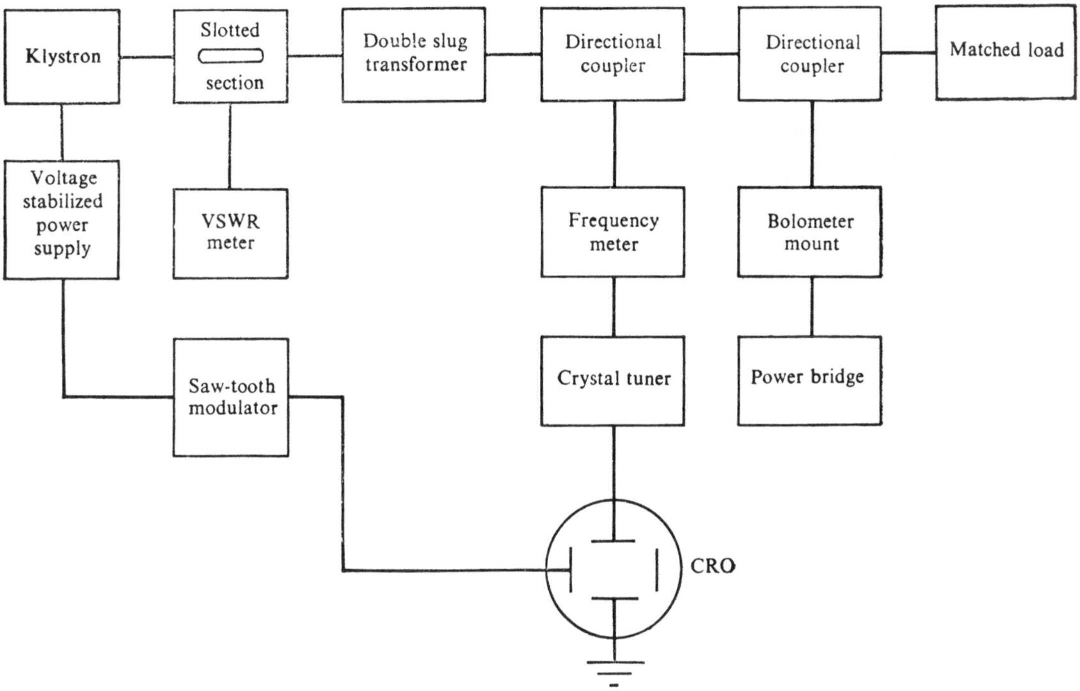

Fig. 11.10 Experimental set-up for mode curve, frequency tuning curve, and Riecke diagram.

quency and power output are noted. The mode curve and electronic tuning curves are shown in Fig. 11.11. The experiment is conducted with different load and beam voltages.

11.32 Riecke Diagram of Reflex Klystron (Experiment 14)

To plot the Riecke diagram, the effect of a changing load impedance on the power output and frequency is measured and contour lines representing constant frequency and constant power output are plotted on a Smith chart. In the experimental set-up (Fig. 11.10), the saw-tooth modulation is switched off. The reflector and beam voltages are fixed so that the operation corresponds to the top of the mode curve. The impedance transformer is changed to vary the load whose impedance is measured by measuring, with the aid of a slotted section, the VSWR and position of voltage minimum. At each load position, the power and frequency are measured. The frequency is measured by noting the d-c crystal output. The VSWR and voltage minima are

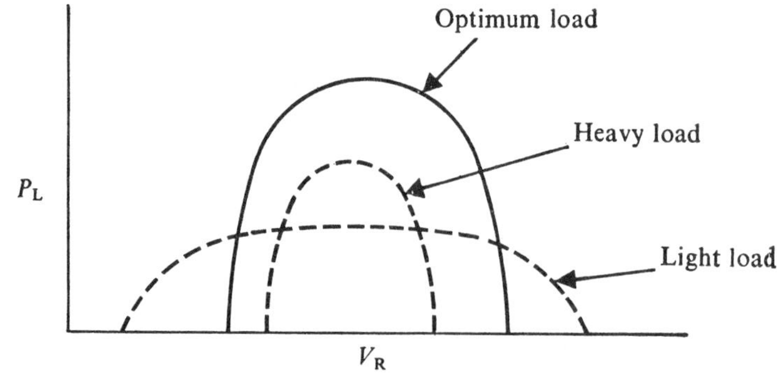

(a) Mode curve

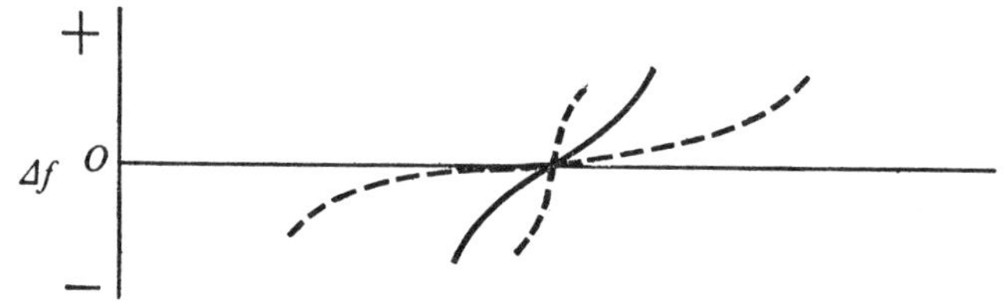

(b) Frequency tuning curve

Fig. 11.11 Klystron characteristics.

converted to susceptance B and conductance G of the load. From the power and frequency for each B and G, one family of constant power contour and another family of constant frequency contour are drawn on the Smith chart for admittance.

The Riecke diagram can be used to determine the frequency pulling figure which is equal

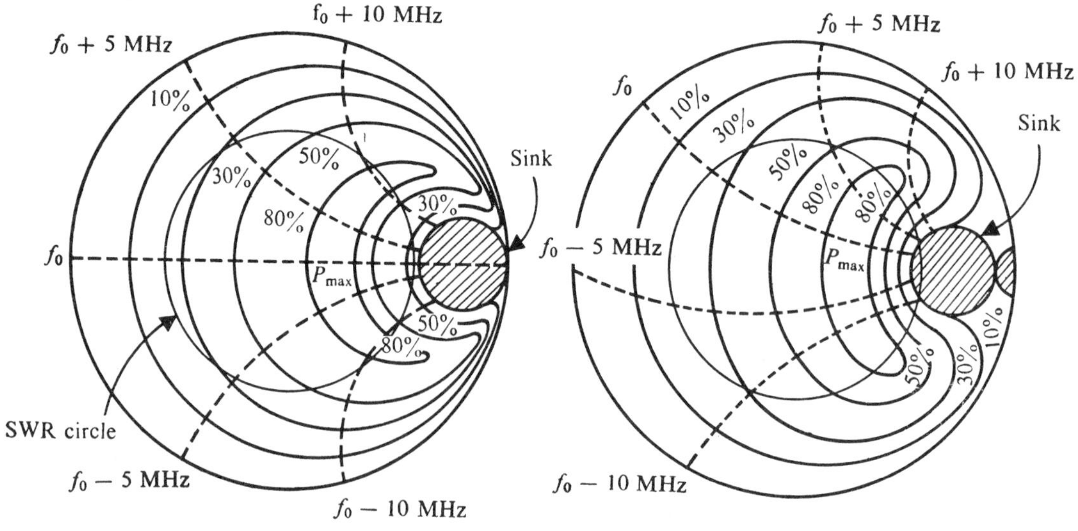

Fig. 11.12 Riecke diagrams of reflex klystron.

to the difference between the maximum and minimum frequencies as the transformer is changed through one half wavelength for a fixed VSWR. Typical Riecke diagrams are shown in Fig. 11.12.

11.33 Measurement of Power

Measurements of power at microwave frequencies are usually done with absorption instruments[1, 2] (see also Sucher and Fox 1963). The power which is thus measured is lost since it is absorbed by the measuring device. The power measuring device may be bolometric or calorimetric. In a bolometric device, the resistance change due to absorption of power and the temperature rise in a calorimeter form the basis of measurement of microwave power. For measurement of small power, generally bolometric elements, which may be of two types, namely, a bolometer having positive temperature coefficient and a thermistor having negative temperature coefficient, are used. The bolometer usually consists of a tungsten filament 0.01 mm diameter mounted in a glass vacuum tube. The thermistor, which may be in the form of bead or disc, normally consists of uranium oxide supported between two wires. The calorimetric method is used for measuring high power. We shall discuss only the measurement of small power, which is generally encountered in laboratory work.

A bridge type of instrument is usually employed for the measurement of small power. Figure 11.13 shows the principle of such measurement. The power sensitive device R_B forms the

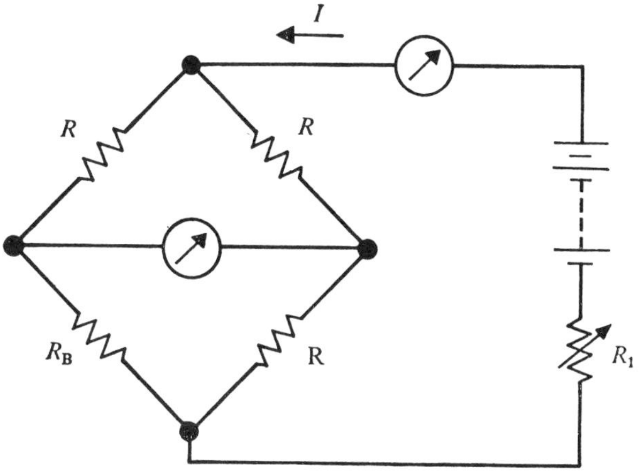

Fig. 11.13 Bolometer bridge.

element of one arm of a four-arm bridge. The element R_B is placed suitably in a waveguide which is connected to the source through an attenuator. The bridge is balanced by adjusting R_1. The total current flowing through the bridge at balance is, say, I_0. The power supplied to the element is $P_{dc} = RI_0^2/4$. The element is then exposed to the microwave power to be measured. The change in the resistance of the element disturbs the balance. The bridge balance is restored by adjusting the total current flow $I = I_1$ to the bridge with R_1. The power supplied to the element is $P_{dc+rf} = I_1^2R/4$. Hence, the microwave power is given by the difference of two powers, i.e.,

[1]W. Rosenberg, A milliwattmeter for power measurements in the superfrequency band of 8700--10,000 Mc/s, *Jour. Sci. Instr.*, **24**, 155, 1947.

[2]G. F. Gainsborough, Some sources of error in microwave milliwattmeters, *J. I.E.E.*, **95**, Pt. III, 229, 1948.

$P_{dc} \approx P_{dc+rf}$, which can be shown to be equal to

$$P_{\mu\,\text{wave}} = \frac{R}{4}(I_0 - I_1)(I_0 + I_1) \approx \frac{RI_0}{2}i \tag{11.48}$$

assuming the difference i between I_1 and I_0 to be very small. This method is simple and useful but is subject to certain errors[1]. The error arises mainly because microwave power and d-c power are not exactly equivalent in their effect on the thermistor and its mounting. This is mostly due to losses contributed by skin effect in the thin lead wires and dielectric losses contributed by the glass envelope containing the thermistor or bolometer. Consequently, the measured value is slightly lower than the actual value.

REFERENCES

Ginzton, E. L., Microwave Measurements, McGraw-Hill, New York, 1957.

Montgomery, C. G. (ed.), Technique of Microwave Measurements, MIT Radiation Laboratory Series, Vol. 11, McGraw-Hill, New York, 1947.

Sucher, M. and Fox, J., Handbook of Microwave Measurements, Polytechnic Press of Polytechnic Institute of Brooklyn, New York, 1963.

[1]G. F. Gainsborough, Some sources of error in microwave milliwattmeters, *J.I.E.E.*, **95**, Pt. III, 229, 1948.

Suggested Reading

CHAPTERS 2 AND 3

Abramowitz, M. and Stegun, I. A., Handbook of Mathematical Functions, Dover, New York, 1970.

Collin, R. E., Foundations for Microwave Engineering, McGraw-Hill, New York, 1966.

Hayt, W. H., Engineering Electromagnetics, 4th edn., McGraw-Hill, New York, 1981.

Johnk, C. T. A., Engineering Electromagnetic Fields and Waves, Wiley, New York, 1975.

Kraus, J. D. and Carver, K. R., Electromagnetics, 2nd edn., McGraw-Hill, New York, 1973.

Kreyszig, E., Advanced Engineering Mathematics, 4th edn., Wiley, New York, 1979.

Mittra, R., Computer Techniques for Electromagnetics, Pergamon Press, Oxford, 1973.

Morse, P. M. and Feshbach, H., Methods of Theoretical Physics, McGraw-Hill, New York, 1953.

Plonsey, R. and Collin, R. E., Principles and Applications of Electromagnetic Fields, McGraw-Hill, New York, 1961.

Schelkunoff, S. A., Applied Mathematics for Engineers and Scientists, 2nd edn., D. Van Nostrand, Princeton, 1965.

Shercliff, J. A., Vector Fields, Cambridge University Press, London, 1977.

Slater, J. C. and Frank, N. H., Electromagnetism, Dover, New York, 1970.

Stratton, J. A., Electromagnetic Theory, McGraw-Hill, New York, 1941.

Watson, G. N., A Treatise on the Theory of Bessel Functions, Cambridge University Press, London, 1944.

CHAPTERS 4, 5, AND 6

Adams, M. J., An Introduction to Optical Waveguides, Wiley, New York, 1981.

Atwater, H. A., Introduction to Microwave Theory, McGraw-Hill, New York, 1962.

Chen, H. C., Theory of Electromagnetic Waves, McGraw-Hill, New York, 1983.

Collin, R. E., Field Theory of Guided Waves, McGraw-Hill, New York, 1966.

Collin, R. E., Foundations for Microwave Engineering, McGraw-Hill, New York, 1966.

Dworsky, L. N., Modern Transmission Line Theory and Applications, Wiley, New York, 1979.

Edwards, T. C., Foundations for Microstrip Circuit Design, Wiley, New York, 1981.

Helszajn, J., Passive and Active Microwave Circuits, Wiley, New York, 1979.

Jordan, E. C. and Balmain, K. G., Electromagnetic Waves and Radiating Systems, 2nd edn., Prentice-Hall, Englewood Cliffs, New Jersey, 1968.

Kurokawa, K., An Introduction to the Theory of Microwave Circuits, Academic Press, New York, 1969.

Liao, S. Y., Microwave Devices and Circuits, Prentice-Hall, Englewood Cliffs, New Jersey, 1980.

Marcuse, D., Theory of Dielectric Optical Waveguides, Academic Press, New York, 1974.

Montgomery, C. G., Dicke, R. H., and Purcell, E. M. (eds.), Principles of Microwave Circuits, Dover, New York, 1965.

Ramo, S., Whinnery, J. R., and Van Duzer, T., Fields and Waves in Communication Electronics, Wiley, New York, 1965.

Sander, K. F. and Reed, G. A. L., Transmission and Propagation of Electromagnetic Waves, Cambridge University Press, London, 1978.

Stratton, J. A., Electromagnetic Theory, McGraw-Hill, New York, 1941.

CHAPTERS 7 AND 8

Atwater, H. A., Introduction to Microwave Theory, McGraw-Hill, New York, 1962.

Baden Fuller, A. J., Microwaves: An introduction to microwave theory and techniques, 2nd edn., Pergamon Press, Oxford, 1979.

Collin, R. E., Foundations for Microwave Engineering, McGraw-Hill, New York, 1966.

Helszajn, J., Non-reciprocal Microwave Junctions, Wiley, New York, 1975.

Helszajn, J., Passive and Active Microwave Circuits, Wiley, New York, 1979.

Kurokawa, K., An Introduction to the Theory of Microwave Circuits, Academic Press, New York, 1969.

Liao, S. Y., Microwave Devices and Circuits, Prentice-Hall, Englewood Cliffs, New Jersey, 1980.

Montgomery, C. G., Dicke, R. H., and Purcell, E. M. (eds.), Principles of Microwave Circuits, Dover, New York, 1965.

Ramo, S., Whinnery, J. R., and Van Duzer, T., Fields and Waves in Communication Electronics, Wiley, New York, 1965.

Sander, K. F. and Reed, G. A. L., Transmission and Propagation of Electromagnetic Waves, Cambridge University Press, London, 1978.

Stratton, J. A., Electromagnetic Theory, McGraw-Hill, New York, 1941.

CHAPTER 9

Brillouin, L., Wave Propagation and Group Velocity, Academic Press, New York, 1960.

Collin, R. E., Field Theory of Guided Waves, McGraw-Hill, New York, 1966.

Collin, R. E., Foundations for Microwave Engineering, McGraw-Hill, New York, 1966.

Liao, S. Y., Microwave Devices and Circuits, Prentice-Hall, Englewood Cliffs, New Jersey, 1980.

Ramo, S., Whinnery, J. R., and Van Duzer, T., Fields and Waves in Communication Electronics, Wiley, New York, 1965.

CHAPTER 10

Collin, R. E., Foundations for Microwave Engineering, McGraw-Hill, New York, 1966.

Dix, C. H. and Aldous, W. H., Microwave Valves, Iliffe, London, 1966.

Howes, M. J. and Morgan, D. V., Microwave Devices, Wiley, New York, 1978.

Liao, S. Y., Microwave Devices and Circuits, Prentice-Hall, Englewood Cliffs, New Jersey, 1980.

Okress, E., Cross-field Microwave Devices, Vols. I and II, Academic Press, New York, 1961.

CHAPTER 11

Hewlett-Packard Co., Microwave Theory and Applications, Prentice-Hall, Englewood Cliffs, New Jersey, 1969.

Liao, S. Y., Microwave Devices and Circuits, Prentice-Hall, Englewood Cliffs, New Jersey, 1980.

Saad, T. S., Microwave Engineers Handbook, Vols. I and II, Artech House, Dedham, Mass., 1971.

Index